D1134694

Fundamentals
of Biomechanics

Fundamentals of Biomechanics

Ronald L. Huston

CRC Press
Taylor & Francis Group
Boca Raton   London   New York

CRC Press is an imprint of the
Taylor & Francis Group, an **informa** business

CRC Press
Taylor & Francis Group
6000 Broken Sound Parkway NW, Suite 300
Boca Raton, FL 33487-2742

© 2013 by Taylor & Francis Group, LLC
CRC Press is an imprint of Taylor & Francis Group, an Informa business

No claim to original U.S. Government works

Printed on acid-free paper
Version Date: 20121203

International Standard Book Number: 978-1-4665-1037-1 (Hardback)

Library of Congress Cataloging-in-Publication Data

Huston, Ronald L., 1937-
    Fundamentals of biomechanics / Ronald L. Huston.
        p. ; cm.
    Rev. ed. of: Principles of biomechanics. 2009.
    Includes bibliographical references and index.
    ISBN 978-1-4665-1037-1 (alk. paper)
    I. Huston, Ronald L., 1937- Principles of biomechanics. II. Title.
    [DNLM: 1. Biomechanics. 2. Models, Biological. WE 103]

    612.7'6--dc23   2012047489

Visit the Taylor & Francis Web site at
http://www.taylorandfrancis.com

and the CRC Press Web site at
http://www.crcpress.com

Contents

Preface

This book was developed from an earlier publication, *Principles of Biomechanics* (CRC Press, Taylor & Francis Group, 2009). Problem sets have been provided for each chapter, and an instructor's manual is also available. The book focuses on the fundamental concepts of biomechanics with an emphasis on biodynamic modeling.

In the last three or four decades, studies of biomechanics have expanded from simple topical applications of elementary mechanics to entire areas of study, drawing the attention of increasing numbers of scientists, engineers, and health-care professionals. Today, studies and research in biomechanics exceed those in basic mechanics itself, even though basic mechanics underlies not only the study of biomechanics but many other fields as well. Consequently, with today's knowledge base, a book or treatise on biomechanics can consider only a few of the many areas on the subject in any depth.

In this book, I have selected a few topics from the fundamentals of solid biomechanics with an emphasis on biodynamic modeling and on the development of human body models. The subject matter is a compilation of material drawn from a sequence of courses taught at the University of Cincinnati for over 40 years.

This book is intended for students, researchers, and practitioners in various fields, with varying backgrounds, who are looking for a basic understanding of the principles of biomechanics analyses. The preparation needed is simply that acquired in the first years of undergraduate science and engineering curricula. It comprises 15 chapters, together with an appendix containing a rather extensive listing of anthropometric data, a large glossary of terms and terminologies, and a bibliography for more in-depth studies.

Following a brief introductory chapter, a review of gross human anatomy and a summary of basic terminology currently in use are provided in Chapter 2. Chapters 3 through 5 describe methods of analysis, from elementary mathematics to elementary mechanics and on to fundamental concepts of the mechanics of materials.

Chapter 6 discusses the modeling of biosystems. Chapter 7 provides a brief overview of tissue biomechanics. Chapters 8 through 10 then introduce concepts of biodynamics and human body modeling, looking at the fundamentals of the kinematics, kinetics, and inertial properties of human body models.

Chapters 11 through 13 present a more detailed analysis of the kinematics, kinetics, and dynamics of human body models. Chapter 14 discusses the numerical procedures for solving the governing dynamical equations.

Finally, Chapter 15 concludes the book by providing a review of a few examples of biodynamic models and their applications, including simple lifting, maneuvering in space, walking, swimming, and crash-victim simulation. References have been provided at the end of each chapter for additional study.

MATLAB® is a registered trademark of The MathWorks, Inc. For product information, please contact:

The MathWorks, Inc.
3 Apple Hill Drive
Natick, MA 01760-2098 USA
Tel: 508-647-7000
Fax: 508-647-7001
E-mail: info@mathworks.com
Web: www.mathworks.com

Acknowledgment

I am deeply appreciative of the encouragement and support of many friends, students, and colleagues in the preparation of this book over the past several years. I am especially appreciative of the vision and inspiration of Alvin Strauss and Chris Passerello, who first brought the subject of biomechanics to my attention over 40 years ago. The subsequent enthusiasm of students and of their focused studies in biomechanics was more than I had ever imagined possible. Their dedication inspired me to proceed with the writing of this book. These students include Roger Adelman, Eric Arthur, Brett Chouinard, John Connelly, Mina Dimov, Fadi El-Khatib, Joe Gallenstein, Cesar Grau, Mark Harlow, Dick Hessel, Stanley Huang, Dan Jones, George Khader, Jim Kamman, Tim King, David Lemmon, Chunghui Li, Fang Li, C.-Q. Liu, Chris Lowell, Sushma Madduri, Soumya Naga, Louise Obergefel, Chris Passerello, Jason Tein, Joe Tzou, Srikant Vallabhajosula, James Wade II, J. T. Wang, Tom Waters, Jim Winget, Michael Wu, and Sharon Yee.

I am appreciative of the inspiration and encouragement of Jerry Rosenbluth of Automotive Consulting Services.

I am grateful to the National Science Foundation and the Office of Naval Research, and to program officers Clifford Astill, Nick Perrone, and Ken Saczalski.

I am also thankful for the patience of the editors at CRC Press and for the work of Charlotte Better in the preparation of the manuscript.

Author

Ronald L. Huston is a professor emeritus of mechanics and a distinguished research professor in the Department of Mechanical Engineering at the University of Cincinnati, Cincinnati, Ohio. He is also a Herman Schneider Chair Professor.

Dr. Huston has been a member of the faculty of the University of Cincinnati since 1962. During his tenure, he was the head of the Department of Engineering Analysis, an interim head of chemical and materials engineering, the director of the Institute for Applied Interdisciplinary Research, and an acting senior vice president and provost. He has also served as a secondary faculty member in the Department of Biomedical Engineering and as an adjunct professor of orthopedic surgery research.

From 1979 to 1980, Dr. Huston was the division director of civil and mechanical engineering at the National Science Foundation. In 1978, he was the visiting professor in applied mechanics at Stanford University. From 1990 to 1996, he was the director of the Monarch Foundation.

Dr. Huston has authored more than 150 journal articles, 150 conference papers, 5 books, and 75 book reviews. He has served as a technical editor of *Applied Mechanics Reviews*, as an associate editor of the *Journal of Applied Mechanics*, and as a book review editor of the *International Journal of Industrial Engineering*.

Dr. Huston is an active consultant in safety, biomechanics, and accident reconstruction. His research interests include multibody dynamics, human factors, biomechanics, and sport mechanics.

1 Introduction

What is biomechanics? Biomechanics is simply mechanics. Mechanics refers to those studies in engineering and applied physics concerned with forces and motion. Biomechanics is mechanics applied with living systems—principally the human body.

While biomechanics is simply mechanics, and while mechanics can be a relatively simple subject (at least conceptually), the application with living systems is usually far from simple. Fabricated and inert systems are much less complex than living systems (or biosystems). With biosystems, the geometry is irregular and not easily represented by elementary figures or shapes. With biosystems, the material properties are inhomogeneous, anisotropic, and nonlinear. Indeed, biosystems are composed of solids, liquids, and gases with nonlinear viscoelastic and non-Newtonian characteristics. Biosystems present students and researchers with an uncountable number of challenging problems in modeling, simulation, and analysis. The aim of this book is to provide methods for simplifying and solving these problems.

1.1 PRINCIPAL AREAS OF BIOMECHANICS

Biomechanics may be conveniently divided into three principal areas: (1) performance, (2) injury, and (3) rehabilitation. Performance refers to the way living systems (primarily human beings) do things. It includes routine movements such as walking, sitting, standing, reaching, throwing, kicking, and carrying objects. It also refers to internal movement and behavior such as blood flow, fluid circulation, heart and muscle mechanics, and skeletal joint kinematics. In addition, performance connotes global activities such as operating vehicles or tools and sport mechanics.

Injury refers to failure and damage of biosystems as in broken bones, torn muscles, ligaments, and tendons, and organ impairment. Injury studies thus include evaluation of tissue properties. They also include studies of accidents and the design of protective devices.

Rehabilitation refers to the recovery from injury and disease. Rehabilitation thus includes all applications of mechanics in the health care industries encompassing such areas as design of corrective and assistive devices, development of implants, design of diagnostic devices, and tissue-healing mechanics.

1.2 APPROACH IN THIS BOOK

Books could be written on each of these topics. Indeed, many have already been written (see Refs. [1–57]). It is thus impossible to encompass biomechanics in a single book. We therefore need to limit our scope to some extent. We have chosen to focus upon gross or whole-body biomechanics and associated analysis methods. That is, we will generally consider the overall system or the system in the large, as opposed to the internal workings of the system. We will also focus upon dynamic as opposed to static phenomena.

As the title suggests, a major portion of this book is devoted to fundamental methods of analysis. While research in biomechanics is closely related to advances in technology, it is believed that individual technological advances are often short-lived and that more long-term benefits are obtained by mastering the fundamental methods. Therefore, we include the text reviews of vector and matrix methods and a summary of the methods of basic mechanics (statics, strength of materials, kinematics, kinetics, inertia, and dynamics). Readers already familiar with these topics may choose to simply skim over them.

We will use these fundamental methods to develop more advanced and computer-oriented methods. These include configuration graphs, lower-body arrays, differentiation algorithms, partial velocity and partial angular velocity vectors, generalized speeds, and Kane's equations.

Finally, although our focus is gross-motion simulation, we will still look at some topics in considerable depth to provide insight into those topics, as well as to illustrate the developed analytical techniques. Throughout the text, we will try to provide references for additional reading.

PROBLEM

From this brief overview, construct a list of potential study areas in biomechanics.

REFERENCES

1. K.-N. An, R. A. Berger, and W. P. Cooney III (Eds.), *Biomechanics of the Wrist Joint*, Springer-Verlag, New York, 1991.
2. C. P. Anthony and N. J. Kolthoff, *Textbook of Anatomy and Physiology*, 9th edn., Mosby, St. Louis, MO, 1975.
3. S. H. Backaitis (Ed.), *Biomechanics of Impact Injury and Injury Tolerances of the Head-Neck Complex*, Publication PT-43, Society of Automotive Engineers, Warrendale, PA, 1993.
4. S. H. Backaitis (Ed.), *Biomechanics of Impact Injury and Injury Tolerances of the Thorax-Shoulder Complex*, Publication PT-45, Society of Automotive Engineers, Warrendale, PA, 1994.
5. S. H. Backaitis (Ed.), *Biomechanics of Impact Injuries and Human Tolerances of the Abdomen, Lumbar Spine, and Pelvis Complex*, Publication PT-47, Society of Automotive Engineers, Warrendale, PA, 1995.
6. S. H. Backaitis (Ed.), *Biomechanics of Impact Injury and Injury Tolerances of the Extremities*, Publication PT-56, Society of Automotive Engineers, Warrendale, PA, 1996.
7. N. Berme and A. Cappozzo (Eds.), *Biomechanics of Human Movement: Applications in Rehabilitation, Sports and Ergonomics*, Bertec Corp., Washington, DC, 1990.
8. J. L. Bluestein (Ed.), *Mechanics and Sport*, American Society of Mechanical Engineers, New York, 1973.
9. R. S. Bridger, *Introduction to Ergonomics*, McGraw-Hill, New York, 1995.
10. P. R. Cavanagh (Ed.), *Biomechanics of Distance Running*, Human Kinetics Books, Champaign, IL, 1990.
11. D. B. Chaffin and G. B. J. Anderson, *Occupational Biomechanics*, John Wiley & Sons, New York, 1984.
12. E. Y. S. Chao, K.-N. An, W. P. Cooney III, and R. L. Linscheid, *Biomechanics of the Hand—A Basic Research Study*, World Scientific Publishing, Singapore, 1989.
13. S. C. Cowin, *Mechanical Properties of Bone*, Publication AMD-45, American Society of Mechanical Engineers, New York, 1981.
14. A. C. Damask, J. B. Damask, and J. N. Damask, *Injury Causation and Analysis—Case Studies and Data Sources*, Vols. 1 and 2, The Michie Company, Charlottesville, VA, 1990.
15. D. Dowson and V. Wright (Eds.), *An Introduction to the Biomechanics of Joints and Joint Replacement*, Mechanical Engineering Publications, London, England, 1981.
16. R. Ducroquet, J. Ducroquet, and P. Ducroquet, *Walking and Limping—A Study of Normal and Pathological Walking*, J. B. Lippincott Co., Philadelphia, PA, 1968.
17. M. Epstein and W. Herzog, *Theoretical Models of Skeletal Muscle*, John Wiley & Sons, Chichester, England, 1998.
18. C. L. Ewing and D. J. Thomas, *Human Head and Neck Response to Impact Acceleration*, Joint Army-Navy Report Nos. NAMRL Monograph 21 and USAARL 73-1, Naval Aerospace Medical Research Laboratory, Pensacola, FL, 1972.
19. R. Ferrari, *The Whiplash Encyclopedia—The Facts and Myths of Whiplash*, Aspen Publishers, Gaithersburg, MD, 1999.
20. V. Frankel and M. Nordin (Eds.), *Basic Biomechanics of the Skeletal System*, Lea & Febiger, Philadelphia, PA, 1980.
21. Y. C. Fung, N. Perrone, and M. Anliker (Eds.), *Biomechanics—It's Foundations and Objectives*, Prentice Hall, Englewood Cliffs, NJ, 1972.
22. Y. C. Fung, *Biomechanics—Motion, Flow, Stress, and Growth*, Springer-Verlag, New York, 1990.
23. M. J. Griffin, *Handbook of Human Vibration*, Academic Press, London, England, 1990.
24. S. J. Hall, *Basic Biomechanics*, 2nd edn., Mosby, St. Louis, MO, 1995.
25. M. B. Harriton, *The Whiplash Handbook*, Charles C. Thomas Publishers, Springfield, IL, 1989.

26. E. F. Hoerner (Ed.), *Head and Neck Injuries in Sports*, Publication STP 1229, American Society for Testing Materials, Philadelphia, PA, 1994.
27. A. S. Hyde, *Crash Injuries: How and Why They Happen*, HAI, Key Biscayne, FL, 1992.
28. A. T. Johnson, *Biomechanics and Exercise Physiology*, John Wiley & Sons, New York, 1991.
29. K. H. E. Kroemer, H. J. Kroemer, and K. E. Kroemer-Elbert, *Engineering Physiology—Bases of Human Factors/Ergonomics*, 2nd edn., Van Nostrand Reinhold, New York, 1990.
30. R. S. Levine (Ed.), *Head and Neck Injury*, Publication P-276, Society of Automotive Engineers, Warrendale, PA, 1994.
31. P. G. J. Maquet, *Biomechanics of the Knee*, 2nd edn., Springer-Verlag, Berlin, Germany, 1984.
32. E. N. Marieb, *Human Anatomy and Physiology*, 3rd edn., The Benjamin/Cummings Publishing Co., Redwood City, CA, 1995.
33. D. I. Miller and R. C. Nelson, *Biomechanics of Sport*, Lea & Febiger, Philadelphia, PA, 1973.
34. A. Mital, A. S. Nicholson, and M. M. Ayoub, *A Guide to Manual Materials Handling*, Taylor & Francis, London, England, 1993.
35. A. Morecki (Ed.), *Biomechanics of Engineering—Modelling, Simulation, Control*, Lecture Notes No. 291, International Centre for Mechanical Sciences, Springer-Verlag, New York, 1987.
36. V. C. Mow and W. C. Hayes (Eds.), *Basic Orthopaedic Biomechanics*, 2nd edn., Lippincott-Raven, Philadelphia, PA, 1997.
37. F. H. Netter, *Atlas of Human Anatomy*, Ciba-Geigy Corp., Summit, NJ, 1989.
38. B. M. Nigg and W. Herzog (Eds.), *Biomechanics of the Musculo-Skeletal System*, John Wiley & Sons, Chichester, England, 1994.
39. M. Nordin and V. H. Frankel (Eds.), *Basic Biomechanics of the Musculoskeletal System*, 2nd edn., Lea & Febiger, Philadelphia, PA, 1989.
40. T. R. Olson, *PDR Atlas of Anatomy*, Medical Economics Co., Montvale, NJ, 1996.
41. N. Ozkaya and M. Nordin, *Fundamentals of Biomechanics—Equilibrium, Motion, and Deformation*, Van Nostrand Reinhold, New York, 1991.
42. J. A. Pike, *Automotive Safety—Anatomy, Injury, Testing, and Regulation*, Society of Automotive Engineers, Warrendale, PA, 1990.
43. V. Putz-Anderson (Ed.), *Cumulative Trauma Disorders—A Manual for Musculoskeletal Diseases of the Upper Limbs*, Taylor & Francis, Bristol, PA, 1994.
44. H. Reul, D. N. Ghista, and G. Rau (Eds.), *Perspectives in Biomechanics*, Harwood Academic Publishers, London, England, 1978.
45. J. A. Roebuck Jr., K. H. E. Kroemer, and W. G. Thompson, *Engineering Anthropometry Methods*, John Wiley & Sons, New York, 1975.
46. J. A. Roebuck Jr., *Anthropometric Methods: Designing to Fit the Human Body*, Human Factors and Ergonomics Society, Santa Monica, CA, 1995.
47. A. Seireg and R. Arvikar, *Biomechanical Analysis of the Musculoskeletal Structure for Medicine and Sports*, Hemisphere Publishing Corporation, New York, 1989.
48. S. L. Stover, J. A. DeLisa, and G. G. Whiteneck (Eds.), *Spinal Cord Injury—Clinical Outcomes from the Model Systems*, Aspen, Gaithersburg, MD, 1995.
49. A. R. Tilley, *The Measure of Man and Woman*, Henry Dreyfuss Associates, New York, 1993.
50. C. L. Vaughan, G. N. Murphy, and L. L. du Toit, *Biomechanics of Human Gait—Annotated Bibliography*, 2nd edn., Human Kinetics Publishers, Champaign, IL, 1987.
51. A. A. White III and M. M. Panjabi, *Clinical Biomechanics of the Spine*, J. B. Lippincott Company, Philadelphia, PA, 1978.
52. W. C. Whiting and R. F. Zernicke, *Biomechanics of Musculoskeletal Injury*, Human Kinetics, Champaign, IL, 1998.
53. D. A. Winter, *Biomechanics of Motor Control of Human Movement*, 2nd edn., John Wiley & Sons, New York, 1990.
54. R. Wirhed, *Athletic Ability and the Anatomy of Motion*, Wolfe Medical Publications, London, England, 1989.
55. W. E. Woodson, B. Tillman, and P. Tillman, *Human Factors Design Handbook*, 2nd edn., McGraw-Hill, New York, 1992.
56. N. Yoganandan, F. A. Pintar, S. J. Larson, and A. Sances Jr. (Eds.), *Frontiers in Head and Neck Trauma—Clinical and Biomechanical*, IOS Press, Amsterdam, the Netherlands, 1998.
57. D. Zacharkow, *Posture: Sitting, Standing, Chair Design and Exercise*, Charles C. Thomas Publishers, Springfield, IL, 1987.

2 Review of Human Anatomy and Some Basic Terminology

Most people are familiar with human anatomy—at least from an intuitive or gross perspective. Since our focus in this book is on gross biomechanics, such a general familiarity is sufficient for most of the discussions and analyses considered herein. Nevertheless, to be consistent in our terminology and to undergird our understanding of anatomical geometry, it is helpful to briefly review some of the terminology and the conventional biomechanics notation.

We begin with a presentation of conventions used in gross (or whole-body) modeling. We follow this with a review of the major bones and segments of the skeletal system. We then take a closer look at the cervical and lumbar spines and the principal connecting/articulating joints (shoulders, hips, elbows, knees, wrists, and ankles). We conclude with a consideration of the major muscle groups and with a presentation of anthropometric data.

2.1 GROSS (WHOLE-BODY) MODELING

Figure 2.1 contains a sketch of the human frame* where the dots represent major connecting joints. Figure 2.2 shows the same sketch with the human frame divided into its major segments or limbs. The resulting figure is a gross model of the human frame. We can further simplify this model by representing the segments by ellipsoids and frustums of elliptical cones as in Figure 2.3.

For analysis purposes, it is convenient to number and label the human model segments as in Figure 2.4. Also, in Figure 2.4, R represents an inertial (or Newtonian) reference frame in the system. It is often convenient to number or label R as body zero.

The human frame modeling in Figure 2.4 is sometimes called finite-segment modeling. The model itself is sometimes called a gross-motion simulator. We will use the model of Figure 2.4 in our analysis of human body kinematics and dynamics (see Table 2.1).

Occasionally we may be interested in a more detailed modeling of the human frame—or more likely, a portion or part of the frame. For example, in injury studies, we may be interested in head/neck motion. Figure 2.5 shows a typical gross-motion model of the head and cervical vertebrae. Adjacent vertebrae can both translate and rotate relative to one another—at least, to some extent. Therefore, the soft tissue connecting the vertebrae is usually modeled by nonlinear springs and dampers. We will explore this further in later chapters.

Similarly, Figure 2.6 shows a model of the hand and wrist, which is useful for studying the gross kinematics (or movement) of the hand and its digits.

On many occasions, it is convenient to combine the use of a gross-motion model with the use of a more detailed model. For example, in neck injury studies of a crash victim, we may use a whole-body model as in Figure 2.4 to obtain the movement of the chest or upper torso. Then, this upper-torso movement may be used to determine more precise movement of the head and vertebrae through the head/neck model of Figure 2.5.

For these gross-motion models to be useful in kinematic and dynamic simulations, it is necessary to have accurate values for the physical (mass/inertia) and geometrical properties of the individual segments of the models. Also, it is necessary to have a good representation of the movement characteristics of the connecting joints. In many simulations, a simple pin (or revolute) joint is

---

* Using a Berol RapiDesign template: R-1050 human figure.

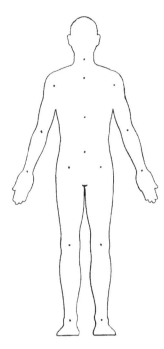

FIGURE 2.1 Sketch of the human frame.

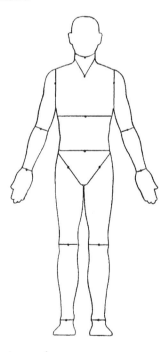

FIGURE 2.2 Major segments of the human frame.

a sufficient model. Other simulations may require a spherical (or ball-and-socket) model, and still others may require full, 6 degrees of freedom movement. For even more precise modeling, it may be necessary to use cam analyses.

The movement and constraints of the joints is governed by the soft tissue connecting the segments—that is, the ligaments, discs, tendons, and muscles. As noted earlier, this soft tissue is often modeled by semilinear and nonlinear springs and dampers.

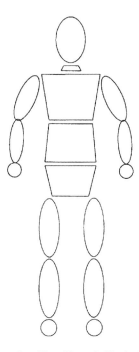

FIGURE 2.3 Modeling the human frame by ellipsoids and elliptical cones.

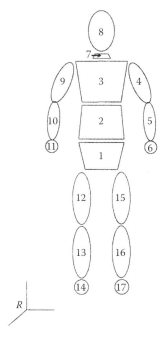

FIGURE 2.4 Numbering and labeling the human frame model.

While it is relatively easy to obtain reasonably accurate values for the physical and geometrical properties of the segments, it is much more difficult to obtain precise values for the coefficients and parameters of the joint spring and damper models. Indeed, improving the accuracy of the values of these coefficients and parameters is a topic of current research of many analysts.

TABLE 2.1
Body Segment Numbers for the
Finite-Segment Model of Figure 2.4

Segment Number	Segment Name
0	Inertial reference frame
1	Pelvis or lower-torso body
2	Midriff or mid-torso body
3	Chest or upper-torso body
4	Left upper arm
5	Left lower arm
6	Left hand
7	Neck
8	Head
9	Right upper arm
10	Right lower arm
11	Right hand
12	Right upper leg or right thigh
13	Right lower leg
14	Right foot
15	Left upper leg or left thigh
16	Left lower leg
17	Left foot

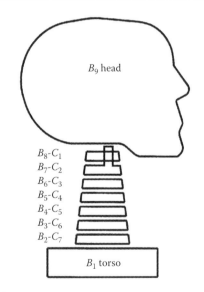

FIGURE 2.5   Head/neck model.

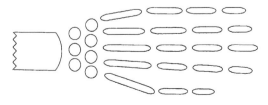

FIGURE 2.6   Model of the hand and wrist.

2.2   POSITION AND DIRECTION TERMINOLOGY

Consider a person in a standing position as in Figure 2.7. If a Cartesian coordinate system is placed in the person's torso, it is common practice to have the X-axis forward, the Z-axis up, and the Y-axis to the person's left, as shown.

These axes define planes, which are also useful in biomechanics analysis (Figure 2.8): The X–Y plane, called the transverse or horizontal plane, divides the body into upper and lower parts;

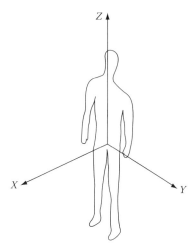

FIGURE 2.7   Coordinate axes for the body.

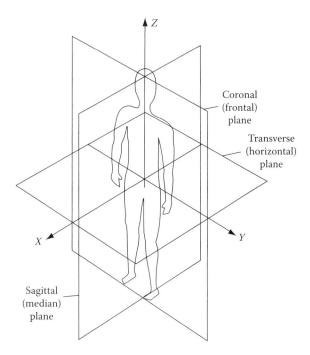

FIGURE 2.8   Principal planes of the human body.

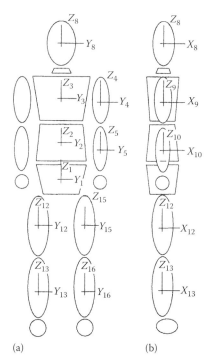

FIGURE 2.9   Coordinate axes of body segments: (a) front view and (b) side view.

the Y–Z plane, called the coronal or frontal plane, divides the body front to rear (anterior to posterior); and the Z–X plane, called the sagittal or median plane, divides the body left to right.

Similarly, X-, Y-, and Z-axes may be affixed to the links and segments of the body selectively as illustrated in Figure 2.9. When the axes of these segments are mutually aligned (parallel) to one another and to the global X-, Y-, and Z-axes of the torso, the body is said to be in the reference configuration.

The reference configuration may vary depending upon the intent of the analysis. For example, if we are interested in studying walking (gait), we may choose a reference configuration as in Figure 2.9. In this regard (for walking), the reference configuration has the planes of the hands facing inward or toward the median plane of the body. Alternatively, if we are interested in studying a vehicle operator, we may choose a reference configuration representing a seated occupant with arms forward and up as in Figure 2.10.

With the torso being the largest segment of the human frame, the position and orientation of the other segments or limbs are usually measured relative to the torso. For example, the orientations of the head and neck are usually measured relative to each other and to the chest, as opposed to measuring their orientation relative to coordinate axes fixed in space. That is, it is usually more convenient to visualize and measure the orientations of the limbs relative to each other and ultimately relative to the chest, as opposed to measuring absolute orientation in space.

The centrality of the torso is an intuitive concept. When people are asked to point to themselves, or to others, they invariably point to the chest.

The torso defines directions for the body: Moving from the torso toward the head is usually regarded as upward (or superior) even if a person is lying down. Similarly, moving from the torso toward the feet is downward (or inferior). Also, limbs or portions of limbs away from the torso (such as fingers or toes) are said to be distal, whereas portions of limbs close to the torso (such as the shoulders) are said to be proximal.

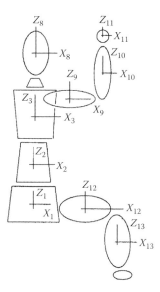

FIGURE 2.10   Reference configuration of a vehicle operator.

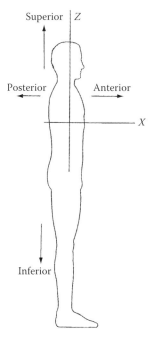

FIGURE 2.11   Superior/inferior and anterior/posterior directions.

Moving forward from the coronal plane is said to be the anterior direction. The rearward direction is called posterior. Similarly, moving away from the mid or sagittal plane is said to be lateral. Moving toward the sagittal plane is the medial direction, or medial side of a limb.

Figures 2.11 and 2.12 show these directions.* Tables 2.2 and 2.3, respectively, provide a summary description of the coordinate planes and direction terminology for the human body.

---

* Again using a Berol template.

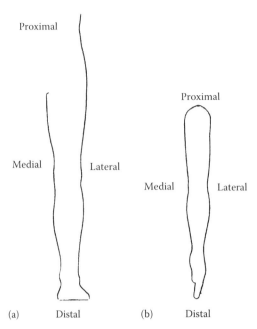

Proximal

Proximal

Medial   Lateral

Medial   Lateral

(a)   Distal   (b)   Distal

FIGURE 2.12   Lateral/medial and distal/proximal directions (Berol template): (a) left leg and (b) left arm.

TABLE 2.2
Coordinate Planes of the Human Body in a Standing Position

Name	Coordinate Axes	Description	Reference
Transverse plane (horizontal plane)	X–Y (normal to Z)	Divides the body into upper and lower parts	Figure 2.8
Coronal plane (frontal plane)	Y–Z (normal to X)	Divides the body front to rear	Figure 2.8
Sagittal plane (medial plane)	Z–X (normal to Y)	Divides the body left to right	Figure 2.8

TABLE 2.3
Direction Terminology for the Human Body

Name	Description	Reference
Superior/inferior	Above/below or upper/lower	Figure 2.11
Anterior/posterior	Front/rear	Figure 2.11
Lateral/medial	Outside/inside	Figure 2.12
Distal/proximal	Away from/near to the chest	Figure 2.12

2.3   TERMINOLOGY FOR COMMON MOVEMENTS

Various movements of the limbs also have special terminology: Perhaps the most frequent of the limb movements is bending the arms at the elbows and the legs at the knees. Such bending is called flexion. Alternatively, straightening the arms or legs is called extension. In general, the bending of any limb or body part is called flexion, and the straightening is called extension (Figures 2.13 and 2.14).

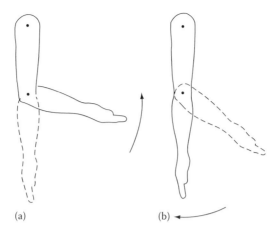

FIGURE 2.13   Arm flexion/extension (Berol template): (a) flexion and (b) extension.

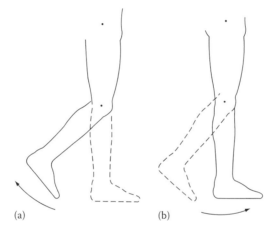

FIGURE 2.14   Leg flexion/extension (Berol template): (a) flexion and (b) extension.

The concepts of flexion and extension are especially important in studying head and neck movement and injury. Bending the head forward, chin to chest, is flexion, while bending the head rearward is called extension (Figure 2.15). The chest restricts the flexion, but there is no comparable restriction to the extension. Thus, extension is generally more harmful than flexion.

The term extension can be misleading in that, in structural mechanics, extension refers to elongation, the opposite of shortening or compression. In body movement (kinesiology), however,

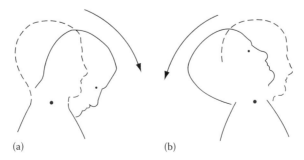

FIGURE 2.15   Head/neck flexion/extension (Berol template): (a) flexion and (b) extension.

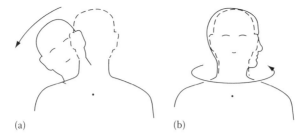

FIGURE 2.16   Head/neck lateral bending and rotation (twisting) (Berol template): (a) lateral bending and (b) rotation (twisting).

extension is simply straightening, the opposite of flexion. With neck extension, there may be either elongation or shortening of the neck [1].

When the head is moved to the side, ear to shoulder, the movement is called lateral bending. When the head is turned left or right, the movement is called axial rotation, or simple rotation, or torsion, or twisting. Figure 2.16 shows these movements.

Some specific movements of the arms and legs are also of interest. When the forearm is rotated so that the palm of the hand faces downward, it is called pronation. When the forearm is rotated so that the palm faces upward, it is called supination. Figure 2.17 shows these movements.

When the legs are brought together, as in clicking one's heels, the movement is called adduction (adding together). When the legs are separated or spread apart, the movement is called abduction. Figure 2.18 depicts these movements.

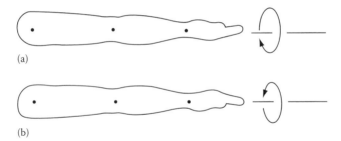

FIGURE 2.17   Forearm rotation (right arm) (Berol template): (a) supination and (b) pronation.

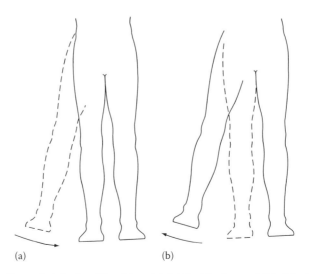

FIGURE 2.18   Adduction and abduction (Berol template): (a) adduction and (b) abduction.

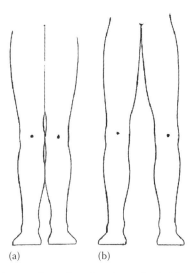

(a) (b)

FIGURE 2.19 Varus and valgus leg configuration (Berol template): (a) varus and (b) valgus.

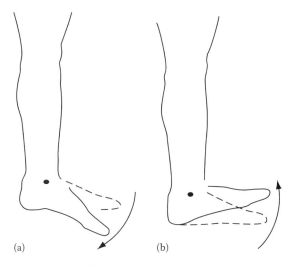

(a) (b)

FIGURE 2.20 Plantarflexion and dorsiflexion foot movement (Berol template): (a) plantarflexion and (b) dorsiflexion.

When a person's legs are together more at the knees than at the feet (as in being knock-kneed), the position is called varus. When a person's legs are spread apart at the knees, more than at the feet (as in being bowlegged), the position is called valgus. Figure 2.19 depicts these positions.

There are also some foot movements of interest. When one pushes the foot downward (as in accelerating a vehicle), the motion is called plantarflexion. The opposite motion, raising the toes upward, is called dorsiflexion. Figure 2.20 shows these movements.

Finally, when the soles of a person's feet are rotated outward, so as to cause a varus leg configuration, the motion is called eversion. Rotation of the feet inward so as to cause a valgus leg configuration is called inversion. Figure 2.21 shows these movements. Table 2.4 summarizes these common movements and their associated terminology.

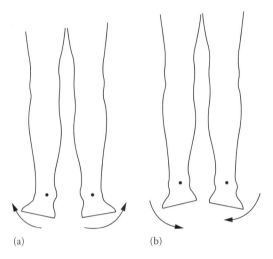

(a)   (b)

FIGURE 2.21   Eversion and inversion of the feet (Berol template): (a) eversion and (b) inversion.

TABLE 2.4
Common Movement Terminology for the Human Body

Name	Description	Reference
Flexion/extension	Bending/straightening	Figures 2.13 through 2.15
Head lateral bending and rotation	Side-to-side movement and axial twisting	Figure 2.16
Supination/pronation	Forearm movement with palm up/palm down	Figure 2.17
Adduction/abduction	Leg bringing together/spreading apart	Figure 2.18
Varus/valgus	Leg positioning knees together/knees apart	Figure 2.19
Plantarflexion/dorsiflexion	Foot pushed down/raised up	Figure 2.20
Eversion/inversion	Foot rotation outward/inward	Figure 2.21

2.4   SKELETAL ANATOMY

Figure 2.22 shows a sketch of the human skeletal system, where the major bones are labeled. The femur (thigh bone) is the largest bone, and the tibia (lower leg) and humerus (upper arm) are the next largest.

Figure 2.23 depicts the shape of the long bones. They are generally cylindrical with enlarged rounded ends. The long shaft is sometimes called the diaphysis and the rounded ends the epiphyses. The diaphysis is similar to a cylindrical shell with the outer wall composed of hard, compact bone (or cortical) and the cavity filled with soft spongy (or cancellous and sometimes called trabecular) bone [2,3]. The epiphyses with their enlarged shapes provide bearing surfaces for the joints and anchoring for the ligaments and tendons. The ligaments connect adjacent bones together, and the tendons connect muscles to the bones.

Referring again to Figure 2.22, the skull is not a single bone but a series of shell-like bones knitted together as represented in Figure 2.24.

Referring yet again to Figure 2.22, the sternum (breast bone) is not a bone at all but is a cartilage, as are those parts of the ribs attached to the sternum and spine.

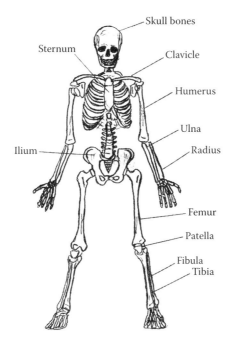

FIGURE 2.22 Human skeleton.

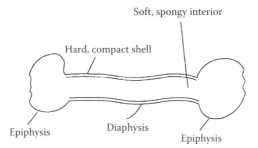

FIGURE 2.23 Sketch of a long bone.

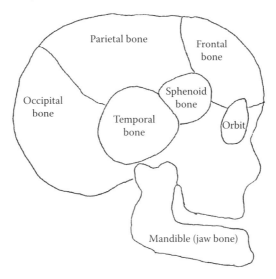

FIGURE 2.24 Skull bones and jaw.

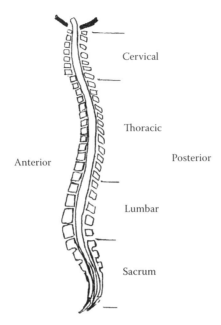

FIGURE 2.25   Sketch of the human spine.

Figure 2.25 shows a sketch of the spine. The spine is the principal supporting structure of the torso. It consists of four major parts: (1) the cervical spine (neck), (2) the thoracic spine (chest), (3) the lumbar spine (lower back), and (4) the sacrum (tail bones).

The spine is composed of annular bones (vertebrae) stacked upon one another and cushioned by discs—spongy, thick-walled annular fibrous structures with fluid interiors [4]. Figure 2.26 has a sketch of a typical vertebra from the cervical spine. The vertebrae are annular structures where the central opening, or foramen, accommodates the spinal cord.

Figure 2.27 provides a sketch of the cervical spine. It consists of seven vertebrae as shown. The cervical spine is the most flexible of all the spine segments, enabling the global movement of the head. This flexibility, however, leaves the neck vulnerable to injury. Aggravating this vulnerability is the relatively fragile nature of the cervical vertebrae. They are small compared with the vertebrae of the thoracic and lumbar spines. But more than this, the foramen of the cervical vertebrae is large enough to accommodate the larger spinal cord of the neck.

The thoracic spine has 12 vertebrae, the lumbar spine has 5 vertebrae, and the sacrum has 5 fused vertebrae. The thoracic spine is supported by the ribs and is thus relatively well protected. The lumbar spine, however, is relatively unprotected and is thus a common source of injury, ailment, and pain.

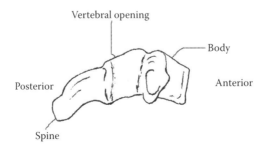

FIGURE 2.26   Typical cervical vertebra.

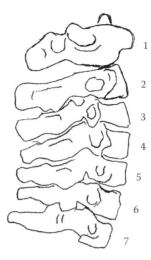

FIGURE 2.27   Sketch of the cervical spine: 1–7.

2.5 MAJOR JOINTS

In machine theory, joints are often classified by their degrees of freedom. The most common machine joint is the pin (the hinge or revolute joint) having 1 degree of freedom and as illustrated in Figure 2.28. Another 1 degree of freedom joint is the slider as in Figure 2.29. The most common 3 degree of freedom joint is the ball-and-socket, or spherical, joint as represented in Figure 2.30.

Bio-joints, or human body joints, are often represented or modeled by these mechanical joints. The elbows and knees are modeled as hinges, and the hips and shoulders are modeled as ball-and-sockets. A close examination of the limb movements at the elbows and knees, however, shows that the joints behave only approximately as hinges. Also, the shoulders and hips are only approximately spherical.

The spine movement may be modeled through a series of joints at the vertebral interfaces. Since the greatest flexibility is in the neck, the cervical joints are best represented by 6 degrees of freedom joints, having both translation and rotation. Since there is less movement and almost no translation in the thoracic and lumbar spines, the movement in these spine segments may be represented through spherical joints.

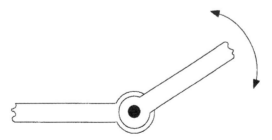

FIGURE 2.28   Pin, or hinge, joint.

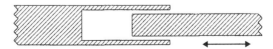

FIGURE 2.29   Slider joint.

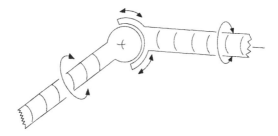

FIGURE 2.30   Ball-and-socket or spherical joint.

2.6   MAJOR MUSCLE GROUPS

The human body has three kinds of muscles: cardiac, smooth, and skeletal. Cardiac muscle is heart muscle, and it occurs only in the heart. Smooth (or visceral) muscle occurs in the intestines, lungs, bladder, and other hollow organs. Skeletal muscle is the prominent visible muscle connected to the bones, which moves the human frame. Skeletal muscle can be voluntarily controlled, whereas cardiac and smooth muscles are involuntary. Skeletal muscles dominate our focus in global biomechanics. Figure 2.31 shows the major skeletal muscles.

Muscles contract and shorten. In this way, they create and exert tension. By lengthening, however, they do not create compression. They pull but they do not push. Instead, they work in pairs: If a muscle causes limb flexion, its counterpart will cause limb extension.

The muscles flexing and extending the arms are the biceps (flexion) and the triceps (extension). For the legs, they are the hamstrings (flexion) and the quadriceps (extension).

Muscles are often classified by the movement they produce—flexors, extensors, pronators, supinators, abductors, adductors, invertors, and evertors [5].

Anatomically, muscles are generally parallel groups of muscle fibers. For example, the biceps are composed of two major muscle groups, the triceps of three groups, and the quadriceps and hamstrings of four groups each.

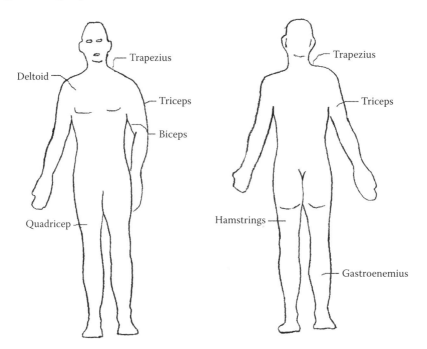

FIGURE 2.31   Major skeletal muscles.

2.7 ANTHROPOMETRIC DATA

For quantitative biomechanical analyses, we need to have numerical values for the geometrical and inertial properties of the human frame and its major segments (see Figures 2.2 and 2.3). The geometric values are frequently called anthropometric data. We summarize the principal values of these data here and in Appendix. We will look at the inertial data in Chapter 10.

While these data can vary considerably from one person to another, there are patterns and averages, which can be useful in most analyses. References [6–11] provide a comprehensive list of anthropometric data for a broad range of statures (U.S. data).

Figures 2.32 and 2.33 and Tables 2.5 and 2.6 summarize these data for the principal human dimensions. Appendix provides a more comprehensive list.

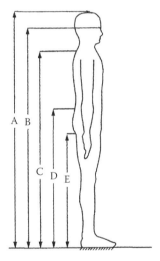

FIGURE 2.32 Standing dimensions: A, stature; B, eye height (standing); C, mid-shoulder height; D, waist height; E, buttocks height.

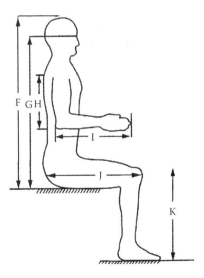

FIGURE 2.33 Sitting dimensions: F, sitting height; G, eye height (sitting); H, upper arm length; I, lower arm/hand length; J, upper leg length; K, lower leg length.

TABLE 2.5

Human Anthropometric Data (in m) (See Figures 2.32 and 2.33)

Name	Figure Dimension	Male 5th%	Male 50th%	Male 95th%	Female 5th%	Female 50th%	Female 95th%
Stature	A	1.649	1.759	1.869	1.518	1.618	1.724
Eye height (standing)	B	1.545	1.644	1.748	1.427	1.520	1.630
Mid-shoulder height	C	1.346	1.444	1.564	1.210	1.314	1.441
Waist height	D	0.993	1.102	1.168	0.907	0.985	1.107
Buttocks height	E	0.761	0.839	0.919	0.691	0.742	0.832
Sitting height	F	0.859	0.927	0.975	0.797	0.853	0.911
Eye height (sitting)	G	0.743	0.800	0.855	0.692	0.743	0.791
Upper arm length	H	0.333	0.361	0.389	0.306	0.332	0.358
Lower arm/hand length	I	0.451	0.483	0.517	0.396	0.428	0.458
Upper leg length	J	0.558	0.605	0.660	0.531	0.578	0.628
Lower leg length	K	0.506	0.553	0.599	0.461	0.502	0.546

Sources: McConville, J.T. and Laubach, L.L., in *Anthropometric Source Book, Anthropometry for Designers*, Vol. 1, J.T. Jackson (Ed.), NASA Reference Publication 1024, National Aeronautics and Space Administration, Washington, DC, 1978, Chapter III; Kroemer, K.H.E.: *Handbook of Human Factors and Ergonomics*, 2nd edn. G. Salvendy (Ed.), John Wiley & Sons, New York, 1997, pp. 219–232; Woodson, W.E., *Human Factors Design Handbook*, McGraw-Hill, New York, 1981, pp. 701–771; Tilley, A.R. and Dreyfuss Associates, H., *The Measure of Man and Woman*, Whitney Library of Design, Watson-Guptill Publishers, New York, 1993; Fung, C.-C., Human factors in aircraft crew systems design, Report no. 20-263-870-457, 1990.

TABLE 2.6

Human Anthropometric Data (in in.) (See Figures 2.32 and 2.33)

Name	Figure Dimension	Male 5th%	Male 50th%	Male 95th%	Female 5th%	Female 50th%	Female 95th%
Stature	A	64.9	69.3	73.6	59.8	63.7	67.9
Eye height (standing)	B	60.8	64.7	68.8	56.2	59.8	64.2
Mid-shoulder height	C	53.0	56.9	61.6	47.6	51.7	56.7
Waist height	D	39.1	43.4	46.0	35.7	38.8	43.6
Buttocks height	E	30.0	33.0	36.2	27.2	29.2	32.7
Sitting height	F	33.8	36.5	38.4	31.4	33.6	35.9
Eye height (sitting)	G	29.3	31.5	33.7	27.2	29.3	31.1
Upper arm length	H	13.1	14.2	15.3	12.0	13.1	14.1
Lower arm/hand length	I	17.8	19.0	20.4	15.6	16.9	18.0
Upper leg length	J	22.0	23.8	26.0	20.9	22.8	24.7
Lower leg length	K	19.9	21.8	23.6	18.1	19.8	21.5

Sources: McConville, J.T. and Laubach, L.L., in *Anthropometric Source Book, Anthropometry for Designers*, Vol. 1, J.T. Jackson (Ed.), NASA Reference Publication 1024, National Aeronautics and Space Administration, Washington, DC, 1978, Chapter III; Kroemer, K.H.E., in *Handbook of Human Factors and Ergonomics*, 2nd edn., G. Salvendy (Ed.), John Wiley & Sons, New York, 1997, pp. 219–232; Woodson, W.E., *Human Factors Design Handbook*, McGraw-Hill, New York, 1981, pp. 701–771; Tilley, A.R. and Dreyfuss Associates, H., in *The Measure of Man and Woman*, Whitney Library of Design, Watson-Guptill Publishers, New York, 1993; Fung, C.-C., Human factors in aircraft crew systems design, Report no. 20-263-870-457, 1990.

PROBLEMS

Section 2.1

P2.1.1 Create a finite-segment model of the right-hand index finger.

P2.1.2 Create a finite-segment model for studying right-hand encirclement of a 1.0 in (2.54 cm) cylinder.

P2.1.3 Create a finite-segment model for left-hand thumb, index finger, and middle finger holding a pen.

Section 2.2

P2.2.1 Using the direction convention of Section 2.2, place X-, Y-, and Z-axes on the models of the left leg and left arm of Figure 2.12.

Section 2.3

P2.3.1 Consider the representations of adduction and abduction in Figure 2.18. Assume this is a view of the front (anterior) of a person—that is, a face-to-face view. Let an axis system be placed on the model in accordance with the convention of Section 2.2.

Let a dextral ("right-hand") rotation about an axis be the rotation direction of a right-hand screw, aligned with the axis and advancing in the positive axis direction, during the rotation. Let a dextral rotation be regarded as positive and a sinistral ("left-hand") rotation be regarded as negative.

Assign positive and negative rotations to the left and right legs in adduction and abduction.

Section 2.5

P2.5.1 Imagine a vehicle operator whose right hand is at the top of the steering wheel, that is, at the "12 o'clock position," using a clock-face analogy. Describe the shoulder and elbow rotations as the operator moves his or her hand to the 3 o'clock position.

P2.5.2 See Problem P2.5.1. Estimate the magnitudes of the joint angle changes.

Section 2.6

P2.6.1 Refer to Figure 2.31. Describe the sequence of contractions of the major muscle groups during (a) standing from a sitting position, (b) walking, and (c) making a simple standing jump.

Section 2.7

P2.7.1 Refer to Tables 2.5 and 2.6. Find a colleague or friend. By using the stature as an entry into the tables, evaluate the accuracy of several other data points in the tables.

REFERENCES

1. B. S. Myers, J. H. McElhaney, and R. Nightingale, Cervical spine injury mechanisms, in *Head and Neck Injury*, R. S. Levine (Ed.), Publication P-276, Society of Automotive Engineers, Warrendale, PA, 1994, pp. 107–155.
2. E. N. Marieb, *Human Anatomy and Physiology*, 3rd edn., Benjamin/Cummings, Redwood City, CA, pp. 248, 293–295.
3. B. M. Nigg and W. Herzog (Eds.), *Biomechanics of the Musculo-Skeletal System*, Wiley, Chichester, England, 1994, pp. 48–50.
4. V. C. Mow and W. C. Hayes, *Basic Orthopaedic Biomechanics*, 2nd edn., Lippincott-Raven, Philadelphia, PA, 1997, p. 356.
5. C. P. Anthony and N. J. Kolthoff, *Textbook of Anatomy and Physiology*, 9th edn., C. V. Mosby, St. Louis, MO, 1975, pp. 5, 54, 60, 62, 63, 84, 121, 123.

6. J. T. McConville and L. L. Laubach, Anthropometry, in *Anthropometric Source Book, Anthropometry for Designers*, Vol. 1, J. T. Jackson (Ed.), NASA Reference Publication 1024, National Aeronautics and Space Administration, Washington, DC, 1978, Chapter III.

7. K. H. E. Kroemer, Engineering anthropometry, in *Handbook of Human Factors and Ergonomics*, 2nd edn., G. Salvendy (Ed.), John Wiley & Sons, New York, 1997, Section 2, Chapter 8, pp. 219–232.

8. W. E. Woodson, *Human Factors Design Handbook*, McGraw-Hill, New York, 1981, Chapter 4, pp. 701–771.

9. A. R. Tilley and H. Dreyfuss, Associates, *The Measure of Man and Woman*, Whitney Library of Design, Watson-Guptill Publishers, New York, 1993.

10. C.-C. Fung, Human factors in aircraft crew systems design, Report No. 20-263-870-457, 1990.

11. J. A. Roebuck, Jr., *Anthropometric Methods: Designing to Fit the Human Body*, Human Factors and Ergonomics Society, Santa Monica, CA, 1995.

3 Methods of Analysis I
Review of Vectors, Dyadics, Matrices, and Determinants

Geometric complexity is the principal hindrance to in-depth biomechanical analyses. When coupled with nonhomogeneous and irregular material properties, this complexity can make even routine analyses virtually intractable. In this chapter, we review elementary methods for organizing and studying complex geometrical systems. These methods include vector, dyadic, and matrix algebra; index notation; and determinants. In Chapter 4, we will look at methods more focused toward biomechanical analyses, including lower-body arrays, configuration graphs, transformation matrices, rotation dyadics, and Euler parameters.

We base our review by stating definitions and results and for the most part, avoiding derivation and proofs. The references at the end of this chapter provide those derivations and proofs, as well as detailed discussions of the methods and related procedures.

3.1 VECTORS

Vectors are usually first encountered in elementary physics in the modeling of forces. In that setting, a force is often described as a push or pull. A little reflection, however, reveals that the effect of the push or pull depends upon (1) how hard one pushes or pulls, (2) the place where one pushes or pulls, and (3) the direction of the push or pull. How hard one pushes or pulls is the magnitude of the force and the direction is the orientation and sense of the force (whether it is a push or a pull).

In like manner, a vector is often described as a directed line segment having magnitude (length), orientation, and sense. In a more formal setting, vectors are defined as elements of a vector space [1–3].

For our purposes, it is sufficient to think of vectors in the simple, intuitive way (i.e., as directed line segments obeying certain geometric and algebraic rules). The directional characteristics then distinguish vectors from scalars. (A scalar is simply a real or complex number.) To maintain this distinction, vectors are usually written in boldface type as $\mathbf{V}$. (The exception is with zero vectors, which are written simply as 0.)

A directed line segment might be called an arrow. In this context, the arrowhead is the head of the directed line segment (or vector), and the opposite end is the tail (Figure 3.1).

We use vectors to represent not only forces but also kinematic quantities such as position, velocity, and acceleration. We also use vectors to designate direction as with unit vectors.

3.2 VECTOR ALGEBRA: ADDITION AND MULTIPLICATION BY SCALARS

3.2.1 VECTOR CHARACTERISTICS

As noted earlier, we will intuitively define vectors as directed line segments, having the characteristics of magnitude and direction (orientation and sense) [4,5]. The magnitude of a vector is simply its length or norm with the same units as the vector itself. If $\mathbf{V}$ is a vector, its magnitude is written as $|\mathbf{V}|$. For example, if $\mathbf{F}$ is a 15 N force, then $|\mathbf{F}| = 15$ N. The magnitude of a vector can be zero but never negative.

FIGURE 3.1 Directed line segment (or arrow).

3.2.2 EQUALITY OF VECTORS

Two vectors **A** and **B** are said to be equal (**A** = **B**) if (and only if) they have equal characteristics, that is, the same magnitude and direction.

3.2.3 SPECIAL VECTORS

Two special and frequently occurring vectors form the basis for vector algebra and analysis. These are zero vectors and unit vectors, defined and described as follows:

1. A zero vector is a vector with magnitude zero and written simply as 0 (not bold). A zero vector has no direction.
2. A unit vector is a vector with magnitude one (1). Unit vectors have no units. Unit vectors are used primarily to designate direction.

3.2.4 MULTIPLICATION OF VECTORS AND SCALARS

If s is a scalar and **V** is a vector, the product s**V** (or **V**s) is a vector whose magnitude is $|s|\,|\mathbf{V}|$ and whose orientation is the same as that of **V**. The sense of s**V** is the same as that of **V** if s is positive and opposite to **V** if s is negative. These ideas are illustrated in Figure 3.2.

Every vector **V** may be represented as a product of a scalar s and a unit vector **n**. Specifically, if s is the magnitude of **V** and **n** is a unit vector with the same direction as **V**, we can express **V** as

$$\mathbf{V} = |\mathbf{V}|\,\mathbf{n}, \quad \text{where } \mathbf{n} = \mathbf{V}/|\mathbf{V}| \tag{3.1}$$

Equation 3.1 is a representation of **V** separated into its characteristics of magnitude ($|\mathbf{V}|$) and direction (**n**).

3.2.5 VECTOR ADDITION

Vectors obey the parallelogram law of addition: If two vectors **A** and **B** are to be added, the sum, **A** + **B**, may be obtained by connecting the vectors "head to tail" and then constructing a vector from the tail of the first (**A**) to the head of the second (**B**), as in Figure 3.3. The sum (resultant) is the constructed vector. It does not matter which vector is used first. The same result is obtained by starting with vector **B** and then adding **A** as in Figure 3.4.

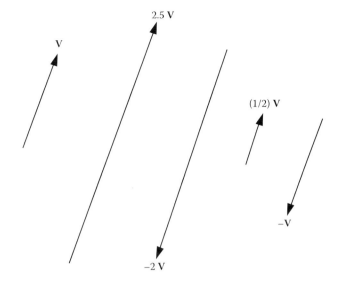

FIGURE 3.2   Multiplication of scalar with vector **V**.

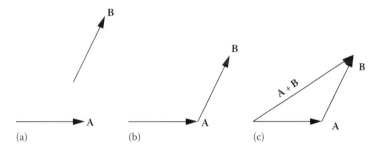

FIGURE 3.3   Vector addition (**A** + **B**). (a) Given vector, (b) connected head to tail, and (c) the sum: tail of **A** to head of **B**.

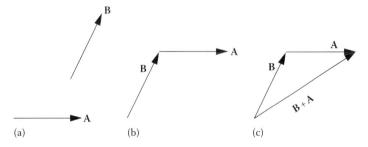

FIGURE 3.4   Vector addition (**B** + **A**). (a) Given vector, (b) connected head to tail, and (c) the sum: tail of **B** to head of **A**.

The superposition of Figures 3.3 and 3.4 shows that the sum **A** + **B** is the diagonal of a parallelogram with sides **A** and **B**, as in Figure 3.5. Hence, the name is parallelogram law. While the sum **A** + **B** is called the resultant, the addends **A** and **B** are called "components."

From Figure 3.5, we see that vector addition is commutative. That is,

$$\mathbf{A} + \mathbf{B} = \mathbf{B} + \mathbf{A} \tag{3.2}$$

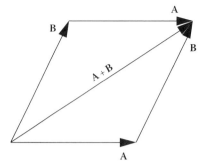

FIGURE 3.5 Parallelogram law of addition.

More generally, for three or more vectors, we also have associative and commutative relations of the form

$$\mathbf{A} + \mathbf{B} + \mathbf{C} = (\mathbf{A} + \mathbf{B}) + \mathbf{C} = \mathbf{A} + (\mathbf{B} + \mathbf{C}) = \mathbf{B} + \mathbf{C} + \mathbf{A}$$

$$= (\mathbf{B} + \mathbf{C}) + \mathbf{A} = \mathbf{B} + (\mathbf{C} + \mathbf{A}) = \mathbf{C} + \mathbf{A} + \mathbf{B} = (\mathbf{C} + \mathbf{A}) + \mathbf{B}$$

$$= \mathbf{C} + (\mathbf{A} + \mathbf{B}) = \mathbf{A} + \mathbf{C} + \mathbf{B} = (\mathbf{A} + \mathbf{C}) + \mathbf{B} = \mathbf{A} + (\mathbf{C} + \mathbf{B})$$

$$= \mathbf{C} + \mathbf{B} + \mathbf{A} = (\mathbf{C} + \mathbf{B}) + \mathbf{A} = \mathbf{C} + (\mathbf{B} + \mathbf{A}) = \mathbf{B} + \mathbf{A} + \mathbf{C}$$

$$= (\mathbf{B} + \mathbf{A}) + \mathbf{C} = \mathbf{B} + (\mathbf{A} + \mathbf{C}) \tag{3.3}$$

3.2.6 ADDITION OF PERPENDICULAR VECTORS

Observe in Figures 3.4 and 3.5 that if we know the magnitude and directions of the components **A** and **B**, we can analytically determine the magnitude and direction of the resultant **A** + **B**. We can do this using the rules of trigonometry, specifically, the law of sines and the law of cosines. In a plane, the procedure is straightforward, although a bit tedious. For nonplanar problems, that is, where there are three or more components, not all in the same plane, the analysis can become lengthy, detailed, and prone to error.

The analysis is greatly simplified, whether in a plane or in three dimensions, if the components are perpendicular or mutually perpendicular. When this happens, the law of cosines reverts to the more familiar and simpler Pythagoras theorem. Consider, for example, the perpendicular vectors **A** and **B** as in Figure 3.6. Let **C** be the resultant of **A** + **B**. The magnitude of the resultant is then determined by the expression

$$|\mathbf{C}|^2 = |\mathbf{A}|^2 + |\mathbf{B}|^2 \tag{3.4}$$

The inclination α of the resultant is

$$\alpha = \tan^{-1}\left(\frac{|\mathbf{B}|}{|\mathbf{A}|}\right) \tag{3.5}$$

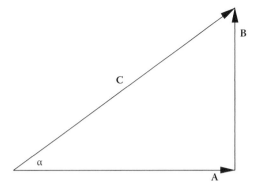

FIGURE 3.6   Addition of perpendicular vectors.

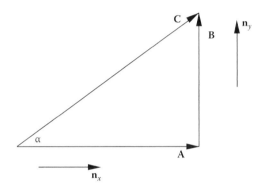

FIGURE 3.7   Unit vectors for vector addition.

Equations 3.3 and 3.4 may be further simplified through the use of unit vectors. Suppose that $\mathbf{n}_x$ and $\mathbf{n}_y$ are horizontal and vertical unit vectors as in Figure 3.7. Then, vectors $\mathbf{A}$ and $\mathbf{B}$ may be expressed as (see Equation 3.1)

$$\mathbf{A} = |\mathbf{A}|\mathbf{n}_x = a_x\mathbf{n}_x \quad \text{and} \quad \mathbf{B} = |\mathbf{B}|\mathbf{n}_y = b_y\mathbf{n}_y \tag{3.6}$$

where a_x and b_x are defined in the equation by inspection. The resultant $\mathbf{C}$ then becomes

$$\mathbf{C} = \mathbf{A} + \mathbf{B} = a_x\mathbf{n}_x + b_y\mathbf{n}_y = c_x\mathbf{n}_x + c_y\mathbf{n}_y \tag{3.7}$$

where c_x and c_y are defined in the equation by inspection. In terms of c_x and c_y, Equations 3.4 and 3.5 become

$$|\mathbf{C}|^2 = c_x^2 + c_y^2 \quad \text{and} \quad \alpha = \tan^{-1}\frac{c_y}{c_x} \tag{3.8}$$

The principal advantage of perpendicular components, however, is not in the simplification of Equation 3.8, but in the simplification seen in three-dimensional (3D) analyses. Indeed, with mutually perpendicular components, 3D analyses are no more complicated than planar analyses. For example, let **A**, **B**, and **C** be mutually perpendicular vectors as in Figure 3.8. Let $\mathbf{n}_x$, $\mathbf{n}_y$, and $\mathbf{n}_z$ be unit vectors parallel to **A**, **B**, and **C** and let **D** be the resultant of **A**, **B**, and **C** as in Figure 3.9. Then **A**, **B**, **C**, and **D** may be expressed as

$$\mathbf{A} = |\,\mathbf{A}\,|\,\mathbf{n}_x = a_x\mathbf{n}_x \quad \mathbf{B} = |\,\mathbf{B}\,|\,\mathbf{n}_y = b_y\mathbf{n}_y \quad \mathbf{C} = |\,\mathbf{C}\,|\,\mathbf{n}_z = c_z\mathbf{n}_z \tag{3.9}$$

and

$$\mathbf{D} = \mathbf{A} + \mathbf{B} + \mathbf{C} = a_x\mathbf{n}_x + b_y\mathbf{n}_y + c_3\mathbf{n}_z \tag{3.10}$$

where, as before, a_x, a_y, and a_z are defined in the equation by inspection in Equation 3.9. The magnitude of the resultant is then given by the simple expression

$$|\,\mathbf{D}\,|^2 = a_x^2 + b_y^2 + c_z^2 \tag{3.11}$$

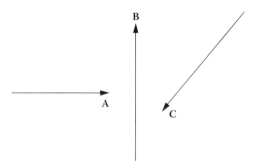

FIGURE 3.8   Mutually perpendicular vectors.

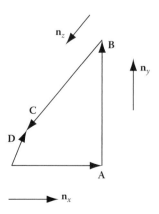

FIGURE 3.9   Addition of vectors **A**, **B**, and **C**.

More generally, if vectors **A**, **B**, and **C** are each expressed in the form of Equation 3.10, their resultant **D** may be expressed as

$$\mathbf{D} = \mathbf{A} + \mathbf{B} + \mathbf{C} = (a_x\mathbf{n}_x + a_y\mathbf{n}_y + a_z\mathbf{n}_z) + (b_x\mathbf{n}_x + b_y\mathbf{n}_y + b_z\mathbf{n}_z) + (c_x\mathbf{n}_x + c_y\mathbf{n}_y + c_z\mathbf{n}_z)$$

$$= (a_x + b_x + c_x)\mathbf{n}_x + (a_y + b_y + c_y)\mathbf{n}_y + (a_z + b_z + c_z)\mathbf{n}_z$$

$$= d_x\mathbf{n}_x + d_y\mathbf{n}_y + d_z\mathbf{n}_z \tag{3.12}$$

where

$$d_x = a_x + b_x + c_x$$

$$d_y = a_y + b_y + c_y \tag{3.13}$$

$$d_z = a_z + b_z + c_z$$

With this evident simplicity in the analysis, it is usually convenient to express all vectors in terms of mutually perpendicular unit vectors. That is, for any given vector **V**, we seek to express **V** in the form

$$\mathbf{V} = v_x\mathbf{n}_x + v_y\mathbf{n}_y + v_z\mathbf{n}_z \tag{3.14}$$

where $\mathbf{n}_x$, $\mathbf{n}_y$, and $\mathbf{n}_z$ are mutually perpendicular unit vectors, which are generally parallel to coordinate axes X, Y, and Z. In this context, the scalars v_x, v_y, and v_z are called the scalar components or simply the components of **V**. **V** is then often expressed in array form as

$$\mathbf{V} = (v_x, v_y, v_z) = \begin{pmatrix} v_x \\ v_y \\ v_z \end{pmatrix} \tag{3.15}$$

3.2.7 USE OF INDEX AND SUMMATION NOTATIONS

Equation 3.14 has a form of vectors continually encountered in biomechanics. That is, the vector is a sum of products of scalars and unit vectors. If the indices x, y, and z are replaced by 1, 2, and 3, Equation 3.14 may be written in the compact form

$$\mathbf{V} = v_1\mathbf{n}_1 + v_2\mathbf{n}_2 + v_3\mathbf{n}_3 = \sum_{i=1}^{3} v_i\mathbf{n}_i \tag{3.16}$$

With the 3D space of biosystems, the sum in the last term is always from 1 to 3. Hence, the summation sign and its limits may be deleted. Therefore, we can express $\mathbf{V}$ in the simplified form

$$\mathbf{V} = v_i\mathbf{n}_i \tag{3.17}$$

where the repeated index i designates a sum from 1 to 3.

Observe in Equation 3.17 that the index i is arbitrary. That is, the same equation is obtained with any repeated index. For example,

$$\mathbf{V} = v_i\mathbf{n}_i = v_j\mathbf{n}_j = v_n\mathbf{n}_n \tag{3.18}$$

It is conventional not to repeat the same index in given equation.

3.3 VECTOR ALGEBRA: MULTIPLICATION OF VECTORS

Vectors may be multiplied with one another in three ways: (1) by a scalar (or dot) product, (2) by a vector (or cross) product, and (3) by a dyadic product. We will review these products in the following sections.

3.3.1 ANGLE BETWEEN VECTORS

The angle between two vectors $\mathbf{A}$ and $\mathbf{B}$ is defined by the following construction: Let the vectors be brought together and connected tail-to-tail. The angle θ between the vectors is then represented in Figure 3.10.

3.3.2 SCALAR PRODUCT

As the name implies, the scalar product of two vectors produces a scalar. The product, often called the dot product, is written with a dot ($\cdot$) between the vectors. For two vectors $\mathbf{A}$ and $\mathbf{B}$, the dot product is defined as

$$\mathbf{A}\cdot\mathbf{B} = |\mathbf{A}\,\|\,\mathbf{B}|\cos\theta \tag{3.19}$$

where θ is the angle between $\mathbf{A}$ and $\mathbf{B}$. From Equation 3.19, we see that the scalar product of perpendicular vectors is zero. Also, we see that the scalar product is commutative. That is,

$$\mathbf{A}\cdot\mathbf{B} = \mathbf{B}\cdot\mathbf{A} \tag{3.20}$$

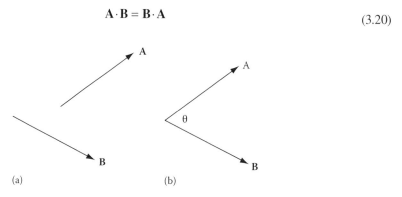

(a)　　　　　(b)

FIGURE 3.10 Angle between two vectors. (a) Given vectors and (b) tail-to-tail construction forming the angle between the vectors.

Further, if s is a scalar multiple of the product, s may be placed on either side of the dot or associated with either of the vectors. That is,

$$s\mathbf{A} \cdot \mathbf{B} = (s\mathbf{A}) \cdot \mathbf{B} = \mathbf{A}s \cdot \mathbf{B} = \mathbf{A} \cdot (s\mathbf{B}) = \mathbf{A} \cdot \mathbf{B}s = s\mathbf{B} \cdot \mathbf{A}$$

$$= (s\mathbf{B}) \cdot \mathbf{A} = \mathbf{B}s \cdot \mathbf{A} = \mathbf{B} \cdot s\mathbf{A} = \mathbf{A} \cdot s\mathbf{B} = \mathbf{B} \cdot s\mathbf{A} \tag{3.21}$$

If $\mathbf{n}_1$, $\mathbf{n}_2$, and $\mathbf{n}_3$ are mutually perpendicular unit vectors, then Equation 3.19 leads to the products

$$\mathbf{n}_1 \cdot \mathbf{n}_1 = 1, \quad \mathbf{n}_1 \cdot \mathbf{n}_2 = 0, \quad \mathbf{n}_1 \cdot \mathbf{n}_3 = 0$$

$$\mathbf{n}_2 \cdot \mathbf{n}_1 = 0, \quad \mathbf{n}_2 \cdot \mathbf{n}_2 = 1, \quad \mathbf{n}_2 \cdot \mathbf{n}_3 = 0 \tag{3.22}$$

$$\mathbf{n}_3 \cdot \mathbf{n}_1 = 1, \quad \mathbf{n}_3 \cdot \mathbf{n}_2 = 0, \quad \mathbf{n}_3 \cdot \mathbf{n}_3 = 1$$

Equation 3.22 may be written in the compact form

$$\mathbf{n}_i \cdot \mathbf{n}_j = \delta_{ij} = \begin{cases} 1 & i = j \\ 0 & i \neq j \end{cases} \tag{3.23}$$

where δ_{ij} is called the Kronecker delta function. Using the summation index notation, it is readily seen that

$$\delta_{kk} = \sum_{k=1}^{3} \delta_{kk} = 3 \tag{3.24}$$

Also, if v_i ($i = 1, 2, 3$), then

$$\delta_{ij} v_j = \sum_{i=1}^{3} \delta_{ij} v_j = v_i \tag{3.25}$$

As a result of Equation 3.25, δ_{ij} is sometimes also called the substitution symbol.

Equations 3.23 through 3.25 are useful in developing another form of the scalar product: If $\mathbf{A}$ and $\mathbf{B}$ are vectors with scalar components a_i and b_i relative to mutually perpendicular unit vectors $\mathbf{n}_i$ ($i = 1, 2, 3$), then $\mathbf{A}$ and $\mathbf{B}$ may be expressed as

$$\mathbf{A} = a_i \mathbf{n}_i \quad \text{and} \quad \mathbf{B} = b_i \mathbf{n}_i = b_j \mathbf{n}_j \tag{3.26}$$

Then $\mathbf{A} \cdot \mathbf{B}$ becomes

$$\mathbf{A} \cdot \mathbf{B} = (a_i \mathbf{n}_i) \cdot (b_j \mathbf{n}_j) = a_i b_j \mathbf{n}_i \cdot \mathbf{n}_j = a_i b_j \delta_{ij}$$

$$= a_i b_i = \sum_{i=1}^{3} a_i b_i = a_1 b_1 + a_2 b_2 + a_3 b_3 \tag{3.27}$$

The scalar product of a vector $\mathbf{V}$ with itself is sometimes written as $\mathbf{V}^2$. Since a vector is parallel to itself, the angle a vector makes with itself is zero. The definition of Equation 3.19 together with Equation 3.27 then shows $\mathbf{V}^2$ to be

$$\mathbf{V}^2 = \mathbf{V} \cdot \mathbf{V} = |\mathbf{V}|^2 = v_i v_i = v_1^2 + v_2^2 + v_3^2 \qquad (3.28)$$

where, as before, the v_i are scalar components of $\mathbf{V}$ relative to mutually perpendicular unit vectors $\mathbf{n}_i$ ($i = 1, 2, 3$).

Taken together, Equations 3.27 through 3.19 lead to the following expression for the cosine of the angle θ between two vectors $\mathbf{A}$ and $\mathbf{B}$:

$$\cos\theta = \frac{\mathbf{A} \cdot \mathbf{B}}{|\mathbf{A}||\mathbf{B}|} = \frac{a_i b_i}{(a_j a_j)^{1/2} (b_k b_k)^{1/2}} = \frac{a_1 b_1 + a_2 b_2 + a_3 b_3}{\left(a_1^2 + a_2^2 + a_3^2\right)^{1/2} \left(b_1^2 + b_2^2 + b_3^2\right)^{1/2}} \qquad (3.29)$$

If two vectors $\mathbf{A}$ and $\mathbf{B}$ are equal and if the vectors are expressed in terms of mutually perpendicular unit vectors $\mathbf{n}_i$ ($i = 1, 2, 3$) as in Equation 3.26, then by taking the scalar product with one of the unit vectors, say $\mathbf{n}_k$, we have

$$\mathbf{A} = \mathbf{B} \Rightarrow \mathbf{n}_k \cdot \mathbf{A} = \mathbf{n}_k \cdot \mathbf{B} \Rightarrow \mathbf{n}_k \cdot (a_i \mathbf{n}_i) = \mathbf{n}_k \cdot (b_j \mathbf{n}_j)$$

$$\Rightarrow a_i \delta_{ki} = b_j \delta_{kj} \quad \text{or} \quad a_k = b_k \qquad (3.30)$$

3.3.3 Vector Product

While the scalar product of two vectors $\mathbf{A}$ and $\mathbf{B}$ produces a scalar, the vector product produces a vector. The vector product, often called the cross product, is written with a cross (×) between the vectors and is defined as

$$\mathbf{A} \times \mathbf{B} = |\mathbf{A}||\mathbf{B}|\sin\theta\,\mathbf{n} \qquad (3.31)$$

where, as before, θ is the angle between the vectors and $\mathbf{n}$ is a unit vector normal to the plane formed by $\mathbf{A}$ and $\mathbf{B}$ when they are brought together and connected tail-to-tail. The sense of $\mathbf{n}$ is the same as the axial advance of a right-hand threaded screw when turned in the same way as when rotating $\mathbf{A}$ toward $\mathbf{B}$, so as to diminish the angle θ.

As with the scalar product, Equation 3.31 determines the properties of the vector product. If s is a scalar multiple of the vector product, then s may be placed at any position in the product. That is,

$$s\mathbf{A} \times \mathbf{B} = \mathbf{A}s \times \mathbf{B} = \mathbf{A} \times s\mathbf{B} = \mathbf{A} \times \mathbf{B}s \qquad (3.32)$$

Also, from Equation 3.31, we see that unlike the scalar product, the vector product is anticommutative. That is,

$$\mathbf{A} \times \mathbf{B} = -\mathbf{B} \times \mathbf{A} \qquad (3.33)$$

If $\mathbf{A}$ and $\mathbf{B}$ are parallel, the angle θ between them is zero, and thus, their vector product is zero. If $\mathbf{A}$ and $\mathbf{B}$ are perpendicular, $\sin\theta$ is unity, and thus, the magnitude of the vector product of $\mathbf{A}$ and $\mathbf{B}$ is equal to the product of the magnitudes of the vectors.

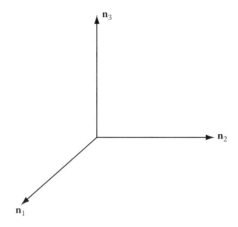

FIGURE 3.11   Mutually perpendicular (dextral) unit vectors.

Let $\mathbf{n}_1$, $\mathbf{n}_2$, and $\mathbf{n}_3$ be mutually perpendicular unit vectors as in Figure 3.11. From Equation 3.31, we obtain the relation

$$\mathbf{n}_1 \times \mathbf{n}_1 = 0 \qquad \mathbf{n}_1 \times \mathbf{n}_2 = \mathbf{n}_3 \qquad \mathbf{n}_1 \times \mathbf{n}_3 = -\mathbf{n}_2$$

$$\mathbf{n}_2 \times \mathbf{n}_1 = -\mathbf{n}_3 \quad \mathbf{n}_2 \times \mathbf{n}_2 = 0 \qquad \mathbf{n}_2 \times \mathbf{n}_3 = \mathbf{n}_1 \qquad (3.34)$$

$$\mathbf{n}_3 \times \mathbf{n}_1 = \mathbf{n}_2 \qquad \mathbf{n}_3 \times \mathbf{n}_2 = -\mathbf{n}_1 \quad \mathbf{n}_3 \times \mathbf{n}_3 = 0$$

Observe that the arrangement of Figure 3.11 produces a positive sign in Equation 3.34 when the index sequence is cyclical (i.e., 1, 2, 3; 2, 3, 1; or 3, 1, 2) and a negative sign when the indices are anticyclic (i.e., 1, 3, 2; 3, 2, 1; or 1, 3, 2). The arrangement of Figure 3.11 is called a right-handed or dextral configuration. For positive signs with anticyclic indices, the vectors need to be configured as in Figure 3.12. Such arrangements are called left-handed or sinistral configurations. In our analyses throughout the text, we will use dextral unit vector sets.

Equation 3.34 may be written in a more compact form as

$$\mathbf{n}_i \times \mathbf{n}_j = e_{ijk}\mathbf{n}_k \qquad (3.35)$$

where the e_{ijk} are elements of the permutation function or permutation symbol [6–8], defined as

$$e_{ijk} = \begin{cases} 1 & i,j,k \text{ distinct and cyclic} \\ -1 & i,j,k \text{ distinct and anticyclic} \\ 0 & i,j,k \text{ not distinct} \end{cases} \qquad (3.36)$$

or as

$$e_{ijk} = (1/2)(i-j)(j-k)(k-i) \qquad (3.37)$$

Consider the cross product of vectors $\mathbf{A}$ and $\mathbf{B}$ where, as before, $\mathbf{A}$ and $\mathbf{B}$ are expressed in terms of mutually perpendicular unit vectors $\mathbf{n}_i$ $(i = 1, 2, 3)$ as

$$\mathbf{A} = a_i\mathbf{n}_i \quad \text{and} \quad \mathbf{B} = b_j\mathbf{n}_j \qquad (3.38)$$

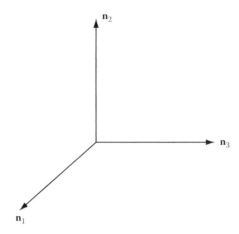

FIGURE 3.12 Mutually perpendicular (sinistral) unit vectors.

From Equations 3.32, 3.33, and 3.35, we have

$$\mathbf{A} \times \mathbf{B} = a_i \mathbf{n}_i \times b_j \mathbf{n}_j = a_i b_j \mathbf{n}_i \times \mathbf{n}_j = e_{ijk} a_i b_j \mathbf{n}_k \tag{3.39}$$

By expanding the expression in the last term of Equation 3.39, we have

$$\mathbf{A} \times \mathbf{B} = (a_2 b_3 - a_3 b_2)\mathbf{n}_1 + (a_3 b_1 - a_1 b_3)\mathbf{n}_2 + (a_1 b_2 - a_2 b_1)\mathbf{n}_3 \tag{3.40}$$

An examination of Equation 3.40 shows that the cross product may also be written in terms of a 3×3 determinant as

$$\mathbf{A} \times \mathbf{B} = \begin{vmatrix} \mathbf{n}_1 & \mathbf{n}_2 & \mathbf{n}_3 \\ a_1 & a_2 & a_3 \\ b_1 & b_2 & b_3 \end{vmatrix} \tag{3.41}$$

3.3.4 DYADIC PRODUCT

The dyadic product of vectors is less well known than the scalar or vector products even though the definition of dyadic product is simpler than the scalar or vector product. For two vectors **A** and **B**, the dyadic product, written simply as **AB**, is defined as

$$
\begin{aligned}
\mathbf{AB} &= (a_i \mathbf{n}_i)(b_j \mathbf{n}_j) \\
&= (a_1 \mathbf{n}_1 + a_2 \mathbf{n}_2 + a_3 \mathbf{n}_3)(b_1 \mathbf{n}_1 + b_2 \mathbf{n}_2 + b_3 \mathbf{n}_3) \\
&= a_1 b_1 \mathbf{n}_1 \mathbf{n}_1 + a_1 b_2 \mathbf{n}_1 \mathbf{n}_2 + a_1 b_3 \mathbf{n}_1 \mathbf{n}_3 \\
&\quad + a_2 b_1 \mathbf{n}_2 \mathbf{n}_1 + a_2 b_2 \mathbf{n}_2 \mathbf{n}_2 + a_2 b_3 \mathbf{n}_2 \mathbf{n}_3 \\
&\quad + a_3 b_1 \mathbf{n}_3 \mathbf{n}_1 + a_3 b_2 \mathbf{n}_3 \mathbf{n}_2 + a_3 b_3 \mathbf{n}_3 \mathbf{n}_3
\end{aligned} \tag{3.42}
$$

where, as before, a_i and b_i are scalar components of **A** and **B** relative to mutually perpendicular unit vectors $\mathbf{n}_i$ ($i = 1, 2, 3$).

The dyadic product of two vectors is simply the multiplication of the vectors following the usual rules of algebra except the commutative rule. That is,

$$\mathbf{AB} \neq \mathbf{BA} \tag{3.43}$$

and more specifically, in Equation 3.42, the relative positions of the individual unit vectors must be maintained. Thus,

$$\mathbf{n_1n_2} \neq \mathbf{n_2n_1}, \quad \mathbf{n_1n_3} \neq \mathbf{n_3n_1}, \quad \mathbf{n_2n_3} \neq \mathbf{n_3n_2} \tag{3.44}$$

3.4 DYADICS

The dyadic product of two vectors (as in Sections 3.4.1 through 3.4.7) and particularly of two unit vectors is often called a dyad. Occasionally, a dyadic product $\mathbf{D}$ may occur in the form

$$\mathbf{D} = \mathbf{An_1} + \mathbf{Bn_2} + \mathbf{Cn_3} \tag{3.45}$$

where, as before, $\mathbf{n_1}$, $\mathbf{n_2}$, and $\mathbf{n_3}$ are mutually perpendicular unit vectors and $\mathbf{A}$, $\mathbf{B}$, and $\mathbf{C}$ are vectors. In this case, the product $\mathbf{D}$ is a sum of dyads and is called a dyadic. The product may also be viewed as a vector whose components are vectors. Thus, dyadics are sometimes called vector–vectors.

Dyadics are useful in continuum mechanics for the representation of stress and strain. In dynamics, dyadics represent inertia properties of bodies. In these applications, dyadics are often expressed in the form

$$\mathbf{D} = d_{ij}\mathbf{n_i}\mathbf{n_j} \tag{3.46}$$

where the d_{ij} are regarded as the scalar components of $\mathbf{D}$ relative to the dyads $\mathbf{n_i}\mathbf{n_j}$. The d_{ij} are conveniently arranged in an array, or matrix, as

$$d_{ij} = \begin{bmatrix} d_{11} & d_{12} & d_{13} \\ d_{21} & d_{22} & d_{23} \\ d_{31} & d_{32} & d_{33} \end{bmatrix} \tag{3.47}$$

The following sections describe several special dyadics that are useful in dynamics and continuum mechanics.

3.4.1 ZERO DYADIC

A dyadic whose scalar components are all zero is called a zero dyadic and written simply as 0 (without boldface).

3.4.2 IDENTITY DYADIC

A dyadic whose scalar components have correspondingly the same values as the Kronecker delta (see Equation 3.23) is called the identity dyadic and is usually designated by $\mathbf{I}$. If, as before, $\mathbf{n_i}$ ($i = 1, 2, 3$) are mutually perpendicular unit vectors, then $\mathbf{I}$ may be written as

$$\mathbf{I} = \delta_{ij}\mathbf{n_i}\mathbf{n_j} = \mathbf{n_1n_1} + \mathbf{n_2n_2} + \mathbf{n_3n_3} \tag{3.48}$$

where, from Equation 3.23, the array of δ_{ij} is

$$\delta_{ij} = \begin{bmatrix} 1 & 0 & 0 \\ 0 & 1 & 0 \\ 0 & 0 & 1 \end{bmatrix} \tag{3.49}$$

3.4.3 DYADIC TRANSPOSE

Let **A** be a dyadic with scalar components a_{ij}, which, relative to mutually perpendicular unit vectors $\mathbf{n}_i$, has the form

$$\mathbf{A} = a_{ij}\mathbf{n}_i\mathbf{n}_j \tag{3.50}$$

The dyadic formed by interchanging the rows and columns of the a_{ij} array is called the transpose of the dyadic or dyadic transpose and is written as $\mathbf{A}^T$. $\mathbf{A}^T$ then has the form

$$\mathbf{A}^T = a_{ji}\mathbf{n}_i\mathbf{n}_j \tag{3.51}$$

3.4.4 SYMMETRIC DYADICS

If a dyadic **A** is equal to its transpose, it is said to be symmetric. That is, **A** is symmetric if (and only if)

$$\mathbf{A} = \mathbf{A}^T \quad \text{or equivalently } a_{ij} = a_{ji} \tag{3.52}$$

where, as before, a_{ij} are the scalar components of **A**.

3.4.5 MULTIPLICATION OF DYADICS

Dyadics, when viewed as vector–vectors, may be multiplied among themselves or with vectors using the dot, cross, or dyadic product. The most useful of these products is the dot product.

As an illustration, if **A** is a dyadic with components a_{ij} and **v** is a vector with components v_k, referred to as mutually perpendicular unit vectors, then the dot product **w** of **A** and **v** is a vector given by

$$\mathbf{w} = \mathbf{A} \cdot \mathbf{v} = a_{ij}\mathbf{n}_i\mathbf{n}_j \cdot v_k\mathbf{n}_k = \mathbf{n}_i a_{ij}\mathbf{n}_j \cdot \mathbf{n}_k v_k$$

$$= \mathbf{n}_i a_{ij}\delta_{jk}v_k = \mathbf{n}_i a_{ij}v_j = w_i\mathbf{n}_i \tag{3.53}$$

where the components w_i of **w** are

$$w_i = a_{ij}v_j \tag{3.54}$$

Similarly, if **B** is a dyadic with components b_{kl}, the dot product **C** of **A** and **B** is

$$\mathbf{C} = \mathbf{A} \cdot \mathbf{B} = a_{ij}\mathbf{n}_i\mathbf{n}_j \cdot b_{k\ell}\mathbf{n}_k\mathbf{n}_\ell$$

$$= \mathbf{n}_i a_{ij}\mathbf{n}_j \cdot \mathbf{n}_k b_{k\ell}\mathbf{n}_\ell = \mathbf{n}_i a_{ij}\delta_{jk}b_{k\ell}\mathbf{n}_\ell$$

$$= \mathbf{n}_j a_{ij}b_{j\ell}\mathbf{n}_\ell = c_{i\ell}\mathbf{n}_i\mathbf{n}_\ell \tag{3.55}$$

where the components $c_{i\ell}$ of $\mathbf{C}$ are

$$c_{i\ell} = a_{ij}b_{j\ell} \tag{3.56}$$

Observe in Equations 3.54 and 3.56 that the products obey the same rules as matrix products (see Section 3.6.11).

3.4.6   INVERSE DYADICS

If the dot product of two dyadics, say $\mathbf{A}$ and $\mathbf{B}$, produces the identity dyadic (see Section 3.4.2), then $\mathbf{A}$ and $\mathbf{B}$ are said to be inverses of each other and are written as $\mathbf{B}^{-1}$ and $\mathbf{A}^{-1}$. Specifically, if a_{ij} and b_{ij} are the components of $\mathbf{A}$ and $\mathbf{B}$ relative to mutually perpendicular unit vectors $\mathbf{n}_i$ and if $\mathbf{A}$ and $\mathbf{B}$ are inverses of each other, we have the relations

$$\mathbf{A} \cdot \mathbf{B} = \mathbf{B} \cdot \mathbf{A} = \mathbf{I} \tag{3.57}$$

$$\mathbf{A} = \mathbf{B}^{-1} \quad \text{and} \quad \mathbf{B} = \mathbf{A}^{-1} \tag{3.58}$$

$$\mathbf{A} = a_{ij}\mathbf{n}_i\mathbf{n}_j \quad \text{and} \quad \mathbf{B} = b_{ij}\mathbf{n}_i\mathbf{n}_j \tag{3.59}$$

$$a_{ij}b_{jk} = \delta_{ij} \quad \text{and} \quad b_{ij}a_{jk} = \delta_{ij} \tag{3.60}$$

3.4.7   ORTHOGONAL DYADICS

If the inverse of a dyadic is equal to its transpose, the dyadic is said to be orthogonal. That is, a dyadic $\mathbf{A}$ is orthogonal if

$$\mathbf{A}^{-1} = \mathbf{A}^T \tag{3.61}$$

When a dyadic is orthogonal, the rows (and columns) of the matrix of components form the components of mutually perpendicular unit vectors, since then $\mathbf{A} \cdot \mathbf{A}^T = \mathbf{A}^T \cdot \mathbf{A} = \mathbf{I}$. (This is the reason for the name "orthogonal.")

3.5   MULTIPLE PRODUCTS OF VECTORS

We can multiply vectors with products of vectors, thus producing multiple products of vectors. The most common and useful of these are the scalar triple product, the vector triple product, and the product of a dyadic and a vector. These are discussed in the following sections.

3.5.1   SCALAR TRIPLE PRODUCT

As the name implies, the scalar triple product is a product of three vectors resulting in a scalar. Let $\mathbf{A}$, $\mathbf{B}$, and $\mathbf{C}$ be vectors and as before, let $\mathbf{n}_i$ ($i = 1, 2, 3$) be mutually perpendicular unit vectors so that $\mathbf{A}$, $\mathbf{B}$, and $\mathbf{C}$ may be expressed in the forms

$$\mathbf{A} = a_i\mathbf{n}_i \quad \mathbf{B} = b_i\mathbf{n}_i \quad \mathbf{C} = c_i\mathbf{n}_i \tag{3.62}$$

where a_i, b_i, and c_i are the scalar components of **A**, **B**, and **C** relative to the $\mathbf{n}_i$. The scalar triple product s of **A**, **B**, and **C** may then be expressed as

$$s = (\mathbf{A} \times \mathbf{B}) \cdot \mathbf{C} \tag{3.63}$$

Recall from Equation 3.41 that the vector product $\mathbf{A} \times \mathbf{B}$ may be expressed in the determinantal form as

$$\mathbf{A} \times \mathbf{B} = \begin{vmatrix} \mathbf{n}_1 & \mathbf{n}_2 & \mathbf{n}_3 \\ a_1 & a_2 & a_3 \\ b_1 & b_2 & b_3 \end{vmatrix} \tag{3.64}$$

Also, recall from Equation 3.27 that the scalar product of two vectors, say **D** and **C**, may be written in the form

$$\mathbf{D} \cdot \mathbf{C} = d_i c_i = d_1 c_1 + d_2 c_2 + d_3 c_3 \tag{3.65}$$

where d_i are the $\mathbf{n}_i$ components of **D**. From an algorithmic perspective, the right side of Equation 3.65 may be viewed as a replacement of the unit vectors of **D** by the scalar components of **C**. In this regard, if **D** represents $\mathbf{A} \times \mathbf{B}$ in Equations 3.63 and 3.64, then by comparing with Equation 3.65, we have

$$s = (\mathbf{A} \times \mathbf{B}) \cdot \mathbf{C} = \begin{vmatrix} c_1 & c_2 & c_3 \\ a_1 & a_2 & a_3 \\ b_1 & b_2 & b_3 \end{vmatrix} \tag{3.66}$$

Recall from the rules for evaluating determinants that we may cyclically permutate the rows without changing the value of the determinant and that by interchanging two rows, we change the sign of the value. That is,

$$s = \begin{vmatrix} c_1 & c_2 & c_3 \\ a_1 & a_2 & a_3 \\ b_1 & b_2 & b_3 \end{vmatrix} = \begin{vmatrix} a_1 & a_2 & a_3 \\ b_1 & b_2 & b_3 \\ c_1 & c_2 & c_3 \end{vmatrix} = \begin{vmatrix} b_1 & b_2 & b_3 \\ c_1 & c_2 & c_3 \\ a_1 & a_2 & a_3 \end{vmatrix} = -\begin{vmatrix} a_1 & a_2 & a_3 \\ c_1 & c_2 & c_3 \\ b_1 & b_2 & b_3 \end{vmatrix}$$

$$= -\begin{vmatrix} c_1 & c_2 & c_3 \\ b_1 & b_2 & b_3 \\ a_1 & a_2 & a_3 \end{vmatrix} = -\begin{vmatrix} b_1 & b_2 & b_3 \\ a_1 & a_2 & a_3 \\ c_1 & c_2 & c_3 \end{vmatrix} \tag{3.67}$$

By comparing Equations 3.66 and 3.67, we see that, in Equation 3.66, the dot and cross may be interchanged and that the vectors may be cyclically permutated without affecting the value of the triple product. Also, by interchanging any two of the vectors, we change the sign of the value of the product. That is,

$$(\mathbf{A} \times \mathbf{B}) \cdot \mathbf{C} = \mathbf{A} \cdot (\mathbf{B} \times \mathbf{C}) = (\mathbf{C} \times \mathbf{A}) \cdot \mathbf{B} = \mathbf{C} \cdot (\mathbf{A} \times \mathbf{B}) = (\mathbf{B} \times \mathbf{C}) \cdot \mathbf{A}$$

$$= \mathbf{B} \cdot (\mathbf{C} \times \mathbf{A}) = -(\mathbf{B} \times \mathbf{A}) \cdot \mathbf{C} = -\mathbf{B} \cdot (\mathbf{A} \times \mathbf{C}) = -(\mathbf{C} \times \mathbf{B}) \cdot \mathbf{A}$$

$$= -\mathbf{C} \cdot (\mathbf{B} \times \mathbf{A}) = -(\mathbf{A} \times \mathbf{C}) \cdot \mathbf{B} = -\mathbf{A} \cdot (\mathbf{C} \times \mathbf{B}) \tag{3.68}$$

Finally, observe that since there is no definition for the vector product of a vector and a scalar, the vector product of a triple scalar product must be evaluated first. Therefore, the parentheses in the foregoing equations are not necessary.

3.5.2 VECTOR TRIPLE PRODUCT

If we have a vector product of two vectors and then take the vector product with a third vector, we produce a vector triple product. As this process and name implies, the result is a vector. To illustrate this, let $\mathbf{A}$, $\mathbf{B}$, and $\mathbf{C}$ be vectors. Then, there are two forms of vector triple products: $(\mathbf{A} \times \mathbf{B}) \times \mathbf{C}$ and $\mathbf{A} \times (\mathbf{B} \times \mathbf{C})$. If we express $\mathbf{A}$, $\mathbf{B}$, and $\mathbf{C}$ in terms of mutually perpendicular unit vectors $\mathbf{n}_i$, as in Equation 3.62, then by using the index notation together with the Kronecker delta function and permutation symbols (see Equations 3.25 and 3.35), we can express these triple products in terms of scalar products as follows:

$$(\mathbf{A} \times \mathbf{B}) \times \mathbf{C} = (\mathbf{A} \cdot \mathbf{C})\mathbf{B} - (\mathbf{B} \cdot \mathbf{C})\mathbf{A} \tag{3.69}$$

and

$$\mathbf{A} \times (\mathbf{B} \times \mathbf{C}) = (\mathbf{A} \cdot \mathbf{C})\mathbf{B} - (\mathbf{A} \cdot \mathbf{B})\mathbf{C} \tag{3.70}$$

Observe that the last terms of Equations 3.69 and 3.70 are different. Therefore, the two products are distinct and thus determined by the position of the parentheses. That is, unlike the scalar triple product, the parentheses are necessary. A development of Equations 3.69 and 3.70 is given in Section 3.8.

3.5.3 DYADIC/VECTOR PRODUCT

Let $\mathbf{A}$ be a dyadic and $\mathbf{x}$ be a vector and, as before, let both $\mathbf{A}$ and $\mathbf{x}$ be referred to as mutually perpendicular unit vectors so that they may be expressed in terms of scalar components in the form

$$\mathbf{A} = a_{ij}\mathbf{n}_i\mathbf{n}_j \quad \text{and} \quad \mathbf{x} = x_i\mathbf{n}_i \tag{3.71}$$

Then, the dot product $\mathbf{y}$ of $\mathbf{A}$ and $\mathbf{x}$ may be defined as

$$\mathbf{y} = \mathbf{A} \cdot \mathbf{x} = a_{ij}\mathbf{n}_i\mathbf{n}_j \cdot x_k\mathbf{n}_k = \mathbf{n}_i a_{ij} x_k \mathbf{n}_j \cdot \mathbf{n}_k$$

$$= \mathbf{n}_i a_{ij} x_k \delta_{jk} = a_{ij} x_j \mathbf{n}_i \tag{3.72}$$

Observe that the product, as defined, produces a vector, $\mathbf{y}$. If $\mathbf{y}$ is expressed in component form as $y_i\mathbf{n}_i$, then from Equation 3.72, we have

$$y_i\mathbf{n}_i = a_{ij} x_j \mathbf{n}_i \quad \text{or} \quad y_i = a_{ij} x_j \quad (i = 1, 2, 3) \tag{3.73}$$

where the individual y_i are

$$y_1 = a_{11}x_1 + a_{12}x_2 + a_{13}x_3$$

$$y_2 = a_{21}x_1 + a_{22}x_2 + a_{23}x_3 \tag{3.74}$$

$$y_3 = a_{31}x_1 + a_{32}x_2 + a_{33}x_3$$

Now, consider the product $\mathbf{x} \cdot \mathbf{B}$ where $\mathbf{B}$ is a dyadic with scalar components b_{ij}:

$$\mathbf{x} \cdot \mathbf{B} = x_i \mathbf{n}_i \cdot b_{jk} \mathbf{n}_j \mathbf{n}_k = x_i b_{jk} \mathbf{n}_i \cdot \mathbf{n}_j \mathbf{n}_k$$

$$= x_k b_{jk} \delta_{ij} \mathbf{n}_k = x_i b_{ik} \mathbf{n}_k \tag{3.75}$$

Observe that, as in the product of Equation 3.72, $\mathbf{x} \cdot \mathbf{B}$ is a vector. If we name this vector $\mathbf{w}$, with scalar components w_k, then $\mathbf{w}$ and the w_k ($k = 1, 2, 3$) may be expressed as

$$\mathbf{w} = \mathbf{x} \cdot \mathbf{B} \quad \text{and} \quad w_k = x_i b_{ik} \quad (k = 1, 2, 3) \tag{3.76}$$

where the individual w_k are

$$w_1 = x_1 b_{11} + x_2 b_{21} + x_3 b_{31}$$

$$w_2 = x_1 b_{12} + x_2 b_{22} + x_3 b_{32} \tag{3.77}$$

$$w_3 = x_1 b_{13} + x_2 b_{23} + x_3 b_{33}$$

By comparing the pattern of the indices in Equations 3.74 and 3.77, we see that the subscripts of $\mathbf{B}$ are the reverse of those of $\mathbf{A}$. Therefore, $\mathbf{w}$ could also be expressed as

$$\mathbf{w} = \mathbf{x} \cdot \mathbf{B} = \mathbf{B}^T \cdot \mathbf{x} \tag{3.78}$$

where, as before, $\mathbf{B}^T$ is the transpose of $\mathbf{B}$ (see Section 3.4.3).

3.5.4 OTHER MULTIPLE PRODUCTS

We list here a few less well-known multiple product identities, which may be of use in advanced biomechanical analyses [9]. Let $\mathbf{A}$, $\mathbf{B}$, $\mathbf{C}$, $\mathbf{D}$, $\mathbf{E}$, and $\mathbf{F}$ be vectors. Then,

$$(\mathbf{A} \times \mathbf{B}) \times (\mathbf{C} \times \mathbf{D}) = (\mathbf{A} \times \mathbf{B} \cdot \mathbf{D})\mathbf{C} - (\mathbf{A} \times \mathbf{B} \cdot \mathbf{C})\mathbf{D}$$

$$= (\mathbf{C} \times \mathbf{D} \cdot \mathbf{A})\mathbf{B} - (\mathbf{C} \times \mathbf{D} \cdot \mathbf{B})\mathbf{A} \tag{3.79}$$

$$(\mathbf{A} \times \mathbf{B} \cdot \mathbf{C})(\mathbf{D} \times \mathbf{E} \cdot \mathbf{F}) = \begin{vmatrix} \mathbf{A} \cdot \mathbf{D} & \mathbf{A} \cdot \mathbf{E} & \mathbf{A} \cdot \mathbf{F} \\ \mathbf{B} \cdot \mathbf{D} & \mathbf{B} \cdot \mathbf{E} & \mathbf{B} \cdot \mathbf{F} \\ \mathbf{C} \cdot \mathbf{D} & \mathbf{C} \cdot \mathbf{E} & \mathbf{C} \cdot \mathbf{F} \end{vmatrix} \tag{3.80}$$

$$(\mathbf{A} \times \mathbf{B}) \cdot (\mathbf{C} \times \mathbf{D}) = (\mathbf{A} \cdot \mathbf{B})(\mathbf{B} \cdot \mathbf{D}) - (\mathbf{A} \cdot \mathbf{D})(\mathbf{B} \cdot \mathbf{C})$$

$$= \begin{vmatrix} \mathbf{A} \cdot \mathbf{C} & \mathbf{A} \cdot \mathbf{D} \\ \mathbf{B} \cdot \mathbf{C} & \mathbf{B} \cdot \mathbf{D} \end{vmatrix} \tag{3.81}$$

$$\mathbf{A} \times \left[\mathbf{B} \times (\mathbf{C} \times \mathbf{D}) \right] = (\mathbf{B} \cdot \mathbf{D})(\mathbf{A} \times \mathbf{C}) - (\mathbf{B} \cdot \mathbf{C})(\mathbf{A} \times \mathbf{D}) \tag{3.82}$$

3.6 MATRICES/ARRAYS

A matrix is simply a structured array of numbers [10–13]. These numbers, which can be represented by variables, are called the "elements" of the matrix. Matrices are usually designated by capital letters and their elements by lowercase subscripted letters. For example, consider an array of 12 numbers arranged in 3 rows and 4 columns. Let this array be called A and let the numbers (whatever their values) be designated by a_{ij} (i = 1, 2, 3; j = 1, …, 4). That is,

$$A = \begin{bmatrix} a_{11} & a_{12} & a_{13} & a_{14} \\ a_{21} & a_{22} & a_{23} & a_{24} \\ a_{31} & a_{32} & a_{33} & a_{34} \end{bmatrix} = [a_{ij}] \tag{3.83}$$

where the subscripts i and j of a_{ij} designate the row (first subscript) and the column (second subscript) of the element.

A matrix with m rows and n columns is said to be an $m \times n$ array or an $m \times n$ matrix. The numbers m and n are said to be the dimensions or order of the matrix. A matrix with only one row is said to be a row array or row matrix. Correspondingly, a matrix with only one column is said to be a column array or column matrix. A matrix with the same number of rows as columns is said to be a square matrix.

There is no limit to the number of rows or columns in a matrix. In our biomechanical analyses, we will mostly use 3×3 arrays and row and column arrays.

In the following sections, we briefly review special matrices and elementary matrix operations.

3.6.1 ZERO MATRICES

If all the elements of a matrix are zero, the matrix is said to be a zero matrix, denoted by 0. A zero matrix may have any dimension, that is, a zero matrix may be a row matrix, a column matrix, or a rectangular array.

3.6.2 IDENTITY MATRICES

For a square matrix, the diagonal is the position of all the elements with the same row and column numbers, that is, $a_{11}, a_{22}, …, a_{nn}$ for an $n \times n$ array. The elements of the diagonal are called diagonal elements, and the remaining elements are called off-diagonal elements.

If all the diagonal elements are 1 and all the off-diagonal elements are zero, the matrix is said to be an identity matrix, usually denoted by I. An identity matrix can have any dimension.

3.6.3 MATRIX TRANSPOSE

If the rows and columns of matrix A are interchanged, the resulting matrix A^T is said to be the transpose of A.

3.6.4 EQUAL MATRICES

Two matrices A and B are said to be equal if they have equal elements, respectively. That is,

$$A = B \quad \text{if, and only if, } a_{ij} = b_{ij} \tag{3.84}$$

Observe that equal matrices must have the same dimensions, that is, the same number of rows and columns.

3.6.5   SYMMETRIC MATRICES

Matrix A is said to be symmetric if it is equal to its transpose A^T. Observe that these symmetric matrices are square and that

$$A = A^T \quad \text{if, and only if } a_{ij} = a_{ji} \tag{3.85}$$

3.6.6   SKEW-SYMMETRIC MATRICES

Matrix A is said to be skew-symmetric if (1) it is square, (2) its diagonal elements are zero, and (3) its off-diagonal elements are negative to their respective elements on the other side of the diagonal. That is, A is skew-symmetric if, and only if,

$$a_{ii} = 0 \text{ (no sum)} \quad \text{and} \quad a_{ij} = -a_{ji} \tag{3.86}$$

3.6.7   DIAGONAL MATRIX

If a square matrix has zero-valued elements off the diagonal but nonzero elements on the diagonal, it is called a diagonal matrix. The identity matrices are diagonal matrices.

3.6.8   MATRIX MEASURES

Two prominent scalar measures of square matrices are the trace and the determinant. The trace is the sum of the elements on the diagonal. The determinant is the sum of products of elements and negatives of elements using minors and cofactors as discussed in elementary algebra courses and as reviewed in Section 3.7. For a diagonal matrix, the determinant is simply the product of the diagonal elements. It follows that for an identity matrix, the determinant is 1.

3.6.9   SINGULAR MATRICES

If the determinant of a matrix is zero, the matrix is said to be singular.

3.6.10   MULTIPLICATION OF MATRICES BY SCALARS

Let s be a scalar and A be a matrix with elements a_{ij}. Then the product sA is a matrix whose elements are sa_{ij}. That is,

$$sA = s[a_{ij}] = [sa_{ij}] \tag{3.87}$$

The negative of a matrix, $-A$, is obtained by multiplying the matrix (A) by -1. That is,

$$-A = (-1)A \tag{3.88}$$

3.6.11   ADDITION OF MATRICES

If two matrices, say A and B, have the same dimensions, they may be added. The sum C is simply the matrix whose elements c_{ij} are the respective sums of the elements a_{ij} and b_{ij} of A and B. That is,

$$C = A + B \quad \text{if, and only if, } c_{ij} = a_{ij} + b_{ij} \tag{3.89}$$

Matrix subtraction is accomplished by adding the negative (see Equation 3.88) of a matrix to be subtracted. That is,

$$A - B = A + (-B) \tag{3.90}$$

3.6.12 MULTIPLICATION OF MATRICES

Matrices may be multiplied using the so-called row–column rule. If C is the product of matrices A and B, written simply as AB, then the element c_{ij}, in the ith row and jth column of C, is the sum of element by element products of the ith row of A with the jth column of B. Specifically,

$$C = AB \quad \text{if, and only if,} \ c_{ij} = a_{ik}b_{kj} \tag{3.91}$$

where, as before, the repeated index k designates a sum over the range of the index.

Observe that the sum of products in Equation 3.91 is meaningful only if the number of elements in the rows of A is the same as the number of elements in the columns of B, or alternatively, the number of columns of A is the same as the number of rows of B. When this happens, the matrices A and B are said to be conformable.

It is readily seen that matrix multiplication obeys the associative and distributive laws:

$$ABC = (AB)C = A(BC) \tag{3.92}$$

and

$$A(B + C) = AB + AC \quad \text{and} \quad (A + B)C = AC + BC \tag{3.93}$$

However, matrix multiplication in general is not commutative. That is,

$$AB \neq BA \tag{3.94}$$

Also, from the definitions of transpose (Section 3.6.3), we have

$$(AB)^T = B^T A^T \tag{3.95}$$

3.6.13 INVERSE MATRICES

Let A and B be square matrices having the same dimensions (thus conformable). If the product AB is an identity matrix I, then A and B are said to be inverses of each other, written as B^{-1} and A^{-1}. That is,

$$\text{If } AB = I, \quad \text{then } A = B^{-1} \quad \text{and} \quad B = A^{-1} \tag{3.96}$$

From this definition, it is also seen that the inverse of a product of matrices is the product of the inverses of the matrices in reverse order. That is,

$$(CD)^{-1} = D^{-1}C^{-1} \tag{3.97}$$

Similarly, it is seen that

$$(A^T)^{-1} = (A^{-1})^T \tag{3.98}$$

3.6.14 ORTHOGONAL MATRICES

If a matrix inverse is equal to its transpose, the matrix is said to be orthogonal. That is,

$$A^T = A^{-1} \tag{3.99}$$

3.6.15 SUBMATRICES

If rows or columns are deleted from matrix A, the array remaining forms a matrix $\bar{A}$ called a submatrix of A.

3.6.16 RANK

The dimension of the largest nonsingular submatrix of a matrix is called the rank of a matrix.

3.6.17 PARTITIONING OF MATRICES, BLOCK MULTIPLICATION

A matrix can be divided into submatrices by positioning, illustrated as follows:

$$A = [A] = \begin{bmatrix} A_{11} & A_{12} & A_{13} \\ A_{21} & A_{22} & A_{23} \\ A_{31} & A_{32} & A_{33} \end{bmatrix} \tag{3.100}$$

where the A_{ij} are submatrices. A partitioned matrix is made up of rows and columns of matrices. If the submatrices of two partitioned matrices are conformable, the matrices may be multiplied using the row–column rule as if the submatrices were elements. For example, if a matrix B is partitioned into three submatrices conformable to the submatrices of the columns of A of Equation 3.100, then the product AB may be expressed as

$$AB = \begin{bmatrix} A_{11} & A_{12} & A_{13} \\ A_{21} & A_{22} & A_{23} \\ A_{31} & A_{32} & A_{33} \end{bmatrix} \begin{bmatrix} B_1 \\ B_2 \\ B_3 \end{bmatrix} = \begin{bmatrix} A_{11}B_1 + A_{12}B_2 + A_{13}B_3 \\ A_{21}B_1 + A_{22}B_2 + A_{23}B_3 \\ A_{31}B_1 + A_{32}B_2 + A_{33}B_3 \end{bmatrix} \tag{3.101}$$

This matrix multiplication is called block multiplication.

3.6.18 PSEUDOINVERSE

If a matrix is singular, it has no inverse. However, a pseudoinverse, useful in certain least squares approximations, can be constructed for singular and even nonsquare matrices. If an $m \times n$ matrix A has rank n, then the pseudoinverse of A, written as A^+, is

$$A^+ = (A^T A)^{-1} A^T \tag{3.102}$$

Similarly, if A has rank m, A^+ is

$$A^+ = A^T (AA^T)^{-1} \tag{3.103}$$

The development of these concepts is beyond our scope, but derivations and details may be found in Refs. [14–16].

3.7   DETERMINANTS

For square matrices, the determinant is a number (or scalar) used as a measure of the matrix. Recall that determinants play a central role in using Cramer's rule [17] where matrices are used in the solution of simultaneous linear algebraic equations.

The determinant of matrix A is usually designated by vertical lines on the sides of A or on the sides of the elements a_{ij} of A. That is,

$$\det A = |A| = \begin{vmatrix} a_{11}a_{12} & \cdots & a_{1n} \\ a_{21}a_{22} & \cdots & a_{2n} \\ \vdots & & \vdots \\ a_{n1}a_{n2} & \cdots & a_{nn} \end{vmatrix} = |a_{ij}| \tag{3.104}$$

If A has only one row (and one column), that is, if it is a single element matrix, the determinant is defined as that single element. Thus,

$$\det A = |a_{11}| = a_{11} \tag{3.105}$$

The determinant of higher-dimension matrices may then be defined in terms of the determinants of submatrices of lower order as follows: let A be an $n \times n$ array and A_{ij} be a submatrix of A formed by deleting the ith row and the jth column of A. Let the determinant of A_{ij} be M_{ij} and M_{ij} be called the minor of element a_{ij}. Then the cofactor C_{ij} of a_{ij} is defined in terms of M_{ij} as

$$C_{ij} = (-1)^{i+j} M_{ij} \tag{3.106}$$

The determinant of A is defined as a sum of products of elements, of any row or column of A, with their cofactors. That is,

$$\det A = |A| = \sum_{i=1}^{n} a_{ij}C_{ij} \quad (j = 1, \ldots, n \text{ with no sum on } j) \tag{3.107}$$

or

$$\det A = |A| = \sum_{i=1}^{n} a_{ij}C_{ij} \quad (i = 1, \ldots, n \text{ with no sum on } i) \tag{3.108}$$

Even though the choice of row or column for evaluating a determinant in Equations 3.107 and 3.108 is arbitrary, the value of the determinant is nevertheless unique (see, e.g., Ref. [18]).

The expansions of Equations 3.107 and 3.108 together with Equation 3.105 may be used to inductively determine the determinant of any size array. To develop this, consider first the 2×2 array A: Using Equations 3.105 and 3.108 and by expanding in the first row of A, we have

$$\det A = \begin{vmatrix} a_{11} & a_{12} \\ a_{21} & a_{22} \end{vmatrix} = a_{11} \left| a_{22} \right| - a_{12} \left| a_{21} \right| = a_{11} a_{22} - a_{12} a_{21} \tag{3.109}$$

By a similar procedure, the determinant of a 3×3 array A may be expressed as

$$\det A = \begin{vmatrix} a_{11} & a_{12} & a_{13} \\ a_{21} & a_{22} & a_{23} \\ a_{31} & a_{32} & a_{33} \end{vmatrix} = a_{11} \begin{vmatrix} a_{22} & a_{23} \\ a_{32} & a_{33} \end{vmatrix} - a_{12} \begin{vmatrix} a_{21} & a_{23} \\ a_{31} & a_{33} \end{vmatrix} + a_{13} \begin{vmatrix} a_{21} & a_{22} \\ a_{31} & a_{32} \end{vmatrix}$$

$$= a_{11}(a_{22}a_{33} - a_{23}a_{32}) - a_{12}(a_{21}a_{33} - a_{23}a_{31}) + a_{13}(a_{21}a_{32} - a_{22}a_{31})$$

$$= a_{11}a_{22}a_{33} - a_{11}a_{23}a_{32} + a_{12}a_{23}a_{31} - a_{12}a_{21}a_{33} + a_{13}a_{21}a_{32} - a_{13}a_{22}a_{31} \tag{3.110}$$

Higher-order determinants may be evaluated using the same procedure. Observe, however, that the number of terms rapidly increases with the order of the array.

In Equation 3.110, we evaluated the determinant by expansion through the first row, that is, by multiplying the elements of the first row by their respective cofactors and then adding the products. To see that the same result is obtained by expansion using a different row or column, consider expansion using the second column:

$$\det A = \begin{vmatrix} a_{11} & a_{12} & a_{13} \\ a_{21} & a_{22} & a_{23} \\ a_{31} & a_{32} & a_{33} \end{vmatrix} = -a_{12} \begin{vmatrix} a_{21} & a_{23} \\ a_{31} & a_{33} \end{vmatrix} + a_{22} \begin{vmatrix} a_{11} & a_{13} \\ a_{31} & a_{33} \end{vmatrix} - a_{32} \begin{vmatrix} a_{11} & a_{13} \\ a_{21} & a_{23} \end{vmatrix}$$

$$= -a_{12}(a_{21}a_{33} - a_{31}a_{23}) + a_{22}(a_{11}a_{33} - a_{31}a_{13}) - a_{32}(a_{11}a_{23} - a_{21}a_{13})$$

$$= -a_{12}a_{21}a_{33} + a_{12}a_{31}a_{23} + a_{22}a_{11}a_{33} - a_{22}a_{31}a_{13} - a_{32}a_{11}a_{23} + a_{32}a_{21}a_{13} \tag{3.111}$$

The results of Equations 3.110 and 3.111 are seen to be the same.

The definition of the determinant also induces the following properties of determinants [18]:

1. The interchange of the rows and columns does not change the value of the determinant. That is, a square matrix A and its transpose A^T have the same determinant value.
2. The interchange of any two rows (or any two columns) produces a determinant with the negative value of the original determinant.
3. The rows (or columns) may be cyclically permutated without affecting the value of the determinant.
4. If all the elements in any row (or column) are zero, the value of the determinant is zero.
5. If the elements in any row (or column) are, respectively, proportional to the elements in any other row (or column), the value of the determinant is zero.
6. If the elements of a row (or column) are multiplied by a constant, the value of the determinant is multiplied by the constant.
7. If the elements of a row (or column) are multiplied by a constant and, respectively, added to the elements of another row (or column), the value of the determinant is unchanged.

3.8 RELATIONSHIP OF 3 × 3 DETERMINANTS, PERMUTATION SYMBOLS, AND KRONECKER DELTA FUNCTIONS

Recall that we already used determinants in our discussion of vector products and vector triple products (see Equations 3.41 and 3.66). From those discussions, we can express properties of 3 × 3 determinants in terms of the permutation symbols (see Equation 3.36). We can then use these results to obtain useful relations between the permutation symbols and Kronecker delta functions.

To develop this, recall from Equation 3.39 that the vector product of two vectors **A** and **B** may be expressed as

$$\mathbf{A} \times \mathbf{B} = e_{ijk} a_i b_j \mathbf{n}_k \tag{3.112}$$

where, as before, $\mathbf{n}_k$ are mutually perpendicular dextral unit vectors and a_k and b_k are the scalar components of **A** and **B** relative to the $\mathbf{n}_k$. Then the scalar product of **A** × **B** with a vector **C** (expressed as $c_\ell \mathbf{n}_\ell$) is simply

$$(\mathbf{A} \times \mathbf{B}) \cdot \mathbf{C} = (e_{ijk} a_i b_j \mathbf{n}_k) \cdot (c_\ell \mathbf{n}_\ell) = e_{ijk} a_i b_j c_\ell (\mathbf{n}_k \cdot \mathbf{n}_\ell)$$

$$= e_{ijk} a_i b_j c_\ell \delta_{k\ell} = e_{ijk} a_i b_j c_k \tag{3.113}$$

where c_k are components of **C** relative to the $\mathbf{n}_k$. But, from Equations 3.66 and 3.67, we see that the triple scalar product may be expressed as

$$\mathbf{A} \times \mathbf{B} \cdot \mathbf{C} = \begin{vmatrix} a_1 & a_2 & a_3 \\ b_1 & b_2 & b_3 \\ c_1 & c_2 & c_3 \end{vmatrix} \tag{3.114}$$

Thus, we have the relation

$$\begin{vmatrix} a_1 & a_2 & a_3 \\ b_1 & b_2 & b_3 \\ c_1 & c_2 & c_3 \end{vmatrix} = e_{ijk} a_i b_j c_k \tag{3.115}$$

Next, suppose that **A**, **B**, and **C** are renamed as $\mathbf{a}_1$, $\mathbf{a}_2$, and $\mathbf{a}_3$, respectively, and that correspondingly, the scalar components a_i, b_i, and c_i are renamed as a_{1i}, a_{2i}, and a_{3i}, respectively. Then Equation 3.115 may be rewritten in the form

$$\begin{vmatrix} a_{11} & a_{12} & a_{13} \\ a_{21} & a_{22} & a_{23} \\ a_{31} & a_{32} & a_{33} \end{vmatrix} = e_{ijk} a_{1i} a_{2j} a_{3k} \tag{3.116}$$

Now, if a_{1i}, a_{2i}, and a_{3i} are viewed as elements of a 3 × 3 matrix A whose determinant is "a," we can write Equation 3.116 in an expanded form as

$$\mathbf{a}_1 \times \mathbf{a}_2 \cdot \mathbf{a}_3 = \begin{vmatrix} a_{11} & a_{12} & a_{13} \\ a_{21} & a_{22} & a_{23} \\ a_{31} & a_{32} & a_{33} \end{vmatrix} = \det A = a = e_{ijk} a_{1i} a_{2j} a_{3k} \tag{3.117}$$

From the rules for rearranging the rows and columns of determinants, as in the foregoing section, if we replace the indices 1, 2, and 3 by variables r, s, and t in Equation 3.117, we can obtain a more general expression:

$$\mathbf{a}_r \times \mathbf{a}_s \cdot \mathbf{a}_t = \begin{vmatrix} a_{r1} & a_{r2} & a_{r3} \\ a_{s1} & a_{s2} & a_{s3} \\ a_{t1} & a_{t2} & a_{t3} \end{vmatrix} = e_{ijk} a_{ri} a_{sj} a_{tk} \tag{3.118}$$

where the last equality follows from a comparison of the rules at the end of Section 3.7 and the definition of the permutation symbol in Equation 3.36.

By similar reasoning, if we let the numeric column indices (1, 2, 3) in Equation 3.118 be replaced by variable indices, say ℓ, m, and n, we obtain

$$\begin{vmatrix} a_{r\ell} & a_{rm} & a_{rn} \\ a_{s\ell} & a_{sm} & a_{sn} \\ a_{t\ell} & a_{tm} & a_{tn} \end{vmatrix} = e_{rst} e_{\ell mn} a \tag{3.119}$$

Suppose now that the matrix A is the identity matrix I so that the elements a_{ij} become δ_{ij}. Then the determinant value is 1.0, and then, Equation 3.119 becomes

$$\begin{vmatrix} \delta_{r\ell} & \delta_{rm} & \delta_{rn} \\ \delta_{s\ell} & \delta_{sm} & \delta_{sn} \\ \delta_{t\ell} & \delta_{tm} & \delta_{tn} \end{vmatrix} = e_{rst} e_{\ell mn} \tag{3.120}$$

Then by expanding the determinant, we obtain

$$\delta_{r\ell}\delta_{sm}\delta_{tn} - \delta_{r\ell}\delta_{tm}\delta_{sn} + \delta_{rm}\delta_{t\ell}\delta_{sn} - \delta_{rm}\delta_{s\ell}\delta_{tn} + \delta_{rn}\delta_{s\ell}\delta_{tm} - \delta_{rn}\delta_{t\ell}\delta_{sm} = e_{rst} e_{\ell mn} \tag{3.121}$$

Next, by setting $r = \ell$ and recalling that $\delta_{rr} = 3$, we have

$$e_{rst} e_{rmn} = \delta_{sm}\delta_{tn} - \delta_{sn}\delta_{tm} \tag{3.122}$$

Further, by setting $s = m$, we have

$$e_{rst} e_{rsn} = 2\delta_{tn} \tag{3.123}$$

Finally, by setting $t = n$, we obtain

$$e_{rst} e_{rst} = 6 = 3! \tag{3.124}$$

Returning to Equation 3.118, by multiplying both sides of the last equality by e_{rst}, we have

$$e_{rst} e_{ijk} a_{ri} a_{sj} a_{tk} = e_{rst} e_{rst} a = 3! a$$

or

$$a = \left(\frac{1}{3!}\right)e_{rst}e_{ijk}a_{ri}a_{sj}a_{tk} \tag{3.125}$$

Recall again the procedure for evaluating the determinant by multiplying the elements in any row or column by their cofactors and then adding the results. In Equation 3.125, let the 3×3 array C be formed with elements C_{ri} defined as

$$C_{ri} = \left(\frac{1}{2!}\right)e_{rst}e_{ijk}a_{si}a_{tk} \tag{3.126}$$

Then Equation 3.125 becomes

$$a = \left(\frac{1}{3}\right)a_{ri}C_{ri} \tag{3.127}$$

Thus, we see that the C_{ri} of Equation 3.126 are the elements of the cofactor matrix of A. (Observe that the 3 in the denominator of Equation 3.127 is needed since the sum over r designates a sum over all three rows of the determinant.)

By a closer examination and comparison of Equations 3.125 and 3.127, we see that

$$a_{ri}C_{si} = \delta_{rs}a \quad \text{or} \quad \delta_{rs} = a_{ri}\left(\frac{C_{si}}{a}\right) \tag{3.128}$$

That is, the elements of the identity matrix (δ_{ij}) are obtained by the matrix product of the elements of A with the transpose of elements of the cofactor matrix divided by the determinant of A. Specifically,

$$I = \frac{AC^T}{a} \quad \text{or} \quad \frac{C^T}{a} = A^{-1} \tag{3.129}$$

For an illustration of the application of Equation 3.122, consider again the triple vector products of Equations 3.69 and 3.70. As before, let $\mathbf{A}$, $\mathbf{B}$, and $\mathbf{C}$ be vectors with scalar components a_i, b_i, and c_i relative to mutually perpendicular unit vectors $\mathbf{n}_i$ ($i = 1, 2, 3$). Then, using Equations 3.35 and 3.122, the triple vector product $(\mathbf{A} \times \mathbf{B}) \times \mathbf{C}$ becomes

$$(\mathbf{A} \times \mathbf{B}) \times \mathbf{C} = (a_i\mathbf{n}_i \times b_j\mathbf{n}_j) \times c_k\mathbf{n}_k = a_ib_jc_k(\mathbf{n}_i \times \mathbf{n}_j) \times \mathbf{n}_k$$

$$= a_ib_jc_ke_{ij\ell}\mathbf{n}_\ell \times \mathbf{n}_k = a_ib_jc_ke_{ij\ell}e_{\ell km}\mathbf{n}_m$$

$$= a_ib_jc_ke_{\ell ij}e_{\ell km}\mathbf{n}_m = a_ib_jc_k(\delta_{ik}\delta_{jm} - \delta_{im}\delta_{jk})\mathbf{n}_m$$

$$= a_kb_mc_k\mathbf{n}_m - a_mb_kc_k\mathbf{n}_m$$

$$= (\mathbf{A} \cdot \mathbf{C})\mathbf{B} - (\mathbf{B} \cdot \mathbf{C})\mathbf{A} \tag{3.130}$$

Similarly,

$$\mathbf{A} \times (\mathbf{B} \times \mathbf{C}) = a_i \mathbf{n}_i \times (b_j \mathbf{n}_j \times c_k \mathbf{n}_k) = a_i b_j c_k \mathbf{n}_i \times (\mathbf{n}_j \times \mathbf{n}_k)$$

$$= a_i b_j c_k \mathbf{n}_i \times (e_{jk\ell} \mathbf{n}_\ell) = a_i b_j c_k e_{jkl} e_{ilm} \mathbf{n}_m$$

$$= a_i b_j c_k e_{ljk} e_{lmi} \mathbf{n}_m = a_i b_j c_k (\delta_{jm}\delta_{ki} - \delta_{ji}\delta_{km}) \mathbf{n}_m$$

$$= a_k b_m c_k \mathbf{n}_m - a_i b_j c_m \mathbf{n}_m$$

$$= (\mathbf{A} \cdot \mathbf{C})\mathbf{B} - (\mathbf{A} \cdot \mathbf{B})\mathbf{C} \qquad (3.131)$$

3.9 EIGENVALUES, EIGENVECTORS, AND PRINCIPAL DIRECTIONS

In the design and analysis of mechanical systems, analysts generally take advantage of geometrical symmetry to assign directions for coordinate axes. It happens that such directions are usually directions for maximum and minimum values of parameters of interest such as stresses, strains, and moments of inertia. With biosystems, however, there is little symmetry, and the geometry is irregular. Thus, directions of maximum/minimum parameter values (principal directions) are not readily apparent.

Fortunately, even for irregular shapes, we can determine the principal direction by solving a 3D eigenvalue problem. In this section, we will briefly review the procedures for solving this problem.

Recall from Section 3.5 that the dot product (or projection) of a dyadic with a vector produces a vector. For an arbitrary dyadic $\mathbf{A}$ and an arbitrary vector $\mathbf{x}$, the vector $\mathbf{y}$ produced by the product $\mathbf{A} \cdot \mathbf{x}$ will also in general appear to be arbitrary having little or no resemblance to the original vector $\mathbf{x}$. If it should happen that $\mathbf{y}$ is parallel to $\mathbf{x}$, then $\mathbf{x}$ and $\mathbf{y}$ are said to be eigenvectors of $\mathbf{A}$ or designators of principal directions. The ratio of the magnitudes of $\mathbf{y}$ and $\mathbf{x}$ ($|y|/|x|$ or $-|y|/|x|$ if y has opposite sense of x) is called an eigenvalue, or principal value, of $\mathbf{A}$. More specifically, let dyadic $\mathbf{A}$ and vectors $\mathbf{x}$ and $\mathbf{y}$ be expressed in terms of mutually perpendicular unit vectors $\mathbf{n}_i$ ($i = 1, 2, 3$) as

$$\mathbf{A} = a_{ij} \mathbf{n}_i \mathbf{n}_j, \quad \mathbf{x} = x_k \mathbf{n}_k, \quad \text{and} \quad \mathbf{y} = y_\ell \mathbf{n}_\ell \qquad (3.132)$$

Then

$$\mathbf{A} \cdot \mathbf{x} = \mathbf{y} \qquad (3.133)$$

or

$$a_{ij} x_j = y_i \qquad (3.134)$$

If $\mathbf{y}$ is parallel to $\mathbf{x}$, say $\mathbf{y} = \lambda \mathbf{x}$ (λ is a scalar multiplier), then Equations 3.133 and 3.134 have the forms

$$\mathbf{A} \cdot \mathbf{x} = \lambda \mathbf{x} \quad \text{and} \quad a_{ij} x_j = \lambda x_i \qquad (3.135)$$

This last expression may be expressed in the form

$$(a_{ij} - \lambda \delta_{ij}) x_j = 0 \qquad (3.136)$$

or more explicitly as

$$(a_{11} - \lambda)x_1 + a_{12}x_2 + a_{13}x_3 = 0$$

$$a_{21}x_1 + (a_{22} - \lambda)x_2 + a_{23}x_3 = 0 \qquad (3.137)$$

$$a_{31}x_1 + a_{32}x_2 + (a_{33} - \lambda)x_3 = 0$$

Equation 3.137 forms a set of homogeneous linear algebraic equations for the components x_i of **x**. Recall from elementary algebra (see, e.g., Refs. [17,19]) that the only solution is $x_i = 0$ ($i = 1, 2, 3$) unless the determinant of the coefficients is zero. That is, **x** = 0 unless

$$\det(a_{ij} - \lambda\delta_{ij}) = \begin{vmatrix} (a_{11} - \lambda) & a_{12} & a_{13} \\ a_{21} & (a_{12} - \lambda) & a_{23} \\ a_{31} & a_{32} & (a_{33} - \lambda) \end{vmatrix} = 0 \qquad (3.138)$$

By expanding the determinant of Equation 3.118, we obtain a cubic equation for λ:

$$\lambda^3 - a_{\mathrm{I}}\lambda^2 + a_{\mathrm{II}}\lambda + a_{\mathrm{III}} = 0 \qquad (3.139)$$

where the coefficients a_{I}, a_{II}, and a_{III} are

$$a_{\mathrm{I}} = a_{11} + a_{22} + a_{33}$$

$$a_{\mathrm{II}} = a_{22}a_{33} - a_{32}a_{23} + a_{11}a_{33} - a_{31}a_{13} + a_{11}a_{22} - a_{12}a_{21} \qquad (3.140)$$

$$a_{\mathrm{III}} = a_{11}a_{22}a_{33} - a_{11}a_{32}a_{23} + a_{12}a_{31}a_{23} - a_{12}a_{21}a_{33} + a_{21}a_{32}a_{13} - a_{31}a_{13}a_{22}$$

By inspection of these terms, we see that if A is the matrix whose elements are a_{ij}, then a_{I} is the trace (sum of diagonal elements) of A, a_{II} is the trace of the matrix of cofactors of A, and a_{III} is the determinant of A.

By solving Equation 3.119 for λ, we obtain three roots: λ_1, λ_2, and λ_3. If A is symmetric, it is seen that these roots are real [6]. In general, they will also be distinct. Thus, there will, in general, be three solution vectors **x** of Equation 3.135, or equivalently, three solution sets of components x_i of **x**. To obtain these components, we can select one of the roots, say λ_1, substitute it into Equation 3.137, and then solve for the corresponding x_i. A problem arising, however, is that if λ_1 is a root of Equation 3.138, the determinant of the coefficients of Equation 3.137 is zero and Equation 3.137 is not independent. Instead, they are dependent, meaning that at most two of the three equations are independent. Therefore, to obtain a unique set of x_i, we need an additional equation. We can obtain this equation by observing that in Equation 3.135, the magnitude of **x** is arbitrary. Hence, if we require that **x** be a unit vector, we have the additional equation

$$x_1^2 + x_2^2 + x_3^2 = 1 \qquad (3.141)$$

Thus, by selecting any two of Equation 3.137 and combining them with Equation 3.141, we have three independent equations for the three x_i. Upon solving for these x_i, we can repeat the process with λ having values λ_2 and λ_3 and thus obtain two other sets of x_i solutions.

The procedure is perhaps best understood by considering a specific illustrative example. To this end, let **A** be a dyadic whose matrix A of components relative to the $\mathbf{n}_i$ is

$$A = [a_{ij}] = \begin{bmatrix} 4 & \sqrt{3}/2 & 1/2 \\ \sqrt{3}/2 & 7/2 & \sqrt{3}/2 \\ 1/2 & \sqrt{3}/2 & 5/2 \end{bmatrix} \tag{3.142}$$

From Equation 3.135, suppose we search for a vector **x** (an eigenvector) with $\mathbf{n}_i$ components x_i such that

$$\mathbf{A} \cdot \mathbf{X} = \lambda \mathbf{x} \quad \text{or} \quad a_{ij} x_j = \lambda x_i \tag{3.143}$$

Equation 3.137 then becomes

$$(4 - \lambda) x_1 + \left(\sqrt{3}/2\right) x_2 + (1/2) x_3 = 0$$

$$\left(\sqrt{3}/2\right) x_1 + (7/2 - \lambda) x_2 + \left(\sqrt{3}/2\right) x_3 = 0 \tag{3.144}$$

$$(1/2) x_1 + \left(\sqrt{3}/2\right) x_2 + (5/2 - \lambda) x_3 = 0$$

A nontrivial solution x_i is obtained only if the determinant of the coefficients is zero (see Equation 3.138). Thus, we have

$$\begin{vmatrix} 4 - \lambda & \sqrt{3}/2 & 1/2 \\ \sqrt{3}/2 & 7/2 - \lambda & \sqrt{3}/2 \\ 1/2 & \sqrt{3}/2 & 5/2 - \lambda \end{vmatrix} = 0 \tag{3.145}$$

By expanding the determinant, we obtain

$$\lambda^3 - 10\lambda^2 + 31\lambda - 30 = 0 \tag{3.146}$$

Solving for λ (the eigenvalues), we have

$$\lambda = \lambda_1 = 2, \quad \lambda = \lambda_2 = 3, \quad \lambda = \lambda_3 = 5 \tag{3.147}$$

From Equation 3.141, if we require that the magnitude of the eigenvectors be unity, we have

$$x_1^2 + x_2^2 + x_3^2 = 1 \tag{3.148}$$

For each of the eigenvalues of Equation 3.147, Equation 3.144 is dependent. Thus, for a particular eigenvalue, if we select two of the equations, say the first two, and combine them with Equation 3.148, we have three equations for three eigenvector components x_i. If $\lambda = \lambda_1 = 2$, we have

$$2x_1 + \left(\sqrt{3}/2\right) x_2 + (1/2) x_3 = 0$$

$$\left(\sqrt{3}/2\right) x_1 + (3/2) x_2 + \left(\sqrt{3}/2\right) x_3 = 0 \tag{3.149}$$

$$x_1^2 + x_2^2 + x_3^2 = 1$$

Solving for x_1, x_2, and x_3, we have

$$x_1 = x_1^{(1)} = 0, \quad x_2 = x_2^{(1)} = -1/2, \quad x_3 = x_3^{(1)} = \sqrt{3}/2 \tag{3.150}$$

where the superscript (1) is used to identify the components with the first eigenvalue λ_1. The corresponding eigenvector $\mathbf{x}^{(1)}$ is

$$\mathbf{x}^{(1)} = -(1/2)\mathbf{n}_2 + \left(\sqrt{3}/2\right)\mathbf{n}_3 \tag{3.151}$$

Similarly, if $\lambda = \lambda_2 = 3$, we obtain the equations

$$x_1 + \left(\sqrt{3}/2\right)x_2 + (1/2)x_3 = 0$$

$$\left(\sqrt{3}/2\right)x_1 + (1/2)x_2 + \left(\sqrt{3}/2\right)x_3 = 0 \tag{3.152}$$

$$x_1^2 + x_2^2 + x_3^2 = 1$$

Solving for x_1, x_2, and x_3, we have

$$x_1 = x_1^{(2)} = -\sqrt{2}/2, \quad x_2 = x_2^{(2)} = \sqrt{6}/4, \quad x_3^{(2)} = \sqrt{2}/4 \tag{3.153}$$

with the eigenvector $\mathbf{x}^{(2)}$, thus being

$$\mathbf{x}^{(2)} = \left(-\sqrt{2}/2\right)\mathbf{n}_1 + \left(\sqrt{6}/4\right)\mathbf{n}_2 + \left(\sqrt{2}/4\right)\mathbf{n}_3 \tag{3.154}$$

Finally, if $\lambda = \lambda_3 = 5$, we have

$$-x_1 + \left(\sqrt{3}/2\right)x_2 + (1/2)x_3 = 0$$

$$\left(\sqrt{3}/2\right)x_1 + (-3/2)x_2 + \left(\sqrt{3}/2\right)x_3 = 0 \tag{3.155}$$

$$x_1^2 + x_2^2 + x_3^2 = 1$$

and then x_1, x_2, and x_3 are

$$x_1 = x_1^{(3)} = -\sqrt{2}/2, \quad x_2 = x_2^{(3)} = -\sqrt{6}/4, \quad x_3 = x_3^{(3)} = -\sqrt{2}/4 \tag{3.156}$$

and the eigenvector $\mathbf{x}^{(3)}$ is then

$$\mathbf{x}^{(3)} = \left(-\sqrt{2}/2\right)\mathbf{n}_1 + \left(-\sqrt{6}/4\right)\mathbf{n}_2 + \left(-\sqrt{2}/4\right)\mathbf{n}_3 \tag{3.157}$$

Observe that $\mathbf{x}^{(1)}$, $\mathbf{x}^{(2)}$, and $\mathbf{x}^{(3)}$ are mutually perpendicular. That is,

$$\mathbf{x}^{(i)} \cdot \mathbf{x}^{(j)} = \delta_{ij} = \begin{cases} 1 & i = j \\ 0 & i \neq j \end{cases} \tag{3.158}$$

It is obvious that when the eigenvalues are distinct, the eigenvectors are mutually perpendicular [6]. Alternatively, if two eigenvalues are equal, then all vectors perpendicular to the eigenvector of the distinct eigenvalue are eigenvectors. That is, when two eigenvalues are equal, all vectors parallel to the plane normal to the eigenvector of the distinct eigenvalue are eigenvectors. Finally, if all three eigenvalues are equal, then all vectors are eigenvectors [6].

In any event, there always exist three mutually perpendicular eigenvectors. Let these vectors be normalized to unit vectors and notationally represented as $\hat{\mathbf{n}}_i$ (i = 1, 2, 3). Then, in the immediate foregoing example, $\hat{\mathbf{n}}_i$ are

$$\hat{\mathbf{n}}_1 = \mathbf{x}^{(1)}, \quad \hat{\mathbf{n}}_2 = \mathbf{x}^{(2)}, \quad \hat{\mathbf{n}}_3 = \mathbf{x}^{(3)} \tag{3.159}$$

With these $\hat{\mathbf{n}}_i$ expressed in terms of the unit vectors $\mathbf{n}_j$ through Equations 3.150, 3.154, and 3.157, we can form a transformation matrix between $\hat{\mathbf{n}}_i$ and $\mathbf{n}_j$. The elements S_{ij} of such a transformation matrix are

$$S_{ij} = \mathbf{n}_i \cdot \hat{\mathbf{n}}_j \tag{3.160}$$

Let $\mathbf{V}$ be any vector. Let $\mathbf{V}$ be expressed in terms of the $\mathbf{n}_i$ and $\hat{\mathbf{n}}_j$ as

$$\mathbf{V} = v_i \mathbf{n}_i \quad \text{and} \quad \mathbf{V} = \hat{v}_j \hat{\mathbf{n}}_j \tag{3.161}$$

From Equations 3.23 and 3.25, we have

$$v_i = \mathbf{V} \cdot \mathbf{n}_i \quad \text{and} \quad \hat{v}_j = \mathbf{V} \cdot \hat{\mathbf{n}}_j \tag{3.162}$$

Thus, from Equation 3.160, we have

$$v_i = S_{ij} \hat{v}_j \quad \text{and} \quad \hat{v}_j = S_{ij} v_i \tag{3.163}$$

Then from Equations 3.161 and 3.162, we have

$$\mathbf{V} = (\mathbf{V} \cdot \mathbf{n}_i) \mathbf{n}_i \quad \text{and} \quad \mathbf{V} = (\mathbf{V} \cdot \hat{\mathbf{n}}_j) \hat{\mathbf{n}}_j \tag{3.164}$$

In the second expression of Equation 3.164, let $\mathbf{V}$ be $\mathbf{n}_i$. Then $\mathbf{n}_i$ may be expressed as

$$\mathbf{n}_i = (\mathbf{n}_i \cdot \hat{\mathbf{n}}_j) \hat{\mathbf{n}}_j = S_{ij} \hat{\mathbf{n}}_j \tag{3.165}$$

Similarly, from the first expression of Equation 3.164, we obtain

$$\hat{\mathbf{n}}_j = S_{ij} \mathbf{n}_i \tag{3.166}$$

Observe that the forms of Equation 3.163 and those of Equations 3.165 and 3.166 are the same. Observe particularly the positioning of the free and repeated indices and the consistency of this positioning relative to whether the indices are associated with $\mathbf{n}_i$ or $\hat{\mathbf{n}}_j$.

Returning now to Equation 3.132, if we substitute for $\mathbf{n}_i$ using Equation 3.165, we obtain

$$\mathbf{A} = a_{ij}\mathbf{n}_i\mathbf{n}_j = a_{ij}S_{ik}S_{j\ell}\hat{\mathbf{n}}_k\hat{\mathbf{n}}_\ell = \hat{a}_{k\ell}\hat{\mathbf{n}}_k\hat{\mathbf{n}}_\ell \tag{3.167}$$

or

$$\hat{a}_{k\ell} = S_{ik}S_{j\ell}a_{ij} = S_{ki}^T a_{ij}S_{j\ell} \tag{3.168}$$

To illustrate the use of Equation 3.168, consider again the numerical example of Equation 3.142. From Equations 3.150, 3.154, and 3.157, we see that the elements of S_{ij} are

$$S_{ij} = \begin{bmatrix} 0 & -\sqrt{2}/2 & -\sqrt{2}/2 \\ -1/2 & \sqrt{6}/4 & -\sqrt{6}/4 \\ \sqrt{3}/2 & \sqrt{2}/4 & -\sqrt{2}/4 \end{bmatrix} \tag{3.169}$$

Then, by substituting into Equation 3.168 for the matrix of Equation 3.142, we have

$$S_{ik}S_{j\ell}a_{ij} = S_{ki}^T a_{ij}S_{j\ell} = \hat{a}_{k\ell} \tag{3.170}$$

or

$$\begin{bmatrix} 0 & -1/2 & \sqrt{3}/2 \\ -\sqrt{2}/2 & \sqrt{6}/4 & \sqrt{2}/4 \\ -\sqrt{2}/2 & -\sqrt{6}/4 & -\sqrt{2}/4 \end{bmatrix} \begin{bmatrix} 4 & \sqrt{3}/2 & 1/2 \\ \sqrt{3}/2 & 7/2 & \sqrt{3}/2 \\ 1/2 & \sqrt{3}/2 & 5/2 \end{bmatrix} \begin{bmatrix} 0 & -\sqrt{2}/2 & -\sqrt{2}/2 \\ -1/2 & \sqrt{6}/4 & -\sqrt{6}/4 \\ \sqrt{3}/2 & \sqrt{2}/4 & -\sqrt{2}/4 \end{bmatrix}$$
$$= \begin{bmatrix} 2 & 0 & 0 \\ 0 & 3 & 0 \\ 0 & 0 & 5 \end{bmatrix} \tag{3.171}$$

Therefore,

$$\hat{a}_{k\ell} = \begin{bmatrix} 2 & 0 & 0 \\ 0 & 3 & 0 \\ 0 & 0 & 5 \end{bmatrix} \tag{3.172}$$

and thus, we see that by using the unit eigenvectors as a basis, the dyadic $\mathbf{A}$ becomes diagonal.

3.10 MAXIMUM AND MINIMUM EIGENVALUES AND THE ASSOCIATED EIGENVECTORS

Consider the quadratic form $x_i a_{ij} x_j$. If x_i are components of a unit vector $\mathbf{n}_x$, and a_{ij} are components of a physically developed dyadic (such as a stress dyadic or an inertia dyadic), then the form $x_i a_{ij} x_j$ represents the projection of the dyadic in the $\mathbf{n}_x$ direction (such as normal stress or moment of inertia).

It is therefore of interest to find the directions of $\mathbf{n}_x$ such that $x_i a_{ij} x_j$ has maximum and minimum values and also to find the maximum and minimum values themselves.

To determine these directions, recall that since $\mathbf{n}_x$ is a unit vector, we have

$$x_i x_i = 1 \tag{3.173}$$

The problem is then a constrained maximum/minimum problem. Specifically, the objective is to find maximum/minimum values of $x_i a_{ij} x_j$ such that $x_i x_i = 1$. This problem is readily solved using Lagrange multipliers [19]. To this end, let a function $h(x_i)$ be defined as

$$h = x_i a_{ij} x_j - \lambda(1 - x_i x_i) \tag{3.174}$$

where λ is a Lagrange multiplier (to be determined in the sequel of the analysis). Then the maximum/minimum values of h are obtained by setting the derivative of h with respect to the x_i ($i = 1$, 2, 3) equal to zero. That is,

$$\frac{\partial h}{\partial x_i} = 0 \tag{3.175}$$

If a_{ij} are elements of a symmetric dyadic, then Equation 3.175 immediately leads to

$$a_{ij} x_j = \lambda x_i \tag{3.176}$$

Equation 3.176 is identical to Equation 3.135, the eigenvalue equation. That is, the directions of $\mathbf{n}_x$ corresponding to the maximum/minimum values of $x_i a_{ij} x_j$ are those of the eigenvectors, and the maximum/minimum values themselves are the eigenvalues.

3.11 USE OF MATLAB®

MATLAB® is the best known and most widely used software for computations involving vector and matrix arrays. Indeed, the name MATLAB is an acronym for MATrix LABoratory. Consequently, MATLAB is ideally suited for biomechanical analyses.

MATLAB was initially written as a Fortran program by Cleve Moler in the late 1970s [20]. It has since then become increasingly updated and expanded to become more versatile and user oriented. MATLAB's widespread use has prompted the writing of numerous books, monographs, and tutorials. References [21,22] are excellent texts for beginning users.

In many ways, MATLAB is like a super, interactive, multifunction calculator. That is, MATLAB has all the commands found on advanced calculators and many more, and unlike most engineering analysis software, MATLAB is used interactively. Moreover, MATLAB functions and commands are very short, enabling a user to get quick and, of course, accurate results.

In exchange for the convenience of brevity, however, a user needs to be careful not to make syntax of typing errors. Nevertheless, if an error occurs, MATLAB will provide an appropriate warning and/or error identification.

As a further aid to the user, MATLAB provides graphing for visual display of the results. There is also a "help" command.

In the following paragraphs, we present a few illustrative computations, typical of those used in biodynamic analyses. In the following sections, we provide a listing of commonly used MATLAB functions.

Fortunately, most of the functions and commands useful in biodynamic analyses are among the simplest. Indeed, once a unit vector basis is established, the vectors and matrices of interest are simply 3×1 and 3×3 arrays of numbers. In MATLAB, these arrays are represented with square brackets: [ ] with the components of vectors and the elements of matrix rows separated by blank spaces or by commas. The matrix rows are separated by semicolons or by using separate lines.

Suppose, for example, that $\mathbf{n}_1$, $\mathbf{n}_2$, and $\mathbf{n}_3$ are mutually perpendicular unit vectors and that $\mathbf{v}$ and $\mathbf{D}$ are a vector and dyadic given by

$$\mathbf{v} = 3\mathbf{n}_1 - 2\mathbf{n}_2 + 7\mathbf{n}_3 \tag{3.177}$$

and

$$\mathbf{D} = 2\mathbf{n}_1\mathbf{n}_1 - 3\mathbf{n}_1\mathbf{n}_2 + 4\mathbf{n}_1\mathbf{n}_3 + 6\mathbf{n}_2\mathbf{n}_1 - \mathbf{n}_2\mathbf{n}_2 + 8\mathbf{n}_2\mathbf{n}_3 - 5\mathbf{n}_3\mathbf{n}_1 + 7\mathbf{n}_3\mathbf{n}_2 + 5\mathbf{n}_3\mathbf{n}_3 \tag{3.178}$$

In index notation, if we express $\mathbf{v}$ and $\mathbf{D}$ as

$$\mathbf{v} = v_i\mathbf{n}_i \quad \text{and} \quad \mathbf{D} = d_{ij}\mathbf{n}_i\mathbf{n}_j \tag{3.179}$$

then the v_i $(i = 1, 2, 3)$ and d_{ij} $(i,j = 1, 2, 3)$ may be regarded as elements of the matrix arrays v and D as

$$v = \begin{bmatrix} 3 \\ -2 \\ 7 \end{bmatrix} \quad \text{and} \quad D = \begin{bmatrix} 2 & -3 & 4 \\ 6 & -1 & 8 \\ -5 & 7 & 5 \end{bmatrix} \tag{3.180}$$

In MATLAB, v and D are written as

$$v = [3 \quad -2 \quad 7] \quad \text{or} \quad v = [3, \quad -2, \quad 7] \tag{3.181}$$

and

$$D = [2 \quad -3 \quad 4; \quad 6 \quad -1 \quad 8; \quad -5 \quad 7 \quad 5]$$

or

$$D = [2, -3, 4; \quad 6, -1, 8; \quad -5, 7, 5] \tag{3.182}$$

Alternatively, D may be written as

$$D = \begin{bmatrix} 2 & -3 & 4 \\ 6 & -1 & 8 \\ -5 & 7 & 5 \end{bmatrix} \tag{3.183}$$

Suppose we are interested in the matrix product: Mv. If the vector w is Mv, then w is

$$w = \begin{bmatrix} w_1 \\ w_2 \\ w_3 \end{bmatrix} = \begin{bmatrix} 2 & -3 & 4 \\ 6 & -1 & 8 \\ -5 & 7 & 5 \end{bmatrix} \begin{bmatrix} 3 \\ -2 \\ 7 \end{bmatrix} \tag{3.184}$$

Using MATLAB, we can immediately obtain the elements of w using the command

$$w = M * v' \tag{3.185}$$

where v' is the transpose of v (in Equation 3.181, v is a row array, and thus, v' is the corresponding column array). The result for w is

$$w = \begin{bmatrix} 40 \\ 76 \\ 6 \end{bmatrix} \tag{3.186}$$

Next, suppose we are interested in solving for x in the matrix equation

$$Mx = b \tag{3.187}$$

where b is the column array.

$$b = \begin{bmatrix} -2 \\ 9 \\ 4 \end{bmatrix} \tag{3.188}$$

In matrix notation, the solution is

$$x = M^{-1}b \tag{3.189}$$

where M^{-1} is the inverse of M.

Using MATLAB, we can immediately obtain x in a couple of ways: First, let M_{inv} be M^{-1}. Then the MATLAB command $inv(M)$ immediately produces M_{inv} as

$$M_{inv} = \begin{bmatrix} -0.2585 & 0.1822 & -0.0847 \\ -0.2966 & 0.1271 & 0.0339 \\ 0.1568 & 0.0042 & 0.0678 \end{bmatrix} \tag{3.190}$$

From Equation 3.189, x is then

$$x = M_{inv} * b = \begin{bmatrix} 0.8178 \\ 1.8729 \\ -0.0042 \end{bmatrix} \tag{3.191}$$

A second, and perhaps more direct, way of determining x is to use the MATLAB command

$$x = \frac{M}{b} \tag{3.192}$$

The result is the same as in Equation 3.191.

MATLAB is especially convenient for finding eigenvalues and eigenvectors: Consider the example in Section 3.9 with the matrix of Equation 3.142. That is,

$$A = \begin{bmatrix} 4 & \sqrt{3}/2 & 1/2 \\ \sqrt{3}/2 & 7/2 & \sqrt{3}/2 \\ 1/2 & \sqrt{3}/2 & 5/2 \end{bmatrix} = \begin{bmatrix} 4.0000 & 0.8660 & 0.5000 \\ 0.8660 & 3.5000 & 0.8660 \\ 0.5000 & 0.8660 & 2.5000 \end{bmatrix} \tag{3.193}$$

In the analysis of Section 3.9, we found the eigenvalues and the corresponding unit eigenvectors to be

$$\lambda_1 = 2, \quad \lambda_2 = 3, \quad \lambda_3 = 5 \tag{3.194}$$

and

$$\mathbf{x}^{(1)} = 0\mathbf{n}_1 - (1/2)\mathbf{n}_2 + \left(\sqrt{3}/2\right)\mathbf{n}_3 \tag{3.195}$$

$$\mathbf{x}^{(2)} = \left(-\sqrt{2}/2\right)\mathbf{n}_1 + \left(\sqrt{6}/4\right)\mathbf{n}_2 + \left(\sqrt{2}/4\right)\mathbf{n}_3 \tag{3.196}$$

$$\mathbf{x}^{(3)} = \left(-\sqrt{2}/2\right)\mathbf{n}_1 + \left(-\sqrt{6}/4\right)\mathbf{n}_2 + \left(-\sqrt{2}/4\right)\mathbf{n}_3 \tag{3.197}$$

In decimal form, the unit eigenvectors are

$$\mathbf{x}^{(1)} = 0.0000\mathbf{n}_1 - 0.5000\mathbf{n}_2 + 0.8600\mathbf{n}_3 \tag{3.198}$$

$$\mathbf{x}^{(2)} = -0.7071\mathbf{n}_1 + 0.6124\mathbf{n}_2 + 0.3536\mathbf{n}_3 \tag{3.199}$$

$$\mathbf{x}^{(3)} = -0.7071\mathbf{n}_1 - 0.6124\mathbf{n}_2 - 0.3536\mathbf{n}_3 \tag{3.200}$$

We can immediately obtain these same results with the simple MATLAB command

$$[U, D] = eig(A) \tag{3.201}$$

where
   U is a matrix whose columns are the components of the unit eigenvectors
   D is a diagonal matrix with the eigenvalues on the diagonal

Upon entering the elements of matrix A (Equation 3.193), and upon entering the MATLAB command of Equation 3.201, the program immediately produces the results in the form

$$U = \begin{bmatrix} 0.7071 & -0.0000 & 0.7071 \\ -0.6124 & -0.5000 & 0.6124 \\ -0.3536 & 0.8660 & 0.3536 \end{bmatrix} \tag{3.202}$$

and

$$D = \begin{bmatrix} 3.0000 & 0 & 0 \\ 0 & 2.0000 & 0 \\ 0 & 0 & 5.0000 \end{bmatrix} \tag{3.203}$$

Equations 3.202 and 3.203 are seen to be the same as those of Equations 3.194, 3.198, 3.199, and 3.200.

Observe that given matrix A, once its elements are entered into MATLAB, the eigenvalues and the corresponding unit eigenvectors are immediately obtained without the necessity of solving a cubic equation and then three sets of simultaneous algebraic equations.

In the following section, we present a brief listing of a few elementary MATLAB operations and functions believed to be useful in biodynamic analyses.

3.12 ELEMENTARY MATLAB® OPERATIONS AND FUNCTIONS

MATLAB provides its users with the usual operations and functions employed in elementary matrix analysis. In this section, we list a few of these and illustrate them with a few example calculations. For this latter end, we use the following square arrays:

$$A = \begin{bmatrix} 8 & -2 & 4 \\ 4 & 0 & 3 \\ 1 & -7 & 5 \end{bmatrix} \quad \text{and} \quad B = \begin{bmatrix} 9 & -1 & 0 \\ 3 & 4 & 6 \\ 6 & 2 & -8 \end{bmatrix} \tag{3.204}$$

and the column array

$$b = \begin{bmatrix} -1 \\ 3 \\ 2 \end{bmatrix} \tag{3.205}$$

3.12.1 ELEMENTARY OPERATIONS

The basic array operations are addition, subtraction, multiplication, division (inverse multiplication), and transpose, using the symbols

+ addition
− subtraction
* multiplication
\ division
N transpose

Using the arrays of Equations 3.204 and 3.205, we have the following results*:

$$A + B = \begin{bmatrix} 17 & -3 & 4 \\ 7 & 4 & 9 \\ 7 & -5 & -3 \end{bmatrix} \quad (3.206)$$

$$A - B = \begin{bmatrix} -1 & -1 & 4 \\ 1 & -4 & -3 \\ -5 & -9 & 13 \end{bmatrix} \quad (3.207)$$

$$A * B = \begin{bmatrix} 90 & -8 & -44 \\ 54 & 2 & -24 \\ 18 & -19 & -82 \end{bmatrix} \quad (3.208)$$

and

$$A * b = \begin{bmatrix} -6 \\ 2 \\ -12 \end{bmatrix} \quad (3.209)$$

Next, suppose x is an array such that

$$B * x = b \quad (3.210)$$

and then x is

$$x = B \backslash b = \begin{bmatrix} -0.0175 \\ 0.8421 \\ -0.0526 \end{bmatrix} \quad (3.211)$$

Finally, the transposes of A and B are

$$A^T = A' = \begin{bmatrix} 8 & 4 & 1 \\ -2 & 0 & -7 \\ 4 & 3 & 5 \end{bmatrix} \quad \text{and} \quad B^T = B' = \begin{bmatrix} 9 & 3 & 6 \\ -1 & 4 & 2 \\ 0 & 6 & -8 \end{bmatrix} \quad (3.212)$$

3.12.2 Elementary Functions

The basic functions are determinant, inverse, and eigenvalues. In MATLAB, these functions are

$\det(A)$ determinant of A
$\text{inv}(A)$ inverse of A
$\text{eig}(A)$ eigenvalues of A
$[U,D] = \text{eig}(A)$ eigenvectors and eigenvalues of A

---

* As an exercise, a reader may want to verify the results using hand calculations and MATLAB software.

With the last function, MATLAB produces an array U whose columns are the unit eigenvectors of A and a diagonal array D with the eigenvalues on the diagonal. (This function was illustrated in Section 3.11.)

For the arrays of Equations 3.204 and 3.205, we have the following results:

$$A^{-1} = inv(A) = \begin{bmatrix} 0.2333 & -0.2000 & -0.0667 \\ -0.1889 & 0.4000 & -0.0889 \\ -0.3111 & 0.6000 & 0.0889 \end{bmatrix} \quad (3.213)$$

$$B^{-1} = inv(B) = \begin{bmatrix} 0.0965 & 0.0175 & 0.0132 \\ -0.1316 & 0.1579 & 0.1184 \\ 0.0395 & 0.0526 & -0.0855 \end{bmatrix} \quad (3.214)$$

Consider again the linear algebraic expression of Equation 3.210. We can solve for x as

$$x = B^{-1} b \quad (3.215)$$

Using MATLAB, we obtain

$$x = B^{-1} * b = \begin{bmatrix} -0.0175 \\ 0.8421 \\ -0.0526 \end{bmatrix} \quad (3.216)$$

Finally, using MATLAB, we can verify the relation

$$(AB)^{-1} = B^{-1} A^{-1} \quad (3.217)$$

or

$$inv(A*B) = \begin{bmatrix} 0.0151 & -0.0044 & -0.0068 \\ -0.0974 & 0.1605 & 0.0053 \\ 0.0259 & -0.0382 & -0.0149 \end{bmatrix} = B^{-1} * A^{-1} \quad (3.218)$$

PROBLEMS

Section 3.1

P3.1.1 Refer to a convenient linear algebra text (see, e.g., Refs. [1–3]). List the properties of a vector space.

Section 3.2

P3.2.1 A person walks 35 ft down a northbound hallway and then turns to the right and walks 50 ft down an eastbound hallway. What is the magnitude of the person's displacement from the original starting position?

P3.2.2 See Problem P3.2.1. Suppose that after walking down the eastbound hallway, the person walks 18 ft down a hallway directed to the southwest. What is the magnitude of the displacement from the original starting position?

P3.2.3 Are zero vectors equal to one another?

P3.2.4 Are equal vectors physically equivalent?

P3.2.5 See Problem P3.2.4. Provide an example to show that equal vectors do not necessarily describe or produce the same effect on a body.

P3.2.6 If **V** has a magnitude of 12 m/s, what are the magnitudes of vectors $\mathbf{V}_x$ and $\mathbf{V}_y$?

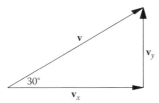

P3.2.7 If **v** has a magnitude of 12 m/s, what are the magnitudes of vectors **a** and **c**?

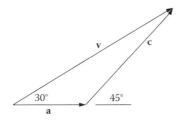

P3.2.8 Consider the vectors **A** and **B** in the following figure. Let **C** = **A** + **B**. Using the law of cosines, find |**C**|, the magnitude of **C**.

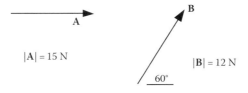

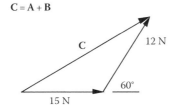

P3.2.9 Repeat Problem P3.2.8 by letting **C** be **B** + **A**.

P3.2.10 Let $\mathbf{n}_x$ and $\mathbf{n}_y$ be horizontal and vertical unit vectors as in the following figure and let **B** be the vector of figure for Problem P3.2.8 (and shown again here). Express **B** in terms of $\mathbf{n}_x$ and $\mathbf{n}_y$.

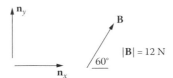

P3.2.11 See Problems P3.2.8 through P3.2.10. With both **A** and **B** expressed in terms of $\mathbf{n}_x$ and $\mathbf{n}_y$, determine the magnitude of **C**.

P3.2.12 See Problems P3.2.8 through P3.2.10. Let $\mathbf{n}_c$ be a unit vector parallel to and with the same sense as **C**. Express $\mathbf{n}_c$ in terms of $\mathbf{n}_x$ and $\mathbf{n}_y$.

P3.2.13 Let $\mathbf{n}_x$ and $\mathbf{n}_y$ be horizontal and vertical unit vectors as in the following figure and let **v** be the velocity vector as shown. Express **v** in terms of $\mathbf{n}_x$ and $\mathbf{n}_y$.

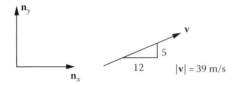

P3.2.14 Let X, Y, and Z be a Cartesian axis system R with origin O as in the following figure. Let P and Q be points in R. Let the X, Y, and Z coordinates of P and Q be P(5 ft, 7 ft, 9 ft) and Q(4 ft, −2 ft, 7 ft). Let $\mathbf{n}_x$, $\mathbf{n}_y$, and $\mathbf{n}_z$ be unit vectors parallel to X, Y, and Z. Let **p** and **q** be the position vectors:

$$\mathbf{p} = \mathbf{OP} \quad \text{and} \quad \mathbf{q} = \mathbf{OQ}$$

a. Express **p** and **q** in terms of $\mathbf{n}_x$, $\mathbf{n}_y$, and $\mathbf{n}_z$.

b. Let **r** and **s** be

$$\mathbf{r} = \mathbf{p} + \mathbf{q} \quad \text{and} \quad \mathbf{s} = \mathbf{p} - \mathbf{q}$$

Express **r** and **s** in terms of $\mathbf{n}_x$, $\mathbf{n}_y$, and $\mathbf{n}_z$.

c. Determine the magnitudes of **r** and **s**.

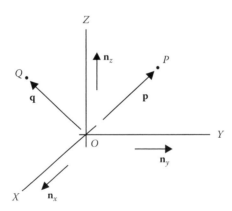

P3.2.15 The following figure shows a rectangular (not to scale) box subjected to a system of eight force vectors $\mathbf{F}_j$, $j = 1, \ldots, 8$, with magnitudes given by

$$|\mathbf{F}_1| = 16 \text{ lb} \qquad |\mathbf{F}_2| = 17 \text{ lb} \qquad |\mathbf{F}_3| = 15 \text{ lb} \qquad |\mathbf{F}_4| = 35 \text{ lb}$$

$$|\mathbf{F}_5| = 26 \text{ lb} \qquad |\mathbf{F}_6| = 8 \text{ lb} \qquad |\mathbf{F}_7| = 20 \text{ lb} \qquad |\mathbf{F}_8| = 30 \text{ lb}$$

The dimensions of the box are also shown in the figure. Finally, $\mathbf{n}_1$, $\mathbf{n}_2$, and $\mathbf{n}_3$ are unit vectors parallel to the edges of the box, as shown.

a. Express the $\mathbf{F}_j$, $j = 1, \ldots, 8$ in terms of the $\mathbf{n}_i$, $i = 1, 2, 3$.

b. Determine the sum (resultant) $\mathbf{R}$ of the $\mathbf{F}_j$, $j = 1, \ldots, 8$, in terms of the $\mathbf{n}_i$, $i = 1, 2, 3$.

c. Determine the magnitude of the resultant: $|\mathbf{R}|$.

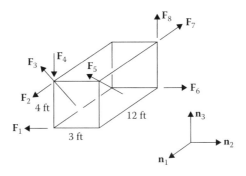

Section 3.3

P3.3.1 What is the angle between parallel vectors having the same sense?

P3.3.2 What is the angle between parallel vectors having opposite sense?

P3.3.3 Consider vectors $\mathbf{A}$ and $\mathbf{B}$ shown in the following figure together with the angles θ, φ, and ψ. Observe that from the definition of Section 3.3 that θ is defined as "the angle between $\mathbf{A}$ and $\mathbf{B}$," as opposed to either φ or ψ.

a. Compare $\cos \varphi$ and $\cos \psi$ with $\cos \theta$.

b. Compare $\sin \varphi$ and $\sin \psi$ with $\sin \theta$.

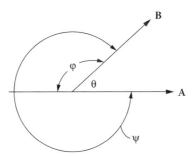

P3.3.4 Imagine two unit vectors $\mathbf{n}_1$ and $\mathbf{n}_2$, along a line and connected "tail-to-tail," as in the following figure.

a. What is the angle between $\mathbf{n}_1$ and $\mathbf{n}_2$?

b. Redundant question: What is the angle between $\mathbf{n}_2$ and $\mathbf{n}_1$?

P3.3.5 See Problem P3.3.4. Imagine three unit vectors $\mathbf{n}_1$, $\mathbf{n}_2$, and $\mathbf{n}_3$ in a plane, connected tail-to-tail and perpendicular to the edges of an equilateral triangle, as in the following figure.

a. What is the angle between $\mathbf{n}_1$ and $\mathbf{n}_2$?

b. What is the angle between $\mathbf{n}_2$ and $\mathbf{n}_3$?

c. What is the angle between $\mathbf{n}_3$ and $\mathbf{n}_1$?

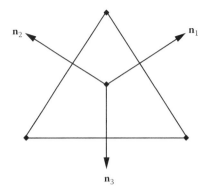

P3.3.6 Refer to figures for Problems P3.9 and P3.10. Observe in the figure for Problem P3.9, there are **two** vectors in **one** dimension (along a line), and in the figure for Problem P3.10, there are **three** vectors in **two** dimensions (a plane). In both cases, each vector makes the same angle with each other vector.

Imagine now a configuration of **four** unit vectors in **three** dimensions aligned tail-to-tail and such that each vector makes the same angle with each other vector. (Hint: Consider the unit vectors to be normal to the faces of the tetrahedron.)

a. What is the common angle theta between the respective vectors?

b. Find $\cos \theta$.

P3.3.7 Consider the vectors **A** and **B** in the following figure. Let the magnitude of A and B be

$$|\mathbf{A}| = 5 \text{ m} \quad \text{and} \quad |\mathbf{B}| = 8 \text{ N}$$

Evaluate $\mathbf{A} \cdot \mathbf{B}$ using the definition of Equation 3.19

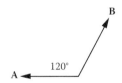

P3.3.8 See Problem P3.3.7 and the figure for that problem. Find

(a) $3\mathbf{A} \cdot \mathbf{B}$, (b) $\mathbf{A} \cdot 3\mathbf{B}$, (c) $\mathbf{A} \cdot (-5)\mathbf{B}$, (d) $3\mathbf{A} \cdot 5\mathbf{B}$

P3.3.9 Suppose $|\mathbf{A}| = 8$ ft and $|\mathbf{B}| = 10$ ft. Suppose further that $\mathbf{A} \cdot \mathbf{B} = 16$ ft². What is the angle θ between **A** and **B**?

P3.3.10 Let $\mathbf{n}_1$, $\mathbf{n}_2$, and $\mathbf{n}_3$ be mutually perpendicular unit vectors and let **A** and **B** be vectors expressed in terms of the $\mathbf{n}_i$ ($i = 1, 2, 3$) as

$$\mathbf{A} = 4\mathbf{n}_1 - 2\mathbf{n}_2 + \mathbf{n}_3 \quad \text{and} \quad \mathbf{B} = -\mathbf{n}_1 + 2\mathbf{n}_2 + 5\mathbf{n}_3$$

Determine $\mathbf{A} \cdot \mathbf{B}$.

P3.3.11 See Problem P3.3.10. Find the magnitudes of **A** and **B**.

P3.3.12 See Problems P3.3.9 and P3.3.10. Determine the angle between **A** and **B**.

P3.3.13 Let $\mathbf{n}_1$, $\mathbf{n}_2$, and $\mathbf{n}_3$ be mutually perpendicular unit vectors and let **p** and **F** be vectors expressed in terms of the $\mathbf{n}_i$ ($i = 1, 2, 3$) as

$$\mathbf{p} = 2\mathbf{n}_1 - 3\mathbf{n}_2 + 6\mathbf{n}_3 \text{ m} \quad \text{and} \quad \mathbf{F} = 3\mathbf{n}_1 + 5\mathbf{n}_2 - 7\mathbf{n}_3 \text{ N}$$

Determine $\mathbf{p} \times \mathbf{F}$.

P3.3.14 See Problem P3.3.13. Find $|\mathbf{p} \times \mathbf{F}|$, the magnitude of $\mathbf{p} \times \mathbf{F}$.

P3.3.15 Review Problems P3.3.4 through P3.3.6. Observe in Problem P3.3.4 that with $\mathbf{n}_1$ and $\mathbf{n}_2$ directed opposite to each other, we have

$$\mathbf{n}_1 + \mathbf{n}_2 = 0 \qquad\qquad\qquad (a)$$

Similarly, observe in the figure for Problem P3.3.5, we have

$$\mathbf{n}_1 + \mathbf{n}_2 + \mathbf{n}_3 = 0 \qquad\qquad\qquad (b)$$

Third, observe in Problem P3.3.6 that we have

$$\mathbf{n}_1 + \mathbf{n}_2 + \mathbf{n}_3 + \mathbf{n}_4 = 0 \qquad\qquad\qquad (c)$$

Observe still further that in each problem with the angles between the respective unit vectors being equal, we have

$$\mathbf{n}_i + \mathbf{n}_j = \cos\theta \qquad i \neq j \qquad\qquad (d)$$

and

$$\mathbf{n}_i \cdot \mathbf{n}_j = 1 \qquad i = j \text{ (no sum)} \qquad\qquad (e)$$

where θ is the common angle.

Use Equations a through e to obtain the solutions to Problems P3.3.4 through P3.3.6.

P3.3.16 Develop Equations a, b, and c of Problem P3.3.15 by observing that in each of Problems P3.3.4 through P3.3.5, the unit vectors are linearly dependent.

P3.3.17 Verify that the expression for e_{ijk} in Equation 3.37 is consistent with the definition given in Equation 3.36.

P3.3.18 Let $\mathbf{n}_i$ ($i = 1, 2, 3$) be mutually perpendicular unit vectors and let $\mathbf{A}$ and $\mathbf{B}$ be the vectors:

$$\mathbf{A} = 3\mathbf{n}_1 - 2\mathbf{n}_2 + 4\mathbf{n}_3 \quad \text{and} \quad \mathbf{B} = 5\mathbf{n}_1 + 7\mathbf{n}_2 - \mathbf{n}_3$$

a. Evaluate the dyadic product $\mathbf{AB}$.
b. Evaluate the dyadic product $\mathbf{BA}$.

P3.3.19 See the results of Problem P3.3.18. Observe that $\mathbf{AB} \neq \mathbf{BA}$. If, however, as before, $\mathbf{A}$ is

$$\mathbf{A} = 3\mathbf{n}_1 - 2\mathbf{n}_2 + 4\mathbf{n}_3$$

find vector $\mathbf{C}$ such that $\mathbf{AC} = \mathbf{CA}$. Is $\mathbf{C}$ unique?

P3.3.20 See Problem P3.3.19. Find a unit vector $\mathbf{n}$ such that $\mathbf{An} = \mathbf{nA}$. Is $\mathbf{n}$ unique?

Section 3.4

P3.4.1 Let $\mathbf{n}_i$ ($i = 1, 2, 3$) be a set of mutually perpendicular unit vectors. Let $\mathbf{u}$ and $\mathbf{v}$ be the vectors:

$$\mathbf{u} = 2\mathbf{n}_1 - 4\mathbf{n}_2 + 5\mathbf{n}_3 \quad \text{and} \quad \mathbf{v} = -\mathbf{n}_1 + 4\mathbf{n}_2 - \mathbf{n}_3$$

Let $\mathbf{A}$ be the dyadic product $\mathbf{uv}$ and let $\mathbf{A}^T$ be the product $\mathbf{vu}$. If $\mathbf{A}$ and $\mathbf{A}^T$ are expressed in the forms

$$\mathbf{A} = a_{ij}\mathbf{n}_i\mathbf{n}_j \quad \text{and} \quad \mathbf{A}^T = a_{ij}^T\mathbf{n}_i\mathbf{n}_j$$

find the matrices of the coefficients a_{ij} and a_{ij}^T.

P3.4.2 See Problem P3.4.1. Let $\mathbf{u}$ and $\mathbf{v}$ be expressed in the general forms

$$\mathbf{u} = u_1\mathbf{n}_1 + u_2\mathbf{n}_2 + u_3\mathbf{n}_3 \quad \text{and} \quad \mathbf{v} = v_1\mathbf{n}_1 + v_2\mathbf{n}_2 + v_3\mathbf{n}_3$$

Let $\mathbf{A}$ and $\mathbf{A}^T$ be the dyadic products $\mathbf{uv}$ and $\mathbf{vu}$, respectively. If, as before, $\mathbf{A}$ and $\mathbf{A}^T$ are expressed in the forms $a_{ij}\mathbf{n}_i\mathbf{n}_j$ and $a_{ij}^T\mathbf{n}_i\mathbf{n}_j$, find the matrices of a_{ij} and a_{ij}^T.

P3.4.3 See Problem P3.4.2. Find values of u_i and v_i ($i = 1, 2, 3$) such that $\mathbf{A}$ and $\mathbf{A}^T$ are *zero* dyadics. Are these values unique?

P3.4.4 See Problems P3.4.2 and P3.4.3. Find values of u_i and v_i such that $\mathbf{A}$ and $\mathbf{A}^T$ are *identity* dyadics. Are these values unique?

P3.4.5 See Problem P3.4.2. Determine the restrictive conditions on u_i and v_i such that $\mathbf{A}$ is a *symmetric* dyadic. (That is, $\mathbf{A} = \mathbf{A}^T$.)

P3.4.6 If dyadic $\mathbf{A}$ is the negative of its transpose, that is, if $\mathbf{A} = -\mathbf{A}^T$, then $\mathbf{A}$ is said to be "skew-symmetric." Referring to Problem P3.4.2, determine the restrictions on u_i and v_i ($i = 1, 2, 3$) so that $\mathbf{A}$ is skew-symmetric.

P3.4.7 Let $\mathbf{n}_i$ ($i = 1, 2, 3$) be mutually perpendicular unit vectors. Let vectors $\mathbf{u}$, $\mathbf{v}$, and $\mathbf{w}$ have the generic forms

$$\mathbf{u} = u_i\mathbf{n}_i, \quad \mathbf{v} = v_i\mathbf{n}_i, \quad \mathbf{w} = w_i\mathbf{n}_i$$

Let $\mathbf{A}$ and $\mathbf{B}$ be dyadics formed by the products

$$\mathbf{A} = \mathbf{uv} + \mathbf{uw} + \mathbf{wu} \quad \text{and} \quad \mathbf{B} = \mathbf{vu} + \mathbf{wv} + \mathbf{uw}$$

Let $\mathbf{C}$ be the dyadic product $\mathbf{C} = \mathbf{A} \cdot \mathbf{B}$. Let $\mathbf{C}$ be expressed in the form $\mathbf{C} = c_{ij}\mathbf{n}_i\mathbf{n}_j$. Determine the c_{ij} in terms of the u_i, v_i, w_i.

P3.4.8 If dyadic $\mathbf{D}$ is written in the form $\mathbf{D} = d_{ij}\mathbf{n}_i\mathbf{n}_j$, the $\mathbf{n}_i$ ($i = 1, 2, 3$) being mutually perpendicular unit vectors, then $\mathbf{D}$ is said to be "diagonal" if

$$d_{ij} = 0 \quad i \neq j \text{ (diagonal } \mathbf{D})$$

Find the components d_{ij}^{-1} of the inverse dyadic $\mathbf{D}^{-1}$.

P3.4.9 Let $\mathbf{n}_i$ ($i = 1, 2, 3$) be mutually perpendicular unit vectors and let $\mathbf{a}_i$ ($i = 1, 2, 3$) be the vectors:

$$\mathbf{a}_1 = -\sqrt{2}/2\mathbf{n}_1 + \sqrt{6}/4\mathbf{n}_2 + \sqrt{2}/4\mathbf{n}_3$$

$$\mathbf{a}_2 = -\sqrt{2}/2\mathbf{n}_1 - \sqrt{6}/4\mathbf{n}_2 - \sqrt{2}/4\mathbf{n}_3$$

$$\mathbf{a}_3 = 0\mathbf{n}_1 - 1/2\mathbf{n}_2 + \sqrt{3}/2\mathbf{n}_3$$

a. Show that $\mathbf{a}_i$ are also mutually perpendicular unit vectors.

b. Let $\mathbf{A}$ be the dyadic

$$\mathbf{A} = \mathbf{n}_1\mathbf{a}_1 + \mathbf{n}_2\mathbf{a}_2 + \mathbf{n}_3\mathbf{a}_3$$

Show that $\mathbf{A}$ is orthogonal ($\mathbf{A}^T = \mathbf{A}^{-1}$).

P3.4.10 Let $\mathbf{A}$ and $\mathbf{B}$ be orthogonal dyadics. Show that the product $\mathbf{C} = \mathbf{A} \cdot \mathbf{B}$ is also orthogonal.

P3.4.11 Let $\mathbf{A}$ be a dyadic and let $\mathbf{x}$ and $\mathbf{b}$ be vectors. Let $\mathbf{A}$, $\mathbf{x}$, and $\mathbf{b}$ be expressed in terms of mutually perpendicular unit vectors $\mathbf{n}_i$ ($i = 1, 2, 3$) as

$$\mathbf{A} = a_{ij}\mathbf{n}_i\mathbf{n}_j, \quad \mathbf{x} = x_i\mathbf{n}_i, \quad \mathbf{b} = b_i\mathbf{n}_i$$

Suppose **A**, **x**, and **b** are related by the expression

$$\mathbf{A} \cdot \mathbf{x} = \mathbf{b}$$

Suppose further that the inverse $\mathbf{A}^{-1}$ of **A** has the form $\mathbf{A}^{-1} = a_{ij}^{-1}\mathbf{n}_i\mathbf{n}_j$. Find the x_i in terms of a_{ij} and b_i.

Section 3.5

P3.5.1 With $\mathbf{n}_i$ (i = 1, 2, 3) being mutually perpendicular unit vectors, let **a**, **b**, and **c** be vectors expressed as

$$\mathbf{a} = a_i\mathbf{n}_i = 2\mathbf{n}_1 - \mathbf{n}_2 - 3\mathbf{n}_3$$

$$\mathbf{b} = b_i\mathbf{n}_i = 4\mathbf{n}_1 + 5\mathbf{n}_2 - 5\mathbf{n}_3$$

$$\mathbf{c} = c_i\mathbf{n}_i = 6\mathbf{n}_1 + 7\mathbf{n}_2 - 8\mathbf{n}_3$$

a. Evaluate $\mathbf{a} \times \mathbf{b}$, $\mathbf{b} \times \mathbf{c}$, and $\mathbf{c} \times \mathbf{a}$.
b. Using the results of (a), evaluate $(\mathbf{a} \times \mathbf{b}) \cdot \mathbf{c}$, $(\mathbf{b} \times \mathbf{c}) \cdot \mathbf{a}$, and $(\mathbf{c} \times \mathbf{a}) \cdot \mathbf{b}$.
c. Evaluate the determinant Δ:

$$\Delta = \begin{vmatrix} 2 & -1 & -3 \\ 4 & 5 & -5 \\ 6 & 7 & -8 \end{vmatrix}$$

d. Compare the results of the computations in (b) and (c).

P3.5.2 With $\mathbf{n}_i$ being mutually perpendicular unit vectors, let **a**, **b**, and **c** be

$$\mathbf{a} = \mathbf{n}_1 - 3\mathbf{n}_2 + 2\mathbf{n}_3$$

$$\mathbf{b} = 2\mathbf{n}_1 - \mathbf{n}_2 + 4\mathbf{n}_3$$

$$\mathbf{c} = -2\mathbf{n}_1 + 5\mathbf{n}_2 + 3\mathbf{n}_3$$

a. Evaluate $\mathbf{a} \times \mathbf{b}$ and $\mathbf{b} \times \mathbf{c}$.
b. Using the results of (a), evaluate $(\mathbf{a} \times \mathbf{b}) \times \mathbf{c}$ and $\mathbf{a} \times (\mathbf{b} \times \mathbf{c})$ and compare these results.
c. Evaluate $\mathbf{a} \cdot \mathbf{b}$, $\mathbf{b} \cdot \mathbf{c}$, and $\mathbf{c} \cdot \mathbf{a}$.
d. Using the results of (a) and (c), evaluate $(\mathbf{a} \cdot \mathbf{c})\mathbf{b} - (\mathbf{a} \cdot \mathbf{b})\mathbf{c}$ and $(\mathbf{a} \cdot \mathbf{c})\mathbf{b} - (\mathbf{b} \cdot \mathbf{c})\mathbf{a}$.
e. Compare the results of (b) and (d).

P3.5.3 Let **A**, **x**, and **b** be the dyadics and vectors:

$$\mathbf{A} = a_{ij}\mathbf{n}_i\mathbf{n}_j, \quad \mathbf{x} = x_i\mathbf{n}_i, \quad \mathbf{b} = b_i\mathbf{n}_i$$

where the $\mathbf{n}_i$ are mutually perpendicular unit vectors and where the components of **A** and **b** are given by the following arrays:

$$[a_{ij}] = \begin{bmatrix} 2 & -1 & 3 \\ 1 & 4 & -2 \\ -5 & 2 & 6 \end{bmatrix} \qquad \{b_i\} = \begin{bmatrix} 2 \\ 1 \\ -3 \end{bmatrix}$$

Determine the components of **x** so that

$$A \cdot x = b$$

P3.5.4 Validate Equation 3.79 using Equations 3.69 and 3.70.
P3.5.5 Validate Equation 3.82 using Equation 3.70.

Section 3.6

Consider the following matrices in Problems P3.6.1 through P3.6.12:

$$
A = \begin{bmatrix} 1 & -2 & 4 \\ 3 & 8 & 5 \\ 1 & -6 & 7 \end{bmatrix} \quad
B = \begin{bmatrix} -2 & 0 & 4 \\ 6 & 1 & 5 \\ -4 & 3 & 1 \end{bmatrix} \quad
C = \begin{bmatrix} 8 & -2 & 4 \\ -2 & 7 & 3 \\ 4 & 3 & -1 \end{bmatrix}
$$

$$
D = \begin{bmatrix} 0 & 0 & 0 \\ 0 & 0 & 0 \\ 0 & 0 & 0 \end{bmatrix} \quad
E = \begin{bmatrix} 0 & 4 & -8 \\ -4 & 0 & 5 \\ 8 & -5 & 0 \end{bmatrix} \quad
F = \begin{bmatrix} 9 & 0 & 0 \\ 0 & 8 & 0 \\ 0 & 0 & 7 \end{bmatrix}
$$

$$
G = \begin{bmatrix} -1 & 2 & -6 \\ 0 & 2 & -3 \\ 5 & 4 & 1 \end{bmatrix} \quad
H = \begin{bmatrix} \sqrt{2}/2 & \sqrt{2}/4 & -\sqrt{6}/4 \\ 0 & \sqrt{3}/2 & 1/2 \\ \sqrt{2}/2 & -\sqrt{2}/4 & \sqrt{6}/4 \end{bmatrix} \quad
I = \begin{bmatrix} 1 & 0 & 0 \\ 0 & 1 & 0 \\ 0 & 0 & 1 \end{bmatrix}
$$

$$
J = \begin{bmatrix} -0.1774 & 0.8894 & 0.1959 \\ 1.0285 & -0.7384 & -0.0024 \\ -0.2726 & -0.8421 & 0.3117 \end{bmatrix} \quad
K = \begin{bmatrix} 0.1015 & -0.8172 & 0.4452 \\ 0.9187 & -0.6242 & 0.1784 \\ -0.6177 & 0.8664 & 0.3911 \end{bmatrix}
$$

$$
L = \begin{bmatrix} 0.5132 & 0.9773 & -0.3150 \\ 0.7071 & 0.0042 & -0.4444 \\ 2.3590 & 0.8660 & 1.7320 \end{bmatrix} \quad
M = \begin{bmatrix} 0.5000 & 0.8660 & 0.0000 \\ -0.6124 & 0.3536 & 0.7071 \\ 0.6124 & -0.3536 & 0.7071 \end{bmatrix}
$$

$$
N = \begin{bmatrix} 0.3846 & 0.0000 & 0.0000 \\ 0.0000 & 0.0000 & 0.0000 \\ 0.0000 & 0.0000 & 0.9231 \end{bmatrix} \quad
O = \begin{bmatrix} 0.5000 & -0.6124 & 0.6124 \\ 0.8660 & 0.3536 & -0.3536 \\ 0.0000 & 0.7071 & 0.7071 \end{bmatrix}
$$

P3.6.1 Which matrices are diagonal?
P3.6.2 Which are orthogonal? (to within 1%)
P3.6.3 Which are inverses? (to within 1%)
P3.6.4 Which are zero?
P3.6.5 Which are identities?
P3.6.6 Which are symmetric?
P3.6.7 Which are skew-symmetric?
P3.6.8 Which are singular?
P3.6.9 Evaluate $(A + B)C$ and $AC + BC$ and compare the results.
P3.6.10 Evaluate $(AB)C$ and $A(BC)$ and compare the results.
P3.6.11 Show that $(AB)^T = B^T A^T$ by direct multiplication.
P3.6.12 Show that $AB \neq BA$ by direct multiplication.

Section 3.7

P3.7.1 Using Equation 3.110, evaluate the determinant a given by

$$a = \begin{vmatrix} 2 & -3 & 2 \\ 4 & -1 & 0 \\ -5 & 5 & 1 \end{vmatrix}$$

P3.7.2 See Problem P3.7.1. Evaluate a by multiplying the elements of the third column by their cofactors and adding. Compare the results and the computational effort.

P3.7.3 Evaluate the determinant a^T:

$$a^T = \begin{vmatrix} 2 & 4 & -5 \\ -3 & -1 & 5 \\ 2 & 0 & 1 \end{vmatrix}$$

Compare the result with that of Problem P3.7.1.

P3.7.4 Evaluate the determinant a^*:

$$a^* = \begin{vmatrix} 4 & 2 & -5 \\ -1 & -3 & 5 \\ 0 & 2 & 1 \end{vmatrix}$$

Compare the result with that of Problem P3.7.3.

Section 3.8

P3.8.1 Validate Equation 3.121 by expanding the determinant of Equation 3.120.

P3.8.2 Review Problem P3.7.1. As before, let a be the determinant,

$$a = \begin{vmatrix} 2 & -3 & 2 \\ 4 & -1 & 0 \\ -5 & 5 & 1 \end{vmatrix}$$

Evaluate a using the terms, multiplications, and additions indicated in Equation 3.125. Compare the result with that of Problem P3.7.1. Compare the computational efforts. Explain the difference in efforts, if any.

P3.8.3 Let A be the matrix.

$$A = \begin{bmatrix} 2 & -1 & -3 \\ 0 & -3 & -5 \\ -5 & 1 & 4 \end{bmatrix}$$

Let C be the matrix of the cofactors of the elements of A. Find the array of the elements of C.

P3.8.4 See Problem P3.8.3. Let a be the determinant of A. Find a.

P3.8.5 See Problems P3.8.3 and P3.8.4.

    a. Form C^T, the transpose of the matrix of cofactors of A.

    b. Multiply C^T by $(1/a)$ forming the inverse A^{-1} of A, as in Equation 3.129. List the array of the elements of A^{-1}.

P3.8.6 Using the result of Problem P3.8.5b, show by multiplication that the matrix products AA^{-1} and $A^{-1}A$ result in the identity matrix I, where A is the matrix of Problem 3.8.3.

P3.8.7 Repeat Problems P3.8.3 through P3.8.6 if A is replaced by the matrix B given by

$$B = \begin{bmatrix} 1 & -2 & 2 \\ -4 & 0 & 3 \\ 3 & 1 & -4 \end{bmatrix}$$

P3.8.8 See Problems P3.8.3 and P3.8.7. Let P be the matrix product AB.

    a. List the elements of P.

    b. List the elements of P^{-1}, the inverse of P.

P3.8.9 See the results of Problems P3.8.5, P3.8.7, and P3.8.8. Show that

$$P^{-1} = (AB)^{-1} = B^{-1}A^{-1}.$$

Section 3.9

P3.9.1 Review the eigenvalue/eigenvector numerical example starting with Equation 3.142. Repeat the steps of the example for the matrix A given by

$$A = (1/16)\begin{bmatrix} 118 & -7\sqrt{6} & 3\sqrt{6} \\ -7\sqrt{6} & 89 & 3 \\ 3\sqrt{6} & 3 & 113 \end{bmatrix}$$

REFERENCES

1. B. Noble, *Applied Linear Algebra*, Prentice Hall, Englewood Cliffs, NJ, 1969.
2. P. C. Shields, *Elementary Linear Algebra*, Worth Publishers, New York, 1968.
3. R. A. Usmani, *Applied Linear Algebra*, Marcel Dekker, New York, 1987.
4. A. J. Pettofrezzo, *Elements of Linear Algebra*, Prentice Hall, Englewood Cliffs, NJ, 1970.
5. F. R. Gantmacher, *The Theory of Matrices*, Vols. I and II, Chelsea, New York, 1977.
6. L. Brand, *Vector and Tensor Analysis*, John Wiley & Sons, New York, 1947.
7. B. Hoffmann, *About Vectors*, Prentice Hall, Englewood Cliffs, NJ, 1966.
8. F. E. Hohn, *Elementary Matrix Algebra*, 2nd edn., Macmillan, New York, 1964.
9. R. C. Wrede, *Introduction to Vector and Tensor Analysis*, John Wiley & Sons, New York, 1963.
10. L. J. Paige and J. D. Swift, *Elements of Linear Algebra*, Ginn & Company, Boston, MA, 1961.
11. M. R. Spiegel, *Vector Analysis and an Introduction to Tensor Analysis*, Schaum, New York, 1959.
12. S. Lipschutz, *Linear Algebra*, Schaum, New York, 1968.
13. R. L. Huston and C. Q. Liu, *Formulas for Dynamic Analysis*, Marcel Dekker, New York, 2001.
14. C. L. Lawson and R. J. Hanson, *Solving Least Squares Problems*, Prentice Hall, Englewood Cliffs, NJ, 1974.
15. T. L. Boullion and P. L. Odell, *Generalized Inverse Matrices*, Wiley-Interscience, New York, 1971.
16. A. Ben-Israel and T. N. E. Greville, *Generalized Inverses: Theory and Applications*, Robert E. Krieger Publishing, Huntington, NY, 1980.

17. T. R. Kane, *Analytical Elements of Mechanics*, Vol. 1, Academic Press, New York, 1959.
18. H. Sharp Jr., *Modern Fundamentals of Algebra and Trigonometry*, Prentice Hall, Englewood Cliffs, NJ, 1961.
19. F. B. Hildebrand, *Advanced Calculus for Applications*, 2nd edn., Prentice Hall, Englewood Cliffs, NJ, 1976.
20. D. J. Higham and N. J. Higham, *MATLAB Guide*, Society for Industrial and Applied Mathematics, Philadelphia, PA, 2000.
21. K. Sigmon, *MATLAB Primer*, 5th edn., CRC Press, Boca Raton, FL, 1998.
22. B. D. Hahn, *Essential MATLAB for Scientists and Engineers*, John Wiley & Sons, Inc., New York, 1997.

4 Methods of Analysis II
Forces and Force Systems

Intuitively, we think of a force as a push or a pull. Then, in elementary mechanics, we represent forces as arrows (or vectors). For rigid bodies, this works well, but for deformable bodies, as human body, it is more appropriate to use force systems, which can represent distributed loadings.

In addition to forces, there are moments and couples (or twistings). Here too, the representation by single vectors, which is effective with rigid bodies, may not be appropriate for living and deformable systems.

In this chapter, we review elementary concepts of forces and force systems. We use these concepts in Chapters 5 and 6.

4.1 FORCES: VECTOR REPRESENTATIONS

If a force is a push or a pull, then the effect of a force depends upon how hard the push or pull is (i.e., the magnitude of the push or pull), the place of the push or pull (the point of application), and the direction of the push or pull (the orientation and sense). As such, forces are ideally represented by vectors lying along specific lines of action. Suppose, for example, that a force is exerted at a point P on a body B as in Figure 4.1. Then the force may be represented by a vector (called $\mathbf{F}$) acting along a line L (the line of action of $\mathbf{F}$), which passes through P. L determines the direction and the location of $\mathbf{F}$.

A vector restricted to a specific line of action is called a bound vector or a sliding vector. A vector not restricted to a specific line or point is called a free vector. (A unit vector is an example of a free vector.)

4.2 MOMENTS OF FORCES

Consider a force $\mathbf{F}$ with line of action L as in Figure 4.2. Let O be an object point (or reference point) and P be any point on L. Then, the moment of $\mathbf{F}$ about O is defined as [1]

$$\mathbf{M}_O = \mathbf{OP} \times \mathbf{F} = \mathbf{p} \times \mathbf{F} \tag{4.1}$$

where $\mathbf{p}$ is the position vector $\mathbf{OP}$ (Figure 4.2).

In the definition of Equation 4.1, the location of point P on L is arbitrary, but the moment $\mathbf{M}_O$ is nevertheless unique. The uniqueness of $\mathbf{M}_O$ is seen by observing that if Q is any other point on L, as in Figure 4.3, then a simple analysis shows that $\mathbf{OQ} \times \mathbf{F}$ is equal to $\mathbf{OP} \times \mathbf{F}$. Specifically,

$$\mathbf{OQ} \times \mathbf{F} = \mathbf{q} \times \mathbf{F} = (\mathbf{OP} + \mathbf{PQ}) \times \mathbf{F} = (\mathbf{p} + \mathbf{r}) \times \mathbf{F} = \mathbf{p} \times \mathbf{F} + \mathbf{r} \times \mathbf{F} \tag{4.2}$$

where
  $\mathbf{q}$ is $\mathbf{OQ}$
  $\mathbf{r}$ is $\mathbf{PQ}$
  $\mathbf{r} \times \mathbf{F}$ is zero since $\mathbf{r}$ is parallel to $\mathbf{F}$

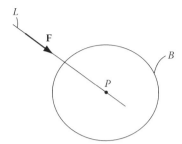

FIGURE 4.1 Force **F** applied on a body *B*.

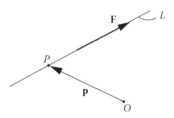

FIGURE 4.2 Force **F**, with line of action *L*, and reference point *O*.

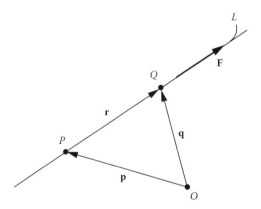

FIGURE 4.3 Two points on the line of action of **F**.

Recall that a judicious choice of the location of *P* can simplify the computation of the moment. Also, observe that if the reference point *O* is on *L* (i.e., if **F** passes through *O*), then $\mathbf{M}_O$ is zero.

4.3 MOMENTS OF FORCES ABOUT LINES

Consider next a force **F** with line of action L_F and a line *L* as in Figure 4.4. Let *O* be a point on *L* (any point of *L*) and let **λ** be a unit vector parallel to *L*. Then, the moment of **F** about *L*, $\mathbf{M}_L$ is defined as the projection of the moment of **F** about *O* ($\mathbf{M}_O$) about *L*. $\mathbf{M}_L$ is

$$\mathbf{M}_L = (\mathbf{M}_L \cdot \boldsymbol{\lambda})\boldsymbol{\lambda} = \left[(\mathbf{p} \times \mathbf{F}) \cdot \boldsymbol{\lambda}\right]\boldsymbol{\lambda} \tag{4.3}$$

where **p** is the position vector locating a point *P* on L_F relative to *O* as in Figure 4.4.

Observe that the specific location of points *P* and *O* on L_F and *L* does not change the value of $\mathbf{M}_L$. Observe further that if L_F intersects *L*, then $\mathbf{M}_L$ is zero.

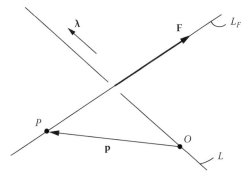

FIGURE 4.4   Force **F** acting about a line L.

4.4   SYSTEMS OF FORCES

Biosystems, as with most mechanical systems, are generally subjected to a number of forces applied simultaneously. To determine the effect of these forces, it is useful to study the assortment (or system) of forces themselves. To this end, consider a set S of N forces $\mathbf{F}_i$ ($i = 1, \ldots, N$) whose lines of action pass through particles $\mathbf{P}_i$ ($i = 1, \ldots, N$) as in Figure 4.5. Two vectors are usually used to characterize (or measure) S. These are (1) the resultant (or sum) of the forces and (2) the sum of the moments of the forces about a reference point O.

The resultant $\mathbf{R}$ of S is simply the sum

$$\mathbf{R} = \mathbf{F}_1 + \mathbf{F}_2 + \cdots + \mathbf{F}_N = \sum_{i=1}^{N} \mathbf{F}_i \tag{4.4}$$

The resultant is a free vector, although with equivalent force systems (see Section 4.5.3), a vector equal to the resultant is given a specific line of action.

The moment of S about a point O, $\mathbf{M}_O$, is simply the sum of the moments of the individual forces of S about O. That is,

$$\mathbf{M}_O = \mathbf{p}_1 \times \mathbf{F}_1 + \mathbf{p}_2 \times \mathbf{F}_2 + \cdots + \mathbf{p}_i \times \mathbf{F}_i + \cdots + \mathbf{p}_N \times \mathbf{F}_N = \sum_{i=1}^{N} \mathbf{p}_i \times \mathbf{F}_i \tag{4.5}$$

where $\mathbf{p}_i$ ($i = 1, \ldots, n$) locates a point on the line of action of $\mathbf{F}_i$ relative to O, as in Figure 4.6.

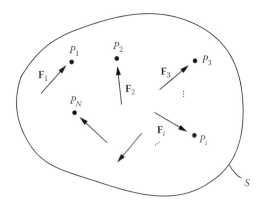

FIGURE 4.5   System of forces.

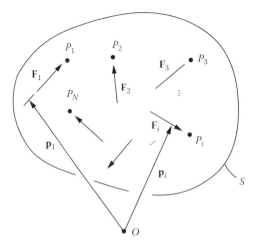

FIGURE 4.6   System of forces and a reference point.

The moment of S about O depends upon the location of O. A question then arises: If Q is a point different from O, how is the moment of S about Q related to the moment of S about O? To answer this question, consider again S with object points O and Q as in Figure 4.7 where $\mathbf{q}_i$ locates a point on the line of action of $\mathbf{F}_i$ relative to Q. Then by inspection in Figure 4.7, we have

$$\mathbf{p}_i = \mathbf{OQ} + \mathbf{q}_i \tag{4.6}$$

By substituting into Equation 4.5, we have

$$\mathbf{M}_O = \sum_{i=1}^{N} \mathbf{p}_i \times \mathbf{F}_i = \sum_{i=1}^{N} (\mathbf{OQ} + \mathbf{q}_i) \times \mathbf{F}_i = \sum_{i=1}^{N} \mathbf{OQ} \times \mathbf{F}_i + \sum_{i=1}^{N} \mathbf{q}_i \times \mathbf{F}_i$$

$$= \mathbf{OQ} \times \sum_{i=1}^{N} \mathbf{F}_i + \sum_{i=1}^{N} \mathbf{q}_i \times \mathbf{F}_i = \mathbf{OQ} \times \mathbf{R} + \mathbf{M}_Q \tag{4.7}$$

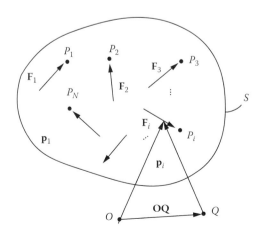

FIGURE 4.7   Force system S with object points O and Q.

where the last equality follows from Equations 4.4 and 4.5 and by inspection of Figure 4.7. Thus, we have

$$\mathbf{M}_O = \mathbf{M}_Q + \mathbf{OQ} \times \mathbf{R} \qquad (4.8)$$

Equation 4.8 is especially useful for studying special force systems as in the following section.

4.5   SPECIAL FORCE SYSTEMS

There are three special force systems, which are useful in the study of mechanical systems and particularly biosystems. These are (1) zero systems, (2) couples, and (3) equivalent systems. The following subsections provide a description of each of these.

4.5.1   ZERO FORCE SYSTEMS

If a force system S has a zero resultant $\mathbf{R}$ and a zero moment about some point O, it is called a zero system. Observe from Equation 4.8 that if $\mathbf{R}$ is zero and if $\mathbf{M}_O$ is also zero, then the moment about any and all other points Q is also zero.

Zero force systems form the basis for static analysis. If a body B (or a collection of bodies C) is in static equilibrium, then, as a consequence of Newton's laws, the force system on B (or C) is a zero system. Then the resultant $\mathbf{R}$ of the forces exerted on B (or C) is zero. Consequently, the sum of the components of the forces on B (or C) in any and all directions is also zero.

Similarly, if a body B (or collection of bodies C) is in static equilibrium, the sum of the moments of the forces exerted on B (or C) about any and all points is zero, and consequently, the sum of the components of these moments in any and all directions is zero.

The foregoing observations form the basis for the construction of free-body diagrams.

4.5.2   COUPLES

If a force system S has a zero resultant but nonzero moment about some point, it is called a couple. Figure 4.8 illustrates such a system.

With the resultant being zero, Equation 4.8 shows that the moment of a couple about all points is the same. This moment is called the torque $\mathbf{T}$ of the couple.

If a couple has only two forces, it is called a simple couple. Figure 4.9 depicts a simple couple. For a simple couple, the forces must have equal magnitudes but opposite directions. The magnitude of the torque of the couple is then simply the magnitude of one of the forces multiplied by the distance d between the forces (Figure 4.9). That is,

$$|\mathbf{T}| = |\mathbf{F}|d \qquad (4.9)$$

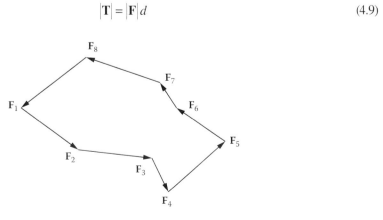

FIGURE 4.8   Example of a couple.

FIGURE 4.9   A simple couple.

4.5.3   Equivalent Force Systems

Two force systems S_1 and S_2 are said to be equivalent if (1) they have equal resultants ($\mathbf{R}_{S_1} = \mathbf{R}_{S_2}$) and (2) they have equal moments about some point O ($\mathbf{M}_O^{S_1} = \mathbf{M}_O^{S_2}$). From Equation 4.8, it is seen that if the resultants of the two systems are equal, and if their moments about one point O are equal, then their moments about all points are equal. Specifically, if Q is any point distinct from O, then Equation 4.8 has the following forms for S_1 and S_2:

$$\mathbf{M}_O^{S_1} = \mathbf{M}_Q^{S_1} + \mathbf{OQ} \times \mathbf{R}_{S_1} \tag{4.10}$$

and

$$\mathbf{M}_O^{S_2} = \mathbf{M}_Q^{S_2} + \mathbf{OQ} \times \mathbf{R}_{S_2} \tag{4.11}$$

Subtracting Equation 4.11 from Equation 4.10, we have

$$\mathbf{O} = \mathbf{M}_Q^{S_1} - \mathbf{M}_Q^{S_2} \quad \text{or} \quad \mathbf{M}_Q^{S_1} = \mathbf{M}_Q^{S_2} \tag{4.12}$$

Using Newton's laws, it can be shown that, for rigid bodies, equivalent force systems have the same physical effect. Thus, for convenience in analysis, for rigid bodies, equivalent force systems may be interchanged. For example, if a force system with many forces is replaced by an equivalent force system with only a few forces, the subsequent analysis effort could be substantially reduced.

Recall that for any given force system, no matter how large, there exists an equivalent force system consisting of a single force whose line of action may be passed through an arbitrary point together with a couple. To see this, let S be a given force system with resultant $\mathbf{R}$ and moment $\mathbf{M}_O$ about some point O. Let $\hat{S}$ be a force system consisting of a single force $\mathbf{F}$ equal to $\mathbf{R}$ with line of action L passing through O together with a couple with torque $\mathbf{T}$ (equal to $\mathbf{M}_O$), as in Figure 4.10. Then, S and $\hat{S}$ are equivalent since (1) the resultants are equal (the resultant of $\hat{S}$ is $\mathbf{R}$ (the couple has zero resultant)) and (2) they have equal moments about O (the moment of $\hat{S}$ about O is $\mathbf{T}$ ($=\mathbf{M}_O$) since the line of action of $\mathbf{F}$ ($=\mathbf{R}$) passes through O, and couples have the same moment about all points).

For nonrigid bodies (such as human limbs), however, equivalent force systems can have vastly different physical effects. Consider, for example, two identical deformable rods acted upon by different but equivalent force systems as in Figure 4.11. In the first case (case (a)), the rod is in compression, and being deformable, it will be shortened. In the second case (case (b)), the rod is in tension and is elongated. The force systems, however, are obviously equivalent: They have equal

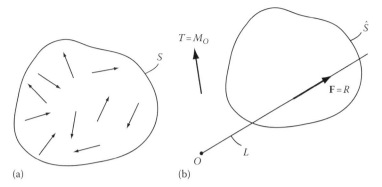

(a)   (b)

FIGURE 4.10   (a) A given force system S and (b) an equivalent force system $\hat{S}$ consisting of a single force and a couple.

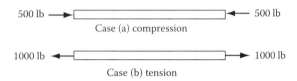

FIGURE 4.11   Equivalent force systems acting on identical deformable rods.

resultants (both are zero) and equal moments about the center point of the rod (both zero). Thus, for deformable bodies, equivalent force systems can have vastly different, even opposite, effects.

This example raises the question—are equivalent force systems of use in reducing computational effort with large force systems acting on deformable bodies? The answer is yes, if we are mindful of Saint Venant's principle when interchanging equivalent force systems. Saint Venant's principle [2] states that, for deformable bodies, the interchange of equivalent force systems produces different effects near the location of force application (i.e., locally) but at locations distant from the points of application, there is no difference in the effects of the equivalent systems. As an illustration, consider the cantilever beam of Figure 4.12.

In case (a), the beam is loaded at its open end with a complex system of forces. In case (b), the beam is loaded with a single force (with axial and transverse components) and a couple. Saint Venant's principle states that near the loaded end of the beam, there will be significant differences in the effects of the equivalent loadings (i.e., stresses, strains, and deformation). At locations away from the loaded end, however (such as at location A), there will be no significant differences in the stresses, strains, or deformation between the equivalent force systems.

In view of this example, a question arises: How far away from the loading application do we need to be for the effects from equivalent force systems to be essentially equal? Unfortunately, there is no precise answer. But, for biosystems, differences between the effects of equivalent

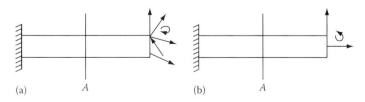

(a)   A   (b)   A

FIGURE 4.12   Cantilever beam with (a) equivalent end loading with many forces and (b) equivalent end loading with few forces.

force systems will generally be within the accuracy of the measurement of physical properties when the distance away from the loading is more than 10 times as great as the characteristic dimension of the loading region. In any event, Saint Venant's principle states that the further away from the loading region we are, the smaller the difference in the effects of equivalent force systems.

4.5.4 SUPERIMPOSED AND NEGATIVE FORCE SYSTEMS

In the study of the mechanics of deformable materials, it is convenient to superpose or add force systems. Suppose, for example, that a body B is subjected to a force system S_1. Suppose that a second force system S_2 is also applied to B. B is then subjected to the forces and couples of both S_1 and S_2, and S_1 and S_2 are said to be superimposed (or superposed) with each other.

If two superimposed force systems result in a zero force system, the force systems are said to be negative of each other. If the superposition of S_1 and S_2 produces a zero force system, then S_1 is the negative of S_2, and S_2 is the negative of S_1.

4.6 PRINCIPLE OF ACTION–REACTION

Newton's laws of motion may be summarized as [3,4]

1. In the absence of a change in the force system exerted on a body B, if B is at rest, it will remain at rest. Further, if B is in motion at a uniform rate, it will remain in motion at a uniform rate.
2. If a particle P with mass m is subjected to a force $\mathbf{F}$, then P will accelerate with an acceleration $\mathbf{a}$ proportional to $\mathbf{F}$ and inversely proportional to m. That is,

$$\mathbf{a} = \frac{\mathbf{F}}{m} \quad \text{or} \quad \mathbf{F} = m\mathbf{a} \tag{4.13}$$

3. If a body B_1 exerts a force $\mathbf{F}$ on a body B_2 along a line of action L, then B_2 exerts a force $-\mathbf{F}$ on B_1 along L (Figure 4.13).

Law 3 is commonly called the law of action and reaction. As a consequence, if a body B_1 exerts a force system S_1 on a body B_2, then B_2 will exert a negative force system S_2 on B_1. That is, taken together, S_1 and S_2 form a zero system.

In a circular argument, Newton's laws are said to be valid in a Newtonian (or inertial) reference frame, where a Newtonian frame is defined as a space where Newton's laws are valid. The question is then as follows: do such reference frames exist? While the answer is physically unresolved, experiments show that a reference frame R fixed on the surface of the earth is an approximate Newtonian,

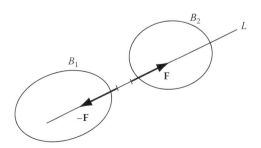

FIGURE 4.13 Equal magnitude, oppositely directed forces between bodies.

or inertial, frame for bodies small relative to the earth and for speeds small compared with the speed of light. Then, for practical purposes, biosystems on earth obey Newton's laws.

Therefore, using the law of action and reaction, we see, for example, with a biosystem, that the forces exerted by one limb (say the upper arm) on another limb (say the lower arm) are counter-acted by equal but oppositely directed forces by the second limb on the first (the lower arm on the upper arm).

PROBLEMS

Section 4.1

P4.1.1 Provide an example of two force systems being equivalent when thought of as vector systems, but having different physical effects.

Section 4.2

P4.2.1 Let **F** be a 12 lb force acting along line L, where L is inclined at $30°$ to the X-axis as shown in the following figure. Express **F** in terms of unit vectors $\mathbf{n}_x$ and $\mathbf{n}_y$, which are parallel to the X- and Y-axes as also shown.

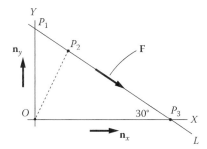

P4.2.2 Refer to the figure for Problem P4.2.1. Let P_3 be the point of intersection of L with the X-axis and let the distance from the origin O to P_3 be 3 inches. Let P_1 be the intersection of L with the Y-axis and let P_2 be the point on L closest to O (i.e., OP_2 is perpendicular to L). Let $\mathbf{p}_1$, $\mathbf{p}_2$, and $\mathbf{p}_3$ be the position vectors $\mathbf{OP}_1$, $\mathbf{OP}_2$, and $\mathbf{OP}_3$, respectively. Express $\mathbf{p}_1$, $\mathbf{p}_2$, and $\mathbf{p}_3$ in terms of the unit vectors $\mathbf{n}_x$ and $\mathbf{n}_y$. Also, determine the magnitudes of $\mathbf{p}_1$, $\mathbf{p}_2$, and $\mathbf{p}_3$.

P4.2.3 Refer to the figure for Problem P4.2.1. and also to the results of Problems P4.2.1 and P4.2.2. Calculate the moment of **F** about O in three ways: (a) as $\mathbf{p}_1 \times \mathbf{F}$, (b) as $\mathbf{p}_2 \times \mathbf{F}$, and (c) as $\mathbf{p}_3 \times \mathbf{F}$. Compare the efficiency and/or ease of computation of the three computations.

Section 4.3

P4.3.1 Consider the sketch of the following figure where L and L_F are lines fixed in a Cartesian reference frame XYZ. Let L and L_F pass through points P, Q and P_F, Q_F as shown. Let the X, Y, and Z coordinates of these points be given by

$$P(3,-2,3), \quad Q(-1,4,4)$$

$$P_F(-2,-2,6), \quad Q_F(1,6,-1)$$

where the units are meters.

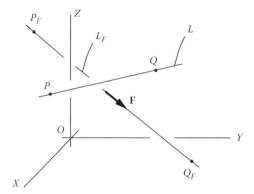

Let **F** be a 35 N force along L_F, directed from P_F to Q_F as shown.
Determine the magnitude and direction of the moment of **F** about L.

P4.3.2 See Problem P4.3.1. Repeat the problem if **F** is 18 lb directed from Q_F to P_F and P, Q, P_F, and Q_F have coordinates given by

$$P(3,2,-1), \quad Q(-2,4,-8)$$

$$P_F(5,-7,4), \quad Q_F(3,-5,-6)$$

Section 4.4

P4.4.1 Consider the system of forces exerted on a box as sketched in the following figure. The box is 3 m long, 2 m deep, and 1 m high, and it is subjected to 10 forces $\mathbf{F}_i$ ($i = 1, \dots ,10$) as shown. For simplicity, each force is directed along one of the edges of the box and thus parallel to one of the coordinate axes X, Y, or Z. Let $\mathbf{n}_x$, $\mathbf{n}_y$, and $\mathbf{n}_z$ be unit vectors parallel to these axes and box edges, as shown.
The magnitudes of the forces are

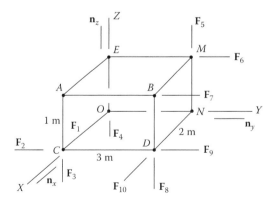

$$|\mathbf{F}_1| = 8\,\text{N}, \quad |\mathbf{F}_2| = 6\,\text{N}, \quad |\mathbf{F}_3| = 7\,\text{N}, \quad |\mathbf{F}_4| = 8\,\text{N}$$

$$|\mathbf{F}_5| = 5\,\text{N}, \quad |\mathbf{F}_6| = 10\,\text{N}, \quad |\mathbf{F}_7| = 10\,\text{N}, \quad |\mathbf{F}_8| = 4\,\text{N}$$

$$|\mathbf{F}_9| = 5\,\text{N}, \quad |\mathbf{F}_{10}| = 12\,\text{N}$$

a. Express the $\mathbf{F}_i$ ($i = 1, ..., 10$) in terms of the unit vectors $\mathbf{n}_x$, $\mathbf{n}_y$, and $\mathbf{n}_z$.

b. Determine the resultant $\mathbf{R}$ of the force system.

c. Determine the moment $\mathbf{M}_O$ of the force system about the origin O of the X-, Y-, and Z-axes.

P4.4.2 See Problem P4.4.1 and its results. Find the moments of the force system about box corners A and B (see the figure for Problem P4.4.1).

P4.4.3 Verify Equation 4.8 for $\mathbf{M}_O$, $\mathbf{M}_A$, and $\mathbf{M}_B$. That is, show that

$$\mathbf{M}_O = \mathbf{M}_A + \mathbf{OA} \times \mathbf{R}$$

$$\mathbf{M}_A = \mathbf{M}_B + \mathbf{AB} \times \mathbf{R}$$

and

$$\mathbf{M}_B = \mathbf{M}_O + \mathbf{BO} \times \mathbf{R}$$

P4.4.4 Consider the box depicted in the following figure together with the system S of eight forces $\mathbf{F}_i$ ($i = 1, ..., 8$) with directions and lines of action as shown. Let the corners of the box be A, B, C, D, O, P, Q, and S; let $\mathbf{n}_1$, $\mathbf{n}_2$, and $\mathbf{n}_3$ be unit vectors parallel to the box edges; and let the edge lengths be 12, 8, and 6 ft. Finally, let the magnitudes of the forces be

$$|\mathbf{F}_1| = 6\,\text{lb}, \quad |\mathbf{F}_2| = 8\,\text{lb}, \quad |\mathbf{F}_3| = 9\,\text{lb}, \quad |\mathbf{F}_4| = 13\,\text{lb}$$

$$|\mathbf{F}_5| = 15\,\text{lb}, \quad |\mathbf{F}_6| = 8\,\text{lb}, \quad |\mathbf{F}_7| = 9\,\text{lb}, \quad |\mathbf{F}_8| = 6\,\text{lb}$$

a. Express the $\mathbf{F}_i$ ($i = 1, ..., 8$) in terms of $\mathbf{n}_1$, $\mathbf{n}_2$, and $\mathbf{n}_3$.

b. Determine the resultant $\mathbf{R}$ of the force system S.

c. Determine the moments of S about points O, Q, and B by direct computation, that is, without using Equation 4.8.

d. Verify the results of (c) by using Equation 4.8.

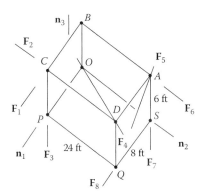

Section 4.5

P4.5.1 Let $\mathbf{F}_i$ ($i = 1, \ldots, N$) be a system S of N forces exerted on a body B, which in turn is fixed in a Cartesian reference frame R as in the following figure. Let X, Y, and Z be the axes of R and let $\mathbf{n}_x$, $\mathbf{n}_y$, $\mathbf{n}_z$ be unit vectors parallel to X, Y, and Z as shown. Finally, let O be the origin of R.

For specificity, let N be 8 and let the forces (in Newtons) be expressed in terms of $\mathbf{n}_x$, $\mathbf{n}_y$, and $\mathbf{n}_z$ as

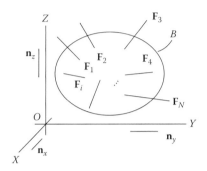

$$\mathbf{F}_1 = 8\mathbf{n}_x + 4\mathbf{n}_y - 3\mathbf{n}_z N, \qquad \mathbf{F}_2 = 1\mathbf{n}_x + 0\mathbf{n}_y + 5\mathbf{n}_z N$$

$$\mathbf{F}_3 = -6\mathbf{n}_x + 4\mathbf{n}_y + 2\mathbf{n}_z N, \quad \mathbf{F}_4 = 3\mathbf{n}_x + 6\mathbf{n}_y - 7\mathbf{n}_z N$$

$$\mathbf{F}_5 = -9\mathbf{n}_x - 8\mathbf{n}_y + 7\mathbf{n}_z N, \quad \mathbf{F}_6 = 10\mathbf{n}_x - 5\mathbf{n}_y - 7\mathbf{n}_z N$$

$$\mathbf{F}_7 = 4\mathbf{n}_x - 3\mathbf{n}_y + 2\mathbf{n}_z N, \qquad \mathbf{F}_8 = 5\mathbf{n}_x + 6\mathbf{n}_y - 9\mathbf{n}_z N$$

Let the forces $\mathbf{F}_i$ pass through points P_i of B where the coordinates of P_i (in meters) are

$$P_1(0,-2,4), \quad P_2(1,3,-7), \quad P_3(2,5,-6), \quad P_4(8,-5,9)$$

$$P_5(7,-6,5), \quad P_6(8,-8,0), \quad P_7(9,1,-2), \quad P_8(5,3,4)$$

a. Compute the resultant $\mathbf{R}$ of S.

b. Compute the moment of S about O, $\mathbf{M}_O$.

P4.5.2 Referring to Problem P4.5.1 and the results, is S a zero system? Why or why not?

P4.5.3 Referring again to Problem P4.5.1, and in view of the result of Problem P4.5.2, if S is not a zero system, envision appending a ninth force $\mathbf{F}_9$ to S so that the resultant is zero. Determine $\mathbf{F}_9$.

P4.5.4 Referring again to Problems P4.5.1 to P4.5.3 and their results, suppose the line of action of $\mathbf{F}_9$ passes through O. Let a couple with torque $\mathbf{T}$ be appended to a system $\hat{S}$ consisting of S and $\mathbf{F}_9$. Find $\mathbf{T}$ such that $\hat{S}$ is a zero system.

P4.5.5 Suppose a couple has a torque $\mathbf{T}$, which, expressed in terms of mutually perpendicular unit vectors $\mathbf{n}_x$, $\mathbf{n}_y$, and $\mathbf{n}_z$, is

$$\mathbf{T} = 18\mathbf{n}_x - 22\mathbf{n}_y + 27\mathbf{n}_z \text{ ft lb}$$

If $\mathbf{n}_x$, $\mathbf{n}_y$, and $\mathbf{n}_z$ are parallel to Cartesian axes X, Y, and Z, construct a set of three simple couples parallel to the Y–Z, Z–X, and X–Y planes, which collectively have the same moment as $\mathbf{T}$ about any and all points.

P4.5.6 Consider the set S of five planar forces represented in the following figure. Let the forces have lines of action passing through points A, B, ..., E with inclinations as shown. Let the XY coordinates of A, B, ..., E be

$$A: (2.0, 0.00) \quad B: (3.732, 1.0) \quad C: (2.232, 3.598)$$
$$D: (0.232, 2.098) \quad E: (0.232, 1.768)$$

where the coordinate units are in feet. Correspondingly, let the magnitudes of the forces be

$$|\mathbf{F}_A| = 30\,\text{lb}, \quad |\mathbf{F}_B| = 45\,\text{lb}, \quad |\mathbf{F}_C| = 37.5\,\text{lb}$$
$$|\mathbf{F}_D| = 4.95\,\text{lb}, \quad |\mathbf{F}_E| = 37.5\,\text{lb}$$

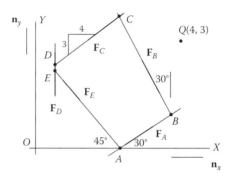

a. Express $\mathbf{F}_A$, $\mathbf{F}_B$, ..., $\mathbf{F}_E$ in terms of unit vectors $\mathbf{n}_x$ and $\mathbf{n}_y$, parallel to axes X and Y.
b. Find the resultant $\mathbf{R}$ of S (consisting of $\mathbf{F}_A$, $\mathbf{F}_B$, ..., $\mathbf{F}_E$), to two decimal places.
c. Find the moment of S about the origin O.
d. Find the moment of S about point Q, having coordinates $(4,3)$.
e. What, if any, conclusions may be stated about S?

P4.5.7 See Problem P4.5.6. Consider a reference frame R having Cartesian axes X, Y, and Z as in the following figure. Let S be a set of five (5) forces $\mathbf{F}_i$ ($i = 1$, ..., 5) with lines of action passing through points P_i ($i = 1$, ..., 5) of R. Let the $\mathbf{F}_i$ be expressed in terms of unit vectors $\mathbf{n}_1$, $\mathbf{n}_2$, and $\mathbf{n}_3$, parallel to X, Y, and Z (see the following figure) as

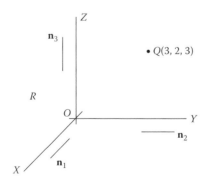

$$\mathbf{F}_1 = 12\mathbf{n}_1 - 8\mathbf{n}_2 - 6\mathbf{n}_3 N$$
$$\mathbf{F}_2 = -9\mathbf{n}_1 + 5\mathbf{n}_2 - 14\mathbf{n}_3 N$$

$$\mathbf{F}_3 = 11\mathbf{n}_1 + 3\mathbf{n}_2 + 0\mathbf{n}_3 N$$

$$\mathbf{F}_4 = -4\mathbf{n}_1 - 7\mathbf{n}_2 + 12\mathbf{n}_3 N$$

$$\mathbf{F}_5 = -10\mathbf{n}_1 + 7\mathbf{n}_2 + 8\mathbf{n}_3 N$$

Let the X, Y, and Z coordinates (in meters) of the P_i be as given by the following table:

I	P_i	X	Y	Z
1	P_1	1	−1	2
2	P_2	4	0	3
3	P_3	3	−5	−7
4	P_4	−8	1	2
5	P_5	−6	−6	5

a. Compute the resultant of S and thereby show that S may be a couple.
b. Compute the moment of S about the origin O of R. (Express the result in terms of $\mathbf{n}_1$, $\mathbf{n}_2$, and $\mathbf{n}_3$.)
c. Compute the moment of S about point Q with X, Y, and Z coordinates (3,2,3).
d. Is S a couple? Why?

P4.5.8 Consider again the box together with its applied force system of Problem P4.4.4 shown in the figure for that problem. For convenience, the system is shown again in the following figure:

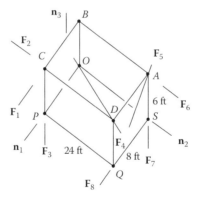

Also, for convenience, the magnitudes of the forces are listed here again as

$$|\mathbf{F}_1| = 6\,\text{lb}, \quad |\mathbf{F}_2| = 8\,\text{lb}, \quad |\mathbf{F}_3| = 9\,\text{lb}, \quad |\mathbf{F}_4| = 13\,\text{lb}$$

$$|\mathbf{F}_5| = 15\,\text{lb}, \quad |\mathbf{F}_6| = 8\,\text{lb}, \quad |\mathbf{F}_7| = 9\,\text{lb}, \quad |\mathbf{F}_8| = 6\,\text{lb}$$

Envision an equivalent force system consisting of a single force $\mathbf{F}_O$, with line of action passing through O, together with a couple with torque $\mathbf{T}_O$. Find $\mathbf{F}_O$ and $\mathbf{T}_O$ and express the results in terms of the unit vectors $\mathbf{n}_1$, $\mathbf{n}_2$, and $\mathbf{n}_3$ shown in the earlier figure.

P4.5.9 See Problem P4.5.8. Repeat Problem P4.5.8 where the equivalent force system consists of a single force $\mathbf{F}_A$, passing through corner A of the box, together with a couple with torque $\mathbf{T}_A$. That is, find $\mathbf{F}_A$ and $\mathbf{T}_A$.

P4.5.10 See Problems P4.5.8 and P4.5.9. Repeat the problems again for corners B and C: Let the single forces be $\mathbf{F}_B$ and $\mathbf{F}_C$ and the corresponding couple torques be $\mathbf{T}_B$ and $\mathbf{T}_C$. Find $\mathbf{F}_B$, $\mathbf{F}_C$, $\mathbf{T}_B$, and $\mathbf{T}_C$.

P4.5.11 Tabulate the results of Problems P4.5.8 through P4.5.10. Evaluate the torque magnitudes $|\mathbf{T}_O|$, $|\mathbf{T}_A|$, $|\mathbf{T}_B|$, and $|\mathbf{T}_C|$.

P4.5.12 For a given force system S, it has been shown (see, e.g., Ref. [1]) that there exists a point P^*, where if the line of action of the single force $\mathbf{F}^*$, of an envisioned equivalent single force/couple system, passes through P^*, the corresponding couple torque $\mathbf{T}^*$ has a magnitude smaller (or as small) as the torque magnitude of any other equivalent force/couple system. That is, $|\mathbf{T}^*|$ is a minimum couple torque magnitude.

If $\mathbf{F}^*$ is the single force passing through P^*, and if $\mathbf{T}^*$ is the torque of the corresponding equivalent force/couple system, then $\mathbf{F}^*$ and the couple with torque $\mathbf{T}^*$ may be thought of as the "smallest" equivalent force system for S.

Specifically, it has been shown that a vector $\mathbf{OP}^*$ locating P^* relative to a reference point O is given by

$$\mathbf{OP}^* = \frac{(\mathbf{R} \times \mathbf{M}_O)}{|\mathbf{R}|^2} \tag{A}$$

where
    $\mathbf{R}$ is the resultant of S
    $\mathbf{M}_O$ is the moment of S about O

a. Find the X, Y, and Z coordinates of P^* for the force system of Problem P4.5.8, where X, Y, and Z are Cartesian axes parallel to $\mathbf{n}_1$, $\mathbf{n}_2$, and $\mathbf{n}_3$ and with origin O.
b. Find the corresponding minimum coupled torque T^* for the force system of Problem P4.5.8.
c. Show that $\mathbf{T}^* = (\mathbf{M}_O \cdot \mathbf{n}_R)\mathbf{n}_R$ where $\mathbf{n}_R$ is a unit vector parallel to $\mathbf{R}$ (i.e., $\mathbf{n}_R = \mathbf{R}/|\mathbf{R}|$).
d. Verify that $|\mathbf{T}^*|$ is smaller than $|\mathbf{T}_O|$, $|\mathbf{T}_A|$, $|\mathbf{T}_B|$, and $|\mathbf{T}_C|$.

REFERENCES

1. T. R. Kane, *Analytical Elements of Mechanics*, Vol. 1, Academic Press, New York, 1959.
2. R. W. Little, *Elasticity*, Prentice Hall, Englewood Cliffs, NJ, 1973.
3. R. L. Huston and C. Q. Liu, *Formulas for Dynamic Analysis*, Marcel Dekker, New York, 2001.
4. J. H. Ginsberg and J. Genin, *Statics and Dynamics*, John Wiley & Sons, New York, 1984.

5 Methods of Analysis III
Mechanics of Materials

In this rather extensive chapter, we briefly review some elementary procedures for studying mechanics of materials with a view toward application with biomaterials. Numerous books have been written on the mechanics of materials with various titles such as *Strength of Materials, Mechanics of Materials, Stress Analysis, Elasticity*, and *Continuum Mechanics*, depending upon the emphasis. The references at the end of this chapter provide a partial list of some of these works [1–27].

Our objective here is to simply review a few fundamental concepts, procedures, and formulas, which could be of interest and use in biomechanic analyses. Readers interested in additional details and more in-depth treatment should consult the references.

We begin with a review of elementary concepts of stress, strain, and Hooke's law. We then generalize these concepts and consider a few illustrative applications primarily in the extension, bending, and torsion of beams and rods, with a view toward loading of long bones and ribs in biosystems.

5.1 CONCEPTS OF STRESS

As noted, a force is like a push or a pull. In this context, a stress is like an average force per area of application.

Consider a system of forces exerted on a small region S of a body B as in Figure 5.1. Let this force system be replaced by a single force $\mathbf{F}$ passing through an arbitrary point P in S together with a couple with torque $\mathbf{T}$ (see Section 4.5.3), as in Figure 5.2.

Suppose the region S is made smaller and shrunk around point P and thus reducing the number of forces applied. Then the magnitude of the equivalent resultant force $\mathbf{F}$ will also decrease. For a reasonably smooth distribution of forces, the magnitude of $\mathbf{F}$ will be proportional to the area A of S. The magnitude of $\mathbf{T}$ will also decrease (even faster), becoming proportional to the product of A and a typical dimension d (characteristic length) of S.

A stress vector $\boldsymbol{\sigma}$ and a couple stress vector $\boldsymbol{\mu}$ at P may be defined as the limiting values of $\mathbf{F}/A$ and $\mathbf{T}/A$. With the magnitude of $\mathbf{T}$ being proportional to Ad, $\boldsymbol{\mu}$ then approaches zero as A becomes increasingly smaller.

Alternatively, with the magnitude of $\mathbf{F}$ being proportional to A, $\boldsymbol{\sigma}$ becomes a finite vector as A becomes smaller. The scalar components of $\boldsymbol{\sigma}$ are called stresses.

In the earlier context, the stress vector is like an average, or normalized, force—a force per unit area. The scalar components, the stresses, are thus like forces per unit area.

As a simple example, consider the rod R depicted in Figure 5.3. Let R have a cross section A and let R be subjected to axial forces with magnitude P as shown. Then the axial component of the stress σ is simply

$$\sigma = \frac{P}{A} \tag{5.1}$$

This stress, being directed axially, is thought of as being exerted on the cross sections normal to the axis of the rod. Also, the forces P of Figure 5.3 tend to elongate the rod, and the rod is

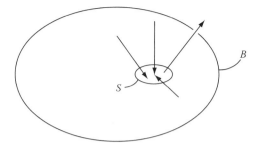

FIGURE 5.1 Body B with a force system exerted over a small region S.

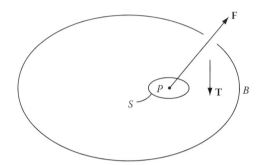

FIGURE 5.2 Equivalent force system to that of Figure 5.1.

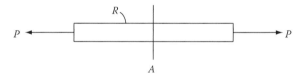

FIGURE 5.3 Rod in tension.

said to be in tension. The stress of Equation 5.1 is then called tension, or tensile stress as well as normal stress, or simple stress.

In like manner, if the forces P of Figure 5.3 were directed axially toward the rod so as to shorten it, the rod would be in compression, and the corresponding stress is called compressive stress as well as normal stress or simple stress.

When forces are directed normal to the surface of a body, they create normal stresses as with the block B of Figure 5.4. If the resultant of these forces is P and if the forces are exerted over a region S of B with area A, the average normal stress σ on B is then

$$\sigma = \frac{P}{A} \tag{5.2}$$

Forces may obviously be directed other than normal to the surface. For example, for the block B of Figure 5.4, let the forces be directed tangent or parallel to the surface as V in Figure 5.5. In this case, the block tends to be distorted into a parallelogram as in Figure 5.6 and is said to be in shear. Again, if the forces have a resultant V, and if they are exerted over a region S of B with area A, the average tangential or shear stress τ is

$$\tau = \frac{V}{A} \tag{5.3}$$

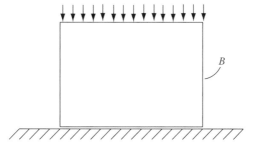

FIGURE 5.4   Forces directed normal to a block *B*.

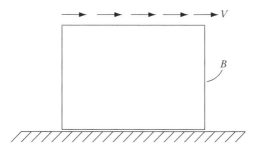

FIGURE 5.5   Tangentially directed forces on block *B*.

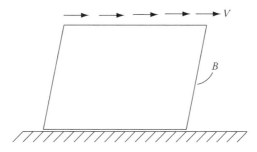

FIGURE 5.6   Block *B* distortion due to shear forces.

In Figures 5.5 and 5.6, the forces are assumed to be continuously distributed over the surface. Although the distribution is continuous, it is not necessarily uniform. Thus, if the surface area was subdivided, the stresses computed by Equations 5.2 and 5.3 will not necessarily be the same. That is, in general, the stress varies across the surfaces where the load is applied. The stresses computed using Equations 5.2 and 5.3 are the average stresses over the loaded surfaces. A more accurate representation of stress may be obtained by letting the loaded area be shrunk around a point. Indeed, as aforementioned, the stress vector $\boldsymbol{\sigma}$ is frequently defined as a limit as

$$\boldsymbol{\sigma} = \lim_{\Delta A \to 0} \frac{\Delta \mathbf{F}}{\Delta A} \qquad (5.4)$$

where
   ΔA is the area of the shrinking region
   $\Delta \mathbf{F}$ is the resultant force, or load, on the region

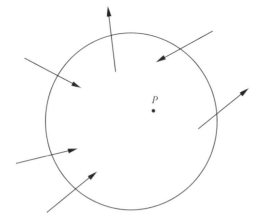

FIGURE 5.7 Deformable body under load.

Consider now the interior of a deformable body *B* where *B* is subjected to a loading on its surface as in Figure 5.7. Let *P* be a typical point in the interior of *B*. Then at *P*, we can visualize three mutually perpendicular directions and three mutually perpendicular surfaces, normal to these directions, where we could evaluate the stress vector, as in Equation 5.4. This results in a net of nine stress components, three of which are normal stresses and the remaining six are shear stresses.

To illustrate and discuss this in more detail, consider a small rectangular element within *B* containing *P* with *X*-, *Y*-, and *Z*-axes normal to the surfaces of the element as in Figure 5.8. Let the vertices of the element be *A*, *B*, *C*, ..., *H* as shown. These vertices define the faces, or sides, of the element. Opposite, parallel faces are perpendicular to the coordinate axes. For example, faces *ABCD* and *EFGH* are perpendicular, or normal, to the *X*-axis. Face *ABCD* is normal to the positive *X*-axis, whereas *EFGH* is normal to the negative *X*-axis. Thus, face *ABCD* is said to be positive, while *EFGH* is said to be negative. In this context, four of the faces are positive and four are negative. These are listed in Table 5.1.

Observe that if a point moves from inside the element to outside the element, it will cross one of the eight faces. The point will be moving in the positive (negative) direction relative to the face's normal axis if the face is positive (negative).

Imagine the element to be subjected to a loading leading to stresses on its surfaces. In general, each face will have three stress components, one for each of the coordinate directions. Thus, in general, there are nine stress components on the element. A convenient notation to account for

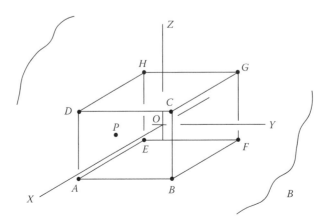

FIGURE 5.8 Element within body *B*.

TABLE 5.1

Positive and Negative Element Faces

Face	Normal Axis	Face Sign
ABCD	+X	Positive
BFGC	+Y	Positive
CGHD	+Z	Positive
EFGH	−X	Negative
AEHD	−Y	Negative
AEFB	−Z	Negative

these components is σ_{ij}, where the first subscript denotes the normal to the face of application and the second subscript denotes the component direction. For example, σ_{xy} represents the stress on a face perpendicular to the X-axis ($ABCD$ or $EFGH$) in the Y direction. A stress component is said to be positive if it is exerted on a positive face in the positive direction or if it is exerted on a negative face in the negative direction. Alternatively, a stress component is said to be negative if it is exerted on a positive face in the negative direction or on a negative face in the positive direction.

Consider now that the element is shrunk to a differential element so that it is essentially shrunk around the point P. Then the stresses on the element are regarded as the stresses at point P.

For the differential element, it is seen* from equilibrium considerations that the normal stresses on opposite faces are equal in magnitude while oppositely directed and that the shear stresses on opposite faces are also equal in magnitude while oppositely directed. This means that of the 18 stress components (three on each of the six faces), there are only six independent components. Specifically, the opposite normal stresses are equal, and the shear stresses obey the following relations:

$$\tau_{xy} = \tau_{yx} \qquad \tau_{yz} = \tau_{zy} \qquad \tau_{zx} = \tau_{xz} \tag{5.5}$$

These results make it convenient to gather the stress components into a 3×3 array Σ (see Chapter 3) as

$$\Sigma = \begin{bmatrix} \sigma_{xx} & \sigma_{xy} & \sigma_{xz} \\ \sigma_{yx} & \sigma_{yy} & \sigma_{yz} \\ \sigma_{zx} & \sigma_{zy} & \sigma_{zz} \end{bmatrix} \tag{5.6}$$

where for convenience in notation, we have replaced τ by σ in the shear stresses. In view of Equation 5.5, Σ is seen to be symmetric.

From the procedures of Chapter 3, it is clear that for general directions within B represented by mutually perpendicular unit vectors $\mathbf{n}_1$, $\mathbf{n}_2$, and $\mathbf{n}_3$ (parallel to say X, Y, and Z), we can obtain a stress dyadic Σ as

$$\Sigma = \sigma_{ij} \mathbf{n}_i \mathbf{n}_j \tag{5.7}$$

5.2 CONCEPTS OF STRAIN

Just as stress is intuitively defined as an average force per unit area, strain is also intuitively defined as an average deformation or as a change in length per unit length.

---

* For example, see references.

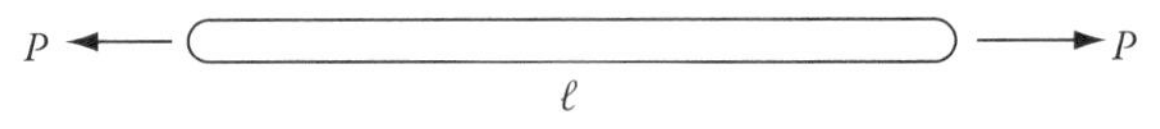

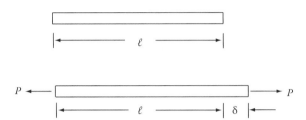

FIGURE 5.9   Elastic rod subjected to tension forces P.

FIGURE 5.10   Elongation of an elastic rod upon exertion of tension forces P.

Consider, for example, an elastic rod of length ℓ being elongated by tensile forces P as represented in Figure 5.9.

Suppose that as a result of the forces the rod is elongated by an amount δ as in Figure 5.10, the strain ε in the rod is defined as

$$\varepsilon = \frac{\delta}{\ell} \tag{5.8}$$

For linearly elastic bodies, the strain is a very small number. Observe also that it is dimensionless.

Equation 5.8 is sometimes referred to as the definition of normal strain or simple strain.

Just as simple stress is augmented by the concept of shear stress (Equation 5.3), so also simple strain is augmented by the concept of shear strain. Consider again the block of Figures 5.5 and 5.6 subjected to a shear loading and distortion as in Figure 5.6 and as shown again in Figure 5.11. If the top of the block is displaced horizontally by a distance δ as represented in Figure 5.11, then the shear strain γ is defined as

$$\gamma = \frac{\delta}{\ell} \tag{5.9}$$

Observe in Figure 5.11 that the ratio δ/ℓ is the tangent of the distortion angle ϕ. That is,

$$\tan \phi = \frac{\delta}{\ell} = \gamma \tag{5.10}$$

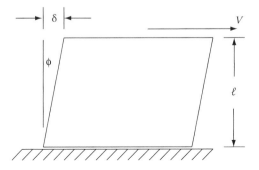

FIGURE 5.11   Block distorted by a shear force.

Observe further that for most linearly elastic bodies, the deformation δ is small compared with a typical dimension ℓ of the body. Thus, in Figure 5.11, if δ is small compared with ℓ, the distortion angle ϕ and consequently the shear strain γ are small. Therefore, Equation 5.11 may be expressed as

$$\tan \phi = \phi = \frac{\delta}{\ell} = \gamma \tag{5.11}$$

Thus, the shear strain is a measure of the distortion of the block due to the shear loading.

To further explore the concepts of strain, consider that the expressions of simple strain in Equation 5.8 and of shear strain in Equations 5.9 and 5.10 are expressions of average strains at points along the rod and in the block. For the rod of Figure 5.10, the strain is uniform along the rod. If, however, the rod does not have a uniform cross section or if there is loading in the interior of the rod, the strain will no longer be uniform and instead will vary from point to point along the rod. In this case, the strain needs to be defined locally or at each point along the rod. To this end, suppose a point P at a distance x from the left end of the rod as in Figure 5.12 is displaced a distance u from its original position. Let Δx be the length of a small element (e) containing P. Then the average strain ε in (e) is simply

$$\varepsilon = \frac{\Delta u}{\Delta x} \tag{5.12}$$

where Δu is the change in length of (e). In the limit as the length Δx of (e) becomes smaller, the strain at P is

$$\varepsilon = \lim_{\Delta x \to 0} \frac{\Delta u}{\Delta x} \tag{5.13}$$

The strain at P of Equation 5.13 is a measure of the local rate of elongation along the axis of the rod. It is sometimes called simple strain or normal strain.

Similarly, consider the shear strain as in Equation 5.10. Consider a small planar rectangular element (e) subjected to a shear loading as represented in Figure 5.13. As a result of the loading, the element will be distorted as depicted in exaggerated form in Figure 5.14. In this case, the shear strain γ is defined as the difference of the included angle θ of the distorted element from the right angle $(\pi/2)$ of the undistorted element. That is, the shear strain γ is defined as

$$\gamma = \left(\frac{\pi}{2}\right) - \theta = \alpha + \beta \tag{5.14}$$

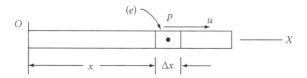

FIGURE 5.12 Elongated element at a typical point P of the rod.

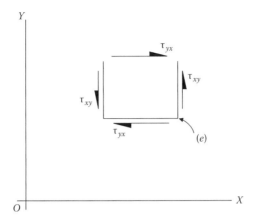

FIGURE 5.13   Small rectangular planar element (*e*) subjected to shear stresses.

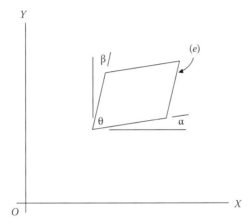

FIGURE 5.14   Element (*e*) distorted by shear stresses.

where α and β measure the inclination of the element sides as in Figure 5.14. If the element size is small and if the horizontal and vertical displacements of the vertex at θ are u and v, then α and β may be approximated as

$$\alpha = \frac{\partial v}{\partial x} \quad \text{and} \quad \beta = \frac{\partial u}{\partial y} \tag{5.15}$$

Thus, the shear strain becomes

$$\gamma = \frac{\partial v}{\partial x} + \frac{\partial u}{\partial y} \tag{5.16}$$

For analytical and computational purposes, it is convenient to redefine the shear strain as the average of α and β in the form

$$\varepsilon_{xy} = \left(\frac{1}{2}\right)\left(\frac{\partial u}{\partial y} + \frac{\partial v}{\partial x}\right) \tag{5.17}$$

where the symbol ε is here also used for shear strain as well as normal strain, and the subscripts (x,y) are used to indicate that the element is in the X–Y plane. In this context, γ of Equation 5.16 may be expressed as

$$\gamma_{xy} = \frac{\partial u}{\partial y} + \frac{\partial v}{\partial x} \tag{5.18}$$

γ_{xy} is sometimes called engineering shear strain, whereas ε_{xy} is called mathematical shear strain. The principal advantage of using ε_{xy} (Equation 5.17) instead of γ_{xy} (Equation 5.18) is that ε_{xy} can be incorporated into a strain tensor or strain dyadic. In this regard, consider a rectangular element as in Figure 5.15. The normal strains on the element may be expressed as

$$\varepsilon_{xy} = \frac{\partial u}{\partial x} \qquad \varepsilon_{yy} = \frac{\partial v}{\partial y} \qquad \varepsilon_{zz} = \frac{\partial w}{\partial z} \tag{5.19}$$

The shear strains may be expressed as

$$\varepsilon_{xy} = \left(\frac{1}{2}\right)\left(\frac{\partial u}{\partial y} + \frac{\partial v}{\partial x}\right)$$

$$\varepsilon_{yz} = \left(\frac{1}{2}\right)\left(\frac{\partial v}{\partial z} + \frac{\partial w}{\partial y}\right) \tag{5.20}$$

$$\varepsilon_{zx} = \left(\frac{1}{2}\right)\left(\frac{\partial w}{\partial x} + \frac{\partial u}{\partial z}\right)$$

Equations 5.19 and 5.20 may be incorporated into a single expression by using index notation as in Section 5.1. Specifically, let x, y, and z be replaced by x_1, x_2, and x_3 and u, v, and w by u_1, u_2, and u_3, respectively. Then a strain matrix E may be expressed as

$$E = E(\varepsilon_{ij}) = \begin{bmatrix} \varepsilon_{11} & \varepsilon_{12} & \varepsilon_{13} \\ \varepsilon_{21} & \varepsilon_{22} & \varepsilon_{23} \\ \varepsilon_{31} & \varepsilon_{32} & \varepsilon_{33} \end{bmatrix} \tag{5.21}$$

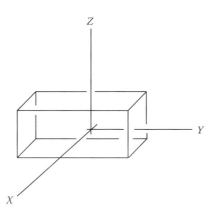

FIGURE 5.15   Rectangular element.

where ε_{ij} are

$$\varepsilon_{ij} = \left(\frac{1}{2}\right)\left(\frac{\partial \dot{u}_i}{\partial x_j} + \frac{\partial u_j}{\partial x_i}\right) \quad \text{or} \quad \varepsilon_{ij} = \left(\frac{1}{2}\right)(u_{i,j} + u_{j,i}) \tag{5.22}$$

where the comma $[(), j]$ is an abbreviation for $\partial()/\partial x_j$.

5.3 PRINCIPAL VALUES OF STRESS AND STRAIN

The stress and strain matrices of Equations 5.6 and 5.21 are both symmetric. That is,

$$\sigma_{ij} = \sigma_{ji} \quad \text{and} \quad \varepsilon_{ij} = \varepsilon_{ji} \tag{5.23}$$

This means that for both stress and strain, there exist mutually perpendicular directions for which there will be maximum and minimum values of the normal stress and strain (Sections 3.9 and 3.10). To find these directions and to determine the corresponding maximum/minimum stress and strain values, we need to simply follow the procedures outlined in Section 3.9. Specifically, from Equations 3.138 and 3.139, we see that the maximum/minimum values are to be found among the solutions of the cubic equation (sometimes called the Hamilton–Cayley equation) of Equation 3.139. For example, for stress, the maximum/minimum values of normal stress are to be found among the eigenvalues σ of the stress matrix given by the determinantal equation (Equation 3.138):

$$\begin{bmatrix} (\sigma_{11} - \sigma) & \sigma_{12} & \sigma_{13} \\ \sigma_{21} & (\sigma_{22} - \sigma) & \sigma_{23} \\ \sigma_{31} & \sigma_{32} & (\sigma_{33} - \sigma) \end{bmatrix} = 0 \tag{5.24}$$

where in view of Equation 5.7, we have replaced the subscripts x, y, and z in Equation 5.6 with 1, 2, and 3, respectively.

By expanding the determinant of Equation 5.24, we obtain (Equation 3.139)

$$\sigma^3 - \sigma_I \sigma^2 + \sigma_{II} \sigma - \sigma_{III} = 0 \tag{5.25}$$

where σ_I, σ_{II}, and σ_{III} are

$$\sigma_I = \sigma_{11} + \sigma_{22} + \sigma_{33}$$

$$\sigma_{II} = \sigma_{22}\sigma_{33} - \sigma_{32}\sigma_{23} + \sigma_{11}\sigma_{33} - \sigma_{31}\sigma_{13} + \sigma_{11}\sigma_{22} - \sigma_{12}\sigma_{21} \tag{5.26}$$

$$\sigma_{III} = \sigma_{11}\sigma_{22}\sigma_{33} - \sigma_{11}\sigma_{32}\sigma_{23} + \sigma_{12}\sigma_{31}\sigma_{23} - \sigma_{12}\sigma_{21}\sigma_{33} + \sigma_{21}\sigma_{32}\sigma_{13} - \sigma_{31}\sigma_{13}\sigma_{22}$$

where

σ_I is the trace (sum of diagonal elements) of the stress matrix
σ_{II} is the trace of its matrix of cofactors
σ_{III} is the stress matrix determinant

The solution of Equation 5.25 produces three real roots σ among which are the maximum and minimum values of the normal stress (see Sections 3.9 and 3.10 and Ref. [18]).

Following the procedures of Section 3.9, we can use the roots to find the directions for the maximum/minimum stresses. Specifically, if $\mathbf{n}_a$ is a unit vector (unit eigenvector) along a direction of maximum or minimum stress, with components a_i relative to mutually perpendicular unit vectors $\mathbf{n}_i$ ($i = 1, 2, 3$), then the a_i are the solutions of the equations

$$(\sigma_{11} - \sigma_a)a_1 + \sigma_{12}a_2 + \sigma_{13}a_3 = 0$$

$$\sigma_{12}a_1 + (\sigma_{22} - \sigma_a)a_2 + \sigma_{23}a_3 = 0 \tag{5.27}$$

$$\sigma_{31}a_1 + \sigma_{32}a_2 + (\sigma_{33} - \sigma_a)a_3 = 0$$

and

$$a_1^2 + a_2^3 + a_3^2 = 1 \tag{5.28}$$

where σ_a is a maximum/minimum root of Equation 5.25. (Recall that since Equation 5.27 is homogeneous with a zero coefficient determinant, as in Equation 5.24, at most two of Equation 5.27 are independent.) Knowing the a_i, $\mathbf{n}_a$ is immediately determined.

Section 3.9 presents a numerical example of the foregoing procedure.

A directly analogous procedure produces the principal values and principal directions for strain.

5.4   TWO-DIMENSIONAL EXAMPLE: MOHR'S CIRCLE

To obtain more insight, recall the 2D analysis of principal values of stress and strain with a geometric visualization provided by Mohr's circle [1]: To this end, let the stress matrix Σ have the form

$$\Sigma = [\sigma_{ij}] = \begin{bmatrix} \sigma_{11} & \sigma_{12} & 0 \\ \sigma_{21} & \sigma_{22} & 0 \\ 0 & 0 & \sigma_{33} \end{bmatrix} \tag{5.29}$$

By inspection, we see that if σ_{ij} are the stress dyadic components relative to mutually perpendicular unit vectors $\mathbf{n}_i$ ($i = 1, 2, 3$), then $\mathbf{n}_3$ is a unit eigenvector. For the matrix of Equation 5.29, Equation 5.24 becomes

$$\begin{vmatrix} (\sigma_{11} - \sigma) & \sigma_{12} & 0 \\ \sigma_{21} & (\sigma_{22} - \sigma) & 0 \\ 0 & 0 & (\sigma - \sigma_{33}) \end{vmatrix} = 0 \tag{5.30}$$

By expanding the determinant, the Hamilton–Cayley equation (Equation 5.31) becomes

$$\left[\sigma^2 - \sigma(\sigma_{11} + \sigma_{22}) + \sigma_{11}\sigma_{22} - \sigma_{11}^2 \right](\sigma - \sigma_{33}) = 0 \tag{5.31}$$

where we have used the property of stress matrix symmetry and assigned σ_{21} to be σ_{12}. The eigenvalues σ are then

$$\sigma_1, \sigma_2 = \frac{\sigma_{11} + \sigma_{22}}{2} \pm \left[\left(\frac{\sigma_{11} + \sigma_{22}}{2} \right)^2 + \sigma_{12}^2 - \sigma_{11}\sigma_{22} \right]^{1/2} \quad \text{and} \quad \sigma_3 = \sigma_{33} \tag{5.32}$$

As before, let $\mathbf{n}_a$ be a unit vector along a principal direction. Let $\mathbf{n}_a$ have components a_i relative to the $\mathbf{n}_i$ ($i = 1, 2, 3$). Then from Equations 5.27 and 5.28, a_i are seen to satisfy the relations

$$(\sigma_{11} - \sigma)a_1 + \sigma_{12}a_2 + 0a_3 = 0$$

$$\sigma_{12}a_1 + (\sigma_{22} - \sigma)0a_3 = 0 \tag{5.33}$$

$$0a_1 + 0a_2 + (\sigma_{33} - \sigma)a_3 = 0$$

and

$$a_1^2 + a_2^2 + a_3^2 = 1 \tag{5.34}$$

(Observe that in view of Equation 5.30, Equation 5.33 is dependent.)

Suppose $\sigma = \sigma_3 = \sigma_{33}$. Then by inspection, a solution of Equations 5.33 and 5.34 is

$$a_1 = a_2 = 0, \quad a_3 = 1 \tag{5.35}$$

The result of Equation 5.35 shows that $\mathbf{n}_3$ is an eigenvector. Thus, for a Cartesian axis system with the Z-axis parallel to $\mathbf{n}_3$, and knowing that there exist three mutually perpendicular unit eigenvectors [18], we can look for two other eigenvectors parallel to the X–Y plane.

To this end, let $\sigma = \sigma_1$. Then we seek a solution of Equations 5.33 and 5.34 in the form

$$a_1 = \cos\theta, \quad a_2 = \sin\theta, \quad a_3 = 0 \tag{5.36}$$

where θ is the angle between $\mathbf{n}_a$ and $\mathbf{n}_1$ as in Figure 5.16. Equation 5.34 and the third of Equation 5.33 are then identically satisfied. The first two of Equation 5.33 are then dependent. Selecting the first of these, we have

$$(\sigma_{11} - \sigma_1)\cos\theta + \sigma_{12}\sin\theta = 0 \tag{5.37}$$

or

$$\tan\theta = -\left(\frac{\sigma_{11} - \sigma_1}{\sigma_{12}}\right) = \frac{\sigma_{11} - \sigma_{22}}{2\sigma_{22}} + \left[\left(\frac{\sigma_{11} - \sigma_{22}}{2\sigma_{12}}\right)^2 + 1\right]^{1/2} \tag{5.38}$$

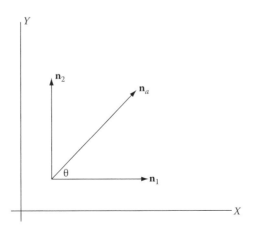

FIGURE 5.16 Inclination of unit eigenvector $\mathbf{n}_a$.

By solving for the radical, by squaring, and thus eliminating the radical, we obtain (after simplification)

$$\tan^2 \theta + \left(\frac{\sigma_{11} - \sigma_{22}}{\sigma_{12}} \right) \tan \theta - 1 = 0 \tag{5.39}$$

or

$$\sin^2 \theta + \left(\frac{\sigma_{11} - \sigma_{22}}{\sigma_{12}} \right) \sin \theta \cos \theta - \cos^2 \theta = 0 \tag{5.40}$$

By converting to double angles and simplifying, we have

$$\left(\frac{\sigma_{11} - \sigma_{22}}{2\sigma_{12}} \right) \sin 2\theta = \cos 2\theta \tag{5.41}$$

or

$$\tan 2\theta = \frac{2\sigma_{12}}{\sigma_{11} - \sigma_{22}} \tag{5.42}$$

By letting $\sigma = \sigma_2$ (using the minus sign of Equation 5.32), we obtain the same result. The solution of Equation 5.42 however produces two values of θ differing by $\pi/2$ rad (or 90°), as can be seen from a graphical representation of $\tan \theta$ and $\tan 2\theta$.

Suppose the stress dyadic Σ is written in the form (Equation 5.7)

$$\Sigma = \sigma_{ij} \mathbf{n}_i \mathbf{n}_j \tag{5.43}$$

where, as before, the $\mathbf{n}_i$ ($i = 1, 2, 3$) are mutually perpendicular unit vectors parallel to a reference X-, Y-, and Z-axes system. Let $\hat{\mathbf{n}}_i$ ($i = 1, 2, 3$) be unit eigenvectors so that

$$\hat{\mathbf{n}}_1 = \mathbf{n}_a^{(1)}, \quad \hat{\mathbf{n}}_2 = \mathbf{n}_a^{(2)}, \quad \hat{\mathbf{n}}_3 = \mathbf{n}_a^{(3)} \tag{5.44}$$

Then since $\hat{\mathbf{n}}_i$ are eigenvectors, we have, by the definition of eigenvectors (Equation 3.135), the results

$$\Sigma \cdot \hat{\mathbf{n}}_1 = \sigma_1 \hat{\mathbf{n}}_1, \quad \Sigma \cdot \hat{\mathbf{n}}_2 = \sigma_2 \hat{\mathbf{n}}_2, \quad \Sigma \cdot \hat{\mathbf{n}}_3 = \sigma_3 \hat{\mathbf{n}}_3 \tag{5.45}$$

Suppose now that Σ is expressed in terms of the $\hat{\mathbf{n}}_i$ as

$$\Sigma = \hat{\sigma}_{ij} \hat{\mathbf{n}}_i \hat{\mathbf{n}}_j \tag{5.46}$$

Then from Equation 5.45, we have

$$\hat{\mathbf{n}}_1 \cdot \Sigma \cdot \hat{\mathbf{n}}_1 = \sigma_1 \quad \hat{\mathbf{n}}_2 \cdot \Sigma \cdot \hat{\mathbf{n}}_2 = \sigma_2 \quad \hat{\mathbf{n}}_3 \cdot \Sigma \cdot \hat{\mathbf{n}}_3 = \sigma_3 \tag{5.47}$$

and

$$\hat{\mathbf{n}}_i \cdot \Sigma \cdot \hat{\mathbf{n}}_j = 0 \quad i \neq j \tag{5.48}$$

Therefore, the stress matrix Σ relative to the unit eigenvector system is diagonal. That is,

$$\hat{\Sigma} = \hat{\Sigma}(\hat{\sigma}_{ij}) = \begin{bmatrix} \sigma_1 & 0 & 0 \\ 0 & \sigma_2 & 0 \\ 0 & 0 & \sigma_3 \end{bmatrix} \tag{5.49}$$

Let S_{ij} be the elements of a transformation matrix between the $\mathbf{n}_i$ and the $\hat{\mathbf{n}}_i$ as in Equation 3.160. That is,

$$S_{ij} = \mathbf{n}_i \cdot \hat{\mathbf{n}}_j \tag{5.50}$$

The stress matrix elements are then related by the expressions

$$\sigma_{ij} = S_{ik} S_{j\ell} \hat{\sigma}_{k\ell} = S_{ik} \hat{\sigma}_{k\ell} S_{\ell j}^{T} \tag{5.51}$$

and

$$\hat{\sigma}_{k\ell} = S_{ik} S_{j\ell} \sigma_{ij} = S_{ki}^{T} \sigma_{ij} S_{j\ell} \tag{5.52}$$

Suppose now that from the foregoing analysis, we have (Equations 5.35 and 5.36)

$$\mathbf{n}_{\mathbf{a}}^{(1)} = \cos\theta \mathbf{n}_1 + \sin\theta \mathbf{n}_2 = \hat{\mathbf{n}}_1 \tag{5.53}$$

$$\mathbf{n}_{\mathbf{a}}^{(2)} = -\sin\theta \mathbf{n}_1 + \cos\theta \mathbf{n}_2 = \hat{\mathbf{n}}_2 \tag{5.54}$$

$$\mathbf{n}_{\mathbf{a}}^{(3)} = \mathbf{n}_3 = \hat{\mathbf{n}}_3 \tag{5.55}$$

Then from Equation 5.50, we have

$$S_{ij} = \begin{bmatrix} \cos\theta & \sin\theta & 0 \\ -\sin\theta & \cos\theta & 0 \\ 0 & 0 & 1 \end{bmatrix} \tag{5.56}$$

By substituting into Equation 5.51, from Equation 5.49, we obtain

$$\Sigma = \begin{bmatrix} \sigma_{11} & \sigma_{12} & \sigma_{13} \\ \sigma_{21} & \sigma_{22} & \sigma_{23} \\ \sigma_{31} & \sigma_{32} & \sigma_{33} \end{bmatrix} = \begin{bmatrix} \cos\theta & \sin\theta & 0 \\ -\sin\theta & \cos\theta & 0 \\ 0 & 0 & 1 \end{bmatrix} \begin{bmatrix} \sigma_1 & 0 & 0 \\ 0 & \sigma_2 & 0 \\ 0 & 0 & \sigma_3 \end{bmatrix} \begin{bmatrix} \cos\theta & -\sin\theta & 0 \\ \sin\theta & \cos\theta & 0 \\ 0 & 0 & 1 \end{bmatrix}$$

$$= \begin{bmatrix} (c^2\sigma_1 + s^2\sigma_2) & (-sc\sigma_1 + sc\sigma_2) & 0 \\ (-sc\sigma_1 + sc\sigma_2) & (-s^2\sigma_1 + c^2\sigma_2) & 0 \\ 0 & 0 & \sigma_3 \end{bmatrix} \tag{5.57}$$

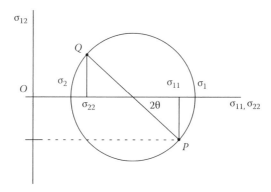

FIGURE 5.17 Mohr's circle for planar stress.

Thus, σ_{11}, σ_{22}, and σ_{12} are

$$\sigma_{11} = c^2\sigma_1 + s^2\sigma_2 = \left(\frac{\sigma_1 + \sigma_2}{2}\right) - \left(\frac{\sigma_1 - \sigma_2}{2}\right)\cos 2\theta \qquad (5.58)$$

$$\sigma_{22} = s^2\sigma_1 + c^2\sigma_2 = \left(\frac{\sigma_1 + \sigma_2}{2}\right) - \left(\frac{\sigma_1 - \sigma_2}{2}\right)\cos 2\theta \qquad (5.59)$$

$$\sigma_{12} = \sigma_{21} = -sc\sigma_1 + sc\sigma_2 = \frac{\sigma_1 - \sigma_2}{2}\sin 2\theta \qquad (5.60)$$

The results of Equations 5.58 through 5.60 may be represented geometrically as in Figure 5.17 where a circle, known as Mohr's circle, can be used to evaluate the normal and shear stresses. In this procedure, the horizontal axis represents normal stress, and the vertical axis represents shear stress. The circle then has its center on the horizontal axis at a distance from the origin equal to the average value of the principal stresses. The circle diameter is equal to the magnitude of the difference of the principal stresses. Then the ordinates and the abscissas of points (P and Q) on the circle, located angularly by 2θ as shown, determine values of the normal and shear stresses.

5.5 ELEMENTARY STRESS–STRAIN RELATIONS

Hooke's law [1] is the earliest and most basic of all stress–strain relations. Simply stated, if an elastic rod is subjected to tension (or compression) forces P, as in Figure 5.18, the resulting elongation (or shortening) δ of the rod is directly proportional to the applied force magnitude. That is,

$$P = k\delta \quad \text{or} \quad \delta = \left(\frac{1}{k}\right)P \qquad (5.61)$$

The rod shown in Figure 5.18 is in simple tension as was the rod we considered earlier in Figure 5.3. Suppose the rod has length ℓ and uniform cross-sectional area A. Then from Equation 5.1, the tensile stress σ in the rod may be defined as follows:

$$\sigma = \frac{P}{A} \qquad (5.62)$$

FIGURE 5.18   Rod in tension with elongation.

As a result of the applied forces P, producing the stress σ and the elongation δ, the rod is in simple strain as the earlier considered rod depicted in Figure 5.10. Then from Equation 5.8, this strain ε may be defined as

$$\varepsilon = \frac{\delta}{\ell} \tag{5.63}$$

By substituting from Equations 5.62 and 5.63 into Equation 5.61, we have

$$\sigma A = k\varepsilon\ell \quad \text{or} \quad \sigma = \left(\frac{k\ell}{A}\right)\varepsilon \tag{5.64}$$

The quantity $k\ell/A$ is usually designated by E and called Young's modulus of elasticity or the elastic modulus. Thus, we have

$$k = \frac{AE}{\ell} \quad \text{and} \quad \sigma = E\varepsilon \tag{5.65}$$

Finally, for the rod of Figure 5.18, from Equations 5.62, 5.63, and 5.65, we have

$$\delta = \frac{P\ell}{AE} \tag{5.66}$$

Equations 5.61 and 5.65 are only approximate expressions in reality. For most materials, and particularly for biological materials, the stress and strain are nonlinearly related. Figure 5.19 shows a typical relation between stress and strain for a large class of biological tissues [26] and also for many inert materials [1]. Observe that for sufficiently small values of the stress and strain, they are linearly related as in Equation 5.65.

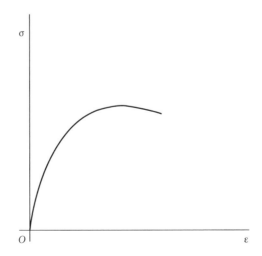

FIGURE 5.19   Typical stress–strain relation.

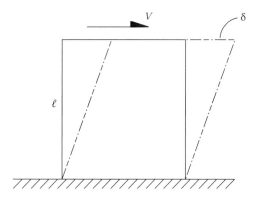

FIGURE 5.20   Block with shear loading and shear deformation.

In like manner, Hooke's law may be extended to shear loading and shear deformation. Consider again the loading and deformation of a block as in Figures 5.5, 5.6, and 5.11 and as shown again in Figure 5.20. Recall from Equations 5.3 and 5.9 that the shear stress τ and the shear strain γ are

$$\tau = \frac{V}{A} \quad \text{and} \quad \gamma = \frac{\delta}{\ell} \tag{5.67}$$

where as before,

  V is the shear loading
  A is the area of the block over which V acts
  δ is the shear deformation
  ℓ is the height of the block, as in Figure 5.20

Hooke's law then relates the shear stress to the shear strain as

$$\tau = G\gamma \tag{5.68}$$

where the proportional constant G is called the shear modulus or modulus of rigidity.

Referring again to Figure 5.18, an axial loading of a rod is sometimes said to be uniaxial or 1D. The resulting stress is 1D and is thus called uniaxial stress. The strain, however, is not 1D. Consider, for example, a rod with a rectangular cross section being elongated as represented in Figure 5.21. As the rod is being stretched, its cross section will be reduced, and thus, there will be strain in directions normal to the loading axis. The magnitude of the ratio of the strain normal to the loading axis relative to the axial strain is called the transverse contraction ratio or Poisson's ratio and is often designated by υ. Specifically, for the rod of Figure 5.21, υ is

$$\upsilon = \frac{-\varepsilon_{yy}}{\varepsilon_{xx}} \tag{5.69}$$

where

  the minus sign occurs since ε_{yy} is negative
  ε_{xx} is positive so that υ is positive

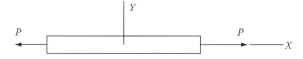

FIGURE 5.21   Rectangular cross-sectional rod being elongated.

For isotropic materials (materials having the same properties in all directions), there are thus two elastic constants (the elastic modulus E and Poisson's ratio υ). The shear modulus G is not independent of E and υ but instead may be expressed in terms of E and υ as [15]

$$G = \frac{E}{2}(1 + \upsilon) \tag{5.70}$$

The value of υ ranges from 0 to 0.5 depending upon the material. For engineering materials, such as metals, υ is typically 0.25–0.35. For incompressible materials (materials maintaining constant volume) such as liquids, υ is 0.5. Since many biological tissues are liquid based, a good approximation for υ for such tissue is also 0.5.

In Section 5.6, we provide a generalization of Equations 5.65 and 5.68 for 3D loadings and deformations.

5.6 GENERAL STRESS–STRAIN (CONSTITUTIVE) RELATIONS

Consider now an arbitrarily shaped elastic body B subjected to an arbitrary loading as in Figure 5.22. Consider a planar surface S within B and let P be a point of S. Due to the arbitrary shape of B and the arbitrary loading, the surface S at P will in general experience a stress vector having both normal stress and shear stress components. Surfaces perpendicular to S at P will also experience such stress vectors so that at P, there will in general be nine stress components (three normal stresses and six shear stresses as in Equation 5.6). Consequently, there will also be nine strain components (three normal strains and six shear strains as in Equation 5.21) at P. Recall that both the stress and strain arrays are symmetric so that there are only three independent shear stress components and only three independent shear strain components or a net of six independent stresses and six independent strains.

For sufficiently small deformations of elastic bodies, these stress and strain components are linearly related as with Equations 5.65 and 5.68, but here, six equations are needed. These equations may be expressed in various forms depending upon the objective of the analysis. For example, the stresses may be expressed in terms of the strains, and conversely, the strains may be expressed in terms of the stresses. Index notation may be used to write the equations in compact form, or alternatively, explicit equations may be written in greater detail. In the following paragraphs, we list some of the more common and useful forms of these equations. (Interested readers may want to consult the references for derivation of these equations and for additional forms of the equations [7,11,20,27].)

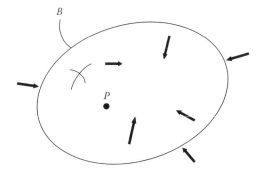

FIGURE 5.22 Arbitrarily shaped elastic body subjected to an arbitrary loading.

First, in a Cartesian XYZ coordinate system, the strains ε_{ij} may be expressed in terms of the stresses σ_{ij} as [1,9,19,20]

$$\varepsilon_{xx} = \left(\frac{1}{E}\right)\left[\sigma_{xx} - \upsilon(\sigma_{yy} + \sigma_{zz})\right]$$

$$\varepsilon_{yy} = \left(\frac{1}{E}\right)\left[\sigma_{yy} - \upsilon(\sigma_{zz} + \sigma_{xx})\right] \tag{5.71}$$

$$\varepsilon_{zz} = \left(\frac{1}{E}\right)\left[\sigma_{zz} - \upsilon(\sigma_{xx} + \sigma_{yy})\right]$$

and

$$\varepsilon_{xy} = \left(\frac{1}{2G}\right)\sigma_{xy}$$

$$\varepsilon_{yz} = \left(\frac{1}{2G}\right)\sigma_{yz} \tag{5.72}$$

$$\varepsilon_{zx} = \left(\frac{1}{2G}\right)\sigma_{zx}$$

where, as before, E, υ, and G are elastic constants (elastic modulus, Poisson's ratio, and shear modulus). By solving Equations 5.71 and 5.72 for the stresses, we obtain

$$\sigma_{xx} = \frac{E}{(1+\upsilon)(1-2\upsilon)}\left[(1-\upsilon)\varepsilon_{xx} + \upsilon(\varepsilon_{yy} + \varepsilon_{zz})\right]$$

$$\sigma_{yy} = \frac{E}{(1+\upsilon)(1-2\upsilon)}\left[(1-\upsilon)\varepsilon_{yy} + \upsilon(\varepsilon_{zz} + \varepsilon_{xx})\right] \tag{5.73}$$

$$\sigma_{zz} = \frac{E}{(1+\upsilon)(1-2\upsilon)}\left[(1-\upsilon)\varepsilon_{zz} + \upsilon(\varepsilon_{xx} + \varepsilon_{yy})\right]$$

and

$$\sigma_{xy} = 2G\varepsilon_{xy}$$

$$\sigma_{yz} = 2G\varepsilon_{yz} \tag{5.74}$$

$$\sigma_{zx} = 2G\varepsilon_{zx}$$

It is often useful to use numerical indices and a summation notation to express these equations in more compact form. Specifically, if x, y, and z are replaced by 1, 2, and 3, respectively, Equations 5.73 and 5.74 may be expressed as [6]

$$\sigma_{ij} = \lambda\delta_{ij}\varepsilon_{kk} + 2G\varepsilon_{ij} \tag{5.75}$$

where λ is an elastic constant (Lamé constant) defined as [19]

$$\lambda = \frac{\upsilon E}{(1+\upsilon)(1-2\upsilon)} \tag{5.76}$$

and δ_{ij} is Kronecker's delta symbol (Equation 3.23).

By solving Equation 5.75 for the strains in terms of the stresses, we obtain [6]

$$\varepsilon_{ij} = -\left(\frac{\upsilon}{E}\right)\delta_{ij}\sigma_{ik} + \left(\frac{1}{2G}\right)\sigma_{ij} \tag{5.77}$$

The sum of the normal strains ε_{kk} is sometimes called dilatation. It can be interpreted as the normalized volume change of a small volume element. Specifically, let V be the volume of a small element without a body B which is being deformed (Figure 5.22). Let ΔV be the change in volume of the element due to the deformation of B. Then the dilatation ε_{kk} may be expressed as

$$\varepsilon_{kk} = \frac{\Delta V}{V} \tag{5.78}$$

Similarly the sum of the normal stresses σ_{kk} is sometimes called the hydrostatic pressure. By adding the normal stresses in Equation 5.75, we obtain

$$\sigma_{kk} = 3K\varepsilon_{kk} \tag{5.79}$$

where K is called the bulk modulus.

Of all these elastic constants (E, υ, G, λ, and K), only two are independent. The most commonly used are E and υ. As in Equations 5.70 and 5.76, G, λ, and K may be expressed as

$$G = \frac{E}{2(1+\upsilon)}, \quad \lambda = \frac{\upsilon E}{(1+\upsilon)(1-2\upsilon)}, \quad K = \frac{E}{3(1-2\upsilon)} \tag{5.80}$$

5.7 EQUATIONS OF EQUILIBRIUM AND COMPATIBILITY

Consider a deformable body B subjected to a loading as represented in Figure 5.23. Let E be a small rectangular element within B with edge lengths Δx, Δy, and Δz in a Cartesian system as in Figure 5.24. Imagine a free-body diagram of E and consider the forces in the X direction. Figure 5.24 shows, for example, the normal stress and its variation in the X direction, using the first term of a Taylor series expansion. By adding the X-directed forces from all the faces, we obtain

$$\left(\sigma_{xx} + \frac{\partial \sigma_{xx}}{\partial x}\Delta x\right)\Delta y\Delta z - \sigma_{xx}\Delta y\Delta z + \left(\sigma_{zx} + \frac{\partial \sigma_{zx}}{\partial z}\Delta z\right)\Delta x\Delta y - \sigma_{zx}\Delta x\Delta y$$

$$+ \left(\sigma_{yx} + \frac{\partial \sigma_{yx}}{\partial y}\Delta y\right)\Delta x\Delta z - \sigma_{yx}\Delta x\Delta z = \rho a_x \Delta x\Delta y\Delta z \tag{5.81}$$

where
ρ is the mass density of E
a_x is the X-directed component of the acceleration of E

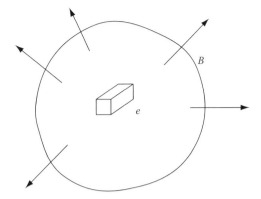

FIGURE 5.23   Loaded deformable body B with an interior rectangular element.

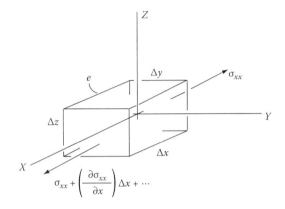

FIGURE 5.24   Interior element of a loaded deformable body (see Figure 5.23).

By canceling terms and simplifying, we have

$$\frac{\partial \sigma_{xx}}{\partial x} + \frac{\partial \sigma_{xy}}{\partial y} + \frac{\partial \sigma_{xz}}{\partial z} = \rho a_x \qquad (5.82)$$

Similarly, by adding forces in the Y and Z directions, we obtain

$$\frac{\partial \sigma_{yx}}{\partial x} + \frac{\partial \sigma_{yy}}{\partial y} + \frac{\partial \sigma_{yz}}{\partial z} = \rho a_y \qquad (5.83)$$

and

$$\frac{\partial \sigma_{zx}}{\partial x} + \frac{\partial \sigma_{zy}}{\partial y} + \frac{\partial \sigma_{zz}}{\partial z} = \rho(g + a_z) \qquad (5.84)$$

where g is the gravity acceleration. (The Z direction is vertical.)

Equations 5.82 through 5.84 may be expressed in more compact form by using index notation as

$$\frac{\partial \sigma_{ij}}{\partial x_j} = \rho a_i \qquad (5.85)$$

where, as before, the indices 1, 2, 3 represent x, y, z, respectively, and a_3 is $g + a_z$.

Equation 5.85 is commonly known as the equilibrium equation. They can be simplified even further by replacing $\partial()/\partial x_i$ by $()$, i. Thus,

$$\sigma_{ij,j} = \rho a_i \tag{5.86}$$

By substituting from Equation 5.75, we can express the equilibrium equation in terms of the strains as

$$\lambda \varepsilon_{kk,i} + \delta G \varepsilon_{ij,j} = \rho a_i \tag{5.87}$$

Similarly by substituting from Equation 5.22, we can express the equilibrium equation in terms of the displacements as

$$(\lambda + G)u_{j,ji} + Gu_{i,jj} = \rho a_i \tag{5.88}$$

Equation 5.88 includes three equations for the three displacements u_i ($i = 1, 2, 3$). Hence, given suitable boundary conditions, we can solve the equations for the displacements, and then knowing the displacements, we can use Equation 5.22 to determine the strains, and finally Equation 5.75 to obtain the stresses.

In many cases, however, the boundary conditions are not conveniently expressed in terms of displacement, but instead in terms of stresses (or loadings). In these cases, we may seek to use Equation 5.87 and thus formulate the problem in terms of the stresses. However, here we have six unknowns but only three equations. Thus, to obtain a unique solution, it is necessary to impose additional conditions. These conditions are needed to keep the stresses compatible with each other. They are developed in elasticity theory and are called compatibility equations [6]. While the development of these equations is beyond our scope, we simply state the results here and direct interested readers to Refs. [5,6,12] for additional details:

$$\sigma_{ij,kk} + \left(\frac{1}{1+\sigma}\right)\sigma_{kk,ij} - \left(\frac{\upsilon}{1-\upsilon}\right)\delta_{ij}(\rho a_k)_{,k} - (\rho a_i)_{,j} - (\rho a_j)_{,i} = 0 \tag{5.89}$$

These expressions are also known as Beltrami–Mitchell compatibility equations.

5.8   USE OF CURVILINEAR COORDINATES

In biosystems, such as the human body, few, if any, parts are rectangular. Instead, the limbs and the other body parts do not have elementary shapes. Nevertheless, in some instances, they may be modeled as cylinders or spheres. Therefore, it is useful to have the foregoing equations expressed in cylindrical and spherical coordinates.

5.8.1   CYLINDRICAL COORDINATES

Recall that the transformation between rectangular Cartesian coordinates (x, y, z) and cylindrical coordinates (r, θ, z) is [21] (Figure 5.25)

$$x = r\cos\theta \qquad y = r\sin\theta \qquad z = z$$

$$r = [x^2 + y^2]^{1/2} \qquad \theta = \tan^{-1}\left(\frac{y}{x}\right) \qquad z = z \tag{5.90}$$

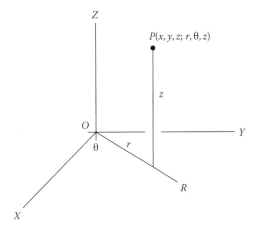

FIGURE 5.25   Cylindrical coordinate system.

In terms of cylindrical coordinates, the displacements are u_r, u_θ, and u_z and the corresponding strain–displacement relations (Equation 5.22) are [6,21]

$$\varepsilon_{rr} = \frac{\partial \mathbf{u}_r}{\partial r}$$

$$\varepsilon_{\theta\theta} = \frac{1}{r}\frac{\partial \mathbf{u}_\theta}{\partial \theta} + \frac{\mathbf{u}_r}{r}$$

$$\varepsilon_{zz} = \frac{\partial \mathbf{u}_z}{\partial z}$$

$$\varepsilon_{r\theta} = \frac{1}{2}\left(\frac{1}{r}\frac{\partial \mathbf{u}_r}{\partial \theta} + \frac{\partial u_\theta}{\partial r} - \frac{u_\theta}{r}\right)$$

$$\varepsilon_{rz} = \frac{1}{2}\left(\frac{\partial u_z}{\partial r} + \frac{1}{4}\frac{\partial \mathbf{u}_z}{\partial z}\right)$$

$$\varepsilon_{\theta z} = \frac{1}{2}\left(\frac{\partial \mathbf{u}_\theta}{\partial z} + \frac{1}{4}\frac{\partial \mathbf{u}_z}{\partial \theta}\right)$$

(5.91)

The equilibrium equations (Equation 5.85) are [6,21]

$$\frac{\partial \sigma_{rr}}{\partial r} + \frac{1}{r}\frac{\partial \sigma_{r\theta}}{\partial \theta} + \frac{\partial \sigma_{rz}}{\partial z} + \frac{\sigma_{rr} - \sigma_{\theta\theta}}{r} = \rho a_\mathbf{r}$$

$$\frac{\partial \sigma_{r\theta}}{\partial r} + \frac{1}{r}\frac{\partial \sigma_{\theta\theta}}{\partial \theta} + \frac{\partial \sigma_{\theta z}}{\partial z} + \frac{2}{r}\sigma_{r\theta} = \rho a_\theta$$

(5.92)

$$\frac{\partial \sigma_{rz}}{\partial r} + \frac{1}{r}\frac{\partial \sigma_{\theta z}}{\partial \theta} + \frac{\partial \sigma_{zz}}{\partial z} + \frac{1}{r}\sigma_{rz} = \rho a_z$$

The stress–strain equations (Equation 5.75) are [21]

$$\sigma_{rr} = \frac{\lambda}{\upsilon}\left[(1-\upsilon)\varepsilon_{rr} + \upsilon(\varepsilon_{\theta\theta} + \varepsilon_{zz})\right]$$

$$\sigma_{\theta\theta} = \frac{\lambda}{\upsilon}\left[(1-\upsilon)\varepsilon_{\theta\theta} + \upsilon(\varepsilon_{rr} + \varepsilon_{zz})\right]$$

$$\sigma_{zz} = \frac{\lambda}{\upsilon}\left[(1-\upsilon)\varepsilon_{zz} + \upsilon(\varepsilon_{rr} + \varepsilon_{\theta\theta})\right] \quad (5.93)$$

$$\sigma_{r\theta} = 2G\varepsilon_{r\theta}$$

$$\sigma_{rz} = 2G\varepsilon_{rz}$$

$$\sigma_{\theta z} = 2G\varepsilon_{rz}$$

5.8.2 SPHERICAL COORDINATES

Recall that the transformation between rectangular Cartesian coordinates (x, y, z) and spherical coordinates (r, θ, ϕ) is [21] (Figure 5.26)

$$x = r\sin\theta\cos\phi \qquad y = r\sin\theta\sin\phi \qquad z = r\cos\theta$$

$$r = \sqrt{x^2 + y^2 + z^2} \qquad \theta = \cos^{-1}\frac{z}{\sqrt{x^2 + y^2 + z^2}} \qquad \phi = \tan^{-1}\left(\frac{y}{x}\right) \quad (5.94)$$

In terms of spherical coordinates, the displacements are u_r, u_θ, and u_ϕ, and the corresponding strain–displacement relations (Equation 5.22) are [6,21]

$$\varepsilon_{rr} = \frac{\partial \mathbf{u}_r}{\partial r}$$

$$\varepsilon_{\theta\theta} = \frac{1}{r}\frac{\partial u_\theta}{\partial \theta} + \frac{\mathbf{u}_r}{r}$$

$$\varepsilon_{\phi\phi} = \frac{1}{r\sin\theta}\frac{\partial u_\phi}{\partial \phi} + \frac{\mathbf{u_r}}{r} + u_\theta\frac{\cot\theta}{r}$$

$$\varepsilon_{r\theta} = \frac{1}{2}\left(\frac{1}{r}\frac{\partial \mathbf{u_r}}{\partial \theta} - \frac{u_\theta}{r} + \frac{\partial u_\theta}{\partial r}\right) \quad (5.95)$$

$$\varepsilon_{r\phi} = \frac{1}{2}\left(\frac{1}{r\sin\theta}\frac{\partial \mathbf{u_r}}{\partial \phi} - \frac{u_\theta}{r} + \frac{\partial \mathbf{u}_\phi}{\partial r}\right)$$

$$\varepsilon_{\theta\phi} = \frac{1}{2}\left(\frac{1}{r}\frac{\partial u_\phi}{\partial \theta} - \frac{u_\phi\cot\theta}{r} + \frac{1}{r\sin\theta}\frac{\partial \mathbf{u}_\theta}{\partial \phi}\right)$$

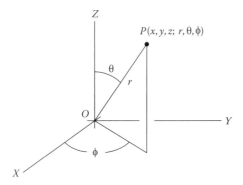

FIGURE 5.26 Spherical coordinate system.

The equilibrium equations (Equation 5.85) are [6,21]

$$\frac{\partial \sigma_{rr}}{\partial r} + \frac{1}{r}\frac{\partial \sigma_{r\theta}}{\partial \theta} + \frac{1}{r\sin\theta}\frac{\partial \sigma_{r\phi}}{\partial \phi} + \frac{2}{r}\sigma_{rr} - \frac{1}{r}\sigma_{\theta\theta} - \frac{1}{r}\sigma_{\phi\phi} + \frac{\cot\theta}{r}\sigma_{r\theta} = \rho a_r$$

$$\frac{\partial \sigma_{r\theta}}{\partial r} + \frac{1}{r}\frac{\partial \sigma_{\theta\theta}}{\partial \theta} + \frac{1}{r\sin\theta}\frac{\partial \sigma_{\theta\phi}}{\partial \phi} + \frac{3}{r}\sigma_{r\theta} + \frac{\cot\theta}{r}\sigma_{\theta\theta} - \frac{\cot\theta}{r}\sigma_{\phi\phi} = \rho a_\theta \qquad (5.96)$$

$$\frac{\partial \sigma_{r\phi}}{\partial r} + \frac{1}{r}\frac{\partial \sigma_{\theta\phi}}{\partial \theta} + \frac{1}{r\sin\theta}\frac{\partial \sigma_{\phi\phi}}{\partial \phi} + \frac{3}{r}\sigma_{r\phi} + \frac{2\cot\theta}{r}\sigma_{\phi\theta} = \rho a_\phi$$

The stress–strain equations (Equations 5.75) are [21]

$$\sigma_{rr} = \frac{\lambda}{\upsilon}\Big[(1-\upsilon)\varepsilon_{rr} + \upsilon\varepsilon_{\theta\theta} + \upsilon\varepsilon_{\phi\phi}\Big]$$

$$\sigma_{\theta\theta} = \frac{\lambda}{\upsilon}\Big[(1-\upsilon)\varepsilon_{\theta\theta} + \upsilon\varepsilon_{rr} + \upsilon\varepsilon_{\phi\phi}\Big]$$

$$\sigma_{\phi\phi} = \frac{\lambda}{\upsilon}\Big[(1-\upsilon)\varepsilon_{\phi\phi} + \upsilon\varepsilon_{rr} + \upsilon\varepsilon_{\theta\theta}\Big] \qquad (5.97)$$

$$\sigma_{r\theta} = 2G\varepsilon_{r\theta}$$

$$\sigma_{\theta\phi} = 2G\varepsilon_{\theta\phi}$$

$$\sigma_{r\phi} = 2G\varepsilon_{r\phi}$$

5.9   REVIEW OF ELEMENTARY BEAM THEORY

Since many parts of skeletal systems are long slender members (arms, legs, limbs), they may often be adequately modeled by beams. Therefore, it is useful to review the concepts of elementary beam theory.

We usually think of a beam as being a long slender number. By "long" and "slender," we mean a structure whose length is at least 10 times its thickness.

5.9.1 Sign Convention

There are a variety of sign conventions we can employ to designate the positive direction for loading and deformation. While each of these conventions has its own advantages, it is important in any analysis to simply be consistent with the rules of the convention. In this section, we outline a commonly used convention.

First, for the beam itself, we position the undeformed beam in a Cartesian axis frame with the origin at the left end, the X-axis along the beam, and the Y-axis down as in Figure 5.27. The positive displacement of a point on the beam is to the right and downward.

Next, for loading transverse to the beam, the positive direction is also downward. For a concentrated bending moment, the positive direction is counterclockwise (or in the negative Z direction) as in Figure 5.28, where $q(x)$ is a distributed loading, P is a concentrated force, and M is a concentrated moment.

Recall in elementary beam analysis that the useful loadings for determining stresses and displacements are the transverse shear V and the bending moment M. The positive directions for shear loading are shown in Figure 5.29 where, as before, the shear is positive if it is exerted on a positive face in the positive direction or on a negative face in the negative direction. (Recall that for an element or section of the beam, a positive face is one which is crossed by going from inside the element to the outside by moving in a positive direction.)

Finally, for bending (or flexion), the positive directions are illustrated in Figure 5.30.

FIGURE 5.27   Axis system of a beam.

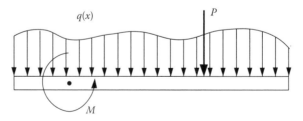

FIGURE 5.28   Positive directions for distributed loading $q(x)$, a concentrated force P, and a concentrated moment M.

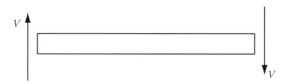

FIGURE 5.29   Positive directions for transverse shear forces.

FIGURE 5.30   Positive bending of a beam element.

5.9.2   Equilibrium Consideration

Consider a small element of a beam subjected to a distributed transverse loading $q(x)$ as in Figure 5.31, where the shear and bending moment reactions of the element ends are shown. The terms on the right end are the first terms of a Taylor series representation of the shear and bending moment where Δx is the element length. By regarding Figure 5.31 as a free-body diagram of the element, we immediately obtain (by adding forces and by taking moments about the left end)

$$\frac{dV}{dx} = -q \quad \text{and} \quad \frac{dM}{dx} = V \tag{5.98}$$

5.9.3   Strain–Curvature Relations

Consider a segment of a beam having positive bending moments as in Figure 5.32. Let e be an element of the segment and let the axes system originate at the left end of the element as shown. As the beam is bent, it will no longer be straight, but instead, it will become slightly curved. Consider an exaggerated view of the curvature as in Figure 5.33, where Q is the center of curvature of the beam at the origin O of element e and where ρ is the radius of curvature at O.

A fundamental assumption of elementary beam theory is that as a beam is bent, planar cross sections normal to the beam axis remain planar and normal to the beam axis. Observe that as a beam is bent, as in Figure 5.33, the upper fibers of the beam, parallel to the X-axis, are shortened and the corresponding lower fibers are lengthened [14,17].

Consider the deformation of element e of the beam segment of Figure 5.33: With the upper longitudinal fibers being shortened and the lower fibers being lengthened, there will occur at some

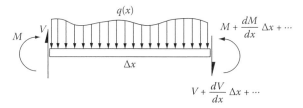

FIGURE 5.31   Distributed loading on a beam element.

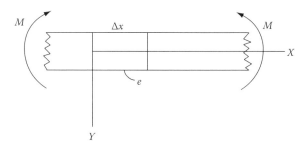

FIGURE 5.32   Beam segment with a positive burden, moment, and element e.

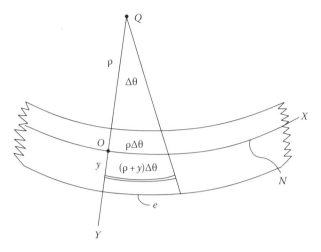

FIGURE 5.33 Exaggerated view of the bending of the beam segment of Figure 5.32.

mid-elevation a fiber which is neither shortened nor lengthened by the bending. Let the X-axis lie along this fiber—also known as the neutral axis, N. If the length of e before bending is Δx (see Figure 5.32), then the length of e along the neutral axis will remain as Δx during the bending. Then also as the beam is bent, the neutral axis of e will form an arc with central angle $\Delta\theta$ as in Figure 5.33 so that in terms of the induced radius of curvature ρ, Δx is

$$\Delta x = \rho \Delta \theta \tag{5.99}$$

From Figure 5.33, we see that a fiber of e at a distance y below the X-axis will have length $(\rho + y)\Delta\theta$. Hence, the strain ε of this fiber is

$$\varepsilon = \frac{(\rho + y)\Delta\theta - \rho\Delta\theta}{\rho\Delta\theta} = \frac{y}{\rho} \tag{5.100}$$

For a plane curve C as in Figure 5.34, it is known [19,25] that the radius of curvature ρ at a point P of C may be expressed as

$$\rho = \frac{\left[1 + (dy/dx)^2\right]^{3/2}}{d^2y/dx^2} \tag{5.101}$$

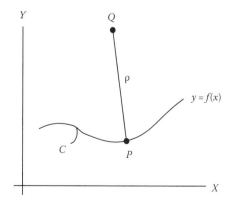

FIGURE 5.34 Plane curve and its radius of curvature.

where $y = f(x)$ is the equation of C. In comparing Figures 5.33 and 5.34, we see that if C is identified with the neutral axis N, of the beam element e, then for small bending, the slope dy/dx, of N, is small. Therefore, to a reasonable approximation, the radius of curvature is

$$\rho = \frac{-1/d^2 y}{dx^2} \quad \text{or} \quad \frac{d^2 y}{dx^2} = \frac{-1}{\rho} \tag{5.102}$$

where the minus sign arises since the Y-axes are in opposite directions in Figures 5.33 and 5.34.

5.9.4 STRESS–BENDING MOMENT RELATIONS

Equations 5.100 and 5.102 provide the means for developing the governing equations of a beam relating the stresses, strains, and displacements to the loading, shear, and bending moment: Consider again the element e of the beam segment of Figure 5.33. With the upper longitudinal fibers being shortened and the lower fibers being lengthened proportional to their distances away from the neutral axis N, we see that the strain, and consequently the stress, in e varies linearly across the cross section of e. Moreover, the stress and strain are uniaxial, or simple, stress and strain (see Sections 5.1 and 5.2).

From the elementary stress–strain equations (Equation 5.65) and from Equation 5.100, the stress in the beam may be expressed as

$$\sigma = E\varepsilon = \frac{Ey}{\rho} \tag{5.103}$$

Equation 5.103 shows that the stress varies linearly in the Y direction across the cross section. Thus, for an element e of the beam with positive bending, the stress distribution at the ends of e may be represented as in Figure 5.35. The bending moment M on e may be expressed as

$$M = \int_A \sigma y \, dA = \left(\frac{E}{\rho}\right) \int_A y^2 \, dA = \frac{EI}{\rho} \tag{5.104}$$

where A is the cross-sectional area and I, called the second moment of area, is defined as

$$I = \int_A y^2 \, dA \tag{5.105}$$

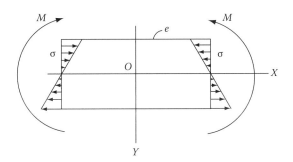

FIGURE 5.35   Stress distribution across the cross section of a beam element (e) with positive bending.

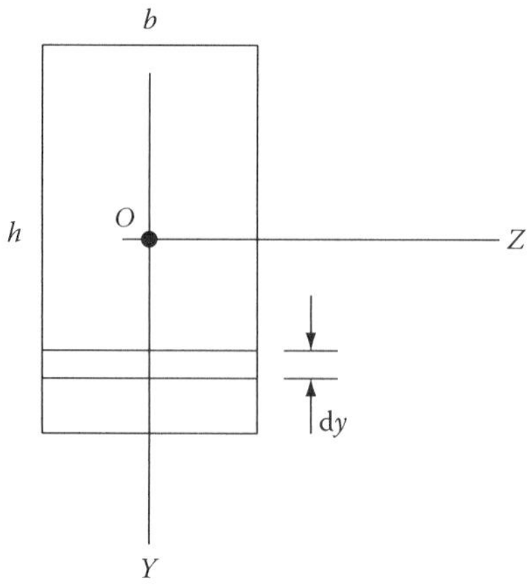

FIGURE 5.36 Rectangular beam cross section.

By comparing Equations 5.103 and 5.104, it is clear that the stress across the cross section is

$$\sigma = \frac{My}{I} \tag{5.106}$$

Also, by comparing Equations 5.102 and 5.104, we have the relation

$$\frac{d^2y}{dx^2} = \frac{-M}{EI} \tag{5.107}$$

Observe that in Equation 5.107, y refers to the displacement of the neutral axis, whereas in Equation 5.106, y represents the coordinate of the cross section.

Finally, recall, for example, that for a rectangular beam cross section, with width b and height h, as in Figure 5.36, dA is seen to be bdy. Then I is immediately seen to be

$$I = \frac{bh^3}{12} \tag{5.108}$$

5.9.5 Summary of Governing Equations

For convenience and quick reference, it is helpful to summarize the foregoing expressions:

1. *Shear load* (Equation 5.98)

$$\frac{dV}{dx} = -q \tag{5.109}$$

2. *Bending moment–shear load* (Equation 5.98)

$$\frac{dM}{dx} = V, \quad \frac{d^2M}{dx^2} = -q \tag{5.110}$$

3. *Bending moment–displacement* (Equation 5.107)

$$M = -EI\frac{d^2 y}{dx^2} \quad \text{or} \quad \frac{d^2 y}{dx^2} = \frac{-M}{EI} \tag{5.111}$$

4. *Shear displacement* (Equations 5.109 and 5.110)

$$V = -EI\frac{d^3 y}{dx^3} \quad \text{or} \quad \frac{d^3 y}{dx^3} = \frac{V}{EI} \tag{5.112}$$

5. *Load displacement* (Equations 5.108 and 5.111)

$$q = EI\frac{d^4 y}{dx^4} \quad \text{or} \quad \frac{d^4 y}{dx^4} = \frac{q}{EI} \tag{5.113}$$

6. *Cross-sectional stress* (Equation 5.106)

$$\sigma = \frac{My}{I} \tag{5.114}$$

Observe again that in Equations 5.111 through 5.113, y refers to the displacement of the neutral axis whereas in Equation 5.114, y represents the coordinate of the cross section.

5.10 THICK BEAMS

The stress of Equation 5.114 is sometimes called flexural stress. As noted earlier, this is a normal stress, acting perpendicular to the beam cross section. If the beam is relatively short or thick, there may also be significant shear stresses along the beam.

Thick beams are useful for modeling skeletal structures such as vertebral pedicles; hand, wrist, fist, and ankle bones; and the proximal femur.

To quantify the shear stresses, consider an element e with length Δx of a thick beam segment as in Figure 5.37. For simplicity, let the beam have a rectangular cross section with width b and height h as also indicated in Figure 5.37. Let the bending moment vary along the beam so that on the left side of e, the bending moment is M and on the right side, by Taylor series approximation, the bending moment is $M + (dM/dx)\Delta x$.

The variation in bending moment along the beam will give rise to shear forces on e due to Equation 5.113. More specifically, the difference in bending moments on the ends of e causes there to be a difference in normal stresses on the ends of e. This difference in normal stresses then

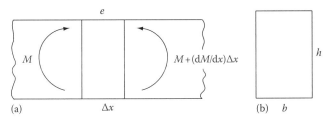

FIGURE 5.37 Element of a segment of a thick beam.

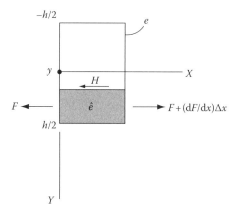

FIGURE 5.38   Lower portion $\hat{e}$ of an element (e) of a thick beam.

requires that there be horizontal shear forces in e to maintain equilibrium of e. To see this, consider the portion $\hat{e}$ of e a distance y below the neutral axis N as in Figure 5.38. Let F and $F + (\mathrm{d}F/\mathrm{d}x)\Delta x$ represent the resultant normal forces on the ends of $\hat{e}$ as shown. Let H be the resultant shearing force on $\hat{e}$ (also shown). Then for equilibrium of $\hat{e}$, we have

$$H = \left(\frac{\mathrm{d}F}{\mathrm{d}x}\right)\Delta x \qquad (5.115)$$

Observe in Equation 5.115, and in Figures 5.37 and 5.38, that H may also be expressed as

$$H = \tau b \Delta x \qquad (5.116)$$

where τ is the horizontal shear stress. Observe further that the resultant normal force F on the left end of $\hat{e}$ is

$$F = \int_{y}^{h/2} \sigma b \mathrm{d}y = \int_{y}^{h/2}\left(\frac{Myb}{I}\right)\mathrm{d}y = \frac{Mb}{I}\left(\frac{h^2}{8} - \frac{y^2}{2}\right) \qquad (5.117)$$

where we have used Equation 5.113 to express the stress in terms of the bending moment. Then $\mathrm{d}F/\mathrm{d}x$ is

$$\frac{\mathrm{d}F}{\mathrm{d}x} = \left(\frac{\mathrm{d}M}{\mathrm{d}x}\right)\left(\frac{b}{I}\right)\left(\frac{h^2}{8} - \frac{y^2}{2}\right) = \left(\frac{Vb}{I}\right)\left(\frac{h^2}{8} - \frac{y^2}{2}\right) \qquad (5.118)$$

where we have used Equation 5.109 to express $\mathrm{d}M/\mathrm{d}x$ in terms of the shear load. Finally, by using Equations 5.115 and 5.116 to solve for the shear stress τ, we obtain

$$\tau = \left(\frac{V}{I}\right)\left(\frac{h^2}{8} - \frac{y^2}{2}\right) \qquad (5.119)$$

Observe in Equation 5.119 that the maximum shear stress will occur on the neutral axis, where y is zero. Observe further that for a rectangular cross section with I being $bh^3/12$ (Equation 5.108), the maximum shear stress τ_{max} is

$$\tau_{max} = \frac{3}{2}\frac{V}{bh} = \frac{3}{2}\frac{V}{A} \text{ (rectangle)} \qquad (5.120)$$

where A is the rectangular cross-sectional area bh.

For beams with an arbitrary cross-sectional profile, Equation 5.119 is often written in the form [19,22]

$$\tau = \frac{VQ}{Ib} \qquad (5.121)$$

where b is now the beam width at the elevation of interest and where Q is

$$Q = \int_{k}^{h/2} y d\hat{A} \qquad (5.122)$$

where $\hat{A}$ is the area of $\hat{e}$. For a circular cross section, the maximum shear stress also occurs at the neutral axis with value [19,22]

$$\tau_{max} = \frac{4}{3}\frac{V}{A} \text{ (circle)} \qquad (5.123)$$

For a thin hollow circular cross section, the maximum shear stress is

$$\tau_{max} = \frac{2V}{A} \text{ (thin hollow circular)} \qquad (5.124)$$

5.11   CURVED BEAMS

In biosystems, some of the slender structures, such as the ribs, are not straight but instead are curved. If the curvature is not large, the behavior under loading is similar to that of a straight beam. If, however, the curvature is significant (radius of curvature less than five times the beam thickness), the stress distribution can measurably be changed. The principal change is that the stress is no longer linearly distributed across the cross section but instead it varies hyperbolically. Also, the neutral axis is displaced away from the centroid of the cross section. The analysis is detailed, but interested readers may consult Refs. [1–4,10,13,16,19,22–24].

To summarize the results, consider a short section of a curved beam as in Figure 5.39. If the section is sufficiently short, the beam profile at the section may be approximated by circles with radii r_o and r_i as shown. Let the beam be subjected to a bending moment M. Then as the beam is deformed, the circumferential fibers will change length creating strain in the beam. Since the fibers near the inside of the beam (near the center of curvature O) are shorter than those near the outside of the beam, the inner fibers will have greater strain and consequently greater stress than the outer fibers.

A bending moment that tends to straighten the beam is said to be positive—as in Figure 5.39. Although the stress varies continuously and monotonically across the cross section, the variation

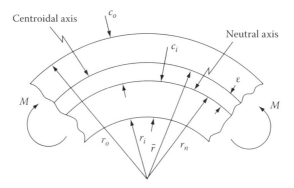

FIGURE 5.39 Section of a curved beam and its geometry.

is not linear. Then the neutral axis (axis of zero stress) is not on the centroidal axis, but instead, it is closer to the center of curvature by a small distance ε as indicated in Figure 5.39 where r_n is the neutral axis radius and $\bar{r}$ is the centroidal axis radius.

Using the geometrical description, it is readily seen that the stresses at the inner and outer surfaces are [19,22]

$$\sigma_i = \frac{Mc_i}{\varepsilon A r_i} \quad \text{and} \quad \sigma_o = \frac{-Mc_o}{\varepsilon A r_o} \tag{5.125}$$

where
  A is the cross-sectional area
  c_i and c_o are the distances from the neutral axis to the inner and outer surfaces
  M is the bending moment, and as in Figure 5.39, r_i and r_o are the inside and outside radii
  ε is the radial separation between the neutral axis and the centroidal axis

For computational purposes, it is convenient to express Equation 5.125 in the forms

$$\sigma_i = K_i \frac{Mc_i}{I} \quad \text{and} \quad \sigma_o = -K_o \frac{Mc_o}{I} \tag{5.126}$$

where
  I is the second moment of area of the cross section relative to the centroid
  K_i and K_o are geometric correction factors [22]

The curvature of the beam is often described in terms of the ratio $\bar{r}/\bar{c}$ where as shown in Figure 5.39 $\bar{r}$ is the radius of the centroidal axis and $\bar{c}$ is the distance from the centroidal axis to the inner surface. For $\bar{r}/\bar{c}$ ratios greater than 10, there is a slight difference between the stress distribution in a curved beam and a straight beam. For small $\bar{r}/\bar{c}$ ratios (say 1.5), the inside factor K_i can become as large as 3, whereas the outside factor K_o does not get smaller than 0.5. See Ref. [22] for additional details.

5.12 SINGULARITY FUNCTIONS

Equation 5.113 is a relatively simple ordinary differential equation. Indeed, given the loading function $q(x)$, we can simply integrate four times to obtain the displacement y as a function of x. Then using Equations 5.111 and 5.112, we can obtain the bending moment and shear force. Equation 5.114 provides the flexural stress.

A difficulty that arises, however, is that the loading function $q(x)$ is often discontinuous, representing concentrated forces and moments, as occurs with traumatic loading of biosystems.

To alleviate this difficulty, many analysts use singularity functions defined using angular brackets as $\langle x - a \rangle^n$ where a is the X-coordinate of any point on the neutral axis and n is any integer (positive, negative, or zero). The angular bracket function is defined as [19]

$$\langle x - a \rangle^n = \begin{cases} 0 & x < a \quad \text{for all } n \\ 0 & x > a \quad \text{for } n < 0 \\ 1 & x = a \quad \text{for } n < 1 \\ 1 & x \geq a \quad \text{for } n = 0 \\ (x-a)^n & x \geq a \quad \text{for } n > 0 \end{cases} \tag{5.127}$$

The derivatives and antiderivatives (integrals) of $\langle x - a \rangle^n$ are defined by the following expressions:

$$\frac{d}{dx} \langle x - a \rangle^{n+1} = \langle x - a \rangle^n \quad n < 0 \tag{5.128}$$

$$\frac{d}{dx} \langle x - a \rangle^n = n \langle x - a \rangle^{n-1} \quad n > 0 \tag{5.129}$$

$$\int_b^x \langle \xi - a \rangle^n \, d\xi = \langle x - a \rangle^{n+1} \quad n < 0 \tag{5.130}$$

and

$$\int_b^x \langle \xi - a \rangle^n \, d\xi = \frac{\langle x - a \rangle^{n+1}}{n+1} \quad n \geq 0 \tag{5.131}$$

where $b < a$.

As noted, these singularity functions are useful for representing discontinuities in the loading function $q(x)$. To develop this, consider a graphical description of selected angular bracket functions as in Figure 5.40. These functions are useful for representing concentrated forces and for uniform

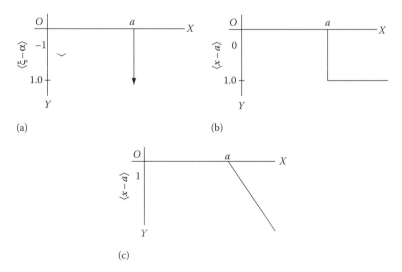

(a)   (b)

(c)

FIGURE 5.40 Angular bracket functions useful for representing discontinuous loadings. (a) Unit impulse function. (b) Unit step function. (c) Ramp function.

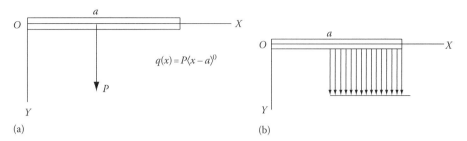

FIGURE 5.41 Concentrated force and uniform loading represented by angular bracket functions. (a) Concentrated load. (b) Uniform loading.

loadings as shown in Figure 5.41. Concentrated moments and ramp loading may be similarly represented by the functions $M_0 \langle x - a \rangle^{-2}$ and $q_0 \langle x - a \rangle^1$.

5.13 ELEMENTARY ILLUSTRATIVE EXAMPLES

By using the singularity functions, we can reduce the analysis of simple beam problems to a routine drill. This involves the following steps:

1. Construct a free-body diagram of the beam determining the support reactions.
2. Using the given loading and the support reactions and using the singularity functions, determine the loading function $q(x)$.
3. Using Equation 5.113, construct an explicit form of the governing equations.
4. Express the boundary conditions in terms of the neutral axis displacement y and its derivatives.*
5. Integrate the governing equations using the boundary conditions to determine the constants of integration.
6. Using the results of step 5, construct shear, bending moment, and displacement diagrams.

The following sections illustrate these steps with a few simple examples.

5.13.1 Cantilever Beam with a Concentrated End Load

Consider first a cantilever beam of length ℓ supported (fixed) at its right end and loaded with a concentrated force at its left end as in Figure 5.42. The objective is to find the beam displacement, the bending moment, and the shear force along the beam.

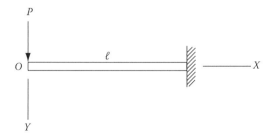

FIGURE 5.42 Cantilever beam loaded at its unsupported end.

---

* The boundary conditions are to be applied on that portion of the beam in the positive X direction, that is, on the right-hand portion of the beam.

FIGURE 5.43   Free-body diagram of the beam of Figure 5.42.

This configuration could provide a gross modeling of a number of common biomechanical activities such as lifting, pushing/pulling, and kicking.

Following the steps listed earlier, we first construct a free-body diagram of the beam as in Figure 5.43 where R and M_ℓ are the reactive force and bending moment from the support. By adding forces and by computing moments about the right end, we obtain

$$R = P \quad \text{and} \quad M = P\ell \tag{5.132}$$

Then by using the singularity functions, the loading $q(x)$ on the beam may be expressed as

$$q(x) = P\langle x - 0 \rangle^{-1} - P\langle x - \ell \rangle^{-1} - P\ell\langle x - \ell \rangle^{-2} \tag{5.133}$$

The governing differential equation (Equation 5.113) is then

$$EI\frac{\mathrm{d}^4 y}{\mathrm{d}x^4} = P\langle x - 0 \rangle^{-1} - P\ell\langle x - \ell \rangle^{-2} \tag{5.134}$$

The boundary (or end) conditions are

At $x = 0$

$$M = 0 \quad \text{and} \quad V = -P$$

or

$$\frac{\mathrm{d}^2 y}{\mathrm{d}x^2} = 0 \quad \text{and} \quad EI\frac{\mathrm{d}^3 y}{\mathrm{d}x^3} = P \tag{5.135}$$

At $x = 1$

$$y = 0 \quad \text{and} \quad \frac{\mathrm{d}y}{\mathrm{d}x} = 0 \tag{5.136}$$

Observe in Equation 5.135 that the shear force V is negative since the load force P is exerted on a negative face in the positive direction.

By integrating Equation 5.135, we obtain

$$EI\frac{\mathrm{d}^3 y}{\mathrm{d}x^3} = -V = P\langle x - 0 \rangle^0 - P\langle x - \ell \rangle^0 - P\ell\langle x - \ell \rangle^{-1} + C_1 \tag{5.137}$$

But since at $x = 0$, $EI\,\mathrm{d}^3 y/\mathrm{d}x^3 = P$, we have

$$P = P + C_1 \quad \text{or} \quad C_1 = 0 \tag{5.138}$$

(Note $\langle 0 \rangle^0 = 1$.)

By integrating again, we have

$$EI\frac{d^2y}{dx^2} = -M = P\langle x-0\rangle^1 - P\langle x-\ell\rangle^1 - P\ell\langle x-\ell\rangle^0 + C_2 \tag{5.139}$$

But since at $x = 0$, $d^2y/dx^2 = 0$, we have

$$0 = 0-0-0+C_2 \quad \text{or} \quad C_2 = 0 \tag{5.140}$$

Integrating again, we obtain

$$EI\frac{dy}{dx} = \frac{P\langle x-0\rangle^2}{2} - \frac{P\langle x-\ell\rangle^2}{2} - P\ell\langle x-\ell\rangle^1 + C_3 \tag{5.141}$$

But since at $x = \ell$, $dy/dx = 0$, we have

$$0 = \frac{P\ell^2}{2} - 0 - 0 + C_3 \quad \text{or} \quad C_3 = \frac{-P\ell^2}{2} \tag{5.142}$$

Finally, integrating for the fourth time, we have

$$EIy = \frac{P\langle x-0\rangle^3}{6} - \frac{P\langle x-\ell\rangle^3}{6} - \frac{P\ell\langle x-\ell\rangle^2}{2} - \frac{P\ell^2 x}{2} + C_4 \tag{5.143}$$

But since at $x = 0$, $y = 0$, we obtain

$$0 = \frac{P\ell^3}{6} - 0 - 0 - \frac{P\ell^3}{2} + C_4 \quad \text{or} \quad C_4 = \frac{P\ell^3}{3} \tag{5.144}$$

Therefore, $EIy(x)$ is seen to be

$$EIy(x) = \frac{P\langle x-0\rangle^3}{6} - \frac{P\langle x-\ell\rangle^3}{6} - \frac{P\ell\langle x-\ell\rangle^2}{2} - \frac{P\ell^2 x}{2} + \frac{P\ell^3}{3} \tag{5.145}$$

Observe in Equation 5.145 that the maximum displacement y_{max} occurs at $x = 0$:

$$y_{max} = \frac{P\ell^3}{3EI} \tag{5.146}$$

From Equations 5.138 and 5.139, we obtain the shear and bending moment diagrams shown in Figures 5.44 and 5.45. Observe from Equation 5.114 that the maximum normal stress in the beam occurs at the position of maximum bending moment. As shown in Figure 5.45, the maximum bending moment and hence the maximum stress occur at the support (left) end.

5.13.2 CANTILEVER BEAM WITH A CONCENTRATED END LOAD ON THE RIGHT END

To illustrate the effect upon the analysis of a change of the direction of the *X*-axis, consider a cantilever beam of length ℓ supported (fixed) at its left end and loaded with a concentrated force at its

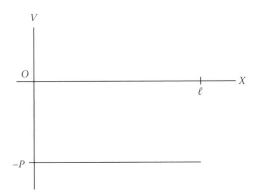

FIGURE 5.44   Shear diagram (note that the positive ordinate is up, as opposed to the positive Y direction being down for the beam itself) for the beam and loading of Figure 5.42.

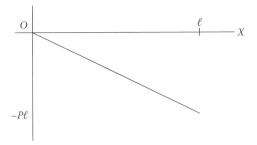

FIGURE 5.45   Bending moment diagram (note that the positive ordinate is up, as opposed to the positive Y direction being down for the beam itself) for the beam and loading of Figure 5.42.

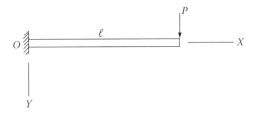

FIGURE 5.46   Cantilever beam loaded at its left end and loaded at its right end.

FIGURE 5.47   Free-body diagram of the beam of Figure 5.46.

right end as in Figure 5.46. As before, the objective is to find the beam displacement, the bending moment, and the shear force along the beam axis.

Following the steps of the foregoing example, we construct a free-body diagram of the beam as in Figure 5.47 where R and M_ℓ are the reactive force and bending moment at the support, respectively. By adding forces and by computing moments about the left end, we obtain

$$R = P \quad \text{and} \quad M_\ell = P\ell \tag{5.147}$$

The loading $q(x)$ on the beam is then

$$q(x) = -P\langle x-0\rangle^{-1} - P\ell\langle x-0\rangle^{-2} + P\langle x-\ell\rangle^{-1} \tag{5.148}$$

The governing differential equation (Equation 5.113) is then

$$EI\frac{d^4y}{dx^4} = -P\langle x-0\rangle^{-1} - P\ell\langle t-0\rangle^{-2} + P\langle x-\ell\rangle^{-1} \tag{5.149}$$

The boundary conditions are

At $x = 0$

$$y = 0 \quad \text{and} \quad \frac{dy}{dx} = 0$$

At $x = \ell$

$$M = 0 \quad \text{and} \quad V = 0$$

$$\tag{5.150}$$

or

$$\frac{d^2y}{dx^2} = 0 \quad \text{and} \quad \frac{d^3y}{dx^3} = 0 \tag{5.151}$$

Observe, according to the footnote in step 4 in Section 5.13, the shear load on the right portion of the beam is zero.

By integrating in Equation 5.134, we obtain

$$EI\frac{d^3y}{dx^3} = -P\langle x-0\rangle^0 - P\ell\langle x-0\rangle^{-1} + P\langle x-\ell\rangle^0 + c_1 \tag{5.152}$$

But since $d^3y/dx^3 = 0$ at $x = \ell$, we have

$$0 = -P - 0 + P + c_1 \quad \text{or} \quad c_1 = 0 \tag{5.153}$$

By integrating again, we have

$$EI\frac{d^2y}{dx^2} = -P\langle x-0\rangle^1 - P\ell\langle x-0\rangle^0 + P\langle x-\ell\rangle^1 + c_2 \tag{5.154}$$

But since $d^2y/dx^2 = 0$ at $x = \ell$, we have

$$0 = -P\ell - P\ell + 0 + c_2 \quad \text{or} \quad c_2 = 2P\ell \tag{5.155}$$

Integrating again, we have

$$EI\frac{dy}{dx} = \frac{-P\langle x-0\rangle^2}{2} - P\ell\langle x-0\rangle^1 + \frac{P\langle x-\ell\rangle^2}{2} + 2P\ell x + c_3 \tag{5.156}$$

But since $dy/dx = 0$ at $x = 0$, we have

$$0 = -0 - 0 + 0 + 0 + c_3 \quad \text{or} \quad c_3 = 0 \tag{5.157}$$

Finally, by integrating for the fourth time, we obtain

$$EIy = \frac{-P\langle x-0\rangle^3}{6} - \frac{P\ell\langle x-0\rangle^2}{2} + \frac{P\langle x-\ell\rangle^3}{6} + P\ell x^2 + c_4 \tag{5.158}$$

But since $y = 0$ at $x = 0$, we have

$$0 = -0 - 0 + 0 + 0 + c_4 \quad \text{or} \quad c_4 = 0 \tag{5.159}$$

Therefore, $y(x)$ is seen to be

$$EIy = \frac{-P\langle x-0\rangle^3}{6} - \frac{P\ell\langle x-0\rangle^2}{2} + \frac{P\langle x-\ell\rangle^3}{6} + P\ell x^2 \tag{5.160}$$

Observe again that the maximum displacement occurs at $x = \ell$ as

$$y_{\max} = \frac{P\ell^3}{3EI} \tag{5.161}$$

As expected, this is consistent with Equation 5.146.

By using Equations 5.152 and 5.154, the shear and bending moment are as shown in Figures 5.48 and 5.49. Observe here that the maximum bending moment again occurs at the support, this time at the right end.

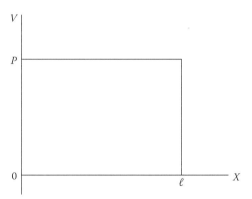

FIGURE 5.48   Shear diagram for the beam and loading of Figure 5.46.

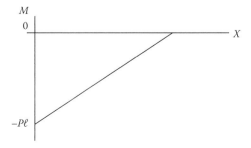

FIGURE 5.49   Bending moment diagram for the beam and loading of Figure 5.46.

5.13.3   Simply Supported Beam with a Concentrated Interior Span Load

Next, consider a beam having simple (pin) supports at its ends and with a concentrated interior load as in Figure 5.50. Let A and B designate the ends of the beam and let the beam have length ℓ. Let the force have magnitude P, located a distance a from the left end A and a distance b from the right end B as shown. (Recall that a simple support exerts no moment on the beam, but instead it provides vertical support (or vertical force) so that the vertical displacement is zero.)

As before, the objective is to determine an expression for the displacement and the shear and bending moment diagrams.

By again following the steps outlined earlier, we construct a free-body diagram of the beam as in Figure 5.51, where R_A and R_B represent the reactive forces at the supports. By adding forces vertically and by evaluating moments about the end A, we find R_A and R_B to be

$$R_A = \frac{Pb}{\ell} \quad \text{and} \quad R_B = \frac{Pa}{\ell} \tag{5.162}$$

The load function $q(x)$ is then

$$q(x) = -P\left(\frac{b}{\ell}\right)\langle x - 0 \rangle^{-1} + P\langle x - a \rangle^{-1} - P\left(\frac{a}{\ell}\right)\langle x - \ell \rangle^{-1} \tag{5.163}$$

and the governing differential equation becomes

$$EI\frac{\mathrm{d}^4 y}{\mathrm{d}x^4} = -P\left(\frac{b}{\ell}\right)\langle x - 0 \rangle^{-1} + P\langle x - a \rangle^{-1} - P\left(\frac{a}{\ell}\right)\langle x - \ell \rangle^{-1} \tag{5.164}$$

The boundary conditions are

At $A\,(x = 0)$

$$y = 0 \quad \text{and} \quad M = 0 \quad \text{or} \quad \frac{\mathrm{d}^2 y}{\mathrm{d}x^2} = 0 \tag{5.165}$$

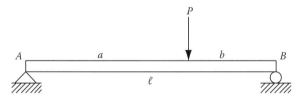

FIGURE 5.50   Simply supported beam with an interior span load.

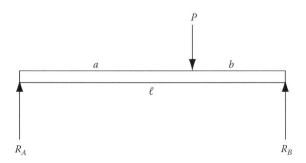

FIGURE 5.51   Free-body diagram for the beam of Figure 5.50.

At $B(x = \ell)$

$$y = 0 \quad \text{and} \quad M = 0 \quad \text{or} \quad \frac{d^2 y}{dx^2} = 0 \tag{5.166}$$

By integrating Equation 5.164, we have

$$EI \frac{d^3 y}{dx^3} = -P\left(\frac{b}{\ell}\right)\langle x - 0 \rangle^0 + P\langle x - a \rangle^0 - P\left(\frac{a}{\ell}\right)\langle x - \ell \rangle^0 + c_1 \tag{5.167}$$

and

$$EI \frac{d^2 y}{dx^2} = -P\left(\frac{b}{\ell}\right)\langle x - 0 \rangle^1 + P\langle x - a \rangle^1 - P\left(\frac{a}{\ell}\right)\langle x - \ell \rangle^1 + c_1 x + c_2 \tag{5.168}$$

Since $d^2 y/dx^2 = 0$ at both $x = 0$ and $x = \ell$, we have

$$c_1 = c_2 = 0 \tag{5.169}$$

By integrating again, we obtain

$$EI \frac{dy}{dx} = -P\left(\frac{b}{\ell}\right)\frac{\langle x - 0 \rangle^2}{2} + \frac{P\langle x - a \rangle^2}{2} - P\left(\frac{a}{\ell}\right)\frac{\langle x - \ell \rangle^2}{2} + c_3 \tag{5.170}$$

and

$$EIy = -P\left(\frac{b}{\ell}\right)\frac{\langle x - 0 \rangle^3}{6} + \frac{P\langle x - a \rangle^3}{6} - P\left(\frac{a}{\ell}\right)\frac{\langle x - \ell \rangle^3}{6} + c_3 x - c_4 \tag{5.171}$$

Since y is zero at both $x = 0$ and $x = \ell$, we have

$$c_3 = \frac{Pb}{6\ell}(\ell^2 - b^2) \quad \text{and} \quad c_4 = 0 \tag{5.172}$$

Therefore, the displacement y is

$$y = -\left(\frac{Pb}{6EI\ell}\right)\langle x - 0 \rangle^3 + \left(\frac{P}{6EI}\right)\langle x - a \rangle^3 - \left(\frac{Pa}{6EI\ell}\right)\langle x - \ell \rangle^3 + \left(\frac{Pb}{6EI\ell}\right)(\ell^2 - b^2)x \tag{5.173}$$

If the load is at midspan ($a = b = \ell/2$), the maximum displacement is also at midspan, under the load, with the value

$$y_{\text{max}} = \frac{P\ell^3}{48EI} \tag{5.174}$$

Finally, from Equations 5.167 and 5.168, the shear and bending moments may be represented by the diagrams in Figures 5.52 and 5.53.

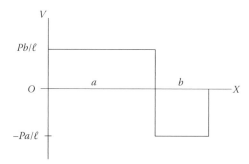

FIGURE 5.52   Shear diagram for the beam and loading of Figure 5.50.

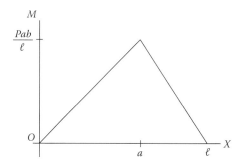

FIGURE 5.53   Bending moment diagram for the beam and loading of Figure 5.50.

5.13.4   SIMPLY SUPPORTED BEAM WITH UNIFORM LOAD

For another example, which can have direct application in biosystems, consider a simply supported beam with a uniform loading as represented in Figure 5.54.

Figure 5.55 shows a free-body diagram of the beam where the uniform loading is q_0 (force per unit length), the beam length is ℓ, and R_A and R_B are the support reactions at ends A and B. R_A and R_B are seen to be

$$R_A = R_B = \frac{q_0\ell}{2} \tag{5.175}$$

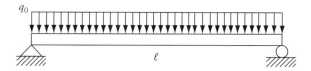

FIGURE 5.54   Uniformly loaded simply supported beam.

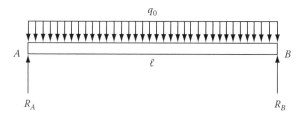

FIGURE 5.55   Free-body diagram of the beam of Figure 5.54.

The loading $q(x)$ on the beam is then

$$q(x) = -\left(\frac{q_0\ell}{2}\right)\langle x-0\rangle^{-1} + q_0\langle x-0\rangle^0 - \left(\frac{q_0\ell}{2}\right)\langle x-\ell\rangle^{-1} \tag{5.176}$$

Therefore, the governing differential equation is

$$EI\frac{d^4y}{dx^4} = -\left(\frac{q_0\ell}{2}\right)\langle x-0\rangle^{-1} + q_0\langle x-0\rangle^0 - \left(\frac{q_0\ell}{2}\right)\langle x-\ell\rangle^{-1} \tag{5.177}$$

The boundary conditions are the same as in Section 5.13.3:

At $A(x=0)$

$$y=0 \quad \text{and} \quad M=0 \quad \text{or} \quad \frac{d^2y}{dx^2}=0 \tag{5.178}$$

At $B(x=\ell)$

$$y=0 \quad \text{and} \quad M=0 \quad \text{or} \quad \frac{d^2y}{dx^2}=0 \tag{5.179}$$

By integrating Equation 5.177, we obtain

$$EI\frac{d^3y}{dx^3} = -\left(\frac{q_0\ell}{2}\right)\langle x-0\rangle^0 + q_0\langle x-0\rangle^1 - \left(\frac{q_0\ell}{2}\right)\langle x-\ell\rangle^0 + c_1 \tag{5.180}$$

and

$$EI\frac{d^2y}{dx^2} = -\left(\frac{q_0\ell}{2}\right)\langle x-0\rangle^1 + q_0\frac{\langle x-0\rangle^2}{2} - \left(\frac{q_0\ell}{2}\right)\langle x-\ell\rangle^1 + c_1 x + c_2 \tag{5.181}$$

Since d^2y/dx^2 is zero at both $x=0$ and $x=\ell$, we have

$$c_1 = c_2 = 0 \tag{5.182}$$

Integrating again, we have

$$EI\frac{dy}{dx} = -\left(\frac{q_0\ell}{2}\right)\frac{\langle x-0\rangle^2}{2} + q_0\frac{\langle x-0\rangle^3}{6} - \left(\frac{q_0\ell}{2}\right)\frac{\langle x-\ell\rangle^2}{2} + c_3 \tag{5.183}$$

and

$$EIy = -\left(\frac{q_0\ell}{2}\right)\frac{\langle x-0\rangle^3}{6} + q_0\frac{\langle x-0\rangle^4}{24} - \left(\frac{q_0\ell}{2}\right)\frac{\langle x-\ell\rangle^3}{6} + c_3 x + c_4 \tag{5.184}$$

Since y is zero at $x = 0$ and $x = \ell$, we have

$$c_3 = \frac{q_0\ell^3}{24} \quad \text{and} \quad c_4 = 0 \tag{5.185}$$

Therefore, the displacement is given by

$$EIy = -\left(\frac{q_0\ell}{2}\right)\frac{\langle x - 0 \rangle^3}{6} + q_0\frac{\langle x - 0 \rangle^4}{24} - \left(\frac{q_0\ell}{2}\right)\frac{\langle x - \ell \rangle^3}{6} + q_0\ell^3\,\frac{x}{24} \tag{5.186}$$

The maximum displacement occurs at midspan and is

$$y_{\text{max}} = \frac{5q_0\ell^4}{384EI} \tag{5.187}$$

By using Equations 5.180 and 5.181, the shear and bending moment diagrams are as shown in Figures 5.56 and 5.57.

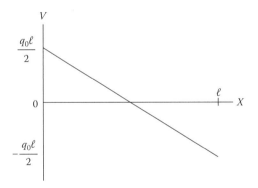

FIGURE 5.56   Shear diagram for the beam of Figure 5.55.

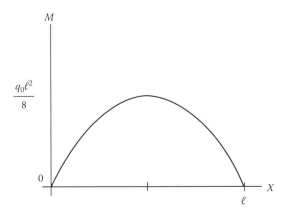

FIGURE 5.57   Bending moment diagram for the beam of Figure 5.55.

5.14 LISTING OF SELECTED BEAM DISPLACEMENT AND BENDING MOMENT RESULTS

Table 5.2 presents a few classical and fundamental results, which may be of use in biomechanical analyses. More extensive lists can be found in Refs. [10,13].

TABLE 5.2
Selected Beam Loading Configurations

Configuration	Maximum Displacement (δ_{max})	Maximum Bending Moment (M_{max})
1. Cantilever beam with concentrated end load	$\delta_{max} = \dfrac{P\ell^3}{3EI}$ (at end B)	$M_{max} = -P\ell$ (at end A)

2. Cantilever beam with a uniform load	$\delta_{max} = \dfrac{q_0\ell^4}{8EI}$ (at end B)	$M_{max} = -\dfrac{q_0\ell^2}{2}$ (at end A)

3. Cantilever beam with moment at the unsupported end	$\delta_{max} = \dfrac{M_B\ell^2}{2EI}$ (at end B)	$M_{max} = -M_B$ (uniform along the beam)

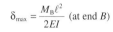

4. Simply supported beam with a concentrated center load	$\delta_{max} = \dfrac{P\ell^4}{48EI}$ (at center span)	$M_{max} = \dfrac{P\ell}{4}$ (at center span)

5. Simply supported beam with a uniform load	$\delta_{max} = \dfrac{5q_0\ell^4}{38EI}$ (at center span)	$M_{max} = \dfrac{q_0\ell^2}{2}$ (at center span)

6. Fixed end beam with a concentrated center load	$\delta_{max} = \dfrac{P\ell^3}{192EI}$ (at center span)	$M_{max} = \dfrac{P\ell}{8}$ (at ends and at the center)

 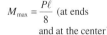 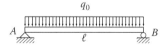 

7. Fixed end beam with a uniform load	$\delta_{max} = \dfrac{q_0\ell^4}{384EI}$ (at center span)	$M_{max} = \dfrac{q_0\ell^2}{12}$ (at ends and center)

5.15 MAGNITUDE OF TRANSVERSE SHEAR STRESS

The foregoing results enable us to estimate the relative magnitude and importance of the transverse shear stress (Section 5.10). Recall from Equation 5.120 that for a rectangular cross-sectional beam with base b and height h, the maximum shearing stress τ_{max} is

$$\tau_{max} = \frac{3V}{2bh} \qquad (5.188)$$

For a simply supported beam with a concentrated midspan load P (see configuration 4 of Table 5.2 and also Section 5.13.3), the maximum shear V_{max} is $P/2$. Hence, the maximum shear stress is

$$\tau_{max} = \frac{3P}{4bh} \qquad (5.189)$$

Also from Section 5.14, we see that the maximum bending moment M_{max} is $P\ell/4$, where ℓ is the beam length. Therefore, the maximum bending stress σ_{max} is

$$\sigma_{max} = \frac{\left[M_{max}(h/2) \right]}{I} = \frac{(P\ell/4)(h/2)}{(bh^3/12)} = \frac{3P\ell}{2bh^2} \qquad (5.190)$$

The ratio of the maximum shear stress to the maximum bending stress is thus

$$\frac{\tau_{max}}{\sigma_{max}} = \frac{(3P/4bh)}{(3P\ell/2bh^2)} = \frac{h}{2\ell} \qquad (5.191)$$

The maximum transverse shear stress will then be less than 10% of the maximum bending stress if $h < \ell/5$.

5.16 TORSION OF BARS

Consider a circular rod subjected to axial twisting by moments (or torques) applied at its ends as in Figure 5.58. Aside from implant components, there are no such simple structures in biosystems. Nevertheless, by reviewing this elementary system, we can obtain qualitative insight and approximation to the behavior of long slender members in the skeletal system.

The symmetry of the circular rod requires that during twisting, planar circular sections normal to the axis remain plane and circular and that radial lines within those sections remain straight [1–3].

Suppose that during the twisting, the ends A and B are rotated relative to one another by an angle θ, as represented in Figure 5.59. Then a point P on the rod surface will be displaced to a point P' as indicated in the figure. The arc length PP' is then both $r\theta$ and $\gamma\ell$ where r is the cylinder radius, ℓ is its length, and γ is the angle between axial surface lines as shown. But γ is also a measure of the shear strain (Section 5.2). Thus, the shear strain at P is

$$\gamma = \frac{r\theta}{\ell} \qquad (5.192)$$

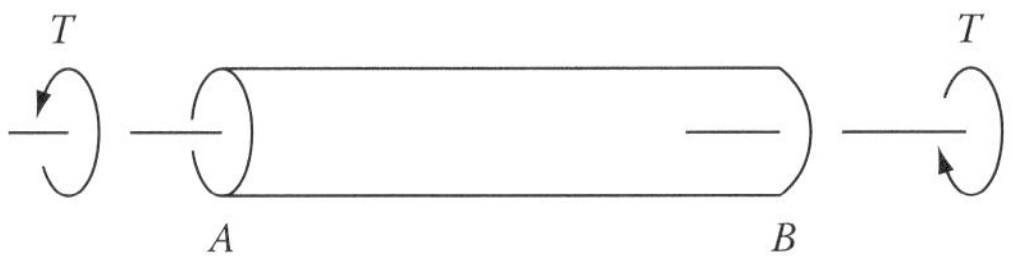

FIGURE 5.58 Rod subjected to torsional moments.

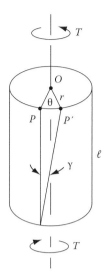

FIGURE 5.59   Twisted circular rod.

Similarly, for a concentric interior circle of radius ρ, the shear strain is

$$\gamma = \frac{\rho\theta}{\ell} \tag{5.193}$$

Therefore, the shear stress τ at an interior point is simply

$$\tau = G\gamma = \frac{G\rho\theta}{\ell} \tag{5.194}$$

where as before, G is the shear modulus.

Equation 5.194 shows that the shear stress varies linearly along a radial line, and consequently, it reaches a maximum value at the outer surface of the rod.

By integrating (summing) the moments of the shear stress over the cross section, we have, by equilibrium, the applied torsioned moment T. To develop this, consider an end view of the cylinder as in Figure 5.60. Consider specifically the shear stress τ on a swell element of the cross section.

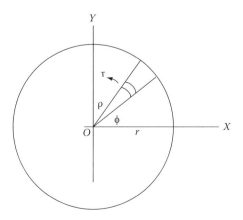

FIGURE 5.60   Shear stress on an element of a cross section.

Using polar coordinates ρ and ϕ as in Figure 5.60, the area of a differential element is $\rho \, d\rho \, d\phi$. Then the resulting force on such an element is $\tau \rho \, d\rho \, d\phi$, and the moment of this force about the center O is $\rho(\tau \rho \, d\rho \, d\phi)$. Thus, the moment T of all such elemental forces about O may be obtained by integration as

$$T = \int_0^{2\pi}\int_0^r \rho \, (\tau \rho \, d\rho \, d\phi) = \int_0^{2\pi}\int_0^r \rho \left(\frac{G\rho\theta}{\ell} \right) \rho \, d\rho \, d\phi = \left(\frac{G\theta}{\ell} \right) \int_0^{2\pi}\int_0^r \rho^3 \, d\rho \, d\phi = \left(\frac{G\theta}{\ell} \right)\left(\frac{\pi r^4}{2} \right) = \left(\frac{G\theta}{\ell} \right) J \quad (5.195)$$

where J, defined as $\pi r^4/2$, is the second polar moment of area (the polar moment of inertia) of the cross section and where we have used Equation 5.194 to obtain an expression for the shear stress in terms of ρ.

From Equations 5.194 and 5.195, we then obtain the fundamental relations:

$$\theta = \frac{T\ell}{JG} \quad \text{and} \quad \tau = \frac{T\rho}{J} \tag{5.196}$$

Observe the similarity of Equations 5.196, 5.66, and 5.106.

5.17 TORSION OF MEMBERS WITH NONCIRCULAR AND THIN-WALLED CROSS SECTIONS

As noted earlier, aside from implants, biosystems do not have long members with circular cross sections. Unfortunately, a torsional analysis of members with noncircular cross sections is somewhat detailed and cumbersome [6,16]. Nevertheless, there are analyses that can give insight into their behavior. The best known of these is the membrane or soapfilm analogy, which relates the shear stresses to the slope of an inflated membrane covering an opening having the shape of the noncircular cross section. Specifically, consider a hollow thin-walled cylinder whose cross section has the same contour as the noncircular member of interest (Figures 5.61 and 5.62). Let a membrane be stretched across the cross section of the thin-walled cylinder and then be inflated by being pressured on the inside as in Figure 5.62. Then the magnitude of the shear stress at any point P of the cross section is proportional to the vertical slope of the membrane at the point Q of the membrane directly above P. The direction of the shear stress is the same as the horizontal tangent at Q [1,6].

FIGURE 5.61 Noncircular cross-sectional member.

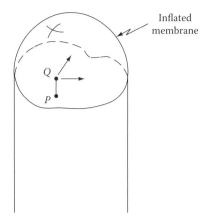

FIGURE 5.62   Membrane atop hollow cylinder with a cross-sectional profile as the cylinder of Figure 5.61.

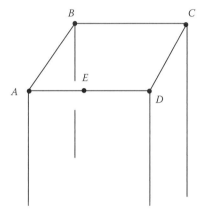

FIGURE 5.63   Bar with a rectangular cross section.

By reflecting about this, we see that the greatest slope of the membrane will occur at the boundary at the point closest to the centroid of the cross section. Thus, for example, for a member with a rectangular cross section, as in Figure 5.63, the maximum shear stress will occur at point E.

Finally, the torsional rigidity, or resistance to twisting, of a bar is proportional to the volume displaced by the expanded membrane [1].

Consider next a hollow cylinder with a thin wall as in Figure 5.64. (Such a cylinder might be a model of a long bone, such as the femur or humerus.) When the cylinder is twisted, the twisting torque produces a shear stress in the thin-walled boundary of the cylinder. Using a relatively simple equilibrium analysis, it is seen [1,3] that the product of the shear stress τ and the wall thickness t is approximately a constant. The shear stress itself is found to be

$$\tau = \frac{T}{2tA} \tag{5.197}$$

where
   T is the magnitude of the externally applied torsional moment
   A is the area enclosed by the center curve of the thin wall

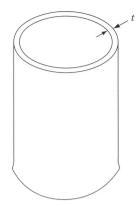

FIGURE 5.64 Thin-walled hollow cylinder.

5.18 ENERGY METHODS

Recall from elementary mechanics analysis how energy methods such as the work–energy method and the concept of potential energy provide a means for quickly obtaining insight into the behavior of mechanical systems. So also with elastic, deformable systems, energy methods can provide useful results with relatively simple analyses. These methods use concepts such as potential energy, complementary energy, and strain energy and procedures such as variational techniques, least squares, and Galerkin analyses. The finite-element and boundary-element methods are also based upon these procedures.

While an exposition of these methods is beyond our scope, it is nevertheless useful to review one of the simplest of these: Castigliano's theorem, which states that the displacement δ under a load P is the derivative of the strain energy U with respect to P [1–3].

To explore this, we define the strain energy U of a body B as the integral over the volume of B of the strain energy density E, which in turn is defined as half the sum of the products of the components of the stress and strain tensors. Specifically,

$$U = \int_B E \, dV \tag{5.198}$$

where

$$E \triangleq \left(\frac{1}{2}\right) \sigma_{ij} \varepsilon_{ij} \tag{5.199}$$

V is the volume of B.

For simple bodies and problems such as extension, bending, or torsion of beams, there is simple straining or unilateral straining, ε, and simple stress, σ, and the strain energy is then simply $(1/2)\sigma\varepsilon$. For example, for a bar B of length ℓ in simple tension or compression due to an axial load P, the strain energy is

$$U = \int_B \left(\frac{1}{2}\right) \sigma \varepsilon \, dV = \int_0^\ell \left(\frac{1}{2}\right)\left(\frac{P}{A}\right) A \, dx = \int_0^\ell \left(\frac{P^2}{2AE}\right) dx = \frac{P^2 \ell}{2AE} \tag{5.200}$$

where
A is the cross-sectional area of B
E is the elastic modulus
x is the axial coordinate

For bending of B, we have

$$U = \int_B \left(\frac{1}{2}\right)\sigma\varepsilon dV = \int_0^\ell \int_A \left(\frac{1}{2}\right)\left(\frac{My}{I}\right)\left(\frac{My}{EI}\right)dA d\ell$$

$$= \int_0^\ell \left(\frac{1}{2}\right)\left(\frac{M^2}{EI}\right)dx \tag{5.201}$$

where
  M is the bending moment
  I is the second moment of area
  y is the transverse coordinate in the cross section
  $\int y^2 dA$ is I

For torsion of B, we have

$$U = \int_B \left(\frac{1}{2}\right)\tau\gamma dV = \int_0^\ell \int_A \left(\frac{1}{2}\right)\left(\frac{T\rho}{J}\right)\left(\frac{T\rho}{JG}\right)dA dx$$

$$= \int_0^\ell \left(\frac{1}{2}\right)\left(\frac{T^2}{JG}\right)dx \tag{5.202}$$

where
  T is the torsional moment
  ρ is the radial coordinate of the cross section
  J is the second axial (polar) moment of area
  G is the shear modulus

As noted earlier, Castigliano's theorem states that if an elastic body B is subjected to a force system, the displacement δ at a point Q of B in the direction of a force P applied at Q is simply [1–3]

$$\delta = \frac{\partial U}{\partial P} \tag{5.203}$$

To illustrate the use of Equation 5.203, consider the simple case of a cantilever beam of length ℓ with an end load P as in Figure 5.65. If the origin of the X-axis is at the loaded end O, the bending moment M along the beam is (see Section 5.9)

$$M = -Px \tag{5.204}$$

FIGURE 5.65   Cantilever beam with a concentrated end load.

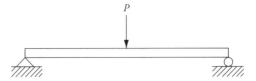

FIGURE 5.66 Simply supported beam with a concentrated center load.

Then from Equation 5.201, the strain energy is

$$U = \int_0^\ell \left(\frac{M^2}{2EI} \right) dx = \frac{P^2 \ell^3}{6EI} \tag{5.205}$$

Thus, from Equation 5.203, the displacement δ at O may be expressed as

$$\delta = \frac{\partial U}{\partial P} = \frac{P\ell^3}{3EI} \tag{5.206}$$

Compare this result and the effort in obtaining it with that of Equation 5.151. Observe that the analysis here is considerably simpler than that in Section 5.13. However, the information obtained is more limited.

Next, consider the simply supported beam of length ℓ with a concentrated load P at its center as in Figure 5.66. From Equation 5.167, the bending moment may be expressed as

$$M = \left(\frac{P}{2} \right) \langle x - 0 \rangle^1 - P \left\langle x - \left(\frac{\ell}{2} \right) \right\rangle^1 + \left(\frac{P}{2} \right) \langle x - \ell \rangle^1 \tag{5.207}$$

Hence, M^2 is

$$M^2 = P^2 \left[\left(\frac{1}{4} \right) \left(\langle x - 0 \rangle^1 \right)^2 + \left(\left\langle x - \left(\frac{\ell}{2} \right) \right\rangle^1 \right)^2 + \left(\frac{1}{4} \right) \left(\langle x - \ell \rangle^1 \right)^2 - \langle x - 0 \rangle^1 \left\langle x - \left(\frac{\ell}{2} \right) \right\rangle^1 \right.$$
$$\left. - \left\langle x - \left(\frac{\ell}{2} \right) \right\rangle^1 \langle x - \ell \rangle^1 + \left(\frac{1}{4} \right) \langle x - 0 \rangle^1 \langle x - \ell \rangle^1 \right] \tag{5.208}$$

The strain energy is then (Equation 5.201)

$$U = \int_0^\ell \left(\frac{M^2}{2EI} \right) dx = \left(\frac{P^2}{2EI} \right) \left[\int_0^\ell \left(\frac{x^2}{4} \right) dx + \int_{\ell/2}^\ell \left(x - \frac{\ell}{2} \right)^2 dx - \int_{\ell/2}^\ell x \left(x - \frac{\ell}{2} \right)^2 dx \right] = \frac{P^2 \ell^3}{96EI} \tag{5.209}$$

Thus, from Equation 5.203, the center displacement δ may be expressed as

$$\delta = \frac{P\ell^3}{48EI} \tag{5.210}$$

(Compare with Equation 5.171.)

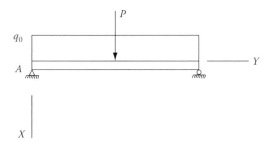

FIGURE 5.67   Simply supported beam with a uniform load and a concentrated center load.

Finally, consider a simply supported beam with a uniformly distributed load and a concentrated center load as in Figure 5.67. We can model this loading by simply superposing the uniform load, the concentrated center load, and the support reactions. That is,

$$q(x) = -R_A\langle x - 0\rangle^{-1} + q_0\langle x - 0\rangle^0 + P\left\langle x - \left(\frac{\ell}{2}\right)\right\rangle^{-1} - R_B\langle x - \ell\rangle^{-1} \tag{5.211}$$

where
R_A and R_B are the support reactions
ℓ is the beam length

From a free-body diagram, with the symmetric loading and end support conditions, we immediately see that R_A and R_B are

$$R_A = R_B = \frac{q_0\ell + P}{2} \tag{5.212}$$

From the governing equations (see Section 5.9.5 and specifically Equation 5.113), we see that $q(x)$ is $EI\, d^ry/dx^4$, and thus by integrating $q(x)$, we can obtain expressions for the shear V and the bending moment M. Specifically, we have

$$EI d^3 y dx^3 = -V$$

$$= -R_A\langle x - 0\rangle^0 + q_0\langle x - 0\rangle^1 + P\left\langle x - \left(\frac{\ell}{2}\right)\right\rangle^0 - R_B\langle x - \ell\rangle^0 + c_1 \tag{5.213}$$

and

$$EI \frac{d^2 y}{dx^2} = -M = -R_A\langle x - 0\rangle^1 + \frac{q_0\langle x - 0\rangle^2}{2} + P\left\langle x - \left(\frac{\ell}{2}\right)\right\rangle^1 - R_B\langle x - \ell\rangle^1 + c_1 x + c_2 \tag{5.214}$$

But since the bending moment M is zero at a simple (pin) support, we have

$$\frac{d^2 y}{dx^2} = 0 \quad \text{at} \quad x = 0 \quad \text{and} \quad x = \ell \tag{5.215}$$

From Equation 5.214, we then have

$$c_1 = c_2 = 0 \tag{5.216}$$

Thus, the bending moment M and its square are

$$M = R_A\langle x-0\rangle^1 - \frac{q_0\langle x-0\rangle^2}{2} - P\left\langle x-\left(\frac{\ell}{2}\right)\right\rangle^1 + R_B\langle x-\ell\rangle^1 \tag{5.217}$$

and

$$M^2 = R_A^2(\langle x-0\rangle^1)^2 + q_0^2\left(\frac{\langle x-0\rangle^2}{2}\right)^2 + P^2\left(\left\langle x-\frac{\ell}{2}\right\rangle^1\right)^2 + R_B^2(\langle x-\ell\rangle^1)^2 - R_A q_0\langle x-0\rangle^1\langle x-0\rangle^2$$

$$- 2R_A P\langle x-0\rangle^1\left\langle x-\frac{\ell}{2}\right\rangle^1 + 2R_A R_B\langle x-0\rangle^1\langle x-\ell\rangle^1 + q_0 P\langle x-0\rangle^2\left\langle x-\frac{\ell}{2}\right\rangle$$

$$- q_0 R_B\langle x-0\rangle^2\langle x-\ell\rangle^1 - 2PR_B\left\langle x-\frac{\ell}{2}\right\rangle^1\langle x-\ell\rangle^1 \tag{5.218}$$

In view of its use in determining the strain energy as in Equation 5.201, the integral of M^2 along the beam is (using the properties of the singularity functions)

$$\int_0^\ell M^2 dx = R_A^2\int_0^\ell x^2 dx + \frac{q_0^2}{4}\int_0^\ell x^4 dx + P^2\int_{\ell/2}^\ell \left(x-\frac{\ell}{2}\right)^2 dx + R_B^2\int_\ell^\ell (x-\ell)^2 dx$$

$$- R_A q_0\int_0^\ell x^3 dx - 2R_A P\int_{\ell/2}^\ell x\left(x-\frac{\ell}{2}\right) dx + 2R_A R_B\int_\ell^\ell x(x-\ell)^2 dx + q_0 P\int_{\ell/2}^\ell x^2\left(x-\frac{\ell}{2}\right) dx$$

$$- q_0 R_B\int_0^\ell x^2(x-\ell)^2 dx - 2PR_B\int_\ell^\ell \left(x-\frac{\ell}{2}\right)(x-\ell) dx \tag{5.219}$$

By evaluating these integrals, we obtain

$$\int_0^\ell M^2 dx = \frac{R_A^2\ell^3}{3} + \frac{q_0^2\ell^5}{20} + \frac{P^2\ell^3}{24} - \frac{5R_A P\ell^3}{24} + \frac{Pq_0\ell^4}{12}$$

$$= \frac{q_0\ell^5}{120} + \frac{5q_0 P\ell^4}{192} + \frac{P^2\ell^3}{48} \tag{5.220}$$

where we have used Equation 5.212 to express R_A in terms of q_0 and P. (Observe that the 4th, 7th, 9th, and 10th integrals of Equation 5.219 are zero.)

From Equation 5.211, the strain energy U for the beam of Figure 5.67 may be expressed as

$$U = \frac{q_0 \ell^5}{240EI} + \frac{5q_0 P \ell^4}{384EI} + \frac{P^2 \ell^3}{96EI} \tag{5.221}$$

From Equation 5.203, the displacement δ at the center of the beam may be expressed as

$$\delta = \frac{\partial U}{\partial P} = \frac{5q_0 \ell^4}{384EI} + \frac{P^2 \ell^3}{48EI} \tag{5.222}$$

This displacement result is seen to be the superposition of the displacement results of Equations 5.171 and 5.184 for a simply supported beam with a concentrated load (Equation 5.171) and with a uniformly distributed load (Equation 5.184). Observe in Equation 5.222 that if P is zero, then we have the center displacement for a uniformly loaded beam. This shows that if we want to determine the displacement at any point Q of a loaded structure, we need simply apply a force P at Q, evaluate the stain energy, use Castigliano's theorem, and then let P be zero. This is sometimes known as the dummy force method.

PROBLEMS

Section 5.1

P5.1.1 Consider a rectangular plate $ABCD$ with a concentrated force applied in the middle, perpendicular to the plate as in the following figure. If the force magnitude is 10,000 lb and if the plate edges have lengths of 16 and 10 in., as shown, what is the average stress on the plate?

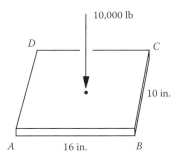

P5.1.2 See Problem P5.1.1. Consider again the plate $ABCD$ of the figure for Problem P5.1.1 with the same loading. Envision a rectangular region $EFGH$ of the plate with dimensions half that of the overall plate and with sides aligned to the plate as in the following figure. What is the average stress on region $EFGH$?

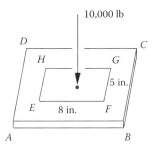

P5.1.3 See Problems P5.1.1 and P5.1.2 and their solutions. Which of the two solutions is more likely to represent the actual physical stress in the plate?

P5.1.4 See Problems P5.1.1 through P5.1.3. Envision a still smaller rectangular region having say half the dimensions of *EFGH* and configured similarly about the 10,000 lb load. What is the average stress on this smaller region? Is this stress more or less representative of the actual physical stress in the plate? Discuss the dilemma of using decreasingly smaller regions about the concentrated load to model the physical stress in the plate.

P5.1.5 Discuss the implications of using equivalent force system to calculate stresses.

P5.1.6 Review Sections 3.6 through 3.10. Suppose a stress matrix Σ is

$$\Sigma = 10,000 \begin{vmatrix} 4 & \sqrt{3}/2 & 1/2 \\ \sqrt{3}/2 & 7/2 & \sqrt{3}/2 \\ 1/2 & \sqrt{3}/2 & 5/2 \end{vmatrix} \text{ psi}$$

In view of Equation 3.142 and the subsequent analysis, what is the value of the maximum stress? What is the direction of the normal to the surface where the maximum stress occurs?

Section 5.2

P5.2.1 Consider Equation 5.21 of the strain matrix E. Verify that E is symmetric.

P5.2.2 Suppose a strain matrix E is

$$E = 10^{-3} \begin{bmatrix} 4 & \sqrt{3}/2 & 1/2 \\ \sqrt{3}/2 & 7/2 & \sqrt{3}/2 \\ 1/2 & \sqrt{3}/2 & 5/2 \end{bmatrix}$$

In view of Equation 3.142 and the subsequent analysis, what are the directions corresponding to zero shear strain?

Section 5.3

P5.3.1 Review Section 3.9. Suppose a stress matrix Σ has the values

$$\Sigma = \begin{bmatrix} 2.700 & -0.4500 & 0.7794 \\ -0.4500 & 2.9250 & 1.2990 \\ 0.7794 & 1.2990 & 2.7750 \end{bmatrix} 10^4 \text{ psi}$$

Determine the principal stresses and the corresponding principal directions.

P5.3.2 Suppose the principal stresses σ_a, σ_b, and σ_c at a point in the interior of a body are 60, 75, and 95 MPa (1 MPa = 10^6 Pa = 10^6 N/m²), respectively. Determine the normal stress on a plane whose normal unit vector **k** is

$$\mathbf{k} = \left(\frac{3}{13} \right) \mathbf{n}_a + \left(\frac{4}{13} \right) \mathbf{n}_b + \left(\frac{12}{13} \right) \mathbf{n}_c$$

where $\mathbf{n}_a$, $\mathbf{n}_b$, and $\mathbf{n}_c$ are along the respective principal directions of the stress.

P5.3.3 See Problem P5.3.2. Let **i** and **j** be unit vectors parallel to the plane whose normal unit vector is **k**. Find the shear stresses on the plane in the directions of **i** and **j** if **i** and **j** are

$$\mathbf{i} = -\left(\frac{4}{5}\right)\mathbf{n}_a + \left(\frac{3}{5}\right)\mathbf{n}_b$$

and

$$\mathbf{j} = -\left(\frac{36}{65}\right)\mathbf{n}_a - \left(\frac{48}{65}\right)\mathbf{n}_b + \left(\frac{5}{13}\right)\mathbf{n}_c$$

Section 5.4

P5.4.1 Suppose the stress matrix Σ at an interior point P of a body is

$$\Sigma = \begin{bmatrix} 8 & 5 & 0 \\ 5 & 10 & 0 \\ 0 & 0 & 4 \end{bmatrix} 10^3 \text{ psi}$$

   a. Determine the principal stresses at P.
   b. Determine the directions at P of the planes where the principal stresses occur.

P5.4.2 See Problem P5.4.1. Construct a Mohr's circle for the data of Problem P5.4.2.

P5.4.3 Repeat Problems P5.4.1 and P5.4.2 if the stress matrix is

$$\Sigma = \begin{bmatrix} -15 & 10 & 0 \\ 10 & 20 & 0 \\ 0 & 0 & 5 \end{bmatrix} \text{ MPa}$$

P5.4.4 Suppose the principal stresses σ_1, σ_2, σ_3 at an interior point P of a body are

$$\sigma_1 = 15 \text{ kips}, \quad \sigma_2 = 12 \text{ kips}, \quad \sigma_3 = 0$$

where 1 kip = 1000 psi. Determine the stresses at P in directions inclined at $30°$ to the directions of the principal stresses.

P5.4.5 See Problem P5.4.4. Repeat Problem P5.4.4 if the principal stresses at P are

$$\sigma_1 = -10 \text{ MPa}, \quad \sigma_2 = 30 \text{ MPa}, \quad \sigma_3 = 0$$

Section 5.5

P5.5.1 Suppose that instead of a linear stress–strain relation

$$\sigma = E\varepsilon \tag{a}$$

(as in Equation 5.65), we have a nonlinear relation as

$$\sigma = E \sin\varepsilon \tag{b}$$

(with ε in radians).
   Compare the differences in stress values provided by (a) and (b) for small values of ε.

P5.5.2 See Problem P5.5.1. Many structural materials are said to have "yielded" when after the material is loaded and then unloaded, there is a residual strain of 0.001 or greater. Another concept of "yield" is that stress level where the stress–strain relation of the material becomes measurably nonlinear.

Combining these two concepts, we have a depiction of a yield point as in the following figure.

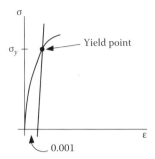

If a stress–strain relation is approximated as Equation (b) of Problem P5.5.1, at what value of ε does the yield stress σ_y occur?

P5.5.3 Verify Equation 5.70.

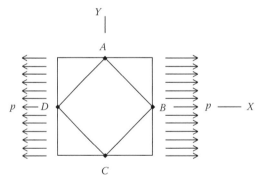

Suggestion: Consider a square plate being stretched by a uniformly distributed load p as in the figure above. As the plate gets longer in the X direction, it gets shorter in the Y direction as modeled by the transverse contraction ratio of Equation 5.69. This lengthening/shortening of the plate distorts the initially square region $ABCD$ into a rhombus. Next, consider a free-body diagram of the rhombus $ABCD$, and evaluate the shear strain γ of $ABCD$. Finally, evaluate the shear stress τ on $ABCD$ and relate it to the tensile stress σ on the overall plate.

P5.5.4 Long bones are occasionally modeled as tapered cylinders or more specifically as frustrums of cones. Consider a conical segment with length P and end cross-sectional areas A_O and A_P, as in the following figure.

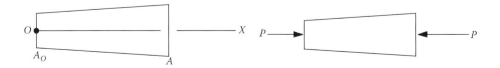

Suppose the segment is subjected to an axial compressive load P as represented in the right-hand side figure shown earlier. Suppose that as a result of the loading, the segment is shortened by an amount δ so that its length is then $\ell - \delta$.

Assuming a linear stress–strain relation as in Equation 5.65, and with a strain–displacement relation as in the first expression of Equation 5.19, determine the relation between P and δ. (*Suggestion*: Express the cross-sectional area A of the segment as a function of the axial coordinate x as

$$A = A_O + (A_1 - A_O)\left(\frac{x}{\ell}\right)$$

and then integrate a differential form of Equation 5.66.)

Section 5.6

P5.6.1 Verify Equation 5.73 by solving Equation 5.71 for σ_{xx}, σ_{yy}, and σ_{zz}.

P5.6.2 By using numerical indices as subscripts, combine Equations 5.73 and 5.74 into a single equation, and thus, verify Equations 5.75 and 5.76.

P5.6.3 Verify Equation 5.77.

P5.6.4 Verify the third expression in Equation 5.80.

Section 5.7

P5.7.1 Using Equations 5.75 and 5.86, verify Equation 5.87.

P5.7.2 Using Equations 5.22 and 5.87, verify Equation 5.88.

Section 5.8

P5.8.1 Suppose with cylindrical coordinates that there is no variation in the θ direction. That is, suppose there is a circular symmetry. Determine the reduced forms of Equations 5.91 through 5.93.

P5.8.2 Suppose with spherical coordinates that there is no variation in the θ and ϕ directions. That is, suppose there is spherical symmetry. Determine the reduced forms of Equations 5.95 through 5.97.

Section 5.9

P5.9.1 The approximation of Equation 5.101 by Equation 5.102 is based upon the assumption that the slope of the centerline of a deflected beam is "small." What is the slope (in degrees) so that the error is less than (a) 1% and (b) 10%?

P5.9.2 Determine the second moment of area I for (a) a circular cross section and (b) an annular region. (*Suggestion*: Consult a Strength of Materials book and/or mechanical handbook.)

P5.9.3 See the solution to Problem P5.9.2. The structural property of long bones may be modeled by hollow (annular cross section) cylinders. Consider a comparison of the second moments of area of the cross sections of solid and hollow circular cylinders: Specifically, suppose a solid circular cylinder has radius r and that an annular circular cylinder has inner and outer radii r_i and r_o as in the following figure.

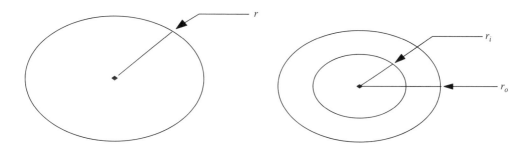

What should be the values of r_i and r_o so that the respective cross-sectional areas of the two cylinders are the same, but such that the second moment of area of the hollow cylinder I_{hollow} is *twice* that of the solid cylinder I_{solid}? (That is, $I_{hollow} = 2I_{solid}$.)

Section 5.10

P5.10.1 Verify Equation 5.123 by using Equations 5.121 and 5.122.

P5.10.2 Verify Equation 5.124.

Section 5.11

P5.11.1 In view of the discussion of Equations 5.125 and 5.126 in Section 5.11, discuss the application to finger grasping.

Section 5.12

P5.12.1 The singularity function $\langle x - a \rangle^{-1}$ is sometimes called the "unit impulse function" and also the "Dirac delta function" (often designated as $\delta(x - a)$). To intuitively conceptualize the function, some have introduced a function $\Delta(x - a, \varepsilon)$ defined geometrically as a "tall" rectangle at $x = a$ as in the following figure, where ε is "small."*

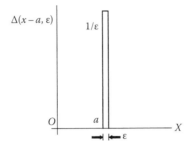

Show that with this definition of Δ

a. $\displaystyle\lim_{\varepsilon \to 0} \Delta(x - a, \varepsilon) = \langle x - a \rangle^{-1}$

b. $\displaystyle\int_{-\infty}^{b} \Delta(x - a, \varepsilon)\mathrm{d}x = \begin{cases} 0 & b \le a \\ m & b = a + m\varepsilon \quad 0 \le m \le 1 \\ 1 & b \ge a + \varepsilon \end{cases}$

c. $\displaystyle\lim_{\sigma \to 0}\int_{-\infty}^{b} f(x)\Delta(x - a, \varepsilon)\mathrm{d}x = \begin{cases} 0 & b \le a \\ f(a) & b > a \end{cases}$

P5.12.2 See Problem P5.12.1. Observe that the function $\Delta(x - a, \varepsilon)$ is not continuous at $x = a$ and at $x = a + \varepsilon$. In Ref. [28], Butkov presents some functions with similar characteristics to the Δ function, but without the points of discontinuity. One of these functions, in the form of a sequence, is

$$\phi_n(x - a) = \frac{n}{\pi}\frac{1}{1 + n^2(x - a)^2}$$

where n is a positive integer.

Compare the properties of $\Delta(x - a, \varepsilon)$ and $\phi_n(x)$ for small ε and large n.

---

* Recall that "small" ε simply means that when approximations are made assuming ε is small, those approximations become increasingly accurate the smaller ε becomes.

P5.12.3 A 2 m long, simply supported beam is loaded over half its span by a uniformly distributed weight of 750 N/m, as represented in the following figure. Determine the magnitudes and directions of the support reactions at A and B.

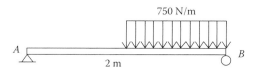

P5.12.4 See Problem P5.12.3. Construct a free-body diagram of the beam of Problem P5.12.3, including the reaction forces. Express the loading on the beam in terms of singularity functions.

P5.12.5 Develop a simply supported beam model of a 50 percentile male doing a push-up.

P5.12.6 See Problem P5.12.5. Construct a free-body diagram of the beam model of the person doing a push-up. Express the loading on the beam model in terms of singularity functions.

Section 5.13

P5.13.1 Consider the uniformly loaded cantilever beam of the following figure. Following the procedures listed in the beginning of Section 5.13 and as illustrated in the examples, determine the shear force, the bending moment, and the displacement along the beam. (Compare the result with that of Table 5.2, Item 2.)

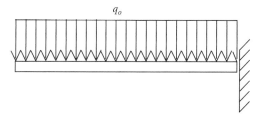

P5.13.2 Consider the uniformly loaded, simply supported beam of the following figure. Following the procedures listed in the beginning of Section 5.13 and as illustrated in the example, determine the shear force, the bending moment, and the displacement along the beam.

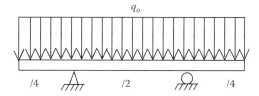

P5.13.3 Consider the model of a diving board and its supports as in the following figure where W represents the weight of a diver standing on the end. Determine the shear force, the bending moment, and the displacement along the board.

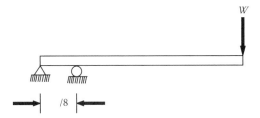

Section 5.14

P5.14.1 Using the listings of Table 5.2 and the principle of superposition, determine the location and values of the maximum bending moment and the maximum displacement for the end-loaded and uniformly loaded cantilever beam of the following figure.

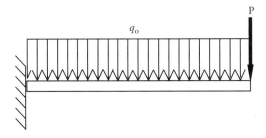

P5.14.2 Consider a simply supported, uniformly loaded beam with a concentrated midspan downward force as in the followiing figure.

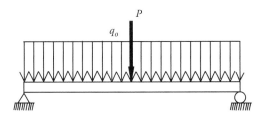

Determine the locations and values of the maximum bending, moment, and displacement of the beam.

P5.14.3 See Problem P5.14.2. What is the magnitude of the upward force P so that the displacement at midspan is zero?

Section 5.15

P5.15.1 Develop an analogous expression to Equation 5.191 for a simply supported beam with a concentrated midspan load, having a *circular* cross section. (See Equation 5.123.)

Section 5.16

P5.16.1 See Problem P5.9.3. The structural property of long bones may be modeled by hollow (annular cross section) cylinders. Consider a comparison of the second polar moments of area of the cross sections of solid and hollow circular cylinders: Specifically, suppose a solid cylinder has radius r and that an annular circular cylinder has inner and outer radii r_i and r_o as in the following figure.

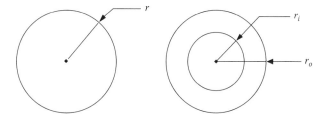

What should be the values of r_i and r_o so that the respective cross-sectional areas of the two cylinders are the same, but such that the second polar moment of area of the hollow cylinder J_{hollow} is *twice* that of the solid cylinder J_{solid}? (That is, $J_{hollow} = 2J_{solid}$.)

Section 5.17

P5.17.1 Verify Equation 5.197 by expressing the second polar moment of area of a hollow circular cylinder in terms of the wall thickness t and then neglecting higher-order terms in t. (That is, let $t = r_o - r_i$ and assume t to be "small.")

Section 5.18

P5.18.1 Consider a vertical bar, or column, whose upper end is just beneath a suspended block as represented in the following left-hand side figure. If the block is initially held by a cable, which is then cut as in the following right-hand side figure, determine the maximum compressive stress in the column.

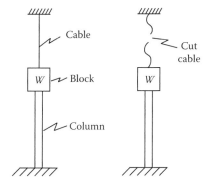

Let the weight of the block be W and let the column have length ℓ, cross-sectional area A, and elastic modulus E. (Neglect the weight of the column.)

Suggestion: Let δ_{max} be the maximum compression (shortening) of the column. Then the change in potential energy of the block, at the maximum downward displacement, is simply $-W\delta_{max}$. Let this energy loss of the block be absorbed by the column as strain energy U expressed as $P_{max}^2 \ell/2AE$ (see Equation 5.200, where P_{max} is now the maximum compressive force in the column). Finally, observing that δ_{max} is $P_{max}\ell/AE$, solve for the maximum stress σ_{max} as P_{max}/A.

P5.18.2 See Problem P5.18.1. Suppose that the column of Problem P5.18.1 is now horizontal and that it is supported on one end and struck on the other end by the block moving toward the column with speed V, as in the following figure. As before, determine the maximum compressive stress σ_{max} in the column.

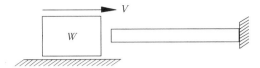

Suggestion: Let K be the kinetic energy of the block given by $(1/2)mV^2$ where m is the block mass W/g (g is gravity acceleration). Then equate the maximum strain energy (corresponding to the maximum compressive stress) with the kinetic energy of the block as it strikes the column.

P5.18.3 See Problem P5.18.2. Suppose the horizontal column and the striking block of the figure for Problem P5.18.2 are used as a model of a baseball catcher catching a high-speed fast ball. That is, suppose the column represents a catcher's forearm, and the moving block is the baseball. Assuming that the force in the forearm is borne primarily by the forearm bones (the radius and ulna), determine the maximum compressive bone stress for the following conditions:

$$V = 100 \text{ mph} = 146.7 \text{ ft/s}$$

$$W = 5.25 \text{ oz} = 0.328 \text{ lb}$$

$$\ell = 12 \text{ in.} = 1.0 \text{ ft}$$

$$A = 0.5 \text{ in}^2 = 3.47 \times 10^{-3} \text{ ft}^2$$

$$E = 29 \times 10^5 \text{ psi} = 4.176 \times 10^8 \text{ lb/ft}^2$$

$$g = 386 \text{ in/s}^2 = 32.2 \text{ ft/s}^2$$

P5.18.4 See Problem P5.18.3. List reasons why the system of the figure for Problem P5.18.2 is *not* a good model for a baseball catcher.

P5.18.5 See Problem P5.18.1. Analogous to the system of the figure for Problem P5.18.1, consider a cantilever beam where a block is suddenly released upon the free end as represented in the following figure. In this case, the stress created is a bending stress with the maximum value occurring at the support.

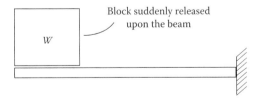

Knowing the weight W of the block and the beam geometry, determine the maximum stress.

Suggestion: Equate the potential energy change of the block with the strain energy U given by Equation 5.205.

P5.18.6 See Problems P5.18.2 and P5.18.5. Analogous to the system of the figure for Problem P5.18.1 and the dynamics of Problem P5.18.2, consider a cantilever beam being struck horizontally by a block moving with speed V as represented in the following figure.

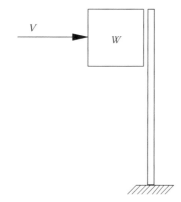

Again, knowing the geometric parameters of the beam, and neglecting the beam mass, determine the maximum bending stress in the beam due to the impact from the block.

Suggestion: As in Problem P5.18.2, equate the kinetic energy of the block with the strain energy of an end-loaded cantilever given by Equation 5.205.

P5.18.7 Discuss the use of the system of the figure for Problem P5.18.6 as a model of a baseball infielder catching a baseball.

REFERENCES

1. F. P. Beer and E. R. Johnston, *Mechanics of Materials*, 2nd edn., McGraw-Hill, New York, 1992.
2. E. P. Popov, *Mechanics of Materials*, 2nd edn., Prentice Hall, Englewood Cliffs, NJ, 1976.
3. F. L. Singer, *Strength of Materials*, 2nd edn., Harper & Row, New York, 1962.
4. J. P. Den Hartog, *Advanced Strength of Materials*, McGraw-Hill, New York, 1952.
5. R. W. Little, *Elasticity*, Prentice Hall, Englewood Cliffs, NJ, 1973.
6. I. S. Sokolnikoff, *Mathematical Theory of Elasticity*, 2nd edn., McGraw-Hill, New York, 1956.
7. A. E. H. Love, *A Treatise on the Mathematical Theory of Elasticity*, 4th edn., Dover, New York, 1944.
8. K. L. Johnson, *Contact Mechanics*, Cambridge University Press, Cambridge, U.K., 1985.
9. E. E. Sechler, *Elasticity in Engineering*, Dover, New York, 1968.
10. W. C. Young, *Roark's Formulas for Stress and Strain*, 6th edn., McGraw-Hill, New York, 1989.
11. H. Reismann and P. S. Pawlik, *Elasticity Theory and Application*, John Wiley & Sons, New York, 1980.
12. S. Timoshenko and J. N. Goodier, *Theory of Elasticity*, McGraw-Hill, New York, 1951.
13. W. D. Pilkey and O. H. Pilkey, *Mechanics of Solids*, Quantum Publishers, New York, 1974.
14. E. Volterra and J. H. Gaines, *Advanced Strength of Materials*, Prentice Hall, Englewood Cliffs, NJ, 1971.
15. S. P. Timoshenko, *Strength of Materials, Part I, Elementary Theory and Problems*, D. Van Nostrand, New York, 1955.
16. S. P. Timoshenko, *Strength of Materials, Part II, Advanced Theory and Problems*, D. Van Nostrand, New York, 1954.
17. R. D. Cook and W. C. Young, *Advanced Mechanics of Materials*, Prentice Hall, Upper Saddle River, NJ, 1985.
18. L. Brand, *Vector and Tensor Analysis*, Wiley, New York, 1947.
19. B. J. Hamrock, B. O. Jacobson, and S. R. Schmid, *Fundamentals of Machine Elements*, McGraw-Hill, New York, 1999.
20. W. F. Hughes and E. W. Gaylord, *Basic Equations of Engineering Science*, Schaum, New York, 1964.
21. S. M. Selby (Ed.), *Standard Mathematical Tables*, The Chemical Rubber Co. (CRC Press), Cleveland, OH, 1972.
22. R. C. Juvinall and K. M. Marshek, *Fundamentals of Machine Component Design*, 2nd edn., Wiley, New York, 1991.
23. J. E. Shigley and C. R. Mischlee, *Mechanical Engineering Design*, 5th edn., McGraw-Hill, New York, 1989.
24. M. F. Spotts, *Design of Machine Elements*, 3rd edn., Prentice Hall, Englewood Cliffs, NJ, 1961.
25. T. R. Kane, *Analytical Elements of Mechanics*, Vol. 2, Academic Press, New York, 1961, p. 40.
26. H. Yamada, *Strength of Biological Materials*, Williams & Wilkins, Baltimore, MD, 1970.
27. W. C. Young, *Roark's Formulas for Stress and Strain*, McGraw-Hill, New York, 1989.
28. E. Butkov, *Mathematical Physics*, Addison Wesley, Reading, MA, 1968, p. 222.

6 Methods of Analysis IV
Modeling of Biosystems

When compared with fabricated mechanical systems, biosystems are extremely complex. The complexity stems from both the geometric and the material properties of the systems. Consider the human frame: Aside from gross symmetry about the sagittal plane (see Chapter 2), there is little if any geometric simplicity. Even the long bones are tapered with noncircular cross sections. The material properties are even more irregular with little or no linearity, homogeneity, or isotropy.

This complexity has created extensive difficulties for analysts and modelers. Indeed, until recently, only very crude models of biosystems have been available, and even with these crude models, the analysis has not been simple. However, with the advent and continuing development of computer hardware and software and with associated advances in numerical procedures, it is now possible to develop and study advanced, comprehensive, and thus more realistic models of biosystems. The objective of this chapter is to consider the modeling procedures.

Our approach is to focus on gross modeling of entire systems, with a vision toward dynamics analyses. This is opposed to a microapproach, which would include modeling at the cellular level. We will also take a finite continuum approach and represent the biosystem (e.g., the human body) as a lumped mass system.

6.1 MULTIBODY (LUMPED MASS) SYSTEMS

As discussed in Chapter 2, we can obtain a gross modeling of a biosystem (the human body) by thinking of the system as a collection of connected bodies—that is, as a multibody system simulating the frame of the biosystem. For a human body, we can globally represent the system as a series of bodies representing the arms, the legs, the torso, the neck, and the head (Figure 6.1). For analysis purposes, the model need not be drawn to scale.

If we regard this system as a multibody system, we can use procedures, which have been developed for studying such systems [1–5]. To briefly review the use of multibody methods, consider the system of Figure 6.2, which is intended to represent a collection of rigid bodies connected by spherical joints and without closed loops. Such a system is often called an open-chain or open-tree system. (More elaborate systems would include those with flexible bodies where adjoining bodies may both translate and rotate relative to one another and where closed loops may occur.)

By inspection of the system depicted in Figure 6.1, we readily recognize it as a multibody system. Then as noted earlier by regarding bio (human) models as multibody systems, we can employ established procedures for kinematic and dynamic analyses as we will do in subsequent chapters.

6.2 LOWER-BODY ARRAYS

Let the system of Figure 6.2 be numbered or labeled as in Figure 6.3 where for the numbering, we arbitrarily select a body, say one of the larger bodies, as a reference body of the system and call it B_1 or simply 1. Next, think of the other bodies of the system (or tree) as being in branches stemming away from B_1. Let these bodies be numbered in ascending progression away from B_1 toward the extremities as in Figure 6.3. Observe that there are several ways we could do this, as illustrated in Figures 6.4 and 6.5. Nevertheless, when the bodies are numbered sequentially through the branches

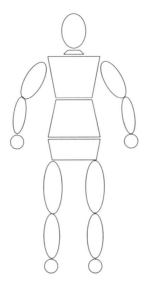

FIGURE 6.1   A human body model.

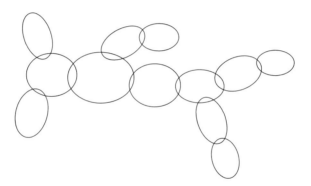

FIGURE 6.2   An open-chain multibody system.

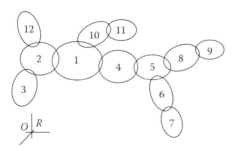

FIGURE 6.3   A numbered (labeled) multibody system (see Figure 6.2).

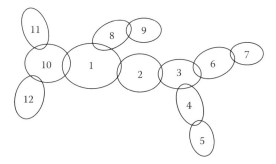

FIGURE 6.4 A second numbering of the system of Figure 6.2.

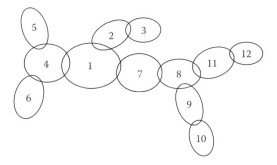

FIGURE 6.5 A third numbering of the system of Figure 6.2.

TABLE 6.1

Lower-Body Array for the System of Figure 6.3

K	1	2	3	4	5	6	7	8	9	10	11	12
$L(K)$	0	1	2	1	4	5	6	5	8	1	10	2

---

toward the extremities, a principal feature of the numbering system is that each body has a unique adjoining lower numbered body. For example, for the numbering sequence of Figure 6.3, if we assign the number 0 to be an inertial reference frame R, we can form an array $L(K)$ listing the adjoining lower numbered bodies for each body B_k (or K) of the system. Specifically, for the system of Figure 6.3, Table 6.1 presents the array $L(K)$ called the lower-body array.

Observe in Figure 6.3 that although each body has a unique adjoining lower numbered body (Table 6.1), some bodies, such as 1, 2, and 5, have more than one adjoining higher numbered body. Observe further in the array $L(K)$ of Table 6.1 some body numbers (3, 7, 9, 11, and 12) do not appear, some body numbers (4, 6, 8, and 10) appear only once in the array, and some body numbers (1, 2, and 5) appear more than once in the array. The non-appearing numbers (3, 7, 9, 11, and 12) are the numbers of the extremity bodies in Figure 6.3. The numbers appearing more than once (1, 2, and 5) are numbers of the branching bodies in Figure 6.3, and hence, they have more than one adjacent higher numbered body. The numbers in $L(K)$ appearing once and only once are the intermediate or connecting bodies of the system of Figure 6.3.

Finally, observe that with all this information contained in $L(K)$, Table 6.1 is seen to be equivalent to Figure 6.3 in terms of the connection configuration of the bodies. $L(K)$ is then like a genetic code for the system.

$L(K)$ may be viewed as an operator, $L(.)$, mapping or transforming the array K into $L(K)$. As such, $L(.)$ can operate on the array $L(K)$ itself forming $L(L(K))$ or $L^2(K)$—the lower-body array of the lower-body array. Specifically, for the system of Figure 6.3, $L^2(K)$ is

$$L^2(K) = (0, 0, 1, 0, 1, 4, 5, 4, 5, 0, 1, 1) \tag{6.1}$$

where we have defined $L(0)$ to be 0.

In like manner, we can construct even higher-order lower-body arrays $L^3(K)$, ..., $L^n(K)$. Table 6.2 lists these higher-order arrays for the system of Figure 6.3 up to the array $L^5(K)$ where all 0s appear.

Observe the columns in Table 6.2. The sequence of numbers in a given column (say column j) represents the numbers of bodies in the branches of the system. Table 6.2 can be generated by repeated use of the $L(K)$ operator; then $L(K)$ may be used to develop algorithms not only to develop the column of Table 6.2 but also to develop the kinematics of the multibody system.

To illustrate the development of such algorithms, consider the fundamental concept of relative velocity: Let R be a reference frame and let P and Q be points moving in R as in Figure 6.6. Then the velocities of P and Q in R (written as $^R\mathbf{V}^P$ and $^R\mathbf{V}^Q$) are often regarded as absolute velocities with respect to R. If R is fixed or understood, then the velocities of P and Q may be written simply as $\mathbf{V}^P$ and $\mathbf{V}^Q$. The difference in the velocities of P and Q is called the relative velocity of P and Q, written as $\mathbf{V}^{P/Q}$. That is,

$$\mathbf{V}^{P/Q} = \mathbf{V}^P - \mathbf{V}^Q \tag{6.2}$$

Then we immediately have the expression

$$\mathbf{V}^P = \mathbf{V}^Q + \mathbf{V}^{P/Q} = \mathbf{V}^{P/Q} + \mathbf{V}^Q \tag{6.3}$$

TABLE 6.2
Higher-Order Lower-Body Arrays for the System of Figure 6.3

K	1	2	3	4	5	6	7	8	9	10	11	12
$L(K)$	0	1	2	1	4	5	6	5	8	1	10	2
$L^2(K)$	0	0	1	0	1	4	5	4	5	0	1	1
$L^3(K)$	0	0	0	0	0	1	4	1	4	0	0	0
$L^4(K)$	0	0	0	0	0	0	1	0	1	0	0	0
$L^5(K)$	0	0	0	0	0	0	0	0	0	0	0	0

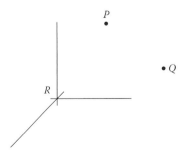

FIGURE 6.6   Points P and Q moving in a reference frame R.

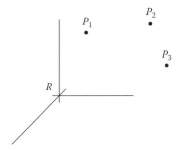

FIGURE 6.7   Three points moving in a reference frame R.

Suppose that we have three points P_1, P_2, and P_3 moving in R as in Figure 6.7. Then as a generalization of Equation 6.3, we have

$$\mathbf{V}^{P_3} = \mathbf{V}^{P_3/P_2} + \mathbf{V}^{P_2/P_1} + \mathbf{V}^{P_1} \tag{6.4}$$

As a further generalization of Equations 6.3 and 6.4, let P_1, P_2, P_3, ..., P_n be n points moving in R. Then Equation 6.4 evolves to

$$\mathbf{V}^{P_n} = \mathbf{V}^{P_n/P_{n-1}} + \mathbf{V}^{P_{n-1}/P_{n-2}} + \cdots + \mathbf{V}^{P_3/P_2} + \mathbf{V}^{P_2/P_1} + \mathbf{V}^{P_1} \tag{6.5}$$

Consider next a set of bodies, or a multibody system, moving in frame R as in Figure 6.8. If the bodies are labeled as $B_1,\ldots, B_N$, we have an expression analogous to Equation 6.5 relating the relative and absolute angular velocities of the bodies. That is,

$$^R\omega^{B_N} = \omega^{B_N} = {}^{B_{N-1}}\omega^{B_N} + \cdots + {}^{B_J}\omega^{B_K} + \cdots + {}^{B_1}\omega^{B_2} + \omega^{B_1} \tag{6.6}$$

where bodies B_J and B_K are bodies in the branches leading out to B_1. Notationally, ${}^{B_J}\omega^{B_K}$ is the angular velocity of B_K in B_J.

Finally, consider the system of Figure 6.3. Consider specifically body 9, or B_9. The angular velocity of B_9 in reference frame R is

$$^R\omega^{B_9} = {}^{B_8}\omega^{B_9} + {}^{B_5}\omega^{B_8} + {}^{B_4}\omega^{B_5} + {}^{B_1}\omega^{B_4} + {}^R\omega^{B_1} \tag{6.7}$$

A convenient notation for absolute angular velocity, such as ${}^R\omega^{B_k}$, is simply

$$^R\omega^{B_k} = {}^D\omega_k \tag{6.8}$$

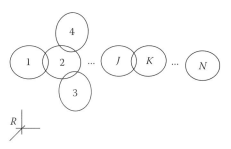

FIGURE 6.8   A multibody system moving in a reference frame R.

Correspondingly, a convenient notation for relative angular velocity of adjoining bodies such as B_j and B_k is

$$^{B_j}\omega^{B_k} = \hat{\omega}_k \tag{6.9}$$

where the overhat designates relative angular velocity. In this notation, Equation 6.7 becomes

$$\omega_9 = \hat{\omega}_9 + \hat{\omega}_8 + \hat{\omega}_5 + \hat{\omega}_4 + \hat{\omega}_1 \tag{6.10}$$

Observe the subscripts on the right-hand side of Equation 6.10: 9, 8, 5, 4, and 1. These are precisely the numbers in the ninth column of Table 6.2.

6.3 WHOLE-BODY, HEAD/NECK, AND HAND MODELS

The lower-body arrays are especially useful for organizing the geometry of human body models. Perhaps the most useful applications are with whole-body models, head/neck models, and with hand models. Figure 6.1 presents a whole-body model, which is sometimes called a gross-motion simulator. It consists of 17 bodies representing the major limbs of the human frame. It is called gross motion since detailed representation of the vertebrae, fingers, and toes is not provided. From a dynamics perspective, the relative movements of the vertebrae are small, and therefore, they can be incorporated into lumped mass segments. Also, from a dynamics perspective, the masses of the fingers and toes are small so that their movement does not significantly affect the gross motion of the system.

Figure 6.9 (see also Figure 2.3) shows a numbering or labeling of the model of Figure 6.1, and Equation 6.11 provides a list of the lower-body array:

K	1	2	3	4	5	6	7	8	9	10	11	12	13	14	15	16	17
$L(K)$	0	1	2	3	4	5	3	7	3	9	10	1	12	13	1	15	16

$$(6.11)$$

Table 6.3 provides a list of the higher-order lower-body arrays for the numbered system of Figure 6.9.

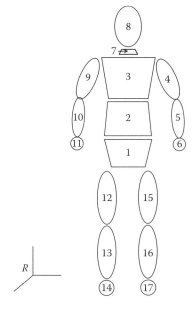

FIGURE 6.9   Numbering and labeling the model of Figure 6.1.

TABLE 6.3
Higher-Order Lower-Body Arrays for the Model of Figure 6.9

K	1	2	3	4	5	6	7	8	9	10	11	12	13	14	15	16	17
$L(K)$	0	1	2	3	4	5	3	7	3	9	10	1	12	13	1	15	16
$L^2(K)$	0	0	1	2	3	4	2	3	2	3	9	0	1	12	0	1	15
$L^3(K)$	0	0	0	1	2	3	1	2	1	2	3	0	0	1	0	0	1
$L^4(K)$	0	0	0	0	1	2	0	1	0	1	2	0	0	0	0	0	0
$L^5(K)$	0	0	0	0	0	1	0	0	0	0	1	0	0	0	0	0	0
$L^6(K)$	0	0	0	0	0	0	0	0	0	0	0	0	0	0	0	0	0

Observe in Figure 6.9 that we have selected the pelvis, or lower torso, as body B_1. This is useful for modeling motor vehicle occupants or machine operators. Sometimes we may think of the chest as the main body of the human frame. (When we point to another person or to ourselves, we usually point to the chest.) Finally, we may be interested in modeling a right-handed baseball pitcher. In this case, it is useful to use the left foot as body B_1.

In neck injury studies, the relative vertebral movement of the neck is important. Figure 6.10 (see also Figure 2.5) presents a model for studying relative vertebral movement. Table 6.4 provides a list of the associated lower-body arrays.

Often in injury studies, such as in motor vehicle accidents, it is useful to initially use the whole-body model to obtain the movement of a crash victim's chest. Then knowing the chest movement, the detailed movement of the head/neck system can be studied using the head/neck model. That is, the output of the whole-body model is used as input for the head/neck model.

Finally, consider the hand model of Figure 6.11 (see also Figure 2.6). Figure 6.12 provides a numerical labeling, and Table 6.5 lists the lower-body arrays. This model is useful for studying grasping and precision finger movements.

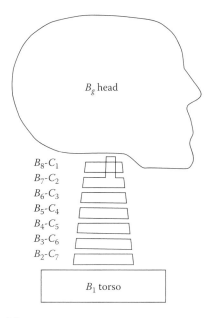

FIGURE 6.10 A head/neck model.

TABLE 6.4
Lower-Body Arrays for the Head/Neck Model of Figure 6.10

K	1	2	3	4	5	6	7	8	9
$L(K)$	0	1	2	3	4	5	6	7	8
$L^2(K)$	0	0	1	2	3	4	5	6	7
$L^3(K)$	0	0	0	1	2	3	4	5	6
$L^4(K)$	0	0	0	0	1	2	3	4	5
$L^5(K)$	0	0	0	0	0	1	2	3	4
$L^6(K)$	0	0	0	0	0	0	1	2	3
$L^7(K)$	0	0	0	0	0	0	0	1	2
$L^8(K)$	0	0	0	0	0	0	0	0	1
$L^9(K)$	0	0	0	0	0	0	0	0	0

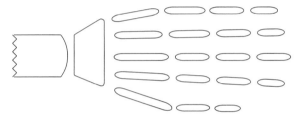

FIGURE 6.11　A model of the hand and wrist.

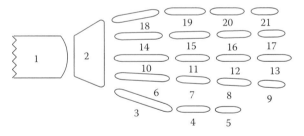

FIGURE 6.12　Numerical labeling of the hand/wrist model of Figure 6.11.

TABLE 6.5
Lower-Body Arrays for Hand/Wrist Model of Figure 6.12

K	1	2	3	4	5	6	7	8	9	10	11	12	13	14	15	16	17	18	19	20	21
$L(K)$	0	1	2	3	4	2	6	7	8	2	10	11	12	2	14	15	16	2	18	19	20
$L^2(K)$	0	0	1	2	3	1	2	6	7	1	2	10	11	1	2	14	15	1	2	18	19
$L^3(K)$	0	0	0	1	2	0	1	2	6	0	1	2	10	0	1	2	14	0	1	2	18
$L^4(K)$	0	0	0	0	1	0	0	1	2	0	0	1	2	0	0	1	2	0	0	1	2
$L^5(K)$	0	0	0	0	0	0	0	0	1	0	0	0	1	0	0	0	1	0	0	0	1
$L^6(K)$	0	0	0	0	0	0	0	0	0	0	0	0	0	0	0	0	0	0	0	0	0

Upon reflection, we see that the whole-body model of Figure 6.9, with its relatively large and massive bodies, is primarily a dynamics model, whereas the hand/wrist model of Figure 6.11 is primarily a kinematics model. We will discuss these concepts in subsequent chapters (Chapters 8, 9, and 11).

6.4 GROSS-MOTION MODELING OF FLEXIBLE SYSTEMS

There is a philosophical difference between the modeling of dynamical systems and the modeling of flexible/elastic systems. Consider a rod pendulum consisting of a thin rod of length ℓ supported at one end by a frictionless pin, allowing the rod to rotate and oscillate in a vertical plane as depicted in Figure 6.13. Consider also a flexible elastic rod with length ℓ with a cantilever support at one end and loaded at its other end by a force with magnitude P as in Figure 6.14.

From a dynamics perspective, the rod pendulum is seen to have one degree of freedom represented by the angle θ. The governing equation of motion is [1]

$$\frac{d^2\theta}{dt^2} + \left(\frac{3}{2}\right)\left(\frac{g}{\ell}\right)\sin\theta = 0 \tag{6.12}$$

For the elastic cantilever beam, the end deflection (under the load P) is seen to be (see Equation 5.5)

$$\delta = \frac{P\ell^3}{3EI} \tag{6.13}$$

where
  E is the elastic modulus
  I is the second moment of area of the beam cross section

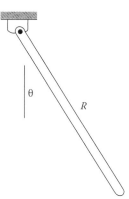

FIGURE 6.13 A rod pendulum.

FIGURE 6.14 A cantilever beam.

Consider the modeling and analysis assumptions made in the development of Equations 6.12 and 6.13. In Equation 6.12, the physical system is idealized for the rod. That is, the pin support is assumed to be frictionless and the rod geometrically simple. The governing equation, however, is based upon exact kinematic equations and on Newton's second law ($F = ma$).

In Equation 6.13, for the cantilever beam, the behavior of the beam is modeled by Hooke's law ($F = kx$)—an approximate expression modeling elastic behavior with an assumption of small deformation.

Thus, for the dynamical system, we have an approximate model (frictionless pin, ideal geometry) with exact kinematic and dynamic equations. Alternatively, for the elastic system, we have a more representative physical model, accounting for deformation, but the underlying governing equations are approximate.

There are, of course, no ideal physical systems such as frictionless pins, and although Newton's laws are postulated as exact, they are really only valid in Newtonian or inertial reference frames, and it is not clear that such frames even exist.* On the other hand, more representative physical models are limited by approximate governing or constitutive equations (linearized Hooke's law) and assumptions of small deformation.

In attempting to model the dynamic phenomena of biosystems where there are large displacements of flexible members, we need to bring together the procedures of dynamic and flexible system modeling. That is, we must blend the procedures of (1) inexact modeling with exact equations and (2) more accurate modeling with less exact equations. This, of course, requires a compromise of both approaches with an attempt to be consistent in the approximations. Being consistent is a difficult undertaking for very flexible systems with large overall motion, such as with the human spine.

In the sequel, in our modeling of biosystems, particularly the human body, we will primarily use the dynamics approach by attempting to obtain accurate governing equations for approximate models. We will then superpose upon this modeling, a representation of the flexibility of the system. Specifically, we will treat the skeletal system of bones as a system of rigid bodies and the connective tissue as flexible semielastic bodies.

PROBLEMS

Section 6.1

P6.1.1 List configurations of a person where with multibody (lumped mass) modeling, it would be advantageous to have closed loops in the model.

P6.1.2 See Problem P6.1.1. List examples in human body modeling via lumped mass (or finite segment) modeling where it would be advantageous to have translation between the bodies.

P6.1.3 See Problems P6.1.1 and P6.1.2. List circumstances where the use of flexible bodies is advantageous.

Section 6.2

P6.2.1 Consider the multibody system of the following figure where a numbering (or labeling) is begun. Complete the numbering of the bodies going through the branches of the system in a counterclockwise procedure.

---

* For most physical systems of interest, particularly biosystems on the Earth's surface, the Earth is an approximate Newtonian reference frame.

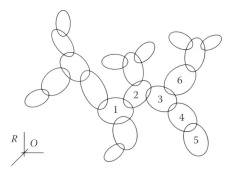

P6.2.2 See Problem P6.2.1. Determine the lower-body array $L(K)$ for the system of the figure for Problem P6.2.1 with its developed numbering system.

P6.2.3 Consider the multibody system of the following figure. Construct the lower-body array $L(K)$ for the given labeling (numbering).

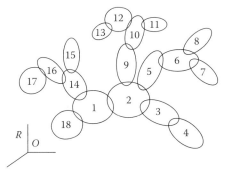

P6.2.4 See Problem P6.2.3 and its solution.
   a. List the extremity bodies.
   b. List the branching bodies.
   c. List the intermediate bodies.

P6.2.5 See Problem P6.2.3 and its solution. Develop a table of higher-order lower-body arrays, analogous to Table 6.2 for the system of the figure for Problem P6.2.3.

P6.2.6 Consider the system of the following figure. Repeat Problems P6.2.3 through P6.2.5 for this system.

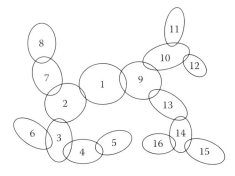

P6.2.7 Given the lower-body array $L(K)$,

$$L(K): 0\ 1\ 2\ 3\ 4\ 5\ 3\ 7\ 7\ 1\ 10\ 11\ 10\ 13$$

sketch a figure representing the connection configuration of the corresponding multibody system.

P6.2.8 Consider the lower-body array $L(K)$,

$$L(K): 0\ 1\ 2\ 3\ 4\ 4\ 2\ 7\ 8\ 8\ 2\ 11\ 1\ 13\ 13$$

Without sketching a connection configuration, develop a table of higher-order lower-body arrays.

P6.2.9 See Problem P6.2.8. Given again $L(K)$ as

$$L(K): 0\ 1\ 2\ 3\ 4\ 4\ 2\ 7\ 8\ 8\ 2\ 11\ 1\ 13\ 13$$

without sketching a connection configuration,
a. List the extremity bodies.
b. List the branching bodies.
c. List the intermediate bodies.

P6.2.10 See Problems P6.2.8 and P6.2.9. Given again $L(K)$ as

$$L(K): 0\ 1\ 2\ 3\ 4\ 4\ 2\ 7\ 8\ 8\ 2\ 11\ 1\ 13\ 13$$

sketch the configuration of the multibody system associated with $L(K)$. Use the sketch to verify the results of Problems P6.2.8 and P6.2.9.

Section 6.3

P6.3.1 Construct a multibody model of an insect (e.g., an ant).
P6.3.2 See Problem P6.3.1. Find a lower-body array for the insect model. Also, determine the corresponding higher-order lower-body arrays.
P6.3.3 Construct a multibody model of a fish.
P6.3.4 See Problem P6.3.3. Find a lower-body array for the fish model. Also, determine the corresponding higher-order lower-body arrays.

Section 6.4

P6.4.1 Construct a multibody model of a chain.
P6.4.2 Construct a multibody model of a rope.
P6.4.3 Construct a multibody model of a snake.

REFERENCES

1. R. L. Huston and C. Q. Liu, *Formulas for Dynamic Analysis*, Marcel Dekker, New York, 2001, p. 445.
2. H. Josephs and R. L. Huston, *Dynamics of Mechanical Systems*, CRC Press, Boca Raton, FL, 2002, Chapter 18.
3. R. L. Huston, *Multibody Dynamics*, Butterworth-Heinemann, Stoneham, MA, 1990.
4. R. L. Huston, Multibody dynamics: Modeling and analysis methods, *Applied Mechanics Reviews*, 44(3), 1991, 109.
5. R. L. Huston, Multibody dynamics since 1990, *Applied Mechanics Reviews*, 49, 1996, 535.

7 Tissue Biomechanics

Tissue biomechanics is a study of the physical properties and the mechanical behavior of biological materials. For almost all biological materials (e.g., bone, muscles, and organs), the properties vary significantly both in location and in direction and often in unexpected ways. Thus, tissue biomechanics is generally a difficult subject.

Most biological materials, except perhaps tooth enamel and cortical bone, are relatively soft and pliable. Thus, most biological materials (or tissue) are nonlinear both in material properties (e.g., stress–strain and stress–strain rate relations) and in geometric response (e.g., strain–displacement relations). These nonlinearities together with inhomogeneity and anisotropy make biological materials significantly more difficult to model than common structural materials such as steel, aluminum, and glass. Similarly, the properties of biofluids (e.g., blood) are more difficult to model than water or oil.

Recent and ongoing advances in computational mechanics (finite-element methods and boundary-element methods) provide a means for more comprehensive and more accurate modeling of biological materials than has previously been possible. Indeed, advances in computational mechanics make it possible to model biomaterials even at the cellular level. The future application of these methods will undoubtedly improve our understanding of biomaterial failure, damage, and injury mechanics, as well as the phenomena of healing and tissue remodeling.

As noted earlier, however, the focus of our studies is on global or gross modeling. Therefore, our focus on tissue mechanics is in understanding gross behavior, particularly of the more rigid components of the human body (bones, cartilage, ligaments, tendons, and muscles). We review their material properties in the following sections.

7.1 HARD AND SOFT TISSUE

From a global perspective, biological material or tissue may be classified as hard or soft. Hard tissue includes bones, cartilage, teeth, and nails. Soft tissue includes fluids (blood, lymph, excretions), muscles, tendons, ligaments, and organ structure. For global modeling of biosystems—particularly human body dynamics—the bones are the most important hard tissue, and the muscles are the most important soft tissue.

7.2 BONES

It is common knowledge that the bones make up the skeletal structure of the human body and of vertebral animals (vertebrates). The bones thus maintain the form and shape of the vertebrates. In the human body, there are 206 bones, of which 26 are in the spine [1]. The bones are held together by ligaments, cartilage, and muscle/tendon groups. The bones vary in size from the large leg, arm, and trunk bones (femur, humerus, and pelvis) to the small ear bones (incus, malleus, and stapes). The bones also have various shapes ranging from long, beam-like bones (as in the arms, legs, and extremities) to concentrated annular structures (vertebrae) to relatively flat plates and shells (scapula and skull bones).

Although bones make up most of the hard tissue of the body, they are not inert. Instead, the bones are living tissue with a vascular network (blood supply). As with muscles, skin, and other soft tissue, the bones are capable of growth, repair, and regrowth (remodeling).

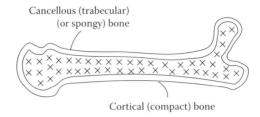

FIGURE 7.1   Sketch of a long bone (femur).

Interestingly, just as biosystem tissue may be globally classified as hard or soft, so also bone tissue may be classified as hard or soft. This is perhaps best seen by considering a long bone such as the femur, which is like a cylindrical shell with a hard outer surface and a soft inner core—as in a tree trunk with bark on the outside and wood in the interior (Figure 7.1). The outer shell or hard bone surface is called compact bone or cortical bone, and the soft, spongy interior is called trabecular or cancellous bone.

We will consider the microstructure of bone in Section 7.3. References [1–9] provide a minibibliography of bone characteristics.

7.3   BONE CELLS AND MICROSTRUCTURE

When we take a close look at bone tissue, it is composed of cells forming an organized or pattern-like structure. These cells, like other tissue cells, are nourished by an elaborate blood supply. The basic bone cell is called an osteocyte. The cells, however, are neither all the same nor do they all have the same function. Some cells, called osteoblasts, are active in bone formation, growth, and healing. Other cells, called osteoclasts, perform bone and mineral absorption or resorption. Bone formation or bone growth is called modeling. Bone resorption and subsequent reformation is called remodeling. Remodeling is sometimes called bone maintenance.

The hardness of bone is due to the presence of mineral salts—principally calcium and phosphorus. These salts are held together by a matrix of collagen (a tough fibrous and flexible material). The salt/collagen compound forms a composite structure analogous to concrete with metal reinforcing rods.

From a different perspective, if we look at the shaft of a long bone, we see that the cortical or hard compact bone is composed of a system of wood-like structures called Haversian systems or osteons as illustrated in Figure 7.2. The central canal houses blood vessels for the nourishment of the system.

The trabecular or soft inner spongy bone is composed of rods and plates with the trabeculae (little beams) generally oriented in the direction of bone loading.

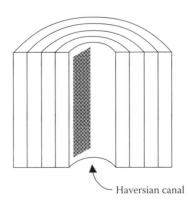

FIGURE 7.2   Haversian/osteon bone representation.

7.4   PHYSICAL PROPERTIES OF BONE

As mentioned earlier, bone tissue, as with all biological tissues, is neither homogeneous nor isotropic. That is, the physical properties of bone vary both in location and in direction. Hence, we can describe only general physical properties with only approximate values for the associated density, strength, and elastic moduli.

The mass density for cortical bone is approximately 1800–1900 kg/m^3 (or 1.8–1.9 g/cm^3) (or 0.065–0.069 lb/in.^{3}). The density of trabecular bone varies considerably depending upon its porosity ranging from 5% to 70% of that of cortical bone [1,2].

Bone is stronger in compression than in tension, and it is the weakest in shear. For cortical bone, the compressive strength is approximately 190 MPa (27,550 psi); for tension, it is 130 MPa (18,855 psi); and for shear, it is 70 MPa (10,152 psi). Correspondingly, the elastic modulus for cortical bone is approximately 20 GPa (29 × 10^5 psi).

For trabecular bone, the strength and modulus depend upon the density. Hayes and Bouxsein [2] model the dependence of the strength σ upon density as being quadratic and the modulus E dependence as being cubic. Specifically,

$$\sigma = 60\rho^2 \text{ MPa} \quad \text{and} \quad E = 2915\rho^3 \text{ MPa} \tag{7.1}$$

where ρ is the mass density (g/cm^3).

7.5   BONE DEVELOPMENT (WOLFF'S LAW)

When we look at the shape of the bones, particularly of the long bones, a question that arises is as follows: Why do they have such shapes? To answer this question, consider that the shape (or form) enables the body to function efficiently: The lumbar and thoracic spines provide for the structure, shape, and stability of the torso. These spines maintain safe operating space for the torso organs. The cervical spine (neck) provides both support and mobility to the head. Indeed, the spine maintains the form of the body so much that humans and other mammals are often called vertebrates. The limbs and extremities, which enable locomotion and other kinematic functions, have long bones shaped like advanced designed beams with high axial strength. The condyles form ideal bearing surfaces. The bones are also protective structures, with the ribs serving as a cage for the heart, lungs, liver, and pancreas; the skull bones forming a helmet for the brain; and the pelvis providing a foundation and base for the lower abdominal organs.

The notion that the shape or form of the bones is determined by their function was advanced by Wolff and others over a century ago [5]. Indeed, the phrase "form follows function" is often known as Wolff's law.

As the bones are loaded (primarily in compression), the trabeculae align themselves and develop along the stress vector directions, in accordance with Wolff's law [2]. Intuitively, the rate and extent of this alignment (through modeling and remodeling) is proportional to the rate and extent of the loading. Specifically, high-intensity loading, albeit only for a short duration, is more determinative of bone remodeling than less intensive longer-lasting loading [10]. Weightlifters are thus more likely to have stronger, higher mineral density bones than light exercise buffs (walkers, swimmers, and recreational bikers).

Some researchers also opine that function follows form [4]. That is, form and function are synergistic. This fascinating concept (mechanobiology) is discussed extensively in the article of van der Meulen and Huiskes [4], which is also an excellent survey for bone development.

7.6 BONE FAILURE (FRACTURE AND OSTEOPOROSIS)

Bones usually fail by fracturing under trauma. There are a variety of fractures ranging from simple microcracks to comminuted fractures (multiple, fragmented disintegration). Whiting and Zernicke [11] provide an excellent summary of bone fractures.

Fractures can occur due to excessive load (usually an impulsive load arising from an impact) or due to fatigue. Fracture due to excessive load often occurs when the load is applied in directions different from the direction of maximum strength of the bone. The extent of load-induced fracture depends, of course, upon the extent to which the load exceeds the bone strength. During impact, as in accidents or in contact sports, the forces and thus the bone loading can be very large, but they usually only occur for a short time (a few milliseconds).

Fatigue fracture occurs from repeated loadings when the combination of the amplitude and frequency of the loadings exceed the ability of the bone to repair (remodeling). That is, fracture will occur when the rate of damage exceeds the rate of remodeling. Fatigue fracture often occurs in sport activities with repeated vigorous motion—as with baseball pitching.

Bone fracture also occurs when the bones are weakened as with osteoporosis. Osteoporosis (or porous bones) occurs when the mineral content of the bones decreases. That is, the activity of the bone resorption cells (the osteoclasts) exceeds the activity of the bone remodeling cells (the osteoblasts). With osteoporosis, the bones become brittle and the geometry frequently changes.

Osteoporosis mostly affects the large bones (femur, humerus, and spine). Osteoporosis in the femoral neck will often lead to spontaneous fracture (hip fracture), causing a person to fall and become immobilized. Osteoporosis is associated with aging and inactivity. Hayes and Bouxsein [2] estimate that by age 90, 17% of men and 33% of women will have hip fractures.

It appears that osteoporosis is best prevented by diet (consuming calcium-rich food) and by vigorous exercise (e.g., running and weight lifting), beginning at an early age and continuing through life.

An observation of human kinematics and dynamics reveals that bones are loaded primarily in compression and seldom in tension. For example, carrying a weight, such as a suitcase, will compress the spine and the leg bones, but the arm bones will be essentially unloaded. (The arm muscles, ligaments, and tendons support the load.)

If strengthening occurs when the bones are loaded to near yielding at the cellular level, then casual exercise (such as slow walking) is not likely to produce significant strengthening. In light and moderate exercise, the muscles become fatigued before the bones are fatigued. Therefore, the best bone-strengthening activity is high-intensity exercise and impact loading (weight lifting and running).

7.7 MUSCLE TISSUE

Muscle tissue forms a major portion of human and animal bodies including fish and fowl. There are three general kinds of muscles: skeletal, smooth, and cardiac. Skeletal muscle moves and stabilizes the body. It is of greatest interest in human body dynamics. Smooth muscle provides a covering of the skeleton and organs of the torso. Cardiac muscle is heart tissue, continually contracting and relaxing. Unlike skeletal and smooth muscles, cardiac muscle does not fatigue.

Cardiac muscle appears to be found only in the heart. Smooth muscle is generally flat and planar, while skeletal muscle tends to be elongated and oval. An illustration distinguishing skeletal and smooth muscle is found in turkey meat with the dark meat being skeletal muscle and the white meat being smooth muscle. In the sequel, we will focus upon skeletal muscle.

The skeletal muscle produces movement by contracting and as a consequence pulling. Muscles do not push. That is, the skeletal muscles go into tension, but not into compression. The skeletal muscle produces movement by pulling on bones through connective tendons. The bones thus serve as levers with the joints acting as fulcrums. To counter (or reverse) the moment on the bone, there are counteracting muscles that pull the bone back. Therefore, the muscles act in groups.

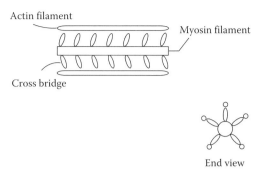

FIGURE 7.3   Sketch of a sarcomere—the basic muscle contracting unit.

In addition to the action and reaction occurring in muscle groups, the muscles often occur in pairs and triplets, forming sets of parallel actuators such as the biceps, triceps, and quadriceps (see Section 2.6).

At the microscopic level, muscle activity is mechanical, electrical, and chemical. Mechanically, forces are generated by microscopic fibers. These fibers are stimulated electrically via nerves, and the energy expended is chemical, being primarily the consumption of sugars and oxygen, although precise details of the process and action are still not fully known.

Figure 7.3 presents a schematic sketch of a microscopic portion of a muscle fiber called a sarcomere [12]. The sarcomere is the basic contractile unit of muscle. It consists of parallel filaments called actin (thin) and myosin (thick) filaments, which can slide (axially) to one another. The contraction occurs when small cross bridges, attached at about a 60° angle to the myosin filament, begin to adhere to the actin filament. The cross filaments then rotate, causing the actin and myosin filaments to slide relative to each other (like oars on a racing scull). The adhering and rotation of the cross bridges are believed to be due to the release of calcium ions and the conversion of adenosine triphosphate (ATP) into adenosine diphosphate (ADP) [1]. The resorption of calcium ions then allows the cross bridges to detach from the actin filament, leading to muscle relaxation.

The contracting of skeletal muscle fibers will not necessarily produce large movement or large length changes. Indeed, a muscle can contract where there may be no significant gross motion—an isometric or constant length contraction. The muscles can have varying forces and movements between these extremes. Muscle tissue contraction can also produce erratic and jerking movement called twitching, fibrillation, tetanic, and convulsion [12].

7.8   CARTILAGE

Next to bones, teeth, and nails, cartilage tissue is the hardest material of the human body. Cartilage provides structure, as in the support of ribs (the sternum). Cartilage also provides bearing surface for joints, as in the knee.

Cartilage is neither supplied with nerves nor is it well supplied with blood vessels. It is primarily a matrix of collagen fibers. (Collagen is a white fibrous protein.) Only about 10% of cartilage is cellular.

As a bearing surface, cartilage is lubricated by synovial fluid—a slippery substance that reduces the coefficient of friction in the joints to about 0.02 [1]. Figure 7.4 is a sketch of a typical cartilage/bone/bearing structure. Near the bone, the cartilage is hard and calcified, and the collagen fibers are directed generally perpendicular to the bone surface. At the joint surface, the cartilage is softer and the collagen fibers are generally parallel to the surface. In between, the fibers are randomly oriented.

Cartilage is neither homogeneous nor isotropic. It can be very hard near bone and relatively soft, wet, and compliant on the sliding surface. When a joint or any surface is loaded by contact forces perpendicular to the surface, the maximum shear stress in the joint will occur just beneath

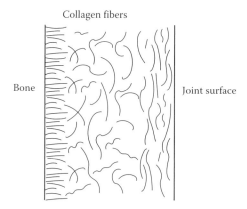

FIGURE 7.4   Collagen fibers of a cartilage surface.

the surface. Interestingly, the collagen fibers, oriented parallel to the cartilage sliding surface, are thus ideally suited to accommodate this shear stress.

On an average, cartilage has a compression strength of approximately 5 MPa (725 psi) and a tensile strength of approximately 25 MPa (3625 psi) if the loading is parallel to the fiber direction [1].

Finally, the presence of the collagen fibers in cartilage means that cartilage may be classified as a fiber-reinforced composite material.

7.9   LIGAMENTS/TENDONS

Ligaments and tendons are like cords or cables, which (1) maintain the stability of the structural system and (2) move the limbs. The ligaments, connecting bone to bone, primarily provide stability and connectedness. The tendons, connecting muscles to bone, provide for the movement. However, as cables, the ligaments and tendons can only support tension. (They can pull but not push.)

Ligaments and tendons have a white, glossy appearance. They have variable cross sections along their lengths, and their physical properties (e.g., their strength) vary along their lengths. At their connection to bone, ligaments and tendons are similar to cartilage, with the connection being similar to plant roots in soil. As with cartilage, ligaments and tendons have a relatively poor blood supply. Their injuries often take a long time to heal, and often deficiencies remain.

Ligaments and tendons are like fibrous rope bands. As such, there is a small amount of looseness in unloaded ligaments and tendons. In this regard, a typical stress–strain relation is like that of Figure 5.19 where the flat horizontal region near the origin is due to the straightening of the fibers as the unstretched member is loaded. The upper portion of the curve represents a yielding, or stretching, of the member—often clinically referred to as sprain.

Water is the major constituent of ligaments and tendons comprising as much as 66% of the material. Elastin is the second major constituent. Elastin is an elastic substance, which helps a ligament or tendon return to its original length after a loading is removed. Ligaments and tendons are also viscoelastic. They slowly creep, or elongate, under loading. Ligaments and tendons are encased in lubricated sheaths (epiligaments and epitendons), which protect the members and allow for their easy movement.

Perhaps the best known of the ligaments are those in the legs and particularly in the knees. Specifically, at the knee, there are ligaments on the sides (the medial collateral and lateral collateral ligaments) connecting the tibia and femur and also crossing between the sides of the tibia and fibula (the anterior cruciate and posterior cruciate ligaments). These ligaments are subject to injury in many sports and even in routine daily activities. Similarly, the familiar tendons are the patellar tendon and the Achilles tendon (or heel cord) providing for leg extension and plantar flexion of the foot.

7.10 SCALP, SKULL, AND BRAIN TISSUE

Of all survivable traumatic injuries, severe brain injury and paralysis are the most devastating. Brain and nerve tissue are very delicate and susceptible to injury, and once the tissue is deformed or avulsed, repair is usually slow and full recovery seldom occurs. Fortunately, however, these delicate tissues have strong natural protection through the skull and the spinal vertebrae. In this section, we briefly review the properties of the skull and the brain/nerve tissue.

Figure 7.5 provides a simple sketch of the scalp, skull, and meningeal tissue, forming a helmet for the brain. On the outside of the scalp, there is a protective matting of hair, which provides for both cushioning and sliding. The hair also provides for temperature modulation. The scalp itself is a multilayered composite consisting of skin, fibrous connective tissue, and blood vessels. The skull is a set of shell-like bones knit together with cortical (hard) outer and inner layers and a trabecular (soft) center layer. Beneath the skull is the dura mater consisting of a relatively strong but inelastic protective membrane. Beneath the dura is another membrane of fibrous tissue called the arachnoid, which in turn covers another fibrous membrane called the pia, which covers the outer surface of the brain.

Goldsmith has conducted, documented, and provided an extensive study and reference of head and brain injury [13]. Even without brain injury, head injury may occur in the face, in the scalp, with skull fracture, and with bleeding (hematoma) beneath the scalp and skull. Scalp injury ranges from minor cuts and abrasions to deep lacerations and degloving. Skull fracture may range from hairline cracks to major crushing. Brain injury may not occur with minor skull fracture. Hematoma can occur above or below the dura matter. Brain injury can occur with or without skull fracture. Brain injury can range from a simple concussion (temporary loss of consciousness) to a contusion (bruising), to diffuse axonal injury (DAI) (scattered microdamage to the tissue), and to laceration of the brain tissue.

Finally, if we consider that brain tissue is stronger in compression than tension, we can visualize brain tissue injury occurring as it is pulled away from the skull membranes. This in turn can occur on the opposite side of the head from a traumatic impact as represented in Figure 7.6, the so-called coup/contrecoup phenomena.

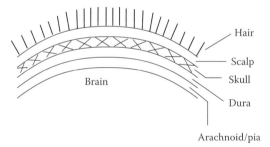

FIGURE 7.5   Protective layers around the brain.

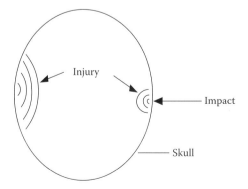

FIGURE 7.6   Impact (coup) and opposite side (contrecoup) head injuries.

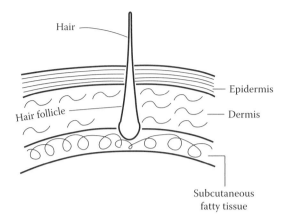

FIGURE 7.7   Skin layers and hair follicle. (From Anthony, C.P. and Kolthoff, N.J., *Textbook of Anatomy and Physiology*, C.V. Mosby Company, St. Louis, MO, 1975. With permission.)

7.11   SKIN TISSUE

Skin is sometimes referred to as the largest organ of the body. Mechanically, it is a membrane serving as a covering for the body. Skin consists of three main layers. The outer layer called the epidermis is the visible palpable part. The epidermis covers a middle layer, called the dermis, which in turn rests upon a subcutaneous (subskin) layer of fatty tissue as represented in Figure 7.7. The skin color is due to melanin contained in the epidermis.

In addition to being a covering, skin serves as a shield or protector of the body. It is also a temperature regulator and a breather (or vent), as well as a valve for gasses and liquids. Like other body tissues, skin varies from place to place on the body. The thickness varies from as thin as 0.05 mm on the eyelids to 2 mm on the soles of the feet. Skin strength varies accordingly. With aging, skin becomes less elastic, more brittle, and thinner. Many regions of the skin contain hair follicles (Figure 7.7), based in the subcutaneous layer. The ensuring hair shaft provides mechanisms for heat transfer, for cushioning, and for friction reduction. Finally, the skin has its own muscle system, allowing it to form goose bumps for heat generation.

PROBLEMS

Section 7.2

P7.2.1   Consider the iron work of an office building and the beams/rafters of a house as being analogous to bones of a person. How are they the same and how are they different?

P7.2.2   What is the principal role of blood supply for the bones? How is this different for muscles?

Section 7.3

P7.3.1   List the ways a long skeletal bone is analogous to a tree branch. How do they differ?

Section 7.4

P7.4.1   Explain intuitively why for trabecular bone the strength is proportional to the square of the mass density, whereas the elastic modulus is proportional to the cube of the mass density.

Section 7.5

P7.5.1   Cite examples in nature illustrating Wolff's law ("form follows function").

P7.5.2   Cite examples in nature illustrating the reverse of Wolff's law—that is, "function follows form."

P7.5.3   Provide examples and illustrations in nature of how "form follows function" and "function follows form" are synergistic (i.e., one stimulates the other).

Section 7.6

P7.6.1 If bones are weakest in shear, what is the likely pattern of fraction for long bones (e.g., the tibia or radius) under compression?

P7.6.2 If bones are weaker in tension than compression, what is the likely fracture pattern of long bones in bending?

P7.6.3 If an elderly person with osteoporosis falls and is then diagnosed with a broken hip, in what ways could the osteoporosis have contributed to the breaking?

P7.6.4 Several theorists have suggested that osteoporosis is systemic, as opposed to being a local phenomena. How might this thesis be tested?

P7.6.5 How might Wolff's law predict a local appearance (or, alternatively, nonappearance) of osteoporosis?

Section 7.7

P7.7.1 Contrast the action of a hydraulic cylinder with a skeletal muscle.

P7.7.2 Contrast the relative advantages (or, alternatively, the disadvantages) of longer and shorter muscles for (a) weight lifting and (b) bodybuilding.

P7.7.3 Discuss the role of blood flow in muscle activity.

P7.7.4 Discuss the role of the nerves in muscle activity.

P7.7.5 See Problem P7.6.3. Explain how cardiac muscle can resist fatigue.

P7.7.6 Discuss the effect of arterial plaque on skeletal and cardiac muscle activity.

P7.7.7 Discuss the use of external electrical stimulation on muscle tissue. How can this be (a) beneficial and (b) deleterious?

Section 7.8

P7.8.1 Use Wolff's law to discuss the geometry and physical properties of joint (articular) cartilage.

P7.8.2 Consider again articular cartilage. Since the cartilage does not have an active blood supply, explain why and how exercise is beneficial for cartilage health.

Section 7.9

P7.9.1 See Problem P7.7.2. Since ligaments and tendons (as with articular cartilage) do not have active blood supply, explain why and how exercise is beneficial for ligament/tendon health.

P7.9.2 Many athletes are strong advocates of stretching both before and after vigorous exercise. Since ligaments and tendons are viscoelastic and can experience creep, explain why and how stretching can be beneficial.

P7.9.3 Since ligament tissue is neither homogeneous nor isotropic, explain why replacement of a torn ligament by a ligament (or even a set of ligaments) from another site of the body (autograft) is not likely to be an "ideal" repair.

P7.9.4 What is the synergistic role of ligaments and bones in burden bearing?

Section 7.10

P7.10.1 The Head Injury Criteria (HIC) is defined as

$$\text{HIC} = \left\{ \left[\frac{1}{(t_2 - t_1)} \right] \int_{t_1}^{t_2} a \, dt \right\}^{2.5} (t_2 - t_1)$$

where t_1 and t_2 are any two points in time, close to each other, during an interval of impact, and a is the resultant head acceleration measured in multiples of the acceleration of gravity.

Lasting injury is a strong possibility if the HIC exceeds 1000.

The Federal Motor Vehicle Safety Standard (FMVSS) 208 sets the upper limit of the HIC to be 700, when $t_2 - t_2 \leq 15$ ms, called HIC_{15} [14].

Discuss why the HIC is both a good and a poor model for head injury.

Section 7.11

P7.11.1 Explain how skin grafts may be less than ideal curatives for skin maladies. Explain further how the benefits of a skin graft generally outweigh the disadvantages.

P7.11.2 In contact stress analysis, it is known that the maximum shear stress occurs slightly below the contact surface [15]. How does this explain the development of blisters in the hands and feet?

REFERENCES

1. B. M. Nigg and W. Herzog, *Biomechanics of the Musculo-Skeletal System*, John Wiley & Sons, New York, 1994.
2. W. C. Hayes and M. L. Bouxsein, Biomechanics of cortical and trabecular bone: Implications for assessment of fracture risk, in *Basic Orthopaedic Biomechanics*, 2nd edn., V. C. Mow and W. C. Hayes (Eds.), Lippincott-Raven, Philadelphia, PA, 1997, 69–111.
3. M. Nordin and V. H. Frankel, Biomechanics of bone, in *Basic Biomechanics of the Musculoskeletal System*, 2nd edn., M. Nordin and V. H. Frankel (Eds.), Lea & Febiger, Philadelphia, PA, 1989, 3–29.
4. M. C. H. van der Meulen and R. Huiskes, Why mechanobiology? A survey article, *Journal of Biomechanics*, 35, 2002, 401–414.
5. J. Wolff, *The Law of Bone Remodeling* (translation by P. Maquet and F. Furlong), Springer-Verlag, Berlin, Germany, 1986 (original in 1870).
6. L. E. Lanyon, Control of bone architecture by functional load bearing, *Journal of Bone and Mineral Research*, 7(Suppl. 2), 1992, S369–S375.
7. J. Cordey, M. Schneider, C. Belendez, W. J. Ziegler, B. A. Rahn, and S. M. Perren, Effect of bone size, not density, on the stiffness of the proximal part of the normal and osteoporotic human femora, *Journal of Bone and Mineral Research*, 7(Suppl. 2), 1992, S437–S444.
8. R. B. Martin and P. J. Atkinson, Age and sex-related changes in the structure and strength of the human femoral shaft, *Journal of Biomechanics*, 10, 1977, 223–231.
9. M. Singh, A. R. Nagrath, P. S. Maini, and R. Hariana, Changes in trabecular pattern of the upper end of the femur as an index of osteoporosis, *Journal of Bone and Joint Surgery*, 52, 1970, 457–467.
10. R. Bozian and R. L. Huston, The skeleton connection, *Nautilus*, 30 (October/November) 1992, 62–63.
11. W. C. Whiting and R. F. Zernicke, *Biomechanics of Musculoskeletal Injury*, Human Kinetics, Champaign, IL, 1998.
12. C. P. Anthony and N. J. Kolthoff, *Textbook of Anatomy and Physiology*, C. V. Mosby Company, St. Louis, MO, 1975.
13. W. Goldsmith, The state of head injury biomechanics: Past, present, and future, Part 1, *Critical Reviews in Biomedical Engineering*, 29(5/6), 2001, 441–600.
14. *Code of Federal Regulations 49*, "Transportation", Section 571.208, S6.3, Office of the Federal Register, 2009, 740.
15. A. P. Boresi and O. M. Sidebottom, *Advanced Mechanics of Materials*, 4th edn., John Wiley & Sons, New York, 1985, 612.

8 Kinematical Preliminaries
Fundamental Equations

In this chapter, we review a few kinematical concepts that are useful in studying the dynamics of biosystems, particularly human body dynamics. Kinematics is one of the three principal subjects of dynamics, with the other two being kinetics (forces) and inertia (mass effects). Kinematics is sometimes described as a study of motion without regard to the cause of the motion.

We will review and discuss such fundamental concepts as position, velocity, acceleration, orientation, angular velocity, and angular acceleration. We will develop the concept of configuration graphs and introduce some simple algorithms for use with biosystems.

In Chapters 9 and 10, we review the fundamentals of kinetics and inertia.

8.1 POINTS, PARTICLES, AND BODIES

The term "particle" suggests a "small body." Thus in biosystems, we might think of a particle as a cell. In dynamic analyses, particularly in kinematics, particles are represented as points. That is, if a particle is a small body, then rotational effects are unimportant, and the kinematics of the particle is the same as the kinematics of a point within the boundary of the particle. Dynamically, we then regard a particle as a point with an associated mass.

A rigid body is a set of particles (or points) at fixed distances from one another. A flexible body is a set of particles whose distances from one another may vary but whose placement (or positioning) relative to one another remains fixed.

Finally, there is the question as to what is "small." The answer depends upon the context. In celestial mechanics, the earth may be thought of as a particle and on the earth an automobile may often be considered as a particle—as in accident reconstruction. The response to the question of smallness is simply that a point-mass modeling of a particle becomes increasingly better if the particle is smaller.

8.2 PARTICLE, POSITION, AND REFERENCE FRAMES

Let P be a point representing a particle. Let R be a fixed, or inertial* reference frame in which P moves. Let X, Y, and Z be rectangular (Cartesian) coordinate axes fixed in R. Let $\mathbf{n}_x$, $\mathbf{n}_y$, $\mathbf{n}_z$ be unit vectors parallel to the X, Y, Z axes as in Figure 8.1. Let (x, y, z) be the coordinates of P measured relative to the origin O of the XYZ axis system. Thus the position vector $\mathbf{p}$ locating P relative to O is simply

$$\mathbf{p} = x\mathbf{n}_x + y\mathbf{n}_y + z\mathbf{n}_z \tag{8.1}$$

---

* A fixed, or inertial reference frame is simply a reference frame in which Newton's laws are valid. There are questions whether such frames exist in an absolute sense, as the entire universe appears to be moving. Nevertheless, from the perspective of biomechanics analysis, it is sufficient to think of the earth as a fixed frame—at least for current problems of practical importance.

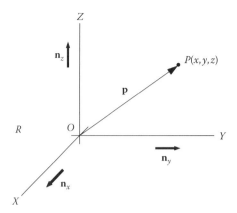

FIGURE 8.1   Position vector locating a particle in a reference frame.

8.3   PARTICLE VELOCITY

See the notation of Section 8.2. Let the coordinates (x, y, z) of particle P be functions of time as

$$x = x(t), \quad y = y(t), \quad z = z(t) \tag{8.2}$$

Thus, P moves in R. Let C be the curve on which P moves. That is, C is the locus of points occupied by P. Then Equations 8.2 are parametric equations defining C. (Figure 8.2). As before, let $\mathbf{p}$ be a position vector locating P relative to the origin O. Then from Equations 8.1 and 8.2, $\mathbf{p}$ is seen to be a function of time. The time derivative of $\mathbf{p}$ is the velocity of P (in reference frame R). That is,

$$\mathbf{v} = \frac{d\mathbf{p}}{dt} = \dot{x}\mathbf{n}_x + \dot{y}\mathbf{n}_y + \dot{z}\mathbf{n}_z \tag{8.3}$$

Several observations may be helpful: first, the time rate of change of P is measured in the inertial frame R, with the unit vectors fixed in R. Next, the velocity $\mathbf{v}$ of P is itself a vector. As such, $\mathbf{v}$ has magnitude and direction. The magnitude of $\mathbf{v}$, written as $|\mathbf{v}|$, is often called the speed and written

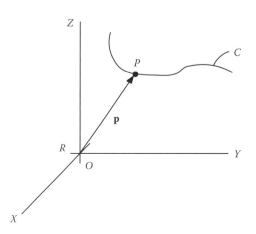

FIGURE 8.2   A particle P moving in a reference frame R and thus defining a curve C.

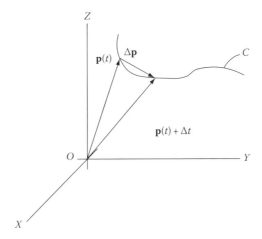

FIGURE 8.3   A chord vector of C.

simply as v. (Note that **v** cannot be negative.) The direction of **v** is tangent to C at P. To see this, recall from elementary calculus that the derivative in Equation 8.3 may be written as

$$\mathbf{v} = \frac{d\mathbf{p}}{dt} = \lim_{\Delta t \to 0} \frac{\mathbf{p}(t + \Delta t) - \mathbf{p}(t)}{\Delta t} = \underset{\Delta t \to 0}{\text{Line}} \frac{\Delta \mathbf{p}}{\Delta t} \tag{8.4}$$

where $\Delta\mathbf{p}$ may be viewed as a chord vector of C as in Figure 8.3. Then as Δt tends to zero, the chord $\Delta\mathbf{p}$ becomes increasingly coincident with C. In the limit, $\Delta\mathbf{p}$ and therefore **v** as well, are tangent to C.

8.4   PARTICLE ACCELERATION

Acceleration of a particle in a reference frame R is defined as the time rate of change of the particle's velocity in R. That is,

$$\mathbf{a} = \frac{d\mathbf{v}}{dt} \tag{8.5}$$

Recall that unlike velocity, the acceleration vector is not, in general, tangent to the curve C on which the particle moves. To see this, let the velocity **v** of a particle P be expressed as

$$\mathbf{v} = v\mathbf{n} \tag{8.6}$$

where
   v is the magnitude of the velocity
   n is a unit vector tangent to C as in Figure 8.4 (see also Section 8.3)

By the product rule of differentiation, the acceleration may be written as

$$\mathbf{a} = \left(\frac{dv}{dt}\right)\mathbf{n} + v\frac{d\mathbf{n}}{dt} \tag{8.7}$$

Thus we see that the acceleration has two components: one tangent to C and one normal to C ($d\mathbf{n}/dt$ is perpendicular to **n**).* The acceleration component tangent to C is due to the speed change, and the component normal to C is due to the direction change of the velocity vector.

---

* Observe that $\mathbf{n}\cdot\mathbf{n} = 1$, and then by differentiation $d(\mathbf{n}\cdot\mathbf{n})dt = 0 = \mathbf{n}\cdot d\mathbf{n}/dt + d\mathbf{n}/dt\cdot\mathbf{n} = 2\mathbf{n}\cdot d\mathbf{n}/dt$, or $\mathbf{n}\cdot d\mathbf{n}/dt = 0$.

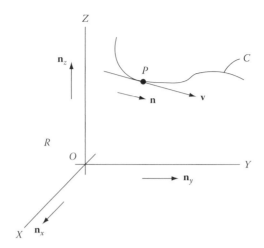

FIGURE 8.4   Velocity of a particle.

Finally, observe from Equation 8.3 that the acceleration of P may also be expressed as

$$\mathbf{a} = \ddot{x}\mathbf{n}_x + \ddot{y}\mathbf{n}_y + \ddot{z}\mathbf{n}_z \tag{8.8}$$

where, as before, $\mathbf{n}_x$, $\mathbf{n}_y$, and $\mathbf{n}_z$ are unit vectors parallel to the fixed X, Y, and Z axes in reference frame R.

8.5   ABSOLUTE AND RELATIVE VELOCITY AND ACCELERATION

Referring again to Figure 8.4, depicting a particle P moving in a reference frame R, we could think of the velocity of P in R as the velocity of P "relative" to R, or alternatively as the velocity of P relative to the origin O of the X, Y, Z frame fixed in R. Observe that the velocity of P is the same relative to all points fixed in R. To see this, consider two points O and $\hat{O}$ fixed in R with position vectors $\mathbf{p}$ and $\hat{\mathbf{p}}$ locating P as in Figure 8.5. Then $\mathbf{p}$ and $\hat{\mathbf{p}}$ are related by the expression

$$\mathbf{p} = \mathbf{O}\hat{\mathbf{O}} + \hat{\mathbf{p}}\overline{\mathbf{p}} \tag{8.9}$$

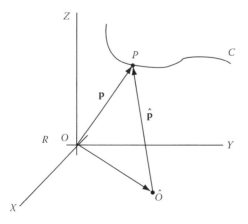

FIGURE 8.5   Location of particle P relative to two reference points.

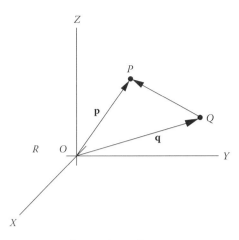

FIGURE 8.6   Two particles moving in a reference frame R.

Since $\mathbf{O\hat{O}}$ is fixed or constant in R, we see that the velocity of P in R is

$$\mathbf{v}^P = \frac{d\mathbf{p}}{dt} = \frac{d\hat{\mathbf{p}}}{dt} \tag{8.10}$$

That is, the velocity of P relative to O is the same as the velocity of P relative to $\hat{O}$, and thus relative to all points fixed in R. As a consequence, $\mathbf{v}^P$ is said to be the "absolute" velocity of P in R.

Consider two particles P and Q moving in R as in Figure 8.6. Let P and Q be located relative to the origin O of an X, Y, Z frame fixed in R. Then the velocities of P and Q in R are

$$\mathbf{v}^P = \frac{d\mathbf{p}}{dt} \quad \text{and} \quad \mathbf{v}^Q = \frac{d\mathbf{q}}{dt} \tag{8.11}$$

From Figure 8.6, we see that $\mathbf{p}$ and $\mathbf{q}$ are related as

$$\mathbf{p} = \mathbf{q} + \mathbf{r} \quad \text{or} \quad \mathbf{r} = \mathbf{p} - \mathbf{q} \tag{8.12}$$

Then by differentiating and comparing with Equation 8.11, we have

$$\frac{d\mathbf{r}}{dt} = \mathbf{v}^P - \mathbf{v}^Q \tag{8.13}$$

The difference $\mathbf{v}^P - \mathbf{v}^Q$ is called the "relative velocity of P and Q in R," or alternatively "the velocity of P relative to Q in R," and is conventionally written as $\mathbf{v}^{P/Q}$. Hence, we have the relation

$$\mathbf{v}^P = \mathbf{v}^Q + \mathbf{v}^{P/Q} \tag{8.14}$$

A similar analysis with acceleration leads to the expression

$$\frac{d^2\mathbf{r}}{dt^2} = \mathbf{a}^P - \mathbf{a}^Q = \mathbf{a}^{P/Q} \quad \text{or} \quad \mathbf{a}^P = \mathbf{a}^Q + \mathbf{a}^{P/Q} \tag{8.15}$$

where
   $\mathbf{a}^{P/Q}$ is the acceleration of P relative to Q in R
   $\mathbf{a}^P$ and $\mathbf{a}^Q$ are absolute accelerations of P and Q in R

8.6   VECTOR DIFFERENTIATION, ANGULAR VELOCITY

The analytical developments of velocity, acceleration, relative velocity, and relative acceleration all involve vector differentiation. If a vector is expressed in terms of fixed unit vectors, then vector differentiation is, in effect, reduced to scalar differentiation—that is, differentiation of the components. Specifically, if a vector $\mathbf{v}(t)$ is expressed as

$$\mathbf{v}(t) = v_1(t)\mathbf{n}_1 + v_2(t)\mathbf{n}_2 + v_3(t)\mathbf{n}_3 \tag{8.16}$$

where $\mathbf{n}_i$ ($i = 1, 2, 3$) are mutually perpendicular unit vectors fixed in a reference frame R, then the derivative of $\mathbf{v}(t)$ in R is simply

$$\frac{d\mathbf{v}}{dt} = \dot{v}_1\mathbf{n}_1 + \dot{v}_2\mathbf{n}_2 + \dot{v}_3\mathbf{n}_3 = \dot{v}_i\mathbf{n}_i \tag{8.17}$$

where the overdot designates time (t) differentiation and the repeated subscript index designates a sum from 1 to 3 over the index. Thus, vector differentiation may be accomplished by simply expressing a vector in terms of fixed unit vectors and then differentiating the components.

It happens, however, particularly with biosystems, that it is often not convenient to express vectors in terms of fixed unit vectors. Indeed, it is usually more convenient to express vectors in terms of "local" unit vectors—that is, in terms of vectors fixed in a limb of the body. For example, to model induced ball rotation in a baseball pitch, it is convenient to use unit vectors fixed in the hand as opposed to unit vectors fixed in the ground frame. This then raises the need to be able to differentiate nonfixed unit vectors—that is, unit vectors with variable orientations.

To explore this, consider a vector $\mathbf{c}$ fixed in a body B, which in turn has a general motion (translation and rotation) in a reference frame R as in Figure 8.7. The derivative of $\mathbf{c}$ relative to an observer fixed in R may be expressed in the relatively simple form

$$\frac{d\mathbf{c}}{dt} = \boldsymbol{\omega} \times \mathbf{c} \tag{8.18}$$

where $\boldsymbol{\omega}$ is the "angular velocity" of B in R.

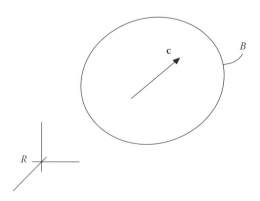

FIGURE 8.7   A vector $\mathbf{c}$, fixed in a body B, moving in a reference frame R.

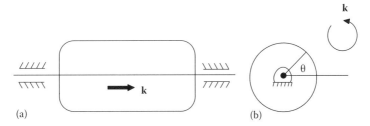

FIGURE 8.8   A body rotating about a fixed axis. (a) Side view and (b) end view.

A remarkable feature of Equation 8.18 is that the derivative is evaluated by a vector multiplication—a useful numerical algorithm.

The utility of Equation 8.18 is, of course, dependent upon knowing or having an expression for the angular velocity vector $\boldsymbol{\omega}$. Recall that in elementary mechanics, angular velocity is usually defined in terms of a rate of rotation of a body about a fixed axis as in Figure 8.8, where $\boldsymbol{\omega}$ may be expressed simply as

$$\boldsymbol{\omega} = \dot{\theta}\mathbf{k} \tag{8.19}$$

where

  $\mathbf{k}$ is a unit vector parallel to the fixed axis of rotation

  θ is the angle between a fixed line perpendicular to $\mathbf{k}$ and a line fixed in the rotating body, also perpendicular to $\mathbf{k}$

The concise form of Equation 8.19 has sometimes led this expression of $\boldsymbol{\omega}$ and the representation in Figure 8.8 to be dubbed: "simple angular velocity."

For more general movement of a body, that is, nonlinear or three-dimensional rotation, the angular velocity is often defined as "the time rate of change of orientation." Thus, for this more general body movement (as occurs with limbs of the human body), to obtain an expression for the angular velocity, we need a means of describing the orientation of the body and then a means of measuring the time rate of change of that orientation.

To develop this, consider again a body B moving in a reference frame R as in Figure 8.9. Let $\mathbf{n}_1$, $\mathbf{n}_2$, and $\mathbf{n}_3$ be mutually perpendicular unit vectors fixed in B and, as before, let $\mathbf{c}$ be a vector, also fixed in B, for which we have to evaluate the derivative $d\mathbf{c}/dt$. Since $\mathbf{n}_1$, $\mathbf{n}_2$, and $\mathbf{n}_3$ are fixed in B, we can think of the orientations of $\mathbf{n}_1$, $\mathbf{n}_2$, $\mathbf{n}_3$ in R as defining the orientation of B in R. The rate of

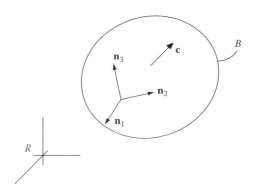

FIGURE 8.9   A body B, with unit vectors $\mathbf{n}_1$, $\mathbf{n}_2$, $\mathbf{n}_3$ fixed in B, and moving in a reference frame R.

change of orientation of B in R (i.e., the angular velocity of B in R) may then be defined in terms of the derivatives of $\mathbf{n}_1$, $\mathbf{n}_2$, and $\mathbf{n}_3$. Specifically, the angular velocity ω of B in R may be defined as

$$\omega = (\dot{\mathbf{n}}_2 \cdot \mathbf{n}_3)\mathbf{n}_1 + (\dot{\mathbf{n}}_3 \cdot \mathbf{n}_1)\mathbf{n}_2 + (\dot{\mathbf{n}}_1 \cdot \mathbf{n}_2)\mathbf{n}_3 \tag{8.20}$$

where, as before, the overdot designates time differentiation.

The definition of Equation 8.20, however, raises several questions: first, as the form for angular velocity of Equation 8.20 is very different from that for simple angular velocity of Equation 8.19, how are they consistent? Next, how does the angular velocity form of Equation 8.20 produce the derivative of a vector $\mathbf{c}$ of B, as desired in Equation 8.18? And finally, what is the utility of Equation 8.20? If we are insightful enough to know the derivatives of $\mathbf{n}_1$, $\mathbf{n}_2$, and $\mathbf{n}_3$, might we not also be insightful enough to know the derivative of $\mathbf{c}$?

Remarkably, each of these questions has a relatively simple and satisfying answer. First, consider the simple angular velocity question: consider a specialization of Equation 8.20, where we have a body B rotating about a fixed axis as in Figure 8.8 and as shown again in Figure 8.10. In this case, let unit vector $\mathbf{n}_3$ be parallel to the axis of rotation as shown. Consequently, $\mathbf{n}_1$ and $\mathbf{n}_2$, being perpendicular to $\mathbf{n}_3$ and fixed in B, will rotate with B. Let the rotation rate be $\dot{\theta}$, as before, as shown where θ is the angle measured in a plane normal to the rotation axis, between a line in the fixed frame and a line in the body as in the end view of B shown in Figure 8.11, where the unit vectors $\mathbf{n}_1$ and $\mathbf{n}_2$ are shown and fixed unit vectors $\mathbf{N}_1$ and $\mathbf{N}_2$ are also shown. Consider the derivatives of $\mathbf{n}_1$, $\mathbf{n}_2$, and $\mathbf{n}_3$ as required for Equation 8.20. For $\mathbf{n}_1$, observe from Figure 8.11 that $\mathbf{n}_1$ may be expressed as

$$\mathbf{n}_1 = \cos\theta \mathbf{N}_1 + \sin\theta \mathbf{N}_2 \tag{8.21}$$

Then $\dot{\mathbf{n}}_1$ is simply

$$\dot{\mathbf{n}}_1 = (-\sin\theta \mathbf{N}_1 + \cos\theta \mathbf{N}_2)\dot{\theta} \tag{8.22}$$

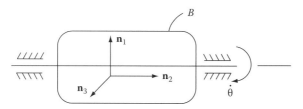

FIGURE 8.10   A body rotating about a fixed axis.

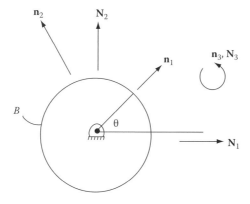

FIGURE 8.11   End view of fixed-axis rotating body.

Observe from Equation 8.21 that the quantity in parenthesis is simply $\mathbf{n}_3 \times \mathbf{n}_1$. Hence, $\dot{\mathbf{n}}_1$ becomes

$$\dot{\mathbf{n}}_1 = \dot{\theta}\mathbf{n}_3 \times \mathbf{n}_1 = \dot{\theta}\mathbf{n}_2 \qquad (8.23)$$

Similarly, for $\mathbf{n}_2$ we find that

$$\dot{\mathbf{n}}_2 = \dot{\theta}\mathbf{n}_3 \times \mathbf{n}_2 = \dot{\theta}\mathbf{n}_1 \qquad (8.24)$$

Since $\mathbf{n}_3$ is fixed, we have

$$\dot{\mathbf{n}}_3 = 0 \qquad (8.25)$$

By substituting from Equations 8.23 through 8.25 into Equation 8.20, we have

$$\boldsymbol{\omega} = \left[(-\dot{\theta}\mathbf{n}_1) \cdot \mathbf{n}_3\right]\mathbf{n}_1 + [0 \cdot \mathbf{n}_1]\mathbf{n}_2 + [\dot{\theta}\mathbf{n}_2 \cdot \mathbf{n}_2]\mathbf{n}_3$$

or

$$\boldsymbol{\omega} = \dot{\theta}\mathbf{n}_3 \qquad (8.26)$$

Equation 8.26 is seen to be identical to Equations 8.19. Therefore, we see that the general angular velocity expression of Equation 8.20 is not inconsistent with the simple angular velocity expression of Equation 8.19. Thus, we can view Equation 8.20 as a generalization of Equation 8.19, or equivalently, Equation 8.19 as a specialization of Equation 8.20.

Consider next the more fundamental issue of how the angular velocity produces a derivative of a body-fixed vector as represented in Equation 8.18. To discuss this, let the vector $\mathbf{c}$ of B be expressed in terms of the unit vectors $\mathbf{n}_i$ $(i = 1, 2, 3)$ as

$$\mathbf{c} = c_1\mathbf{n}_1 + c_2\mathbf{n}_2 + c_3\mathbf{n}_3 = c_i\mathbf{n}_i \qquad (8.27)$$

Since, like $\mathbf{c}$, the $\mathbf{n}_i$ are fixed in B, the components c_i will be constants. Therefore, the derivative of $\mathbf{c}$ (in R) is simply

$$\frac{d\mathbf{c}}{dt} = \dot{\mathbf{c}} = c_1\dot{\mathbf{n}}_1 + c_2\dot{\mathbf{n}}_2 + c_3\dot{\mathbf{n}}_3 \qquad (8.28)$$

Consider the $\dot{\mathbf{n}}_i$. As vectors, the $\dot{\mathbf{n}}_i$ may be expressed in terms of the $\mathbf{n}_i$ as

$$\dot{\mathbf{n}}_i = (\dot{\mathbf{n}}_i \cdot \mathbf{n}_1)\mathbf{n}_1 + (\dot{\mathbf{n}}_i \cdot \mathbf{n}_2)\mathbf{n}_2 + (\dot{\mathbf{n}}_i \cdot \mathbf{n}_3)\mathbf{n}_3 \qquad (8.29)$$

Hence for $\dot{\mathbf{n}}_1$, we have

$$\dot{\mathbf{n}}_1 = (\dot{\mathbf{n}}_1 \cdot \mathbf{n}_1)\mathbf{n}_1 + (\dot{\mathbf{n}}_1 \cdot \mathbf{n}_2)\mathbf{n}_2 + (\dot{\mathbf{n}}_1 \cdot \mathbf{n}_3)\mathbf{n}_3 \qquad (8.30)$$

Note, however, that $\dot{\mathbf{n}}_1 \cdot \mathbf{n}_1$ is zero: that is, as $\mathbf{n}_1$ is a unit vector, we have $\mathbf{n}_1 \cdot \mathbf{n}_1 = 1$ and then

$$\frac{d(\mathbf{n}_1 \cdot \mathbf{n}_1)}{dt} = 0 = \dot{\mathbf{n}}_1 \cdot \mathbf{n}_1 + \mathbf{n}_1 \cdot \dot{\mathbf{n}}_1 = 2\mathbf{n}_1 \cdot \dot{\mathbf{n}}_1 \quad \text{or} \quad \mathbf{n}_1 \cdot \dot{\mathbf{n}}_1 = 0 \tag{8.31}$$

Note, with $\mathbf{n}_1$ perpendicular to $\mathbf{n}_3$, we have $\mathbf{n}_1 \cdot \mathbf{n}_3 = 0$ and then

$$\frac{d(\mathbf{n}_1 \cdot \mathbf{n}_3)}{dt} = 0 = \dot{\mathbf{n}}_1 \cdot \mathbf{n}_3 + \mathbf{n}_1 \cdot \dot{\mathbf{n}}_3 \quad \text{or} \quad \dot{\mathbf{n}}_1 \cdot \mathbf{n}_3 = -\dot{\mathbf{n}}_3 \cdot \mathbf{n}_1 \tag{8.32}$$

Therefore, from Equation 8.30, $\dot{\mathbf{n}}_1$ becomes

$$\dot{\mathbf{n}}_1 = 0\mathbf{n}_1 + (\dot{\mathbf{n}}_1 \cdot \mathbf{n}_2)\mathbf{n}_2 - (\dot{\mathbf{n}}_3 \cdot \mathbf{n}_1)\mathbf{n}_3 \tag{8.33}$$

Observe, however, from Equation 8.20 that $\boldsymbol{\omega} \times \mathbf{n}_1$ is

$$\boldsymbol{\omega} \times \mathbf{n}_1 = -(\dot{\mathbf{n}}_3 \cdot \mathbf{n}_1)\mathbf{n}_3 + (\dot{\mathbf{n}}_1 \cdot \mathbf{n}_2)\mathbf{n}_2 \tag{8.34}$$

Thus by comparing Equations 8.33 and 8.34, we see that

$$\dot{\mathbf{n}}_1 = \boldsymbol{\omega} \times \mathbf{n}_1 \tag{8.35}$$

Similarly, we have

$$\dot{\mathbf{n}}_2 = \boldsymbol{\omega} \times \mathbf{n}_2 \quad \text{and} \quad \dot{\mathbf{n}}_3 = \boldsymbol{\omega} \times \mathbf{n}_3 \tag{8.36}$$

Finally, by substituting from Equations 8.35 and 8.36 into Equation 8.28, $\dot{\mathbf{c}}$ becomes

$$\dot{\mathbf{c}} = c_1 \boldsymbol{\omega} \times \mathbf{n}_1 + c_2 \boldsymbol{\omega} \times \mathbf{n}_2 + c_3 \boldsymbol{\omega} \times \mathbf{n}_3 = \boldsymbol{\omega} \times c_1 \mathbf{n}_1 + \boldsymbol{\omega} \times c_2 \mathbf{n}_2 + \boldsymbol{\omega} \times c_3 \mathbf{n}_3$$

$$= \boldsymbol{\omega} \times (c_1 \mathbf{n}_1 + c_2 \mathbf{n}_2 + c_3 \mathbf{n}_3) = \boldsymbol{\omega} \times \mathbf{c} \tag{8.37}$$

as expected.

The third question about the ease of obtaining expressions for ω is the most difficult to address: before discussing the issue, it is helpful to first consider a couple of additional kinematic and geometric procedures as outlined in Section 8.7.

8.7 TWO USEFUL KINEMATIC PROCEDURES

8.7.1 DIFFERENTIATION IN DIFFERENT REFERENCE FRAMES

An important feature of vector differentiation is the reference frame in which the derivative is evaluated. For example, in Section 8.6 the derivatives of vectors fixed in body B were evaluated in reference frame R. Had there been a reference frame fixed in B, say R_B, the derivatives of vectors fixed in B, evaluated in R_B, would of course be zero. The reference frame where a derivative is evaluated, is often obvious and thus it (the reference frame) need not be explicitly mentioned. However, when

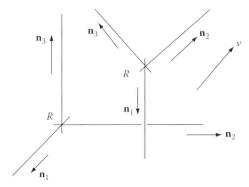

FIGURE 8.12   Reference frames R and $\hat{R}$ moving relative to each other and a variable vector $\mathbf{v}$.

there are several reference frames used in a given analysis, as there are several bodies involved with each body having an embedded reference frame,* it is necessary to know which reference frame is to be used for evaluation of the derivative.

To develop this concept, consider two reference frames R and $\hat{R}$ moving relative to each other and a vector $\mathbf{v}$, which is not fixed in either of the frames, as in Figure 8.12. (i.e., let $\mathbf{v}$ be a time-varying vector relative to observers in both R and $\hat{R}$) A fundamental issue then is how the derivatives of $\mathbf{v}$ evaluated in R and $\hat{R}$ differ from each other. To explore this, let $\mathbf{n}_i$ and $\hat{\mathbf{n}}_i$ ($i = 1, 2, 3$) be mutually perpendicular unit vector sets fixed in R and $\hat{R}$, respectively. Let $\mathbf{v}$ be expressed in terms of the $\hat{\mathbf{n}}_i$ as

$$\mathbf{v} = \hat{v}_1\hat{\mathbf{n}}_1 + \hat{v}_2\hat{\mathbf{n}}_2 + \hat{v}_3\hat{\mathbf{n}}_3 = \hat{v}_i\hat{\mathbf{n}}_i \tag{8.38}$$

Thus for an observer in $\hat{R}$, where $\hat{\mathbf{n}}_i$ are fixed, the derivative of $\mathbf{v}$ is simply

$$\frac{{}^{\hat{R}}d\mathbf{v}}{dt} = \dot{\hat{v}}_1\hat{\mathbf{n}}_1 + \dot{\hat{v}}_2\hat{\mathbf{n}}_2 + \dot{\hat{v}}_3\hat{\mathbf{n}}_3 \tag{8.39}$$

where we have introduced the super-prefix $\hat{R}$ to designate that the derivative is evaluated in $\hat{R}$.

For an observer in R, however, the $\hat{\mathbf{n}}_i$ are not constant. Thus, in R, the derivative of $\mathbf{v}$ is

$$\frac{{}^{R}d\mathbf{v}}{dt} = \dot{\hat{v}}_1\hat{\mathbf{n}}_1 + \hat{v}_1\frac{{}^{R}d\hat{\mathbf{n}}_1}{dt} + \dot{\hat{v}}_2\hat{\mathbf{n}}_2 + \hat{v}_2\frac{{}^{R}d\hat{\mathbf{n}}_2}{dt} + \dot{\hat{v}}_3\hat{\mathbf{n}}_3 + \hat{v}_3\frac{{}^{R}d\hat{\mathbf{n}}_3}{dt}$$

$$= (\dot{\hat{v}}_1\hat{\mathbf{n}}_1 + \dot{\hat{v}}_2\hat{\mathbf{n}}_2 + \dot{\hat{v}}_3\hat{\mathbf{n}}_3) + \left(\hat{v}_1\frac{{}^{R}d\hat{\mathbf{n}}_1}{dt} + \hat{v}_2\frac{{}^{R}d\hat{\mathbf{n}}_2}{dt} + \hat{v}_3\frac{{}^{R}d\hat{\mathbf{n}}_3}{dt}\right)$$

$$= \frac{{}^{\hat{R}}d\mathbf{v}}{dt} + \left(\hat{v}_1\frac{{}^{R}d\hat{\mathbf{n}}_1}{dt} + \hat{v}_2\frac{{}^{R}d\hat{\mathbf{n}}_2}{dt} + \hat{v}_3\frac{{}^{R}d\hat{\mathbf{n}}_3}{dt}\right) \tag{8.40}$$

where the last equality is seen in view of Equation 8.39.

---

* Observe that from a purely kinematic perspective, if a body B has an embedded reference frame, say R_B, then the movement of B is completely determined by the movement of R_B and vice versa. Thus, kinematically (i.e., aside from the mass), there is no difference between a rigid body and a reference frame.

Equation 8.40 provides the sought after relation between the derivatives in the two frames. Observe further, however, that the last set of terms can be expressed in terms of the relative angular velocity of the frames as the unit vector derivatives are simply

$$\frac{^{R}d\hat{\mathbf{n}}_{i}}{dt} = \boldsymbol{\omega} \times \hat{\mathbf{n}}_{i} \tag{8.41}$$

(See Equation 8.18), where $\boldsymbol{\omega}$ is the angular velocity of $\hat{R}$ relative to R, which may be written more explicitly as $^{R}\boldsymbol{\omega}^{\hat{R}}$. Hence, we have

$$\hat{\mathbf{v}}_{1}\frac{^{R}d\hat{\mathbf{n}}_{1}}{dt} + \hat{\mathbf{v}}_{2}\frac{^{R}d\hat{\mathbf{n}}_{2}}{dt} + \hat{\mathbf{v}}_{3}\frac{^{R}d\hat{\mathbf{n}}_{3}}{dt} = \hat{\mathbf{v}}_{i}\frac{^{R}d\hat{\mathbf{n}}_{i}}{dt} = \hat{\mathbf{v}}_{i}{}^{R}\boldsymbol{\omega}^{\hat{R}} \times \hat{\mathbf{n}}_{i}$$

$$= {}^{R}\boldsymbol{\omega}^{\hat{R}} \times (\hat{\mathbf{v}}_{i}\hat{\mathbf{n}}_{i}) = {}^{R}\boldsymbol{\omega}^{\hat{R}} \times \mathbf{v} \tag{8.42}$$

Therefore, Equation 8.40 becomes

$$\frac{^{R}d\mathbf{v}}{dt} = \frac{^{\hat{R}}d\mathbf{v}}{dt} + {}^{R}\boldsymbol{\omega}^{\hat{R}} \times \mathbf{v} \tag{8.43}$$

Since $\mathbf{v}$ could represent any vector, we can write Equation 8.43 as

$$\frac{^{R}d()}{dt} = {}^{\hat{R}}d()/dt + {}^{R}\boldsymbol{\omega}^{\hat{R}} \times () \tag{8.44}$$

where () represents any vector.

Equation 8.44 shows that vector derivatives in different reference frames are related to each other by the relative angular velocity of the frames.

8.7.2 Addition Theorem for Angular Velocity

Perhaps the most important use of Equation 8.44 is in the development of the addition theorem for angular velocity: consider a body B moving relative to two reference frames R and $\hat{\mathbf{R}}$ as in Figure 8.13. Let $\mathbf{c}$ be a vector fixed in B. Then from Equation 8.18, the derivatives of $\mathbf{c}$ with respect to R and $\hat{\mathbf{R}}$ are

$$\frac{^{R}d\mathbf{c}}{dt} = {}^{R}\boldsymbol{\omega}^{B} \times \mathbf{c} \tag{8.45}$$

and

$$\frac{^{\hat{R}}d\mathbf{c}}{dt} = {}^{\hat{R}}\boldsymbol{\omega}^{B} \times \mathbf{c} \tag{8.46}$$

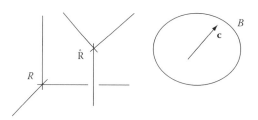

FIGURE 8.13   A body B with a fixed vector $\mathbf{c}$ moving relative to two references frame.

Observe, however, from Equation 8.44 that these derivatives are related by

$$\frac{^R d\mathbf{c}}{dt} = \frac{^{\hat{R}} d\mathbf{c}}{dt} + {}^R\boldsymbol{\omega}^{\hat{R}} \times \mathbf{c} \tag{8.47}$$

Then by substituting from Equations 8.45 and 8.46, we have

$$^R\boldsymbol{\omega}^B \times \mathbf{c} = {}^{\hat{R}}\boldsymbol{\omega}^R \times \mathbf{c} + {}^R\boldsymbol{\omega}^{\hat{R}} \times \mathbf{c}$$

$$= ({}^{\hat{R}}\boldsymbol{\omega}^R + {}^R\boldsymbol{\omega}^{\hat{R}}) \times \mathbf{c} \tag{8.48}$$

Finally, as $\mathbf{c}$ is any vector fixed in B, Equation 8.48 must hold for all such vectors. That is,

$$^R\boldsymbol{\omega}^B \times (\,) = ({}^{\hat{R}}\boldsymbol{\omega}^B + {}^R\boldsymbol{\omega}^{\hat{R}}) \times (\,) \tag{8.49}$$

where ( ) represents any (or all) vectors fixed in B. Then for Equation 8.49 to be valid, we must have

$$^R\boldsymbol{\omega}^B = {}^{\hat{R}}\boldsymbol{\omega}^B + {}^R\boldsymbol{\omega}^{\hat{R}} \tag{8.50}$$

Equation 8.50 is the "addition theorem for angular velocity." It is readily extended to n intermediate reference frames as (see Figure 8.14)

$$^R\boldsymbol{\omega}^B = {}^R\boldsymbol{\omega}^{R_0} + {}^{R_0}\boldsymbol{\omega}^{R_1} + \cdots + {}^{R_{n-1}}\boldsymbol{\omega}^{R_n} + {}^{R_n}\boldsymbol{\omega}^B \tag{8.51}$$

Regarding notation in biosystems, such as with human body models, it is convenient to simplify the notation of Equation 8.51. A way of doing this is to observe that the terms on the right side are angular velocities of frames relative to their adjacent lower numbered frame (i.e., relative angular velocities). A simple index with an overhat can be used to designate the terms. That is,

$$\overset{D}{\hat{\boldsymbol{\omega}}_k} = {}^{R_j}\boldsymbol{\omega}^{R_k} \tag{8.52}$$

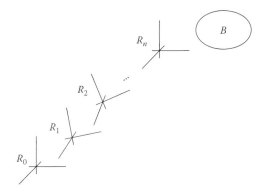

FIGURE 8.14 A body B moving relative to a series of reference frames.

where j is the numerical index below k. Also in this context, we can designate absolute angular velocities by deleting the overhat. That is,

$$^R\boldsymbol{\omega}^B \overset{D}{=} \boldsymbol{\omega}_B \tag{8.53}$$

Equation 8.51 then becomes

$$\boldsymbol{\omega}_B = \hat{\boldsymbol{\omega}}_0 + \hat{\boldsymbol{\omega}}_1 + \hat{\boldsymbol{\omega}}_2 + \cdots + \hat{\boldsymbol{\omega}}_n + \hat{\boldsymbol{\omega}}_B \tag{8.54}$$

8.8   CONFIGURATION GRAPHS

We can establish a simple and reliable procedure for obtaining angular velocity vectors by using configuration graphs, which are simply diagrams for relating unit vector sets to one another. To this end, consider two typical adjoining bodies of a multibody system such as chest and upper arm of a human body model as in Figure 8.15. Let these bodies be represented by generic bodies B_j and B_k, which can have arbitrary orientation relative to each other as depicted in Figure 8.16. Let $\mathbf{n}_{ji}$ and $\mathbf{n}_{ki}$ ($i = 1, 2, 3$) be mutually perpendicular unit vector sets fixed in B_j and B_k as shown. A fundamental geometric question is then: given the orientations of the $\mathbf{n}_{ji}$ and the $\mathbf{n}_{ki}$, what are the relations expressing the vectors relative to each other? To answer this question, consider again the movement of a body B in a reference frame R as in Figure 8.17. Let $\mathbf{n}_i$ and $\mathbf{N}_i$ ($i = 1, 2, 3$) be mutually perpendicular unit vector sets fixed in B and R as shown. Then the objective is to express the $\mathbf{N}_i$ and the $\mathbf{n}_i$ in terms of each other as

$$\mathbf{N}_1 = ()\mathbf{n}_1 + ()\mathbf{n}_2 + ()\mathbf{n}_3$$

$$\mathbf{N}_2 = ()\mathbf{n}_1 + ()\mathbf{n}_2 + ()\mathbf{n}_3 \tag{8.55}$$

$$\mathbf{N}_3 = ()\mathbf{n}_1 + ()\mathbf{n}_2 + ()\mathbf{n}_3$$

and

$$\mathbf{n}_1 = ()\mathbf{N}_1 + ()\mathbf{N}_2 + ()\mathbf{N}_3$$

$$\mathbf{n}_2 = ()\mathbf{N}_1 + ()\mathbf{N}_2 + ()\mathbf{N}_3 \tag{8.56}$$

$$\mathbf{n}_3 = ()\mathbf{N}_1 + ()\mathbf{N}_2 + ()\mathbf{N}_3$$

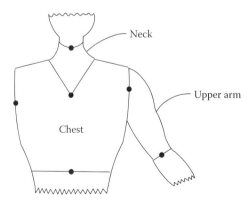

FIGURE 8.15   Human body model: chest and upper arm.

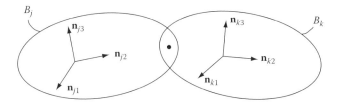

FIGURE 8.16   Two typical adjoining bodies each with fixed unit vector sets.

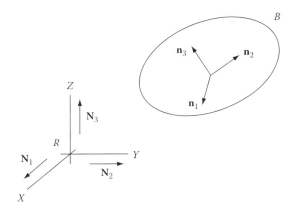

FIGURE 8.17   A body B moving in a reference frame R and associated unit vectors.

The question then is: what quantities should be placed in the parentheses in these equations? To answer this question, we need a means of defining the orientation of B in R. To this end, imagine B to initially be oriented in R so that the unit vector sets are mutually aligned as in Figure 8.18. Next imagine B to be brought into any desired orientation relative to R, by successive rotations of B about the directions of the unit vectors $\mathbf{n}_1$, $\mathbf{n}_2$, and $\mathbf{n}_3$ through angles α, β, and γ, respectively. The values of α, β, and γ then determine the orientation of B in R.

Returning now to Equations 8.55 and 8.56, the parenthetical quantities are readily obtained by introducing two intermediate reference frames, say $\hat{R}$ and $\hat{B}$, with unit vector sets $\hat{\mathbf{N}}_i$ and $\hat{\mathbf{n}}_i$ $(i = 1, 2, 3)$ parallel to the $\mathbf{n}_i$ after B is rotated through the angles α and β, respectively. Consider first the rotation of B about the $\mathbf{n}_1$ direction through the angle α: the unit vectors $\hat{\mathbf{N}}_i$ of $\hat{R}_i$ will be oriented relative to

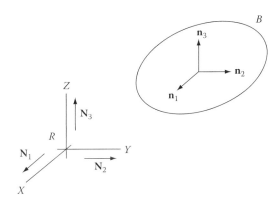

FIGURE 8.18   Orientation of a body B with unit vectors aligned with those of a reference frame R.

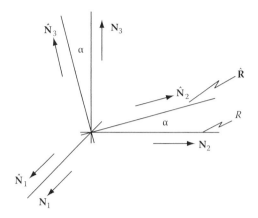

FIGURE 8.19   Relative orientation (or inclination) of unit vector sets $\mathbf{N}_i$, and $\hat{\mathbf{N}}_i$, (i = 1, 2, 3).

the $\mathbf{N}_i$ of R as in Figure 8.19. Then, as in Equations 8.55 and 8.56, we seek to express the $\mathbf{N}_i$ in terms of the $\hat{\mathbf{N}}_i$ (and vice versa) as

$$\mathbf{N}_1 = (\ )\hat{\mathbf{N}}_1 + (\ )\hat{\mathbf{N}}_2 + (\ )\hat{\mathbf{N}}_3$$

$$\mathbf{N}_2 = (\ )\hat{\mathbf{N}}_1 + (\ )\hat{\mathbf{N}}_2 + (\ )\hat{\mathbf{N}}_3 \qquad (8.57)$$

$$\mathbf{N}_3 = (\ )\hat{\mathbf{N}}_1 + (\ )\hat{\mathbf{N}}_2 + (\ )\hat{\mathbf{N}}_3$$

and

$$\hat{\mathbf{N}}_1 = (\ )\mathbf{N}_1 + (\ )\mathbf{N}_2 + (\ )\mathbf{N}_3$$

$$\hat{\mathbf{N}}_2 = (\ )\mathbf{N}_1 + (\ )\mathbf{N}_2 + (\ )\mathbf{N}_3 \qquad (8.58)$$

$$\hat{\mathbf{N}}_3 = (\ )\mathbf{N}_1 + (\ )\mathbf{N}_2 + (\ )\mathbf{N}_3$$

Since $\mathbf{N}_1$ and $\hat{\mathbf{N}}_1$ are equal, these parenthetical expressions are considerably simpler than those of Equations 8.55 and 8.56. Indeed, observe from Figure 8.19 that $\mathbf{N}_2$, $\mathbf{N}_3$, $\hat{\mathbf{N}}_2$, and $\hat{\mathbf{N}}_3$ are all parallel to the same plane and that $\mathbf{N}_1$ and $\hat{\mathbf{N}}_1$ are perpendicular to that plane. Thus we can immediately identify 10 of the 18 unknown coefficients in Equations 8.57 and 8.58 as ones and zeros. Specifically, the terms involving $\mathbf{N}_1$ and $\hat{\mathbf{N}}_1$ have ones and zeros as

$$\mathbf{N}_1 = (1)\hat{\mathbf{N}}_1 + (0)\hat{\mathbf{N}}_2 + (0)\hat{\mathbf{N}}_3$$

$$\mathbf{N}_2 = (0)\hat{\mathbf{N}}_1 + (\ )\hat{\mathbf{N}}_2 + (\ )\hat{\mathbf{N}}_3 \qquad (8.59)$$

$$\mathbf{N}_3 = (0)\hat{\mathbf{N}}_1 + (\ )\hat{\mathbf{N}}_2 + (\ )\hat{\mathbf{N}}_3$$

and

$$\hat{\mathbf{N}}_1 = (1)\mathbf{N}_1 + (0)\mathbf{N}_2 + (0)\mathbf{N}_3$$

$$\hat{\mathbf{N}}_2 = (0)\mathbf{N}_1 + (\ )\mathbf{N}_2 + (\ )\mathbf{N}_3 \qquad (8.60)$$

$$\hat{\mathbf{N}}_3 = (0)\mathbf{N}_1 + (\ )\mathbf{N}_2 + (\ )\mathbf{N}_3$$

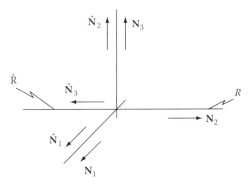

FIGURE 8.20 Rotation of $\hat{R}$ relative to R by 90°.

This leaves eight coefficients to be determined—that is, those relating $\mathbf{N}_2$ and $\mathbf{N}_3$ to $\hat{\mathbf{N}}_2$ and $\hat{\mathbf{N}}_3$. From Figure 8.19 we readily see that these coefficients involve sines and cosines of α. To determine these coefficients, observe that when α is zero, the unit vector sets are mutually aligned. Thus we conclude that unknown "diagonal" coefficients of Equations 8.59 and 8.60 are $\cos \alpha$. That is, the terms having equal subscript values on both sides of an equation are equal to one another when α is zero, and they vary relative to one another as $\cos \alpha$. Also, observe that when α is 90°, the unit vector sets have the configuration shown in Figure 8.20. Here, we see that

$$\mathbf{N}_2 = -\hat{\mathbf{N}}_3 \quad \text{and} \quad \hat{\mathbf{N}}_2 = \mathbf{N}_3 \tag{8.61}$$

The "off-diagonal" coefficients of Equations 8.59 and 8.60 are thus $\pm \sin \alpha$ with the minus signs occurring with the coefficients of $\mathbf{N}_2$ and $\hat{\mathbf{N}}_3$. That is, Equations 8.59 and 8.60 become

$$\mathbf{N}_1 = 1\hat{\mathbf{N}}_1 + 0\hat{\mathbf{N}}_2 + 0\hat{\mathbf{N}}_3$$

$$\mathbf{N}_2 = 0\hat{\mathbf{N}}_1 + \cos \alpha \hat{\mathbf{N}}_2 - \sin \alpha \hat{\mathbf{N}}_3 \tag{8.62}$$

$$\mathbf{N}_3 = 0\hat{\mathbf{N}}_1 + \sin \alpha \hat{\mathbf{N}}_2 + \cos \alpha \hat{\mathbf{N}}_3$$

and

$$\hat{\mathbf{N}}_1 = 1\mathbf{N}_1 + 0\mathbf{N}_2 + 0\mathbf{N}_3$$

$$\hat{\mathbf{N}}_2 = 0\mathbf{N}_1 + \cos \alpha \mathbf{N}_2 + \sin \alpha \mathbf{N}_3 \tag{8.63}$$

$$\hat{\mathbf{N}}_3 = 0\mathbf{N}_1 - \sin \alpha \mathbf{N}_2 + \cos \alpha \mathbf{N}_3$$

Equations 8.62 and 8.63 provide the complete identification of the unknown coefficients of Equations 8.57 and 8.58. In view of the foregoing analysis, we see that these coefficients may be obtained by following a few simple rules: If we think of the 18 unknowns in Equations 8.57 and 8.58 as arranged into two 3 × 3 arrays, then the rows and columns of these arrays corresponding to the two equal unit vectors ($\mathbf{N}_1$ and $\hat{\mathbf{N}}_1$) have ones at the common element and zeros at all the other elements (of rows 1 and columns 1). This leaves only two 2 × 2 arrays to be determined: the diagonal elements of these arrays are cosines and the off-diagonal elements are sines. There is only one minus sign in each of the arrays. The minus sign occurs with the sine located as follows: consider a view of the four vectors parallel to the plane normal to $\mathbf{N}_1$ and $\hat{\mathbf{N}}_1$ as in Figure 8.21. For acute angle α, $\mathbf{N}_3$ and $\hat{\mathbf{N}}_2$ may be regarded as being "inside" $\hat{\mathbf{N}}_3$ and $\mathbf{N}_2$, and

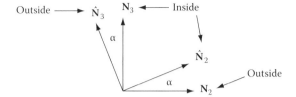

FIGURE 8.21   "Inside" and "outside" unit vectors.

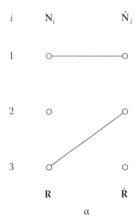

FIGURE 8.22   Configuration graph for the unit vectors of Figure 8.19.

conversely, $\hat{\mathbf{N}}_3$ and $\mathbf{N}_2$ may be regarded as being "outside" $\mathbf{N}_3$ and $\hat{\mathbf{N}}_2$. The minus sign then occurs with the elements of the outside vectors (inside is positive and outside is negative).

These rules may be represented by a simple diagram as in Figure 8.22. In this diagram, called a configuration graph, each dot (or node) represents a unit vector as indicated in the figure. The horizontal line denotes equality and the inclined line denotes inside vectors. The unconnected dots are thus outside vectors. That is, the horizontal line between $\mathbf{N}_1$ and $\hat{\mathbf{N}}_1$ means that $\mathbf{N}_1$ equals $\hat{\mathbf{N}}_1$. The inclined line connecting $\mathbf{N}_3$ and $\hat{\mathbf{N}}_2$ means that $\mathbf{N}_3$ and $\hat{\mathbf{N}}_2$ are "inside vectors." With no lines connecting $\mathbf{N}_2$ and $\hat{\mathbf{N}}_3$, we identify them as "outside vectors."

Equations 8.62 and 8.63 may be written in matrix form as

$$\begin{bmatrix} \mathbf{N}_1 \\ \mathbf{N}_2 \\ \mathbf{N}_3 \end{bmatrix} = \begin{bmatrix} 1 & 0 & 0 \\ 0 & c_\alpha & -s_\alpha \\ 0 & s_\alpha & c_\alpha \end{bmatrix} \begin{bmatrix} \hat{\mathbf{N}}_1 \\ \hat{\mathbf{N}}_2 \\ \hat{\mathbf{N}}_3 \end{bmatrix} \quad \text{and} \quad \begin{bmatrix} \hat{\mathbf{N}}_1 \\ \hat{\mathbf{N}}_2 \\ \hat{\mathbf{N}}_3 \end{bmatrix} = \begin{bmatrix} 1 & 0 & 0 \\ 0 & c_\alpha & s_\alpha \\ 0 & -s_\alpha & c_\alpha \end{bmatrix} \begin{bmatrix} \mathbf{N}_1 \\ \mathbf{N}_2 \\ \mathbf{N}_3 \end{bmatrix} \qquad (8.64)$$

where s_α and c_α are abbreviations for $\sin \alpha$ and $\cos \alpha$. These equations may be written in the simplified forms

$$\mathbf{N} = A\hat{\mathbf{N}} \quad \text{and} \quad \hat{\mathbf{N}} = A^T \mathbf{N} \qquad (8.65)$$

where $\mathbf{N}$, $\hat{\mathbf{N}}$, A, and A^T are defined by inspection and comparison of the equations.

Next, consider the rotation of B about $\mathbf{n}_2$ (or $\hat{\mathbf{N}}_2$) through the angle β, bringing $\mathbf{n}_1$, $\mathbf{n}_2$, and $\mathbf{n}_3$ into the orientation of $\hat{\mathbf{n}}_1$, $\hat{\mathbf{n}}_2$, and $\hat{\mathbf{n}}_3$ as in Figure 8.23. Here, we are interested in the relation of the $\hat{\mathbf{N}}_i$

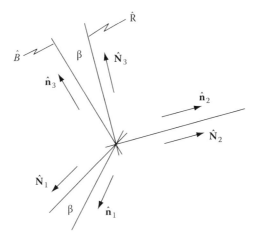

FIGURE 8.23   Relative orientation of unit vector sets $\hat{\mathbf{N}}_i$ and $\mathbf{n}_i$ ($i = 1, 2, 3$).

and $\hat{\mathbf{n}}_i$. With $\hat{\mathbf{n}}_2$ and $\hat{\mathbf{N}}_2$ being equal, we see by following the same procedures as with the α-rotation that the vector sets are related as

$$\hat{\mathbf{N}}_1 = c_\beta \hat{\mathbf{n}}_1 + s_\beta \hat{\mathbf{n}}_3 \qquad\qquad \mathbf{n}_1 = c_\beta \hat{\mathbf{N}}_1 - s_\beta \hat{\mathbf{N}}_3$$

$$\hat{\mathbf{N}}_2 = \hat{\mathbf{n}}_2 \qquad \text{and} \quad \mathbf{n}_2 = \hat{\mathbf{N}}_2 \qquad\qquad (8.66)$$

$$\hat{\mathbf{N}}_3 = -s_\beta \hat{\mathbf{n}}_1 + c_\beta \hat{\mathbf{N}}_3 \qquad\qquad \mathbf{n}_3 = s_\beta \hat{\mathbf{N}}_1 + c_\beta \hat{\mathbf{N}}_3$$

where s_β and c_β represent $\sin \beta$ and $\cos \beta$. In matrix form, these equations are

$$\begin{bmatrix} \hat{\mathbf{N}}_1 \\ \hat{\mathbf{N}}_2 \\ \hat{\mathbf{N}}_3 \end{bmatrix} = \begin{bmatrix} c_\beta & 0 & s_\beta \\ 0 & 1 & 0 \\ -s_\beta & 0 & c_\beta \end{bmatrix} \begin{bmatrix} \hat{\mathbf{n}}_1 \\ \hat{\mathbf{n}}_2 \\ \hat{\mathbf{n}}_3 \end{bmatrix} \quad \text{and} \quad \begin{bmatrix} \hat{\mathbf{n}}_1 \\ \hat{\mathbf{n}}_2 \\ \hat{\mathbf{n}}_3 \end{bmatrix} = \begin{bmatrix} c_\beta & 0 & -s_\beta \\ 0 & 1 & 0 \\ s_\beta & 0 & c_\beta \end{bmatrix} \begin{bmatrix} \hat{\mathbf{N}}_1 \\ \hat{\mathbf{N}}_2 \\ \hat{\mathbf{N}}_3 \end{bmatrix} \qquad (8.67)$$

or simply

$$\hat{\mathbf{N}} = B\hat{\mathbf{n}} \quad \text{and} \quad \hat{\mathbf{n}} = B^T \hat{\mathbf{N}} \qquad\qquad (8.68)$$

where, as before, $\hat{\mathbf{N}}$, $\hat{\mathbf{n}}$, B, and B^T are defined by inspection and comparison of the equations.

Consider the vectors parallel to the plane normal to $\hat{\mathbf{N}}_2$ and $\hat{\mathbf{n}}_2$ as in Figure 8.24. We see that the inside vectors are $\hat{\mathbf{N}}_2$ and $\hat{\mathbf{n}}_3$ and that the outside vectors are $\hat{\mathbf{N}}_3$ and $\hat{\mathbf{n}}_2$. Then analogous to Figure 8.22, we obtain the configuration graph of Figure 8.25. By following the rules as before (the horizontal line designates equality and the inclined line identifies the inside vectors), we see that the graph of Figure 8.25 is equivalent to Equation 8.67.

Finally, consider the rotation of B about $\mathbf{n}_3$ (or $\hat{\mathbf{n}}_3$) through the angle γ bringing $\mathbf{n}_1$, $\mathbf{n}_2$, and $\mathbf{n}_3$, and consequently B, into their final orientation as represented in Figure 8.26. In this case, the equations relating the $\mathbf{n}_i$ and the $\hat{\mathbf{n}}_i$ are

$$\hat{\mathbf{n}}_1 = c_\gamma \mathbf{n}_1 - s_\gamma \mathbf{n}_2 \qquad\qquad \mathbf{n}_1 = c_\gamma \hat{\mathbf{n}}_1 - s_\gamma \hat{\mathbf{n}}_2$$

$$\hat{\mathbf{n}}_2 = s_\gamma \mathbf{n}_1 - c_\gamma \mathbf{n}_2 \quad \text{and} \quad \mathbf{n}_2 = s_\gamma \hat{\mathbf{n}}_1 - c_\gamma \hat{\mathbf{n}}_2 \qquad (8.69)$$

$$\hat{\mathbf{n}}_3 = \mathbf{n}_3 \qquad\qquad \mathbf{n}_3 = \hat{\mathbf{n}}_3$$

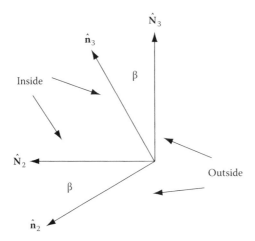

FIGURE 8.24 Inside and outside unit vectors.

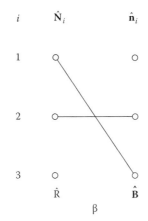

FIGURE 8.25 Configuration graph for the unit vectors of Figure 8.23.

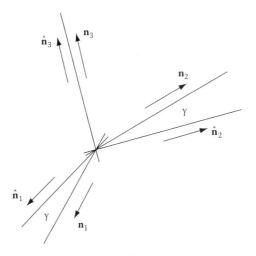

FIGURE 8.26 Relative orientation of unit vector sets $\hat{\mathbf{n}}_i$ and $\mathbf{n}_i$ ($i = 1, 2, 3$).

where s_γ and c_γ are $\sin \gamma$ and $\cos \gamma$. In matrix form, these equations are

$$\begin{bmatrix} \hat{\mathbf{n}}_1 \\ \hat{\mathbf{n}}_2 \\ \hat{\mathbf{n}}_3 \end{bmatrix} = \begin{bmatrix} c_\gamma & -s_\gamma & 0 \\ s_\gamma & c_\gamma & 0 \\ 0 & 0 & 1 \end{bmatrix} \begin{bmatrix} \mathbf{n}_1 \\ \mathbf{n}_2 \\ \mathbf{n}_3 \end{bmatrix} \quad \text{and} \quad \begin{bmatrix} \mathbf{n}_1 \\ \mathbf{n}_2 \\ \mathbf{n}_3 \end{bmatrix} = \begin{bmatrix} c_\gamma & s_\gamma & 0 \\ -s_\gamma & c_\gamma & 0 \\ 0 & 0 & 1 \end{bmatrix} \begin{bmatrix} \hat{\mathbf{n}}_1 \\ \hat{\mathbf{n}}_2 \\ \hat{\mathbf{n}}_3 \end{bmatrix} \tag{8.70}$$

or simply

$$\hat{\mathbf{n}} = C\mathbf{n} \quad \text{and} \quad \mathbf{n} = C^T \hat{\mathbf{n}} \quad \text{and} \quad \mathbf{n} = C^T \hat{\mathbf{n}} \tag{8.71}$$

where, as before, $\hat{\mathbf{n}}$, $\mathbf{n}$, C, and C^T are defined by inspection and comparison of the equations. The vectors parallel to the plane normal to $\hat{\mathbf{n}}_3$ and $\mathbf{n}_3$ are shown in Figure 8.27, where $\mathbf{n}_1$ and $\hat{\mathbf{n}}_2$ are seen to be the inside vectors and $\hat{\mathbf{n}}_1$ and $\mathbf{n}_2$ are the outside vectors. Analogous to Figures 8.22 and 8.25, we obtain the configuration graph of Figure 8.28.

Observe that the configuration graphs of Figures 8.22, 8.25, and 8.28 may be combined into a single diagram as in Figure 8.29.

Observe further that Equations 8.65, 8.68, and 8.71 may also be combined, leading to the following expressions:

$$\mathbf{N} = ABC\mathbf{n} \quad \text{and} \quad \mathbf{n} = C^T B^T A^T \mathbf{N} \tag{8.72}$$

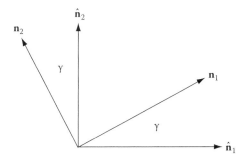

FIGURE 8.27 Inside and outside unit vectors.

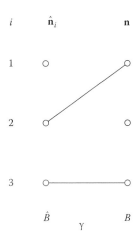

FIGURE 8.28 Configuration graph for the unit vectors of Figure 8.26.

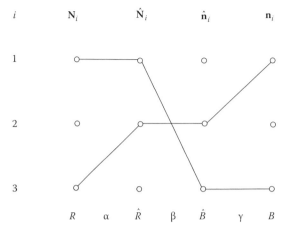

FIGURE 8.29 Combined configuration graph.

or

$$\mathbf{N} = S\mathbf{n} \quad \text{and} \quad \mathbf{n} = S^{T}\mathbf{N} \tag{8.73}$$

where by inspection, S is ABC and the transpose S^{T} is $C^{T}B^{T}A^{T}$. By multiplying the matrices in Equations 8.69, we have

$$S = ABC \begin{bmatrix} c_{\beta}c_{\gamma} & -c_{\beta}s_{\gamma} & s_{\beta} \\ (c_{\alpha}s_{\gamma} + s_{\alpha}s_{\beta}c_{\gamma}) & (c_{\alpha}c_{\gamma} - s_{\alpha}s_{\beta}s_{\gamma}) & -s_{\alpha}c_{\beta} \\ (s_{\alpha}s_{\gamma} - c_{\alpha}s_{\beta}c_{\gamma}) & (s_{\alpha}c_{\gamma} + c_{\alpha}s_{\beta}s_{\gamma}) & c_{\alpha}c_{\beta} \end{bmatrix} \tag{8.74}$$

and

$$S^{T} = C^{T}B^{T}A^{T} = \begin{bmatrix} c_{\beta}c_{\gamma} & (c_{\alpha}s_{\gamma} + s_{\alpha}s_{\beta}c_{\gamma}) & (s_{\alpha}s_{\gamma} - c_{\alpha}s_{\beta}c_{\gamma}) \\ -c_{\beta}s_{\gamma} & (c_{\alpha}c_{\gamma} - s_{\alpha}s_{\beta}s_{\gamma}) & (s_{\alpha}c_{\gamma} + c_{\alpha}s_{\beta}s_{\gamma}) \\ s_{\beta} & -s_{\alpha}c_{\beta} & c_{\alpha}c_{\beta} \end{bmatrix} \tag{8.75}$$

By multiplication of the expressions of Equations 8.74 and 8.75, we see that SS^{T} is the identity matrix and therefore S is orthogonal, and the inverse is the transpose.

Finally, observe that Equations 8.74 and 8.75 provide the following relations between the $\mathbf{N}_i$ and the $\mathbf{n}_i$:

$$\mathbf{N}_1 = c_{\beta}c_{\gamma}\mathbf{n}_1 - c_{\beta}c_{\gamma}\mathbf{n}_2 + s_{\beta}\mathbf{n}_3 \tag{8.76}$$

$$\mathbf{N}_2 = (c_{\alpha}s_{\gamma} + s_{\gamma}s_{\beta}c_{\gamma})\mathbf{n}_1 + (c_{\alpha}c_{\gamma} - s_{\alpha}s_{\beta}s_{\gamma})\mathbf{n}_2 - s_{\alpha}c_{\chi}\mathbf{n}_3 \tag{8.77}$$

$$\mathbf{N}_3 = (s_{\alpha}s_{\gamma} - c_{\alpha}s_{\beta}c_{\gamma})\mathbf{n}_1 + (s_{\alpha}c_{\gamma} + c_{\alpha}s_{\beta}s_{\gamma})\mathbf{n}_2 + c_{\alpha}c_{\beta}\mathbf{n}_3 \tag{8.78}$$

and

$$\mathbf{n}_1 = c_{\beta}c_{\gamma}\mathbf{N}_1 + (c_{\alpha}s_{\gamma} + s_{\alpha}s_{\beta}c_{\gamma})\mathbf{N}_2 + (s_{\alpha}s_{\gamma} - c_{\alpha}s_{\beta}c_{\gamma})\mathbf{N}_3 \tag{8.79}$$

$$\mathbf{n}_2 = -c_\beta s_\gamma \mathbf{N}_1 + (c_\alpha c_\gamma - s_\alpha s_\beta s_\gamma)\mathbf{N}_2 + (s_\alpha c_\gamma + c_\alpha s_\beta s_\gamma)\mathbf{N}_3 \tag{8.80}$$

$$\mathbf{n}_3 = s_\beta \mathbf{N}_1 - s_\alpha c_\beta \mathbf{N}_2 + c_\alpha c_\beta \mathbf{N}_3 \tag{8.81}$$

The foregoing analysis uses dextral (or Bryant) rotation angles (rotation of B about $\mathbf{n}_1$, $\mathbf{n}_2$, and $\mathbf{n}_3$ through the angles α, β, and γ, respectively). These angles are convenient and useful in biomechanical analyses. There are occasions, however, where different rotation sequences may also be useful. The most common of these are the so-called Euler angles developed by a rotation sequence of B about $\mathbf{n}_3$, $\mathbf{n}_1$ and then about $\mathbf{n}_3$ again, through the angles θ_1, θ_2, and θ_3. Figure 8.29 shows the configuration graph for this rotation sequence.

To further illustrate the use of configuration graphs, consider expressing $\mathbf{N}_1$ of the graph in Figure 8.30, in terms of the $\mathbf{n}_1$, $\mathbf{n}_2$, and $\mathbf{n}_3$: we can accomplish this by moving from left to right through the columns of dots (or nodes). Specifically, by focusing on the $\mathbf{N}_1$ node in the first column, we can express $\mathbf{N}_1$ in terms of the unit vectors of the second column (the $\hat{\mathbf{N}}_i$) as

$$\mathbf{N}_1 = c_1 \hat{\mathbf{N}}_1 - s_1 \hat{\mathbf{N}}_2 + 0 \hat{\mathbf{N}}_3 \tag{8.82}$$

where s_1 and c_1 are abbreviations for $\sin \theta_1$ and $\cos \theta_1$. Observe that we have a cosine coefficient where the unit vector indices are the same, and a sine coefficient where the unit vector indices are different. Also, as $\mathbf{N}_1$ and $\hat{\mathbf{N}}_2$ are "outside" vectors (no connecting lines), there is a negative sign for the $\sin \theta_1$ coefficient. Observe further that there is no $\hat{\mathbf{N}}_3$ component of $\mathbf{N}_1$ as $\mathbf{N}_1$ is perpendicular to $\mathbf{N}_3$, and $\mathbf{N}_3$ is equal to $\hat{\mathbf{N}}_3$ due to the horizontal connector.

By proceeding from the second column of dots to the third column, we have

$$\mathbf{N}_1 = c_1 \hat{\mathbf{N}}_1 - s_1 \hat{\mathbf{N}}_2$$
$$= c_1 \hat{\mathbf{n}}_1 - s_1(c_2 \hat{\mathbf{n}}_2 - s_2 \hat{\mathbf{n}}_3) \tag{8.83}$$

where s_2 and c_2 are $\sin \theta_2$ and $\cos \theta_2$, and we have followed the same procedure related to the same and different indices and orthogonal vectors.

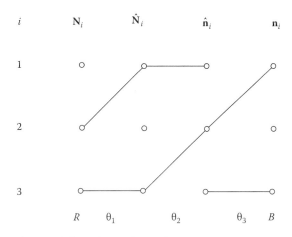

FIGURE 8.30 Configuration graph for Euler angle rotation sequence.

Finally, by proceeding to the fourth column, we have

$$\mathbf{N}_1 = c_1\hat{\mathbf{n}}_1 - s_1c_2\hat{\mathbf{n}}_2 + s_1s_2\hat{\mathbf{n}}_3$$

$$= c_1(c_3\mathbf{n}_1 - s_3\mathbf{n}_2) - s_1c_2(c_3\mathbf{n}_2 + s_3\mathbf{n}_1) + s_1s_2\mathbf{n}_3$$

or

$$\mathbf{N}_1 = (c_1c_3 - s_1c_2s_3)\mathbf{n}_1 + (-c_1s_3 - s_1c_2c_3)\mathbf{n}_2 + s_1s_2\mathbf{n}_3 \qquad (8.84)$$

In like manner, we can express $\mathbf{n}_1$ in terms of $\mathbf{N}_1$, $\mathbf{N}_2$, and $\mathbf{N}_3$ by proceeding from right to left in the columns of dots in the graph. Specifically, referring again to the graph of Figure 8.30, we have

$$\mathbf{n}_1 = c_3\hat{\mathbf{n}}_1 + s_3\hat{\mathbf{n}}_2 \qquad (8.85)$$

Then in going from the third column to the second column, we have

$$\mathbf{n}_1 = c_3\hat{\mathbf{N}}_1 + s_3(c_2\hat{\mathbf{N}}_2 + s_2\hat{\mathbf{N}}_3) \qquad (8.86)$$

Finally, by proceeding to the first column of unit vectors, we obtain

$$\mathbf{n}_1 = c_3(c_1\mathbf{N}_1 + s_1\mathbf{N}_2) + s_3c_2(c_1\mathbf{N}_2 - s_1\mathbf{N}_1) + s_3s_2\mathbf{N}_3$$

or

$$\mathbf{n}_1 = (c_3c_1 - s_3c_2s_1)\mathbf{N}_1 + (c_3s_1 + s_3c_2c_1)\mathbf{N}_2 + s_3s_2\mathbf{N}_3 \qquad (8.87)$$

By using these procedures, we find the remaining vectors ($\mathbf{n}_2$, $\mathbf{n}_3$, $\mathbf{N}_2$, and $\mathbf{N}_3$) to have the following forms:

$$\mathbf{n}_2 = (-s_3c_1 - c_3c_2s_1)\mathbf{N}_1 + (-s_3s_1 + c_3c_2c_1)\mathbf{N}_2 + c_3s_2\mathbf{N}_3 \qquad (8.88)$$

$$\mathbf{n}_3 = s_2s_1\mathbf{N}_1 - s_2c_1\mathbf{N}_2 + c_2\mathbf{N}_3 \qquad (8.89)$$

and

$$\mathbf{N}_2 = (c_1c_2s_3 + s_1c_3)\mathbf{n}_1 + (c_1c_2c_3 - s_1s_3)\mathbf{n}_2 - c_1s_2\mathbf{n}_3 \qquad (8.90)$$

$$\mathbf{N}_3 = s_2s_3\mathbf{n}_1 + s_2c_3\mathbf{n}_2 + c_2\mathbf{n}_3 \qquad (8.91)$$

8.9 USE OF CONFIGURATION GRAPHS TO DETERMINE ANGULAR VELOCITY

Recall that a principal concern about angular velocity in Section 8.6 was the question of its utility and particularly, the questions about the ease of obtaining expressions for the angular velocity for bodies having general movement (specifically arbitrary orientation changes). The configuration graphs of Section 8.8 and the addition theorem provide favorable answers to these questions.

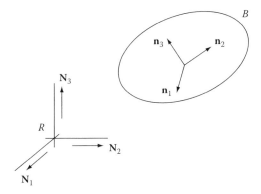

FIGURE 8.31 A body B moving in a reference frame R.

To see this, consider again a body B moving in a reference frame R as in Figure 8.31. (This models the movement of a limb of a body B of the human frame relative to a supporting body such as an upper arm relative to the chest.) As before, let the $\mathbf{n}_i$ and the $\mathbf{N}_i$ ($i = 1, 2, 3$) be mutually perpendicular unit vector sets fixed in B and R. Then the angular velocity of B in R is the same as the rate of change of orientation of the $\mathbf{n}_i$ relative to the $\mathbf{N}_i$.

Consider the configuration graph of Figure 8.32, where we can envision the orientation of B in R as arising from the successive rotations of B about $\mathbf{n}_1$, $\mathbf{n}_2$, and $\mathbf{n}_3$ through the angles α, β, and γ, where $\mathbf{n}_i$ are initially aligned with the $\mathbf{N}_i$ (see Section 8.8). As before, $\hat{R}$ and $\hat{B}$ are intermediate reference frames with unit vectors $\hat{\mathbf{N}}_i$ and $\hat{\mathbf{n}}_i$, defined by the orientation of B after the $\mathbf{n}_1$ and $\mathbf{n}_2$ rotations, respectively. Then during the orientation change of B in R, the rate of change of orientation (i.e., the angular velocity) of B in R is, by the addition theorem (Equation 8.51):

$$^{R}\boldsymbol{\omega}^{B} = {}^{R}\boldsymbol{\omega}^{\hat{R}} + {}^{\hat{R}}\boldsymbol{\omega}^{\hat{B}} + {}^{R}\boldsymbol{\omega}^{B} \tag{8.92}$$

Since each term on the right side of Equation 8.92 is an expression of simple angular velocity (see Section 8.6, Equation 8.19), we can immediately obtain explicit expressions for them by referring to the configuration graph of Figure 8.32. Specifically,

$$^{R}\boldsymbol{\omega}^{\hat{R}} = \dot{\alpha}\mathbf{N} = \dot{\alpha}\hat{\mathbf{N}}_{1}, \quad {}^{\hat{R}}\boldsymbol{\omega}^{\hat{B}} = \dot{\beta}\hat{\mathbf{N}}_{2} = \dot{\beta}\hat{\mathbf{n}}_{2}, \quad {}^{R}\boldsymbol{\omega}^{\hat{B}} = \dot{\gamma}\hat{\mathbf{n}}_{3} = \dot{\gamma}\mathbf{n}_{3} \tag{8.93}$$

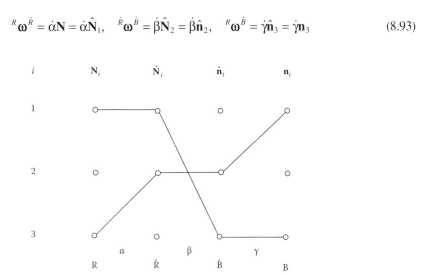

FIGURE 8.32 Configuration graph for the orientation of B in R with dextral orientation angles.

Then $^R\omega^B$ is

$$^R\omega^B = \dot\alpha\mathbf{N}_1 + \dot\beta\hat{\mathbf{N}}_2 + \dot\gamma\hat{\mathbf{n}}_3 = \dot\alpha\hat{\mathbf{N}}_1 + \dot\beta\hat{\mathbf{n}}_2 + \dot\gamma\mathbf{n}_3 \tag{8.94}$$

Observe the simplicity of Equation 8.94: indeed, the terms are each associated with a horizontal line of the configuration graph with the angle derivative being the derivative of the angle beneath the line (compare Equation 8.94 with Figure 8.32). This simplicity may be a bit misleading, however, as the unit vectors in Equation 8.94 are all from different unit vector sets. But this is not a major problem as the configuration graph itself can be used to express $^R\omega^B$ in terms of single set of unit vectors, say the $\mathbf{N}_i$ or the $\mathbf{n}_i$. That is, from Equation 8.94 and the diagram of Figure 8.32, we have

$$^R\omega^B = \dot\alpha\mathbf{N}_1 + \dot\beta(c_\alpha\mathbf{N}_2 + s_\alpha\mathbf{N}_3) + \dot\gamma(c_\beta\mathbf{N}_3 + s_\beta\mathbf{N}_1)$$

$$= \dot\alpha\mathbf{N}_1 + \dot\beta c_\alpha\mathbf{N}_2 + \dot\beta s_\alpha\mathbf{N}_3 + \dot\gamma c_\beta(c_\alpha\mathbf{N}_3 + s_2\mathbf{N}_2) + \dot\gamma s_\beta\mathbf{N}_1$$

or

$$^R\omega^B = (\dot\alpha + \dot\gamma s_\beta)\mathbf{N}_1 + (\dot\beta c_\alpha - \dot\gamma s_\alpha c_\beta)\mathbf{N}_2 + (\dot\beta s_\alpha + \dot\gamma c_\alpha c_\beta)\mathbf{N}_3 \tag{8.95}$$

Similarly, in terms of the $\mathbf{n}_i$, we have

$$^R\omega^B = (\dot\alpha c_\beta c_\gamma + \dot\beta s_\gamma)\mathbf{n}_1 + (-\dot\alpha c_\beta s_\gamma + \dot\beta c_\gamma)\mathbf{n}_2 + (\dot\alpha s_\beta + \dot\gamma)\mathbf{n}_3 \tag{8.96}$$

The form of the angular velocity components depends, of course, upon the rotation sequence. If, for example, Euler angles such as in the diagram of Figure 8.30 are used, the angular velocity has the forms:

$$^R\omega^B = (\dot\theta_2 c_1 - \dot\theta_3 s_1 s_2)\mathbf{N}_1 + (\dot\theta_2 s_1 - \dot\theta_3 c_1 s_2)\mathbf{N}_2 + (\dot\theta_1 + \dot\theta_3 c_2)\mathbf{N}_3 \tag{8.97}$$

and

$$^R\omega^B = (\dot\theta_2 c_3 + \dot\theta_1 s_2 s_3)\mathbf{n}_1 + (\dot\theta s_2 c_3)\mathbf{n}_2 + (\dot\theta_3 + \dot\theta c_2)\mathbf{n}_3 \tag{8.98}$$

References [1,2] provide extensive and comprehensive listings of the angular velocity components for the various possible rotation sequences.

Finally, observe that since the angular velocity components are known, as in Equations 8.95 and 8.96, they do not need to be derived again, but instead, they may be directly incorporated into numerical algorithms.

8.10 APPLICATION WITH BIOSYSTEMS

Equation 8.97 is applicable with biosystems such as the human body model of Figure 8.33 as with the upper arm and the chest, or with any two adjoining bodies. Consider, for example, two typical adjoining bodies such as B_j and B_k of Figure 8.34. Let $\mathbf{n}_{ji}$ and $\mathbf{n}_{ki}$ ($i = 1, 2, 3$) be unit vector sets fixed in B_j and B_k. B_k may be brought into a general orientation relative to B_j by aligning the unit vector sets and then performing successive rotations of B_k about $\mathbf{n}_{k1}$, $\mathbf{n}_{k2}$, and $\mathbf{n}_{k3}$ through the angles α_k, β_k, and γ_k—as we did with body B in R in the foregoing section. The resulting relative orientations of

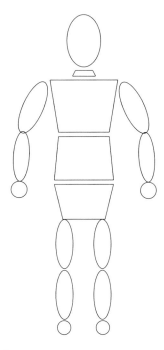

FIGURE 8.33 A human body model.

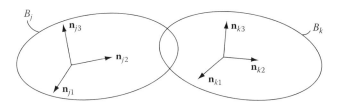

FIGURE 8.34 Two typical adjoining bodies.

the unit vectors may be defined by the configuration graph of Figure 8.35, where $\hat{\mathbf{n}}_{ji}$ and $\hat{\mathbf{n}}_{ki}$ ($i = 1, 2, 3$) are unit vector sets corresponding to the intermediate positions of the $\mathbf{n}_{ki}$ during the successive rotations of B_k. By following the analysis of the foregoing section, the angular velocity of B_k relative to B_j may then be expressed as (see Equation 8.95)

$$
^{B_j}\boldsymbol{\omega}^{B_k} = \dot{\alpha}_k \mathbf{n}_{j1} + \dot{\beta}_k \hat{\mathbf{n}}_{j2} + \dot{\gamma}_k \hat{\mathbf{n}}_{k3}
$$

$$
= (\dot{\alpha}_k + \dot{\gamma}_k s_{\beta_k}\mathbf{n}_{j1} + (\dot{\beta}_k c_{\alpha_k} - \dot{\gamma}_k s_{\alpha_k} c_{\beta_k})\mathbf{n}_{j2} + (\dot{\beta}_k s_{\alpha_k} + \dot{\gamma}_k c_{\alpha_k} c_{\beta_k})\hat{\mathbf{n}}_{j3} \tag{8.99}
$$

Equation 8.99 together with the addition theorem for angular velocity (Equation 8.51) enables us to obtain an expression for the angular velocity of a typical body of the human model, say B_k, with respect to an inertial (or fixed) frame R—the absolute angular velocity of B_k in R. To see this, let the bodies of the human body model be numbered as in Figure 8.36. Consider body B_{11}, the right hand. By inspection of Figure 8.36, the angular velocity of B_{11} in a fixed frame R is

$$
^{R}\boldsymbol{\omega}^{B_{11}} = {}^{R}\boldsymbol{\omega}^{B_1} + {}^{B_1}\boldsymbol{\omega}^{B_2} + {}^{B_2}\boldsymbol{\omega}^{B_3} + {}^{B_3}\boldsymbol{\omega}^{B_9} + {}^{B_9}\boldsymbol{\omega}^{B_{10}} + {}^{B_{10}}\boldsymbol{\omega}^{B_{11}} \tag{8.100}
$$

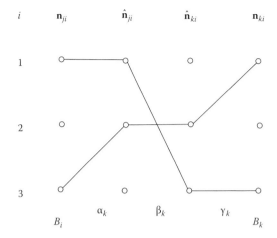

FIGURE 8.35   Configuration graph for unit vector sets of two adjoining bodies.

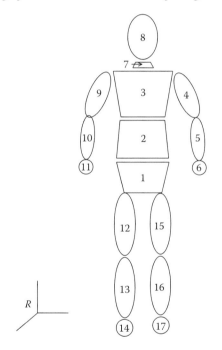

FIGURE 8.36   Numbering and labeling the human frame model.

Observe that each term on the right side is a relative angular velocity as in Equation 8.99. That is, by using Equation 8.99 we have an explicit expression for the angular velocity of the right hand in terms of the relative orientation angles of the links and their derivatives.

Observe further that whereas Equation 8.100 is of the same form as Equation 8.51, it may similarly be simplified as in Equation 8.49. Specifically, let $^{R}\boldsymbol{\omega}^{Bk}$ and $^{Rj}\boldsymbol{\omega}^{Bk}$ be written as

$$^{R}\boldsymbol{\omega}^{Bk} = \boldsymbol{\omega}_k \quad \text{and} \quad ^{B_i}\boldsymbol{\omega}^{Bk} = \hat{\boldsymbol{\omega}}_k \qquad (8.101)$$

where, as before, the overhat designates relative angular velocity of a body with respect to its adjacent lower-numbered body. Then Equation 8.100 becomes

$$\boldsymbol{\omega}_{11} = \boldsymbol{\omega}_1 + \boldsymbol{\omega}_2 + \boldsymbol{\omega}_3 + \boldsymbol{\omega}_9 + \boldsymbol{\omega}_{10} + \boldsymbol{\omega}_{11} \qquad (8.102)$$

TABLE 8.1
Higher-Order Lower-Body Arrays for the System of Figure 8.36

K	1	2	3	4	5	6	7	8	9	10	11	12	13	14	15	16	17
$L(K)$	0	1	2	3	4	5	3	7	3	9	10	1	12	13	1	15	16
$L^2(K)$	0	0	1	2	3	4	2	3	2	3	9	0	1	12	0	1	15
$L^3(K)$	0	0	0	1	2	3	1	2	1	2	3	0	0	1	0	0	1
$L^4(K)$	0	0	0	0	1	2	0	1	0	1	2	0	0	0	0	0	0
$L^5(K)$	0	0	0	0	0	1	0	0	0	0	1	0	0	0	0	0	0
$L^6(K)$	0	0	0	0	0	0	0	0	0	0	0	0	0	0	0	0	0

Finally, observe in Equation 8.102 that the indices correspond to the column of indices in the table of lower-body arrays for the human body model of Figure 8.36 (and of Figures 2.4 and 6.9), as developed in Table 6.3 and as listed again in Table 8.1. Observe the indices in the column for B_{11}: 11, 10, 9, 3, 2, and 1. These are precisely the indices on the right side of Equation 8.102. Therefore, we can express Equation 8.102 in the compact form:

$$\omega_{11} = \sum_{p=0}^{5} \hat{\omega}_q, \quad q = L^p(11) \tag{8.103}$$

In view of these results, we can express the angular velocity of any of the compact forms [3]:

$$\omega_k = \sum_{p=0}^{r} \hat{\omega}_q, \quad q = L^p(k) \tag{8.104}$$

where r is the index such that

$$L^r(k) = 1 \tag{8.105}$$

8.11 ANGULAR ACCELERATION

Angular acceleration is simply the derivative of the angular velocity: consider again a body B moving in a reference frame R as in Figure 8.37. If the angular velocity of B in R is ${}^R\omega^B$, then the angular acceleration of B in R is defined as

$$ {}^R\alpha^B \overset{D}{=} \frac{{}^R\mathrm{d}^R\omega^B}{\mathrm{d}(t)} \tag{8.106}$$

The derivative in Equation 8.106 may be obtained by using the procedures of the foregoing sections: specifically, if ${}^R\omega^B$ is expressed in terms of unit vectors $\mathbf{N}_i$ ($i = 1, 2, 3$) fixed in R, then ${}^R\alpha^B$ is obtained by simply differentiating the components. That is, let ${}^R\omega^B$ be

$$ {}^R\omega^B = \Omega_1\mathbf{N}_1 + \Omega_2\mathbf{N}_2 + \Omega_3\mathbf{N}_3 \tag{8.107}$$

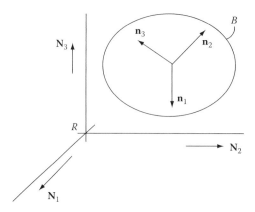

FIGURE 8.37 A body B moving in a reference frame.

then

$$^{R}\boldsymbol{\alpha}^{B} = \dot{\Omega}_{1}\mathbf{N}_{1} + \dot{\Omega}_{2}\mathbf{N}_{2} + \dot{\Omega}_{3}\mathbf{N}_{3} \tag{8.108}$$

If $^{R}\boldsymbol{\omega}^{B}$ is expressed in terms of unit vectors $\mathbf{n}_{i}$ ($i = 1, 2, 3$) fixed in B, then it also happens that $^{R}\boldsymbol{\alpha}^{B}$ may be determined by simply differentiating the components. That is, let $^{R}\boldsymbol{\omega}^{B}$ be

$$^{R}\boldsymbol{\omega}^{B} = \omega_{1}\mathbf{n}_{1} + \omega_{2}\mathbf{n}_{2} + \omega_{3}\mathbf{n}_{3} \tag{8.109}$$

then

$$^{R}\dot{\boldsymbol{\alpha}}^{B} = \dot{\omega}_{1}\mathbf{n}_{1} + \dot{\omega}_{2}\mathbf{n}_{2} + \dot{\omega}_{3}\mathbf{n}_{3} \tag{8.110}$$

The validity of Equation 8.110 is seen by referring to Equation 8.44. That is,

$$^{R}\boldsymbol{\alpha}^{B} = \frac{^{R}\mathrm{d}^{R}\boldsymbol{\omega}^{B}}{\mathrm{d}t} = \frac{^{B}\mathrm{d}^{R}\boldsymbol{\omega}^{B}}{\mathrm{d}t} + {}^{R}\boldsymbol{\omega}^{B} \times {}^{R}\boldsymbol{\omega}^{B} = \frac{^{B}\mathrm{d}^{R}\boldsymbol{\omega}^{B}}{\mathrm{d}t} \tag{8.111}$$

where the last equality holds as the vector product of a vector with itself is zero. The result of Equation 8.111 then leads directly to Equation 8.110.

If, however, $^{R}\boldsymbol{\omega}^{B}$ is expressed in terms of unit vectors other than those in R or B, then $^{R}\boldsymbol{\alpha}^{B}$ cannot, in general, be obtained by simply differentiating the scalar components of $^{R}\boldsymbol{\omega}^{B}$. Therefore, for computational purposes, with biosystems, it is best to express the angular velocity in terms of unit vectors fixed either in the body B or the fixed frame R.

With large human body models, it is usually preferable to express the angular velocities in terms of unit vectors in the fixed frame R instead of those of the body segments. We can obtain vector components of the fixed frame unit vectors by using the transformation matrices developed in the foregoing section. Having made such transformations, however, when we attempt to differentiate the resulting components we will still need to differentiate the transformation matrices. It happens that just as the angular velocity is useful in differentiating vectors (see Equations 8.18 and 8.44), it is also useful in differentiating transformation matrices. We will develop expressions for these derivatives in the following section.

Finally, we observe that whereas we have an addition theorem for angular velocity, as in Equation 8.50, there is not a corresponding simple expression for the addition of angular accelerations. That is, for angular velocity, we have

$$^{R}\boldsymbol{\omega}^{B} = {}^{R}\boldsymbol{\omega}^{\hat{R}} + {}^{\hat{R}}\boldsymbol{\omega}^{B} \tag{8.112}$$

But for angular acceleration, we do not have a corresponding expression. That is,

$$^R\boldsymbol{\alpha}^B \neq {}^R\boldsymbol{\alpha}^{\hat{R}} + {}^{\hat{R}}\boldsymbol{\alpha}^B \tag{8.113}$$

To see this, observe that by differentiating in R in Equation 8.112, we have

$$\frac{^R\mathrm{d}^R\boldsymbol{\omega}^B}{\mathrm{d}t} = \frac{^R\mathrm{d}^R\boldsymbol{\omega}^{\hat{R}}}{\mathrm{d}t} + \frac{^R\mathrm{d}^{\hat{R}}\boldsymbol{\omega}^B}{\mathrm{d}t} \tag{8.114}$$

or

$$^R\boldsymbol{\alpha}^B = {}^R\boldsymbol{\alpha}^{\hat{R}} + \frac{^R\mathrm{d}^{\hat{R}}\boldsymbol{\omega}^B}{\mathrm{d}t} \tag{8.115}$$

In general, however, $^{\hat{R}}\boldsymbol{\alpha}^B$ is not equal to $^R\mathrm{d}^{\hat{R}}\boldsymbol{\omega}^B/\mathrm{d}t$.

8.12 TRANSFORMATION MATRIX DERIVATIVES

Consider again a body B moving in a reference frame R as in Figure 8.38. As before, let $\mathbf{n}_i$ and $\mathbf{N}_i$ be unit vectors fixed in B and R. Let B move in R with angular velocity $\boldsymbol{\omega}$ given by

$$\boldsymbol{\omega} = \Omega_1\mathbf{N}_1 + \Omega_2\mathbf{N}_2 + \Omega_3\mathbf{N}_3 = \omega_1\mathbf{n}_1 + \omega_2\mathbf{n}_2 + \omega_3\mathbf{n}_3 \tag{8.116}$$

In Section 8.8, we saw that the $\mathbf{N}_i$ and $\mathbf{n}_i$ may be related by the transformation matrix S as (see Equation 8.70)

$$\mathbf{N} = S\mathbf{n} \quad \text{and} \quad \mathbf{n} = S^T\mathbf{N} \tag{8.117}$$

or in index notation as

$$\mathbf{N}_j = S_{ij}\mathbf{n}_j \quad \text{and} \quad \mathbf{n}_j = S_{ij}\mathbf{N}_i \tag{8.118}$$

where we have employed the summation convention as in Chapter 3, and where the elements S_{ij} of S are obtained using the configuration graphs of Section 8.8.

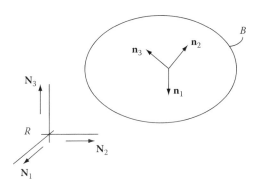

FIGURE 8.38   A body B moving in a reference frame R.

Observe that by taking the scalar (dot) product of the first equation of Equation 8.118 with $\mathbf{n}_k$, we obtain

$$S_{ik} = \mathbf{N}_i \cdot \mathbf{n}_k \tag{8.119}$$

See also Equation 3.160.

Since the $\mathbf{N}_i$ are fixed in R and the $\mathbf{n}_i$ are fixed in B, by differentiating in Equation 8.119, we obtain

$$
\begin{aligned}
\frac{dS_{ik}}{dt} &= \frac{d}{dt}(\mathbf{N}_i \cdot \mathbf{n}_k) = \mathbf{N}_i \cdot \frac{d\mathbf{n}_k}{dt} \\
&= \mathbf{N}_i \cdot \boldsymbol{\omega} \times \mathbf{n}_k = \mathbf{N}_i \cdot \Omega_\ell \mathbf{N}_\ell \times S_{mk}\mathbf{N}_m \\
&= e_{i\ell m}\Omega_\ell S_{mk} = -e_{im\ell}\Omega_\ell S_{mk} \\
&= W_{im}S_{mk}
\end{aligned}
\tag{8.120}
$$

where W_{im} are defined by inspection as

$$W_{im} \overset{D}{=} -e_{im\ell}\Omega_\ell \tag{8.121}$$

or as

$$W_{im} = \begin{bmatrix} 0 & -\Omega_3 & \Omega_2 \\ \Omega_3 & 0 & -\Omega_1 \\ -\Omega_2 & \Omega_1 & 0 \end{bmatrix} \tag{8.122}$$

In matrix form, Equation 8.120 may be written as

$$\frac{dS}{dt} = WS \tag{8.123}$$

W is sometimes called the angular velocity matrix or the matrix whose "dual vector" is $\boldsymbol{\omega}$ [2]. Thus, we see the central role $\boldsymbol{\omega}$ plays in the computation of derivatives.

8.13   RELATIVE VELOCITY AND ACCELERATION OF TWO POINTS FIXED ON A BODY

Consider a body B moving in a reference frame R as in Figure 8.39. Consider B as representing a typical body B_k of a human body model. Let P and Q be particles of B (represented by points P and Q). Let $\mathbf{p}$ and $\mathbf{q}$ be position vectors locating P and Q relative to the origin O of R. Then from Equation 8.4, the velocities of P and Q in R are

$$\mathbf{V}^P = \frac{d\mathbf{p}}{dt} \quad \text{and} \quad \mathbf{V}^Q = \frac{d\mathbf{q}}{dt} \tag{8.124}$$

where the derivatives are calculated in R. Next, let $\mathbf{r}$ locate P relative to Q as in Figure 8.39. (Observe that as P and Q are fixed in B, $\mathbf{r}$ is also fixed in B.) Then from simple vector addition, we have

$$\mathbf{p} = \mathbf{q} + \mathbf{r} \tag{8.125}$$

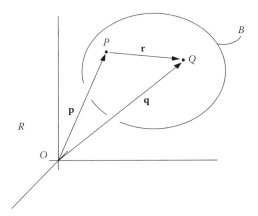

FIGURE 8.39 A body B with points P and Q moving in a reference frame R.

By differentiating, we have

$$\mathbf{V}^P = \mathbf{V}^Q + \frac{d\mathbf{r}}{dt}$$

$$= \mathbf{V}^Q + \boldsymbol{\omega} \times \mathbf{r} \qquad (8.126)$$

where $\boldsymbol{\omega}$ is the angular velocity of B in R, and the last term occurs as $\mathbf{r}$ is fixed in B. Then from Equation 8.14 the relative velocity of P and Q is

$$\mathbf{V}^{P/Q} = \mathbf{V}^P - \mathbf{V}^Q = \boldsymbol{\omega} \times \mathbf{r} \qquad (8.127)$$

These results are extended to accelerations by differentiating again. That is,

$$\mathbf{a}^P = \frac{d\mathbf{V}^P}{dt}, \quad \mathbf{a}^Q = \frac{d\mathbf{q}}{dt}, \quad \mathbf{a}^{P/Q} = \frac{d\mathbf{V}^{P/Q}}{dt} \qquad (8.128)$$

and

$$\mathbf{a}^P = \mathbf{a}^Q + \boldsymbol{\alpha} \times \mathbf{r} + \boldsymbol{\omega} \times (\boldsymbol{\omega} \times \mathbf{r}) \qquad (8.129)$$

where
   $\mathbf{a}^P$ and $\mathbf{a}^Q$ are the accelerations of P and Q in R
   $\boldsymbol{\alpha}$ is the angular acceleration of B in R
   the derivatives are calculated in R

8.14 SINGULARITIES OCCURRING WITH ANGULAR VELOCITY COMPONENTS AND ORIENTATION ANGLES

Consider again Equations 8.95 and 8.96 expressing the angular velocity of a body B in a reference frame R, in terms of unit vectors fixed in R ($\mathbf{N}_i$) and in B ($\mathbf{n}_i$):

$$^R\boldsymbol{\omega}^B = (\dot{\alpha} + \dot{\gamma}s_\beta)\mathbf{N}_1 + (\dot{\beta}c_\alpha - \dot{\gamma}s_\alpha c_\beta)\mathbf{N}_2 + (\dot{\beta}s_\alpha + \dot{\gamma}c_\alpha c_\beta)\mathbf{N}_3 \qquad (8.130)$$

and

$$^{R}\boldsymbol{\omega}^{B} = (\dot{\alpha}c_{\beta}c_{\gamma} + \dot{\beta}s_{\gamma})\mathbf{n}_1 + (-\dot{\alpha}c_{\beta}s_{\gamma} + \dot{\beta}c_{\gamma})\mathbf{n}_2 + (\dot{\alpha}s_{\beta} + \dot{\gamma})\mathbf{n}_3 \tag{8.131}$$

where
   α, β, and γ are dextral orientation angles
   s_θ and c_θ are $\sin\theta$ and $\cos\theta$ for $\theta = \alpha$, β, γ

These equations may be written as

$$^{R}\boldsymbol{\omega}^{B} = \Omega_1\mathbf{N}_1 + \Omega_2\mathbf{N}_2 + \Omega_3\mathbf{N}_3 \tag{8.132}$$

and

$$^{R}\boldsymbol{\omega}^{B} = \omega_1\mathbf{n}_1 + \omega_2\mathbf{n}_2 + \omega_3\mathbf{n}_3 \tag{8.133}$$

where by inspection, the Ω_i and ω_i are

$$\Omega_1 = \dot{\alpha} + \dot{\gamma}s_{\beta} \qquad \omega_1 = \dot{\alpha}c_{\beta}c_{\gamma} + \dot{\beta}s_{\gamma}$$

$$\Omega_2 = \dot{\beta}c_{\alpha} - \dot{\gamma}s_{\alpha}c_{\beta} \qquad \omega_1 = \dot{\alpha}c_{\beta}s_{\gamma} + \dot{\beta}c_{\gamma} \tag{8.134}$$

$$\Omega_3 = \dot{\beta}s_{\alpha} + \dot{\gamma}c_{\alpha}c_{\beta} \qquad \omega_3 = \dot{\alpha}s_{\beta} + \dot{\gamma}$$

Equations 8.134 may be viewed as systems of first-order ordinary differential equations for the orientation angles. They are linear in the orientation angle derivatives, but nonlinear in the angles. These nonlinear terms can lead to singularities in the solutions of the equations, resulting in disruption of numerical solution procedures. To see this, consider solving Equations 8.134 for the orientation angle derivatives, as one would do in forming the equations for numerical solutions. By doing this, we obtain

$$\dot{\alpha} = \Omega_1 + \frac{s_{\beta}(s_{\alpha}\Omega_2 - c_{\alpha}\Omega_3)}{c_{\beta}}$$

$$\dot{\beta} = c_{\alpha}\Omega_2 + s_{\alpha}\Omega_3 \tag{8.135}$$

$$\dot{\gamma} = \frac{-s_{\alpha}\Omega_2 + c_{\alpha}\Omega_3}{c_{\beta}}$$

and

$$\dot{\alpha} = \frac{c_{\gamma}\omega_1 - s_{\gamma}\omega_2}{c_{\beta}}$$

$$\dot{\beta} = s_{\gamma}\omega_1 + c_{\gamma}\omega_2 \tag{8.136}$$

$$\dot{\gamma} = \omega_3 + \frac{s_{\beta}(-c_{\gamma}\omega_1 + s_{\gamma}\omega_2)}{c_{\beta}}$$

Observe in each of these sets of equations, there is a division by zero, or singularity, where β is 90° or 270°. It happens that such singularities occur no matter how the orientation angles are chosen [2].

These singularities and the accompanying problems in numerical integration can be avoided through the use of Euler parameters—a set of four variables, or parameters, defining the orientation of a body in a reference frame. The use of the four variables, as opposed to three orientation angles, creates a redundancy and ultimately an additional differential equation to be solved. To overcome this disadvantage, however, the equations take a linear form and the singularities are avoided.

We explore these concepts in detail in the following sections.

8.15   ROTATION DYADICS

The development of Euler parameters depends upon a classic orientation problem in rigid body kinematics: If a body B is changing orientation in a reference frame R, then for any two orientations of B in R, say orientation one ($O1$) and orientation two ($O2$), B may be moved, or changed, from $O1$ to $O2$ by a single rotation about a line fixed in both B and R. Expressed another way: B may be brought into a general orientation in R from an initial reference orientation by a single rotation through an appropriate angle θ about some line L, where L is fixed in both B and R. The rotation angle θ and the rotation axis line L depend upon the initial and final orientations of B in R.

To see this, we follow the analysis of Ref. [3]: consider the rotation of a body B about a line L in a reference frame R as represented in Figure 8.40. Let λ be a unit vector parallel to L and let θ be the rotation angle. Let P be a point within a particle P of B, not on L, as in Figure 8.41. Let A be a reference point on L.

Next, let B rotate about L through an angle θ. Then point P will rotate to a point $\hat{P}$ in R as represented in Figure 8.42. Let $\mathbf{p}$ and $\hat{\mathbf{p}}$ be position vectors locating P and $\hat{P}$ relative to A and similarly, let $\mathbf{r}$ and $\hat{\mathbf{r}}$ locate P and $\hat{P}$ relative to Q. That is,

$$\mathbf{p} = \mathbf{AP}, \quad \hat{\mathbf{p}} = \mathbf{A\hat{P}}, \quad \mathbf{r} = \mathbf{QP}, \quad \hat{\mathbf{r}} = \mathbf{Q\hat{P}} \tag{8.137}$$

From Figure 8.42, we see that $\mathbf{AQ}$ is the projection of $\mathbf{p}$ along L and that $\mathbf{r}$ is the projection of $\mathbf{p}$ perpendicular to L. That is,

$$\mathbf{p} = \mathbf{AQ} + \mathbf{r} \tag{8.138}$$

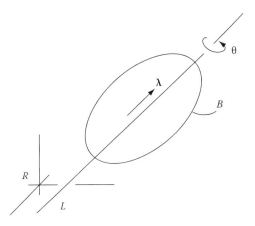

FIGURE 8.40   Rotation of a body about a line.

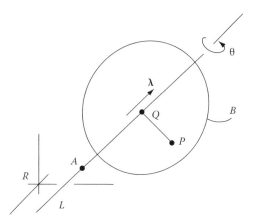

FIGURE 8.41 Point P of B and Q of L with PQ being perpendicular to L.

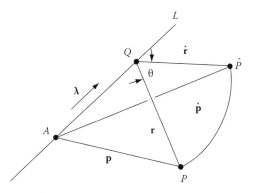

FIGURE 8.42 Points A, Q, P, and $\hat{P}$ and associated position vectors.

where **AQ** and **r** are perpendicular components of **p**, with **AQ** being along L. Thus with λ as a unit vector parallel to L, we have

$$\mathbf{AQ} = (\mathbf{p}\cdot\boldsymbol{\lambda})\boldsymbol{\lambda} \quad \text{and} \quad \mathbf{r} = \mathbf{p} - (\mathbf{p}\cdot\boldsymbol{\lambda})\boldsymbol{\lambda} \tag{8.139}$$

By continuing to follow the analysis of Ref. [3], consider a view of the plane of Q, P, and $\hat{P}$ as in Figure 8.43, where F is at the foot of the line through $\hat{\mathbf{P}}$ perpendicular to QP as shown. Then in view of Figure 8.43, we can establish the following position vector expressions:

$$\mathbf{Q}\hat{\mathbf{P}} = \hat{\mathbf{r}} = \mathbf{QF} + \mathbf{F}\hat{\mathbf{P}} \tag{8.140}$$

$$\mathbf{QF} = |\mathbf{r}|\cos\theta\boldsymbol{\mu} = r\cos\theta\boldsymbol{\mu} \tag{8.141}$$

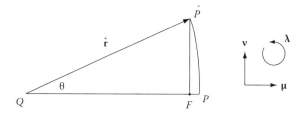

FIGURE 8.43 Planar view of points Q, P, and $\hat{P}$.

$$\mathbf{F}\hat{\mathbf{P}} = |\hat{\mathbf{r}}|\sin\theta\boldsymbol{v} = r\sin\theta\boldsymbol{v} \tag{8.142}$$

where r is the magnitude of $\hat{\mathbf{r}}$, and also of $\mathbf{r}$, then $\mathbf{r}$ is simply

$$\mathbf{r} = \mathbf{QP} = r\boldsymbol{\mu} \tag{8.143}$$

and thus $\mathbf{QF}$ and $\mathbf{F}\hat{\mathbf{P}}$ may be expressed as

$$\mathbf{QF} = r\boldsymbol{\mu}\cos\theta = \mathbf{r}\cos\theta \tag{8.144}$$

and

$$\mathbf{F}\hat{\mathbf{P}} = r\boldsymbol{v}\sin\theta = r\boldsymbol{\lambda}\times\boldsymbol{\mu}\sin\theta = \boldsymbol{\lambda}\times\mathbf{r}\sin\theta \tag{8.145}$$

Then from Equation 8.140 $\hat{\mathbf{r}}$ becomes

$$\hat{\mathbf{r}} = \mathbf{r}\cos\theta + \boldsymbol{\lambda}\times\mathbf{r}\sin\theta \tag{8.146}$$

Recall from Equation 8.139 that $\mathbf{r}$ may be expressed in terms of $\mathbf{p}$ as $\mathbf{p} - (\mathbf{p}\cdot\boldsymbol{\lambda})\boldsymbol{\lambda}$. Therefore, using Equation 8.146, $\hat{\mathbf{r}}$ may be expressed in terms of $\mathbf{p}$ as

$$\hat{\mathbf{r}} = \left[\mathbf{p} - (\mathbf{p}\cdot\boldsymbol{\lambda})\boldsymbol{\lambda}\right]\cos\theta + \boldsymbol{\lambda}\times\mathbf{p}\sin\theta \tag{8.147}$$

Finally, from Figure 8.42 we see that $\hat{\mathbf{p}}$ may be expressed in terms of $\hat{\mathbf{r}}$ as

$$\hat{\mathbf{p}} = \mathbf{AQ} + \hat{\mathbf{r}} \tag{8.148}$$

or in terms of $\mathbf{p}$ as

$$\hat{\mathbf{p}} = (\mathbf{p}\cdot\boldsymbol{\lambda})\boldsymbol{\lambda} + \left[\mathbf{p} - (\mathbf{p}\cdot\boldsymbol{\lambda})\boldsymbol{\lambda}\right]\cos\theta + \boldsymbol{\lambda}\times\mathbf{p}\sin\theta \tag{8.149}$$

Equation 8.149 may be written in the compact form:

$$\hat{\mathbf{p}} = R\cdot\mathbf{p} \tag{8.150}$$

where by inspection, the dyadic $\mathbf{R}$ is

$$R = (1 - \cos\theta)\boldsymbol{\lambda}\boldsymbol{\lambda} + \cos\theta I + \sin\theta\boldsymbol{\lambda}\times I \tag{8.151}$$

where $\mathbf{I}$ is the identity dyadic. If we express $\mathbf{R}$ in terms of unit vectors $\mathbf{N}_1$, $\mathbf{N}_2$, and $\mathbf{N}_3$ fixed in reference frame R, as

$$R = R_{ij}\mathbf{N}_i\mathbf{N}_j \tag{8.152}$$

then the R_{ij} are

$$R_{ij} = (1 - \cos\theta)\lambda_i\lambda_j + \delta_{ij}\cos\theta - e_{ijk}\lambda_k\sin\theta \tag{8.153}$$

where
  δ_{ij} and e_{ijk} are Kronecker's delta function and the permutation function
  the λ_i are the $\mathbf{N}_i$ components of λ

In view of Equation 8.150, we see that $\mathbf{R}$ is a "rotation" dyadic. That is, as an operator, $\boldsymbol{R}$ transforms $\mathbf{p}$ into $\hat{\mathbf{p}}$. Specifically, it rotates $\mathbf{p}$ about line L through the angle θ. Moreover, as P is an arbitrary point of B, $\mathbf{p}$ is an arbitrary vector fixed in B. Therefore, if $\mathbf{V}$ is any vector fixed in B then $\mathbf{R} \cdot \mathbf{V}$ is a vector $\hat{\mathbf{V}}$ whose magnitude is the same as the magnitude of $\mathbf{V}$ and whose direction is the same as that of $\mathbf{V}$ rotated about L through θ. That is,

$$\hat{\mathbf{V}} = \boldsymbol{R} \cdot \mathbf{V} \quad \text{and} \quad |\hat{\mathbf{V}}| = |\mathbf{V}| \tag{8.154}$$

In this context, suppose that $\mathbf{n}_1$, $\mathbf{n}_2$, and $\mathbf{n}_3$ are mutually perpendicular unit vectors fixed in B. Then as B rotates about L through angle θ, the $\mathbf{n}_i$ rotate into new orientations characterized by unit vectors $\hat{\mathbf{n}}_i$ $(i = 1, 2, 3)$. Then in view of Equation 8.150, we have

$$\hat{\mathbf{n}}_i = \boldsymbol{R} \cdot \mathbf{n}_i \tag{8.155}$$

Suppose that the $\mathbf{n}_i$ of Equation 8.155 are initially aligned with the $\mathbf{N}_i$ of reference frame R, so that $\boldsymbol{R}$ may be expressed as

$$\boldsymbol{R} = R_{ij}\mathbf{n}_i n_j \equiv R_{mn}\mathbf{n}_m\mathbf{n}_n \tag{8.156}$$

Then Equation 8.155 becomes

$$\hat{\mathbf{n}}_i = R_{mn}\mathbf{n}_m\mathbf{n}_n \cdot \mathbf{n}_i = R_{mn}\mathbf{n}_m\delta_{ni} = R_{\min}\mathbf{n}_m$$

or

$$\hat{\mathbf{n}}_j = R_{ij}\mathbf{n}_i \tag{8.157}$$

Interestingly, Equation 8.157 has exactly the same form as Equation 3.166 for the transformation matrix:

$$\hat{\mathbf{n}}_j = S_{ij}\mathbf{n}_i \tag{8.158}$$

Therefore, the transformation matrix elements S_{ij} between $\mathbf{n}_i$ and $\hat{\mathbf{n}}_j$ may be expressed in terms of the λ_i and θ by identifying the S_{ij} with the rotation dyadic components. That is,

$$S_{ij} = R_{ij} \tag{8.159}$$

The result of Equation 8.159, however, does not mean that the rotation dyadic is same as the transformation matrix. As operators, they are greatly different: the rotation dyadic rotates and thus changes the vector, whereas the transformation matrix does not change the vector but instead simply expresses the vector relative to different unit vector systems. Indeed, from the definition of the transformation matrix elements of Equation 3.160, we have

$$S_{ij} = \mathbf{n}_i \cdot \hat{\mathbf{n}}_j \tag{8.160}$$

This expression shows that the S_{ij} are referred to different unit vector bases (the first subscript with the $\mathbf{n}_i$ and the second with the $\hat{\mathbf{n}}_j$), whereas the R_{ij} are referred to a single unit vector basis (both subscripts referred to the $\mathbf{n}_i$).

Despite this distinction, the S_{ij} and the R_{ij} have similar properties: for example in Ref. [3] it is shown that, as with the S_{ij}, the R_{ij} are elements of an orthogonal matrix. That is,

$$R \cdot R^T = R^T \cdot R = I \quad \text{or} \quad R^T = R^{-1} \tag{8.161}$$

where I is the identity dyadic. Alternatively,

$$R_{ik}R_{jk} = \delta_{ij} = R_{ki}R_{kj} \tag{8.162}$$

Therefore, we have

$$\det(\delta_{ij}) = 1 = (\det R_{ij})^2 \quad \text{or} \quad \det R_{ij} = 1 \tag{8.163}$$

In view of Equations 8.154 and 8.161, we also have

$$\mathbf{V} = R^T \cdot \mathbf{V}^T \tag{8.164}$$

From Equation 8.152, we see that the elements of the rotation dyadic matrix are

$$R_{ij} = \begin{bmatrix} \lambda_1^2(1-\cos\theta)+\cos\theta & \lambda_1\lambda_2(1-\cos\theta)-\lambda_3\sin\theta & \lambda_1\lambda_2(1-\cos\theta)-\lambda_3\sin\theta \\ \lambda_2\lambda_1(1-\cos\theta)-\lambda_3\sin\theta & \lambda_2^2(1-\cos\theta)+\cos\theta & \lambda_2\lambda_3(1-\cos\theta)-\lambda_1\sin\theta \\ \lambda_3\lambda_1(1-\cos\theta)-\lambda_2\sin\theta & \lambda_3\lambda_2(1-\cos\theta)-\lambda_1\sin\theta & \lambda_3^2(1-\cos\theta)+\cos\theta \end{bmatrix}$$

$$\tag{8.165}$$

Observe in Equation 8.165 that if we know $\lambda_1, \lambda_2, \lambda_3$, and θ, we can immediately obtain the rotation matrix R_{ij}. Then knowing the R_{ij}, we have the rotation dyadic R. Conversely, if initially we have the rotation dyadic, so that we know the R_{ij}, we can use the values of the R_{ij} to obtain the rotation angle θ and the unit vector components λ_i, parallel to the rotation axis.* This last observation shows that given any orientation of a body B, we can bring B into that orientation from any other given orientation by a single rotation about a fixed line—as noted earlier (see Refs. [2,3] for additional details).

8.16 EULER PARAMETERS

Consider again a body B moving in a reference frame R, changing its orientation in R. Suppose that at an instant of interest, B has attained an orientation, say O^*, relative to a reference orientation O in R. Then from the foregoing section, we see that B can be brought from O to O^* by a single rotation of B about a line L fixed in both B and R through an angle θ as

---

* It might appear that the nine R_{ij} producing the four variable $\lambda_1, \lambda_2, \lambda_3$, and θ that the system is overdetermined. But recall that the R_{ij} are not independent. Indeed, as being elements of an orthogonal matrix, the elements forming the rows and columns of the matrix must be components of mutually perpendicular unit vectors.

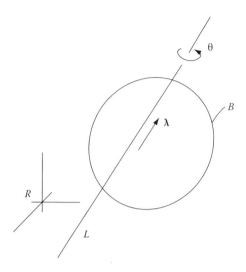

FIGURE 8.44   Orientation change of a body B in a reference frame R.

represented in Figure 8.44. The orientation O^* may be represented in terms of four Euler parameters ε_i ($i = 1, \ldots, 4$) defined as

$$\varepsilon_1 = \lambda_1 \sin\left(\frac{\theta}{2}\right)$$

$$\varepsilon_2 = \lambda_2 \sin\left(\frac{\theta}{2}\right)$$

$$\varepsilon_3 = \lambda_3 \sin\left(\frac{\theta}{2}\right) \qquad (8.166)$$

$$\varepsilon_4 = \cos\left(\frac{\theta}{2}\right)$$

where λ_i ($i = 1, 2, 3$) are the components of a unit vector λ parallel to L referred to mutually perpendicular unit vectors fixed in R.

Observe in Equation 8.166 that the Euler parameters are not independent. Indeed from their definitions, we see that

$$\varepsilon_1^2 + \varepsilon_2^2 + \varepsilon_3^2 + \varepsilon_4^2 = 1 \qquad (8.167)$$

This equation shows that only three parameters are needed to define the orientation of a body such as the customarily used orientation angles α, β, and γ. The use of Euler parameters, however, while redundant, leads to linear and homogeneous quadratic expressions as seen in the following paragraphs.

To see all this, consider again the transformation matrix S whose elements S_{ij} are defined in terms of the relative inclinations of unit vectors fixed in B and R as in Equations 3.160 and 8.160. From Equation 8.159, we see that these elements are the same as the components of the rotation dyadic $\mathbf{R}$. Thus from Equation 8.165, we see that in terms of the λ_i and θ, the S_{ij} are

$$S_{ij} = \begin{bmatrix} \lambda_1^2(1-\cos\theta)+\cos\theta & \lambda_1\lambda_2(1-\cos\theta)-\lambda_3\sin\theta & \lambda_1\lambda_2(1-\cos\theta)-\lambda_3\sin\theta \\ \lambda_2\lambda_1(1-\cos\theta)-\lambda_3\sin\theta & \lambda_2^2(1-\cos\theta)+\cos\theta & \lambda_2\lambda_3(1-\cos\theta)-\lambda_1\sin\theta \\ \lambda_3\lambda_1(1-\cos\theta)-\lambda_2\sin\theta & \lambda_3\lambda_2(1-\cos\theta)-\lambda_1\sin\theta & \lambda_3^2(1-\cos\theta)+\cos\theta \end{bmatrix}$$

$$(8.168)$$

By using the definitions of Equations 8.166, we can express the S_{ij} in terms of the Euler parameters. To illustrate this, consider first S_{11}:

$$S_{11} = \lambda_1^2(1 - \cos\theta) + \cos\theta \qquad (8.169)$$

We can replace $1 - \cos\theta$ and $\cos\theta$ by half-angle functions with the trigonometric identities:

$$1 - \cos\theta \equiv 2\sin^2\left(\frac{\theta}{2}\right) \quad \text{and} \quad \cos\theta = 2\cos^2\left(\frac{\theta}{2}\right) - 1 \qquad (8.170)$$

Then by substitution into Equation 8.169 and in view of Equations 8.166, we have

$$S_{11} = 2\varepsilon_1^2 + 2\varepsilon_4^2 - 1 \qquad (8.171)$$

Then by using Equation 8.167, we obtain

$$S_{11} = \varepsilon_1^2 - \varepsilon_2^2 - \varepsilon_3^2 + \varepsilon_4^2 \qquad (8.172)$$

Consider next S_{12}:

$$S_{12} = \lambda_1\lambda_2(1 - \cos\theta) - \lambda_3\sin\theta \qquad (8.173)$$

In addition to the identity for $1 - \cos\theta$ of Equation 8.170, we can use the following identity for $\sin\theta$:

$$\sin\theta = 2\sin\left(\frac{\theta}{2}\right)\cos\left(\frac{\theta}{2}\right) \qquad (8.174)$$

Then by substituting into Equation 8.173 and again in view of Equation 8.166, we have

$$S_{12} = 2\varepsilon_1\varepsilon_2 - 2\varepsilon_3\varepsilon_4 \qquad (8.175)$$

By similar analyses, we see that the transformation matrix elements are

$$S_{ij} = \begin{bmatrix} (\varepsilon_1^2 - \varepsilon_2^2 - \varepsilon_3^2 + \varepsilon_4^2) & 2(\varepsilon_1\varepsilon_2 - \varepsilon_3\varepsilon_4) & 2(\varepsilon_1\varepsilon_3 + \varepsilon_2\varepsilon_4) \\ 2(\varepsilon_1\varepsilon_2 - \varepsilon_3\varepsilon_4) & -(\varepsilon_1^2 + \varepsilon_2^2 - \varepsilon_3^2 + \varepsilon_4^2) & 2(\varepsilon_2\varepsilon_3 - \varepsilon_1\varepsilon_4) \\ 2(\varepsilon_1\varepsilon_3 - \varepsilon_2\varepsilon_4) & 2(\varepsilon_2\varepsilon_3 - \varepsilon_1\varepsilon_4) & -(\varepsilon_1^2 - \varepsilon_2^2 + \varepsilon_3^2 + \varepsilon_4^2) \end{bmatrix} \qquad (8.176)$$

8.17 EULER PARAMETERS AND ANGULAR VELOCITY

Consider again a body B moving in a reference frame R as in Figure 8.45. As before, let $\mathbf{N}_i$ $(i = 1, 2, 3)$ be unit vectors fixed in R and let $\mathbf{n}_i$ be unit vectors fixed in B. Let $\boldsymbol{\omega}$ be the angular velocity of B in R, expressed as

$$\boldsymbol{\omega} = \Omega_1\mathbf{N}_1 + \Omega_2\mathbf{N}_2 + \Omega_3\mathbf{N}_3 = \Omega_i\mathbf{N}_i \qquad (8.177)$$

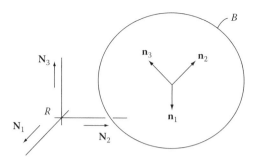

FIGURE 8.45 A body B moving in a reference frame R.

As before, let S be the transformation matrix between the $\mathbf{N}_i$ and the $\mathbf{n}_i$ with the elements S_{ij} of S defined as

$$\mathbf{S}_{ij} = \mathbf{N}_i \cdot \mathbf{n}_j \tag{8.178}$$

Recall from Equation 8.123 that the transformation matrix derivative $\mathrm{d}S/\mathrm{d}t$ may be expressed simply as

$$\frac{\mathrm{d}S}{\mathrm{d}t} = WS \tag{8.179}$$

where W is the angular velocity matrix whose elements W_{ij} are defined in Equation 8.121 as

$$W_{ij} = -e_{ijk}\Omega_k \tag{8.180}$$

Then W is (see Equation 8.122):

$$W = \begin{bmatrix} 0 & -\Omega_3 & \Omega_2 \\ \Omega_3 & 0 & -\Omega_1 \\ -\Omega_2 & \Omega_1 & 0 \end{bmatrix} \tag{8.181}$$

By using Equation 8.180 to express Equation 8.179 in terms of element derivatives, we have

$$\frac{\mathrm{d}S_{ij}}{\mathrm{d}t} = W_{im}S_{mj} = -e_{imk}\Omega_k S_{mj} \tag{8.182}$$

or specifically,

$$\frac{\mathrm{d}S_{11}}{\mathrm{d}t} = -\Omega_3 S_{21} + \Omega_2 S_{31}$$

$$\frac{\mathrm{d}S_{12}}{\mathrm{d}t} = -\Omega_3 S_{22} + \Omega_2 S_{32} \tag{8.183}$$

$$\frac{\mathrm{d}S_{13}}{\mathrm{d}t} = -\Omega_3 S_{32} + \Omega_2 S_{33}$$

$$\frac{dS_{21}}{dt} = -\Omega_1 S_{31} + \Omega_3 S_{11}$$

$$\frac{dS_{22}}{dt} = -\Omega_1 S_{32} + \Omega_3 S_{12} \tag{8.184}$$

$$\frac{dS_{23}}{dt} = -\Omega_1 S_{33} + \Omega_3 S_{13}$$

$$\frac{dS_{31}}{dt} = -\Omega_2 S_{11} + \Omega_1 S_{21}$$

$$\frac{dS_{32}}{dt} = -\Omega_2 S_{12} + \Omega_1 S_{22} \tag{8.185}$$

$$\frac{dS_{33}}{dt} = -\Omega_2 S_{13} + \Omega_1 S_{23}$$

The transformation matrix S is orthogonal; that is, its inverse is equal to its transpose (see Section 8.8). This leads to the expressions:

$$S_{ij}S_{kj} = \delta_{ik} \quad \text{and} \quad S_{ji}S_{jk} = \delta_{ik} \tag{8.186}$$

where δ_{ik} is Kronecker's delta function.

We can use Equations 8.186 to solve Equations 8.183 through 8.185 for Ω_1, Ω_2, and Ω_3. For example, if we multiply the expressions of Equations 8.183 by S_{31}, S_{32}, and S_{33} respectively and add, we obtain

$$\Omega_2 = S_{31}\frac{dS_{11}}{dt} + S_{32}\frac{dS_{12}}{dt} + S_{33}\frac{dS_{13}}{dt} \tag{8.187}$$

Similarly, from Equations 8.184 and 8.185, we obtain

$$\Omega_3 = S_{11}\frac{dS_{21}}{dt} + S_{12}\frac{dS_{22}}{dt} + S_{13}\frac{dS_{23}}{dt} \tag{8.188}$$

and

$$\Omega_1 = S_{21}\frac{dS_{31}}{dt} + S_{22}\frac{dS_{32}}{dt} + S_{23}\frac{dS_{33}}{dt} \tag{8.189}$$

Finally, by substituting from Equation 8.176 into Equations 8.187 through 8.189, we have

$$\Omega_1 = 2(\varepsilon_4\dot{\varepsilon}_1 - \varepsilon_3\dot{\varepsilon}_2 + \varepsilon_2\dot{\varepsilon}_3 - \varepsilon_1\dot{\varepsilon}_4) \tag{8.190}$$

$$\Omega_2 = 2(\varepsilon_3\dot{\varepsilon}_1 + \varepsilon_4\dot{\varepsilon}_2 - \varepsilon_1\dot{\varepsilon}_3 - \varepsilon_2\dot{\varepsilon}_4) \tag{8.191}$$

$$\Omega_3 = 2(-\varepsilon_2\dot{\varepsilon}_1 + \varepsilon_1\dot{\varepsilon}_2 + \varepsilon_4\dot{\varepsilon}_3 - \varepsilon_3\dot{\varepsilon}_4) \tag{8.192}$$

Observe the homogeneity and linearity of these expressions.

8.18 INVERSE RELATIONS BETWEEN ANGULAR VELOCITY AND EULER PARAMETERS

Observe that the linearity in Equations 8.190 through 8.192 is in stark contrast to the nonlinearity in Equation 8.95. Specifically, from Equation 8.95, the angular velocity components Ω_i ($i = 1, 2, 3$) are

$$\Omega_1 = \dot{\alpha} + \dot{\gamma} \sin\beta \tag{8.193}$$

$$\Omega_2 = \dot{\beta} \cos\alpha - \dot{\gamma} \sin\alpha \cos\beta \tag{8.194}$$

$$\Omega_3 = \dot{\beta} \sin\alpha + \dot{\gamma} \cos\alpha \cos\beta \tag{8.195}$$

where we recall that α, β, and γ are dextral orientation angles. While these equations are nonlinear in α, β, and γ, they are nevertheless linear in $\dot{\alpha}$, $\dot{\beta}$, and $\dot{\gamma}$. Thus, we can solve for $\dot{\alpha}$, $\dot{\beta}$, and $\dot{\gamma}$ obtaining (see Equation 8.135)

$$\dot{\alpha} = \Omega_1 + \frac{[\Omega_2 \sin\beta \sin\alpha - \Omega_3 \sin\beta \cos\alpha]}{\cos\beta} \tag{8.196}$$

$$\dot{\beta} = \Omega_2 \cos\alpha + \Omega_3 \sin\alpha \tag{8.197}$$

$$\dot{\gamma} = \frac{[-\Omega_2 \sin\alpha + \Omega_3 \cos\alpha]}{\cos\beta} \tag{8.198}$$

From these expressions, we see that the trigonometric nonlinearities thus produce singularities, where β is either 90° or 270°.

With the linearities of Equations 8.190 through 8.192, such singularities do not occur. To see this, observe first that these equations form but three equations for the four Euler parameter derivatives. A fourth equation may be obtained by recalling that the Euler parameters are redundant, related by Equation 8.167 as

$$\varepsilon_1^2 + \varepsilon_2^2 + \varepsilon_3^2 + \varepsilon_4^2 = 1 \tag{8.199}$$

By differentiating, we have

$$2\varepsilon_1\dot{\varepsilon}_1 + 2\varepsilon_2\dot{\varepsilon}_2 + 2\varepsilon_3\dot{\varepsilon}_3 + 2\varepsilon_4\dot{\varepsilon}_4 = 0 \tag{8.200}$$

Observe that this expression has a similar form to Equations 8.190 through 8.192. Indeed we can cast Equation 8.200 into the same form as Equations 8.190 through 8.192 by introducing an identically zero parameter Ω_4 defined as

$$\Omega_4 \overset{D}{=} 2\varepsilon_1\dot{\varepsilon} + 2\varepsilon_2\dot{\varepsilon} + 2\varepsilon_3\dot{\varepsilon}_3 + 2\varepsilon_4\dot{\varepsilon}_4 = 0 \tag{8.201}$$

Then by appending this expression to Equations 8.190 through 8.192, we have the four equations:

$$\Omega_1 = 2(\varepsilon_4\dot{\varepsilon}_1 - \varepsilon_3\dot{\varepsilon}_2 + \varepsilon_2\dot{\varepsilon}_3 - \varepsilon_1\dot{\varepsilon}_4) \tag{8.202}$$

$$\Omega_2 = 2(\varepsilon_3\dot{\varepsilon}_1 + \varepsilon_4\dot{\varepsilon}_2 - \varepsilon_1\dot{\varepsilon}_3 - \varepsilon_2\dot{\varepsilon}_4) \tag{8.203}$$

$$\Omega_3 = 2(-\varepsilon_2\dot{\varepsilon}_1 + \varepsilon_1\dot{\varepsilon}_2 + \varepsilon_4\dot{\varepsilon}_3 - \varepsilon_3\dot{\varepsilon}_4) \tag{8.204}$$

$$\Omega_4 = 2(\varepsilon_1\dot{\varepsilon}_1 + \varepsilon_2\dot{\varepsilon}_2 + \varepsilon_3\dot{\varepsilon}_3 + \varepsilon_4\dot{\varepsilon}_4) \tag{8.205}$$

These equations may be cast into the matrix form:

$$\begin{bmatrix} \Omega_1 \\ \Omega_2 \\ \Omega_3 \\ \Omega_4 \end{bmatrix} = 2 \begin{bmatrix} \varepsilon_4 & -\varepsilon_3 & \varepsilon_2 & -\varepsilon_1 \\ \varepsilon_3 & \varepsilon_4 & -\varepsilon_1 & -\varepsilon_2 \\ -\varepsilon_2 & \varepsilon_1 & \varepsilon_4 & -\varepsilon_3 \\ \varepsilon_1 & \varepsilon_2 & \varepsilon_3 & \varepsilon_4 \end{bmatrix} \begin{bmatrix} \dot{\varepsilon}_1 \\ \dot{\varepsilon}_2 \\ \dot{\varepsilon}_3 \\ \dot{\varepsilon}_4 \end{bmatrix} \tag{8.206}$$

or simply as

$$\Omega = 2E\dot{\varepsilon} \tag{8.207}$$

where the arrays Ω, E, and ε are defined by inspection.

Equation 8.207 may be solved for $\dot{\varepsilon}$ as

$$\dot{\varepsilon} = \left(\frac{1}{2}\right)E^{-1}\Omega \tag{8.208}$$

Interestingly, it happens that E is an orthogonal array so that its inverse is equal to its transpose, that is,

$$E^{-1} = E^T = \begin{bmatrix} \varepsilon_4 & \varepsilon_3 & -\varepsilon_2 & \varepsilon_1 \\ -\varepsilon_3 & \varepsilon_4 & \varepsilon_1 & \varepsilon_2 \\ \varepsilon_2 & -\varepsilon_1 & \varepsilon_4 & \varepsilon_3 \\ -\varepsilon_1 & -\varepsilon_2 & -\varepsilon_3 & \varepsilon_4 \end{bmatrix} \tag{8.209}$$

Thus, from Equation 8.208, $\dot{\varepsilon}_i$ are seen to be

$$\dot{\varepsilon}_1 = \left(\frac{1}{2}\right)(\varepsilon_4\Omega_1 + \varepsilon_3\Omega_2 - \varepsilon_2\Omega_3) \tag{8.210}$$

$$\dot{\varepsilon}_2 = \left(\frac{1}{2}\right)(-\varepsilon_3\Omega_1 + \varepsilon_4\Omega_2 + \varepsilon_1\Omega_3) \tag{8.211}$$

$$\dot{\varepsilon}_3 = \left(\frac{1}{2}\right)(\varepsilon_2\Omega_1 - \varepsilon_1\Omega_2 + \varepsilon_4\Omega_3) \tag{8.212}$$

$$\dot{\varepsilon}_4 = \left(\frac{1}{2}\right)(-\varepsilon_1\Omega_1 - \varepsilon_2\Omega_2 - \varepsilon_3\Omega_3) \tag{8.213}$$

where in view of Equation 8.201 we have assigned Ω_4 as zero.

Observe that in contrast to Equations 8.196 through 8.198, there are no singularities in Equations 8.210 through 8.213.

8.19   NUMERICAL INTEGRATION OF GOVERNING DYNAMICAL EQUATIONS

In the numerical integration of the governing dynamical equations of multibody systems, such as a human body model, we need to solve equations of the form

$$M\ddot{x} = f \quad \text{or} \quad \ddot{x} = M^{-1}f \qquad (8.214)$$

where
   M is an $n \times n$ symmetric, nonsingular, generalized mass array
   x is an $n \times 1$ array of dependent variables
   f is an $n \times 1$ forcing function array, with n being the number of degrees of freedom

The arrays M and f are themselves functions of x, $\dot{x}$, and time t as

$$M = M(x) \quad \text{and} \quad f = f(x, \dot{x}, t) \qquad (8.215)$$

In commonly used numerical integration procedures, the equations need to be cast into a first-order form such as

$$\dot{z} = \phi(z, t) \qquad (8.216)$$

Equation 8.214 may be obtained in this form by simply introducing an $n \times 1$ array y defined as $\dot{x}$. That is,

$$\dot{x} = y \quad \text{and} \quad \dot{y} = M^{-1}f \qquad (8.217)$$

Thus, we have exchanged a set of n second-order differential equations for $2n$ first-order equations, which may be expressed in the matrix form of Equation 8.216.

   In Equation 8.214, the majority of the dependent variables are usually orientation angles such as the α, β, and γ of Section 8.8. If, however, we use Euler parameters in place of the orientation angles, the governing dynamical equations may be expressed in terms of derivatives of angular velocity components. In this case, the governing equations are already in first-order form as in Equation 8.216. The Euler parameters may then be determined using Equations 8.210 through 8.213, which replace the first of Equations 8.217.

   We will explore and develop these concepts in Chapters 13 and 14.

PROBLEMS

Section 8.1

P8.1.1   In celestial mechanics, stars, planets, and moons are usually modeled as spherical bodies, which in turn are modeled by particles at the center of the sphere and with mass equal to that of the astronomical body. The particle is then represented by a point mass.
   List examples of how such modeling can be useful in biomechanics, and specifically, in human body modeling.

P8.1.2   An object is considered to be "small" if its dimensions are small compared with those of pertinent surrounding objects or with pertinent distances. For example, a grain of sand in a concrete beam has small dimensions compared with those of the beam—thus, in this context, the sand grain is "small." Suppose now that a point P, with a "strength" m, is used as a mathematical model of the sand particle, with m being the mass of the sand particle.

Suppose further that the only criterion for P is that P must be a point within the volume occupied by the sand particle. Since there are an infinite number of such points, how can P be a reliable model of the sand particle? Specifically, if Q is a different point within the volume of the sand particle, how might the results of some mechanical analysis be different for P and Q?

P8.1.3 See Problem P8.1.2. Repeat Problem P8.1.2 for a cell, say an osteocyte (or bone cell) in a long bone.

Section 8.2

P8.2.1 Suppose the location of a point P is defined by cylindrical coordinates: r, θ, z as in Section 5.8.1. Develop an expression for the position vector **OP** (or simple, **P**) from the origin O to P analogous to Equation 8.1.

P8.2.2 See Problem P8.2.1. Repeat Problem P8.2.1 for spherical coordinates: ρ, ϕ, θ.

Section 8.3

P8.3.1 Suppose a particle P, modeled by a point P, moves on a circle with radius r and center O at the origin O of an X, Y Cartesian frame as in the following figure.

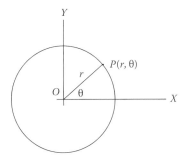

If the polar coordinates of P are: (r,θ) as shown, express the position vector **OP** (designated simply as **P**) in the form of Equation 8.1.

P8.3.2 See Problem P8.3.1. From the result of Problem P8.3.1, determine the Cartesian coordinates of P as in Equation 8.2.

P8.3.3 See Problems P8.3.1 and P8.3.2. What are the corresponding parametric equations of the circle?

P8.3.4 See Problems P8.3.1through P8.3.3. Repeat Problems P8.3.1 through P8.3.3 for the circle with radius r but located by its center Q whose Cartesian coordinates are: (a,b) as in the following figure.

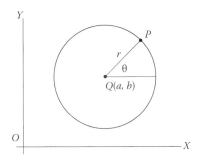

P8.3.5 Suppose the position vector **p** of a point P moving on a curve C is

$$\mathbf{p} = t\mathbf{n}_x + 2\mathbf{n}_y + 4t\mathbf{n}_z$$

where
   p locates P relative to the origin O of a Cartesian (X, Y, Z) frame
   $\mathbf{n}_x$, $\mathbf{n}_y$, and $\mathbf{n}_z$ are unit vectors parallel to the X, Y, and Z axes

a. What are the corresponding parametric equations of C?
b. Identify C as an elementary conic section curve.

P8.3.6 See Problem P8.3.5. Suppose **p** is

$$\mathbf{p} = a\cos t\,\mathbf{n}_x + b\sin t\,\mathbf{n}_y + c\mathbf{n}_z$$

where a, b, and c are positive constants. As before,
a. What are the corresponding parametric equations of curve C?
b. Identify C as an elementary conic section curve.

P8.3.7 See Problems P8.3.5 and P8.3.6. Suppose **p** is

$$\mathbf{p} = a\cos t\,\mathbf{n}_x + b\sin t\,\mathbf{n}_y + ct\mathbf{n}_z$$

where a, b, and c are positive constants.
a. What are the parametric equations of C?
b. Identify C.

P8.3.8 A particle P moves in the X–Y plane on a curve C as represented in the following figure.

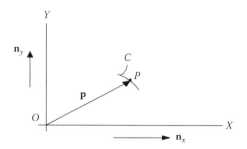

where, as before, **p** is a position vector locating P relative to the origin O.
   Suppose **p** is expressed as:

$$\mathbf{p} = r\cos\omega t\,\mathbf{n}_x + r\sin\omega t\,\mathbf{n}_y$$

where
   r and ω are constants
   $\mathbf{n}_x$ and $\mathbf{n}_y$ are unit vectors parallel to X and Y as in the earlier figure

a. Identify C.
b. Find a unit vector $\mathbf{n}_x$ parallel to **p**.
c. Find the velocity **V** of P.
d. What is the magnitude of **V** (the "speed") of P?
e. Find a unit vector $\mathbf{n}_\theta$ parallel to **V**.

f. Show that $\mathbf{n}_r$ and $\mathbf{n}_\theta$ are perpendicular.

g. Verify that $\mathbf{V}$ is tangent to C.

P8.3.9 See Problem P8.3.7. Suppose P is a particle moving on the curve C of Problem P8.3.7 with the position vector $\mathbf{p}$, locating P relative to the origin O (as before):

a. Find the velocity $\mathbf{V}$ of P.

b. Find the speed of P.

c. Find a unit vector $\mathbf{n}_t$ tangent to C.

Section 8.4

P8.4.1 Let L be a radial line in the X–Y plane as in the following figure where θ measures the inclination of L relative to the X-axis, and where, as before, $\mathbf{n}_x$ and $\mathbf{n}_y$ are unit vector parallel to the X–Y axes. Let $\mathbf{n}_r$ and $\mathbf{n}_\theta$ be unit vectors parallel and perpendicular to L as shown.

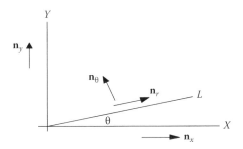

a. Express $\mathbf{n}_r$ and $\mathbf{n}_\theta$ in terms of $\mathbf{n}_x$ and $\mathbf{n}_y$.

b. Using the results of (a), verify that $\mathbf{n}_r \cdot \mathbf{n}_\theta = 0$.

c. Show that $d\mathbf{n}_r/d\theta = \mathbf{n}_\theta$.

d. Show that $d\mathbf{n}_\theta/d\theta = -\mathbf{n}_r$.

P8.4.2 See Problems P8.3.8 and P8.4.1. Suppose as in Problem P8.3.8 that a particle P moves in the X–Y plane with its position vector $\mathbf{p}$ relative to the origin O, being:

$$\mathbf{p} = r\cos\omega t\,\mathbf{n}_x + r\sin\omega t\,\mathbf{n}_y$$

where, as before, r and ω are constants.

a. Using Equation 8.3 evaluate the velocity $\mathbf{V}$ of P. Express $\mathbf{V}$ in terms of $\mathbf{n}_x$ and $\mathbf{n}_y$.

b. Using the results of (a) use Equation 8.5 to determine the acceleration $\mathbf{a}$ of P. Express $\mathbf{a}$ in terms of $\mathbf{n}_x$ and $\mathbf{n}_y$.

c. Using the results of Problems P8.3.8 and P8.4.1 express $\mathbf{V}$ in terms of $\mathbf{n}_r$ and $\mathbf{n}_\theta$.

d. Express the acceleration $\mathbf{a}$ in terms of $\mathbf{n}_r$ and $\mathbf{n}_\theta$ in two ways: first, by using the result of (b) and by expressing $\mathbf{n}_x$ and $\mathbf{n}_y$ in terms of $\mathbf{n}_r$ and $\mathbf{n}_\theta$; and second, by using the result of (c) and the results of Problem P8.4.1 for the derivatives of $\mathbf{n}_r$ and $\mathbf{n}_\theta$.

e. Show that, unlike $\mathbf{V}$, $\mathbf{a}$ is **not** tangent to C.

P8.4.3 See Problem P8.4.2. Repeat Problem P8.4.2 if the position vector $\mathbf{p}$ of P relative to O is

$$\mathbf{p} = \mathbf{p}(t) = r\cos\theta\,\mathbf{n}_x + r\sin\theta\,\mathbf{n}_y$$

where, as before, r is a constant, but θ is a function of time t given by:

$$\theta = \left(\frac{1}{2}\right)\alpha t^2 + \omega t$$

where α and ω are constants.

P8.4.4 See Problem P8.4.1. Suppose a particle P moves in the X–Y plane with its position vector $\mathbf{p}$ expressed as:

$$\mathbf{p} = \mathbf{p}(t) = r(t)\mathbf{n}_r$$

where
    $\mathbf{n}_r$ is a radial unit vector
    r, the distance OP, is now a function of time t

    The following figure provides a representation of P, $\mathbf{p}$, and $\mathbf{n}_r$. The figure also shows a unit vector $\mathbf{n}_\theta$ perpendicular to $\mathbf{n}_r$ and the inclination angle θ (a function of time t) defining the angle between $\mathbf{n}_r$ and $\mathbf{n}_x$, and between $\mathbf{n}_\theta$ and $\mathbf{n}_y$ with $\mathbf{n}_x$ and $\mathbf{n}_y$ being unit vectors parallel to the X and Y axes, as usual.

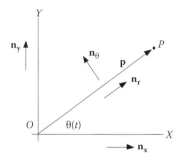

    With r and θ defined as stated, they (r,θ) may be regarded as **polar coordinates** of P.
    Using the results of Problem P8.4.1, showing that

$$\frac{d\mathbf{n}_r}{d\theta} = \mathbf{n}_\theta \quad \text{and} \quad d\mathbf{n}_\theta = -\mathbf{n}_r$$

determine the acceleration $\mathbf{a}$ of P in the X–Y plane. Express the results in terms of r, θ, $\mathbf{n}_r$ and $\mathbf{n}_\theta$. (Note by the chain rule of differentiation: $d\mathbf{n}_r/dt = (d\mathbf{n}_r/d\theta)(d\theta/dt) = \dot{\theta}\,d\mathbf{n}_r/d\theta = \dot{\theta}\mathbf{n}_\theta$ and similarly, $d\mathbf{n}_\theta/dt = -\dot{\theta}\mathbf{n}_r$.)

Section 8.5

P8.5.1 Let P and Q be points representing particles of a rigid body B as indicated in the following figure. Let $\mathbf{r}$ be a vector locating P relative to Q, and let $\mathbf{n}_r$ be a unit vector parallel to $\mathbf{r}$.

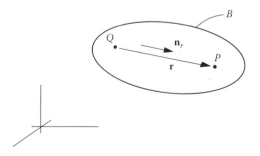

Show that $\mathbf{V}^{P/Q}$ is perpendicular to $\mathbf{n}_r$ but that $\mathbf{a}^{P/Q}$ is not necessarily perpendicular to $\mathbf{n}_r$.

Section 8.6

P8.6.1 Consider a line L in the X–Y plane, passing through the origin, with a variable inclination $\theta(t)$ relative to the X-axis as represented in the following figure. Let $\mathbf{n}_r$ be a unit vector parallel to L and let $\mathbf{n}_\theta$ be a unit vector perpendicular to $\mathbf{n}_r$ (and hence also perpendicular to L), and parallel to the X–Y plane. Finally, as before, let $\mathbf{n}_x$ and $\mathbf{n}_y$ be unit vectors parallel to the X and Y-axes.

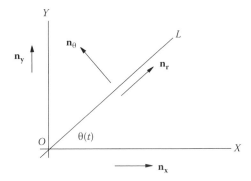

a. Express $\mathbf{n}_r$ and $\mathbf{n}_\theta$ in terms of $\mathbf{n}_x$ and $\mathbf{n}_y$.
b. Calculate the derivatives of $\mathbf{n}_r$ and $\mathbf{n}_\theta$ with respect to θ—that is, find $d\mathbf{n}_r/d\theta$ and $d\mathbf{n}_\theta/d\theta$.
c. Using the chain rule for differentiation (e.g., $df/dt = (df/d\theta)(d\theta/dt)$) calculate $d\mathbf{n}_r/dt$ and $d\mathbf{n}_\theta/dt$.
d. Let $\mathbf{n}_z$ be a unit vector normal to the X–Y plane such that $\mathbf{n}_x \times \mathbf{n}_y = \mathbf{n}_z$. Calculate: $\mathbf{n}_z \times \mathbf{n}_r$ and $\mathbf{n}_z \times \mathbf{n}_\theta$ and express the results in terms of $\mathbf{n}_x$ and $\mathbf{n}_y$.
e. Repeat (d) but express the results in terms of $\mathbf{n}_r$ and $\mathbf{n}_\theta$.
f. Review the results of (d) and (e). Let $\boldsymbol{\omega}$ be: $\dot{\theta}\mathbf{n}_z$. Calculate $\boldsymbol{\omega} \times \mathbf{n}_r$ and $\boldsymbol{\omega} \times \mathbf{n}_\theta$.
g. Compare the results of (c) and (f) and then express $d\mathbf{n}_r/dt$ and $d\mathbf{n}_\theta/dt$ in terms of $\boldsymbol{\omega}$.

P8.6.2 Consider a body B moving freely in a reference frame R as in the following figure. Let $\mathbf{a}$ and $\mathbf{b}$ be non zero, non-colinear vectors fixed in B.

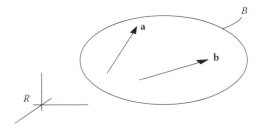

a. Show that

$$\mathbf{a} \cdot \frac{d\mathbf{a}}{dt} = 0 \quad \text{and} \quad \mathbf{b} \cdot \frac{d\mathbf{b}}{dt} = 0$$

b. Show that

$$\mathbf{a} \cdot \frac{d\mathbf{b}}{dt} = -\mathbf{b} \cdot \frac{d\mathbf{a}}{dt}$$

c. Show that

$$\frac{\left[(d\mathbf{a}/dt)\times(d\mathbf{b}/dt)\right]}{\left[(d\mathbf{a}/dt)\cdot\mathbf{b}\right]} = \frac{\left[(d\mathbf{b}/dt)\times(d\mathbf{a}/dt)\right]}{\left[(d\mathbf{b}/dt)\cdot\mathbf{a}\right]}$$

Let ω be defined as

$$\omega = \frac{\left[(d\mathbf{a}/dt)\times(d\mathbf{b}/dt)\right]}{\left[(d\mathbf{a}/dt)\cdot\mathbf{b}\right]}$$

d. Show that $d\mathbf{a}/dt = \omega \times \mathbf{a}$
e. Show that $d\mathbf{b}/dt = \omega \times \mathbf{b}$
f. Show that $d(\mathbf{a} \times \mathbf{b})/dt = \omega \times (\mathbf{a} \times \mathbf{b})$

P8.6.3 See Problem P8.6.2 and review the stated expressions. Let **c** be a third non zero vector fixed in B with the sole restriction being that **a**, **b**, and **c** are non-coplanar, as represented in the following figure. (**c** could be: $\mathbf{a} \times \mathbf{b}$.)

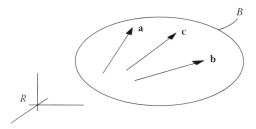

Let V be any other vector fixed in B. Show that **V** may be uniquely expressed in the form

$$\mathbf{V} = p\mathbf{a} + q\mathbf{b} + r\mathbf{c}$$

where p, q, and r are scalar constants. Hint: by evaluating the scalar (dot) product of the expression with $\mathbf{b} \times \mathbf{c}$, determine p as $(\mathbf{b} \times \mathbf{c})/(\mathbf{b} \times \mathbf{c} \cdot \mathbf{V})$. Then find similar expressions for q and r.

P8.6.4 Review Problem P8.6.3. Show that, whether or not **c** is $\mathbf{a} \times \mathbf{b}$, **c** may be uniquely expressed as

$$\mathbf{c} = \alpha\mathbf{a} + \beta\mathbf{b} + \gamma\mathbf{a} \times \mathbf{b}$$

where α, β, and γ are scalar constants.

P8.6.5 Review Problems P8.6.2 through P8.6.4. In view of the assertions of tasks (d), (e), and (f) of Problem P8.6.2, show that

$$\frac{d\mathbf{c}}{dt} = \omega \times \mathbf{c}$$

where, as in Problem P8.6.2, ω is

$$\omega = \frac{\left[(d\mathbf{a}/dt)\times(d\mathbf{b}/dt)\right]}{\left[(d\mathbf{a}/dt)\cdot\mathbf{b}\right]}$$

P8.6.6 See again Problems P8.6.2 through P8.6.5. Show that in view of the result of Problem P8.6.5, if **V** is *any* vector fixed in B d**V**/d*t* is

$$\frac{d\mathbf{V}}{dt} = \boldsymbol{\omega} \times \mathbf{V}$$

where, as before, ω is

$$\omega = \frac{\left[(d\mathbf{a}/dt) \times (d\mathbf{b}/dt)\right]}{\left[(d\mathbf{a}/dt) \cdot \mathbf{b}\right]}$$

P8.6.7 Show that angular velocity as in Equation 8.18 is unique. Hint: let ω_1 and ω_2 be two assumed to be distinct vectors satisfying Equation 8.18. Then by comparison show that ω_1 and ω_2 must be the same.

Section 8.7

P8.7.1 Consider a pair of reference frames R_1 and R_2 moving in a reference frame R as in the following figure.

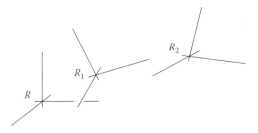

Show that

$$^{R_2}\boldsymbol{\omega}^{R_1} = -\,^{R_1}\boldsymbol{\omega}^{R_2}$$

Section 8.8

P8.8.1 In the single-panel configuration graph of the following figure, determine the relation between the $\mathbf{N}_i$ and the $\hat{\mathbf{N}}_i$. Specifically, express the $\mathbf{N}_i$ in terms of the $\hat{\mathbf{N}}_i$ and conversely, express the $\hat{\mathbf{N}}_i$ in terms of the $\mathbf{N}_i$.

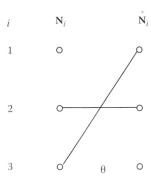

P8.8.2 For the configuration graph of the following figure, express the $\mathbf{N}_i$ in terms of the $\mathbf{n}_i$ and, conversely, express the $\mathbf{n}_i$ in terms of the $\mathbf{N}_i$.

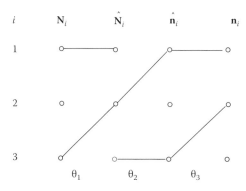

P8.8.3 See Problem P8.8.2 and particularly the configuration graph of the figure for that problem. Let the unit vectors $\mathbf{N}_i$ and $\mathbf{n}_i$ be arranged (or placed) in 3×1 arrays $\mathbf{N}$ and $\mathbf{n}$. Suppose further that $\mathbf{N}$ and $\mathbf{n}$ are related by the expression

$$\mathbf{N} = S\mathbf{n} = ABC\mathbf{n}$$

where S, A, B, and C are matrices.
   Find S, A, B, and C.

P8.8.4 Repeat Problems P8.8.2 and P8.8.3 for the configuration graph of the following figure.

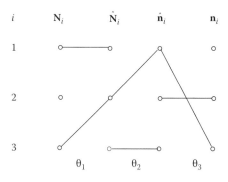

Section 8.9

P8.9.1 See Problem P8.8.2. Let the unit vectors $\mathbf{n}_i$ ($i = 1, 2, 3$) be fixed in a body B moving in a reference frame R and let the $\mathbf{N}_i$ ($i = 1, 2, 3$) be unit vectors fixed in R. By using the procedures of Section 8.9, determine the angular velocity of B in R, ${}^R\omega^B$ for the configuration graph of the figure for Problem P8.8.2. Express the results both in terms of the $\mathbf{n}_i$ and the $\mathbf{N}_i$.

P8.9.2 See Problems P8.8.4 and P8.9.1. Repeat Problem P8.9.1 for the configuration graph of the figure for Problem P8.8.4.

P8.9.3 Determine approximate expressions for Equations 8.95 and 8.96 by assuming that α, β, γ, $\dot{\alpha}$, $\dot{\beta}$, and $\dot{\gamma}$ are small.

P8.9.4 Determine approximate expressions for Equations 8.97 and 8.98 by assuming that θ_1, θ_2, θ_3, $\dot{\theta}_1$, $\dot{\theta}_2$, and $\dot{\theta}_3$ are small.

Section 8.10

P8.10.1 Referring to Table 8.1 and Figure 8.36, determine the angular velocity of the head B_8 in R. Express the result in a form analogous to Equation 8.102.

P8.10.2 See Problem P8.10.1. Use Equations 8.104 and 8.105 to determine the angular velocity of B_8 in $R(\omega_8)$.

P8.10.3 Compare the "labor" (the level of effort) of obtaining ω_8 using (a) Table 8.1 and (b) Equations 8.104 and 8.105.

Section 8.11

P8.11.1 Suppose a body B is moving in a reference frame R as represented in the following figure, where, as before, the $\mathbf{n}_i$ are fixed in B and the $\mathbf{N}_i$ are fixed in R.

By differentiating in Equation 8.95 and 8.96, determine expressions for the angular acceleration of B in R, $^R\alpha^B$, in terms of (a) the $\mathbf{N}_i$ and (b) the $\mathbf{n}_i$.

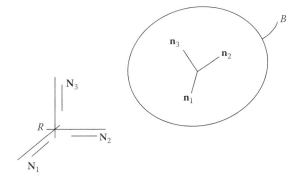

P8.11.2 Suppose a multibody system has purely planar motion such as a rod–pendulum system restricted to movement in say the X–Y plane as represented in the following figure.

Show that for planar motion systems there exists an addition formula for angular acceleration analogous to that for angular velocity (Equation 8.112). That is, show that Equation 8.113 is valid for planar motion systems.

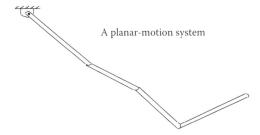

A planar-motion system

Section 8.12

P8.12.1 Consider the configuration graph of Figure 8.29 and shown again here in the following figure. Verify that the associated transformation matrix S, as expressed in Equation 8.74, is

$$S = \begin{bmatrix} c_\beta c_\gamma & -c_\beta s_\gamma & s_\beta \\ (c_\alpha s_\gamma + s_\alpha s_\beta c_\gamma) & (c_\alpha c_\gamma - s_\alpha s_\beta s_\gamma) & -s_\alpha c_\beta \\ (s_\alpha s_\gamma - c_\alpha s_\beta c_\gamma) & (s_\alpha c_\lambda + c_\alpha s_\beta s_\lambda) & c_\alpha c_\beta \end{bmatrix}$$

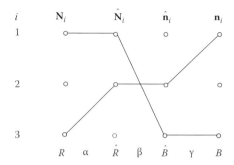

(see also Figures 8.29 and 8.32).

P8.12.2 See Problem P8.12.1. By differentiation of the elements of the transformation matrix S, determine the transformation matrix derivative: dS/dt (or simply, $\dot{S}$).

P8.12.3 See Problem P8.12.1. Review the analysis of Section 8.11 and verify the expression for the angular velocity ${}^R\omega^B$ given by Equation 8.95. List the $\mathbf{N}_i$ components Ω_i of ${}^R\omega^B$.

P8.12.4 See Problem P8.12.3. Determine the angular velocity matrix W.

P8.12.5 See Problems P8.12.1 and P8.12.4. Using the result of Problem P8.12.4, evaluate the matrix product WS.

P8.12.6 See Problems P8.12.2 and P8.12.5. Compare the results of those problems. Discuss the level of effort in obtaining the respective results. What does the level of effort imply about the practicality of numerical evaluation of the transformation matrix derivative.

P8.12.7 See Problems P8.12.1 through P8.12.6. Consider the configuration graph of Figure 8.30, and shown again in the following figure, for the Euler angle rotation sequence. Determine the associated transformation matrix S. Verify the result via Equations 8.84, 8.90, and 8.91.

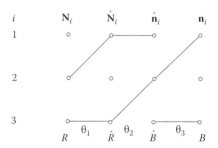

(See also Figure 8.30).

P8.12.8 See the result of Problem P8.12.7. By differentiation of the elements of the transformation matrix S, determine the transformation matrix derivative: $\dot{S}$.

P8.12.9 See Problem P8.12.7. Verify the expression of Equation 8.97 for the angular velocity of B in R, ${}^R\omega^B$. List the $\mathbf{N}_i$ components $\Omega_i (i = 1,2,3)$ of ${}^R\omega^B$.

P8.12.10 See Problem P8.12.9. Determine the angular velocity matrix W.

P8.12.11 See Problems P8.12.7 and P8.12.10. Using the results of Problems P8.12.7 and P8.12.10 evaluate the matrix product: WS.

P8.12.12 See Problems P8.12.8 and P8.12.11. Compare the results of these problems. Discuss the level of effort in obtaining the respective results. See the discussion for Problem P8.12.6.

Section 8.13

P8.13.1 Refer to Figure 8.39. Suppose point Q is fixed in R. Show that $\mathbf{V}^P$ and $\mathbf{a}^P$ are then the velocity and acceleration of a point moving on a sphere with center at Q and radius $r = |\mathbf{r}|$.

P8.13.2 See Problem P8.13.1. Suppose now that Q is not fixed in R. Provide a simple interpretation of Equations 8.126 and 8.129.

Section 8.14

P8.14.1 Review again Equations 8.97 and 8.98, expressing angular velocity in terms of Euler angles θ_1, θ_2, and θ_3. If these equations are expressed in the forms

$$^R\boldsymbol{\omega}^B = \Omega_i \mathbf{N}_i \quad \text{and} \quad ^R\boldsymbol{\omega}^B = \omega_i \mathbf{n}_i$$

solve for $\dot{\theta}_1$, $\dot{\theta}_2$, and $\dot{\theta}_3$ in terms of the Ω_i and the $\boldsymbol{\omega}_i$ ($i = 1,2,3$).

P8.14.2 See Problem P8.14.1. Identify the singularities occurring with the Euler angles.

Section 8.15

P8.15.1 For the matrix of Equation 8.165, determine the "trace"—that is, the sum of the diagonal elements.

P8.15.2 Using the result of Problem P8.15.1, determine the rotation angle θ in terms of the rotation dyadic elements.

P8.15.3 See Problem P8.15.2. What are the geometric interpretations when θ has the values (a) 0, (b) π, (c) 2π, (d) $-\pi$, and (e) -2π?

P8.15.4 See Problems P8.15.2 and P8.15.3. Suppose θ has none of the values $\pm\pi$, 0, nor $\pm 2\pi$. Referring again to Equation 8.165, determine λ_1, λ_2, and λ_3 in terms of the off-diagonal elements of the rotation matrix.

P8.15.5 Using the rotation matrix elements of Equation 8.165, show that the rotation dyadic is orthogonal. That is, show that

$$\mathbf{R} \cdot \mathbf{R}^T = \mathbf{I} \quad \text{and} \quad \mathbf{R}^T \cdot \mathbf{R} = \mathbf{I}$$

where $\mathbf{I}$ is the identity dyadic.

Section 8.16

P8.16.1 Following the procedure used to obtain S_{11} of the matrix of Euler parameters of Equation 8.176, determine S_{22} and S_{33} in terms of the Euler parameters.

P8.16.2 Following the procedure used to obtain S_{12} of the matrix of Euler parameters of Equation 8.176, determine S_{13}, S_{21}, S_{23}, S_{31}, and S_{32}.

Section 8.17

P8.17.1 Validate Equations 8.190 through 8.192 by substituting from Equations 8.176 into Equations 8.187 through 8.189.

Section 8.18

P8.18.1 With the matrix E being defined by inspection of Equations 8.206 and 8.207, show that E is orthogonal. That is, show that

$$ET^T = I \quad \text{and} \quad E^T E \neq I$$

where I is the 4×4 identity matrix.

Section 8.19

P8.19.1 What are possible disadvantages of using Euler parameters when numerically integrating governing dynamical equations?

REFERENCES

1. T. R. Kane, P. W. Likins, and D. A. Levinson, *Spacecraft Dynamics*, McGraw-Hill, New York, 1983, pp. 422–431.
2. R. L. Huston and C.-Q. Liu, *Formulas for Dynamic Analysis*, Marcel Dekker, New York, 2001, pp. 279–288, 292, 293.
3. H. Josephs and R. L. Huston, *Dynamics of Mechanical Systems*, CRC Press, Boca Raton, FL, 2002, pp. 617–628, 637.

9 Kinematic Preliminaries
Inertia Force Considerations

Kinetics refers to forces and force systems. In Chapter 4, we discussed force systems and their characteristics: forces, line of action, moments, resultant, couple, equivalent systems, and replacement. Here we will consider how these concepts may be applied with biosystems—particularly human body models. As in the foregoing chapter, we will consider a typical segment of the model to develop our analysis.

9.1 APPLIED FORCES AND INERTIA FORCES

Consider a typical body B of a human body model, and imagine B to be moving in a fixed frame R, as represented in Figure 9.1. In general, there will be two kinds of forces exerted on B: (1) those arising externally, so-called applied or active forces, and (2) those arising internally, so-called inertia or passive forces. Applied forces are due to gravity; to be in contact with other bodies, with fabric (e.g., garments), and with structural/component surfaces; and due to muscle/tendon activity. Inertia forces are due to the motion of B in R.

As in Chapter 4, for modeling purposes, we will regard B as being rigid so that the active forces may be represented by a single resultant force $\mathbf{F}$ passing through an arbitrary point Q of B together with a couple having a torque $\mathbf{T}$. Although Q is arbitrary, we will usually select the mass center (or center of gravity) G of B for the placement of $\mathbf{F}$ and then evaluate $\mathbf{T}$ accordingly.

To model the inertia forces, we will use d'Alembert's principle—a variant of Newton's second law. Specifically, consider a particle (or cell) P of B, with mass m and acceleration $\mathbf{a}$ in a fixed frame R. Then from Newton's second law, the force $\mathbf{F}$ on P needed to produce the acceleration $\mathbf{a}$ in R is simply

$$\mathbf{F} = m\mathbf{a} \tag{9.1}$$

A fixed frame R such that Equation 9.1 is valid is called an inertial reference frame or alternatively a Newtonian reference frame. There have been extensive discussions about the existence (or nonexistence) of Newtonian reference frames, often resulting in such circular statements as, "A Newtonian reference frame is a reference frame where Newton's laws are valid, and Newton's laws are valid in a Newtonian reference frame." Nevertheless, for most cases of practical importance in biomechanics, the Earth is a good approximation to a Newtonian reference frame (even though the Earth is rotating and also is in orbit about the sun, which itself is moving in the galaxy).

Analytically, d'Alembert's principle is often stated instead of Equation 9.1 as

$$\mathbf{F} - m\mathbf{a} = 0 \quad \text{or} \quad \mathbf{F} + \mathbf{F}^* = 0 \tag{9.2}$$

where $\mathbf{F}^*$ is called an inertia force and is defined as

$$\mathbf{F}^* \overset{\mathrm{D}}{=} -m\mathbf{a} \tag{9.3}$$

In analytical procedures, $\mathbf{F}^*$ is treated in the same manner as applied forces (or active forces).

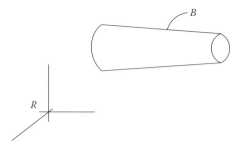

FIGURE 9.1   A human body model segment moving in an inertial reference frame.

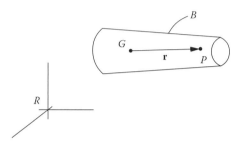

FIGURE 9.2   Position vector of a human body model segment.

From the definition of Equation 9.3, we see that inertia forces arise due to the movement, specifically the acceleration, of the particles. Consider again the typical human body segment B of Figure 9.1 and as shown again in Figure 9.2. Let P be a typical particle of B and let G be the mass center (center of gravity) of B. Then from Equation 8.129, the acceleration of P is

$$\mathbf{a}^P = \mathbf{a}^G + \boldsymbol{\alpha} \times \mathbf{r} + \boldsymbol{\omega} \times (\boldsymbol{\omega} \times \mathbf{r}) \tag{9.4}$$

where
   $\boldsymbol{\omega}$ and $\boldsymbol{\alpha}$ are the angular velocity and angular acceleration of B in R
   $\mathbf{r}$ is a position vector locating P relative to G, as in Figure 9.2

In Equation 9.4, we see that unless $\mathbf{a}^G$, $\boldsymbol{\alpha}$, and $\boldsymbol{\omega}$ are zero, $\mathbf{a}^P$ will not be zero, and thus from Equation 9.3, P will experience an inertia force $\mathbf{F}_P\!*$ as

$$\mathbf{F}_P\!* = -m_p \mathbf{a}^P = -m_p \left[\mathbf{a}^G + \boldsymbol{\alpha} \times \mathbf{r} + \boldsymbol{\omega} \times (\boldsymbol{\omega} \times \mathbf{r}) \right] \tag{9.5}$$

Next, let B be modeled as a set of particles (or cells) P_i ($i = 1, \ldots, N$) (such as a stone made up of particles of sand). Then from Equation 9.4, each of these particles will experience an inertia force as B moves. The set of all these inertia forces (a large set) then constitutes an inertia force system on B. Since this is a very large system of forces, it is needful for analysis purposes to use an equivalent set of forces to represent the system (see Chapter 4 for a discussion of equivalent force systems).

As with the active forces on B, it is convenient to represent the inertia force system by a single force $\mathbf{F}^*$ passing through mass center G together with a couple with torque $\mathbf{T}^*$. Before evaluating $\mathbf{F}^*$ and $\mathbf{T}^*$, however, it is convenient to review the concept of mass center (or center of gravity).

9.2 MASS CENTER

Let P be a particle with mass m and let O be a reference point as in Figure 9.3. Let $\mathbf{p}$ be a position vector locating P relative to O. The product $m\mathbf{p}$, designated as $\boldsymbol{\phi}^{P/O}$, is called the first moment of P relative to O. That is,

$$\boldsymbol{\phi}^{P/O} = m\mathbf{p} \tag{9.6}$$

Next, consider a set S of N particles P_i with masses m_i ($i = 1, ..., N$) as in Figure 9.4. As before, let O be a reference point and let $\mathbf{p}_i$ locate typical particle P_i relative to O. The first moment $\boldsymbol{\phi}^{S/G}$ of S relative to O is simply the sum of the first moments of the individual particles relative to O. That is,

$$\boldsymbol{\phi}^{P/O} = \sum_{i=1}^{N} \boldsymbol{\phi}^{P_i/O} = \sum_{i=1}^{N} m_i \mathbf{p}_i \tag{9.7}$$

Observe in Equations 9.6 and 9.7 that the first moment is dependent upon the position of the reference point O. In this regard, if there is a reference point G such that the first moment of S for G is zero, then G is called the mass center of S. Specifically, G is the mass center of S if

$$\boldsymbol{\phi}^{S/G} = 0 \tag{9.8}$$

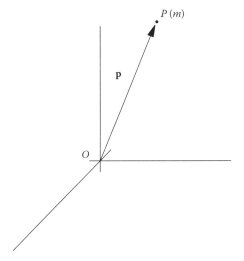

FIGURE 9.3   Particle P with mass m on a reference point O.

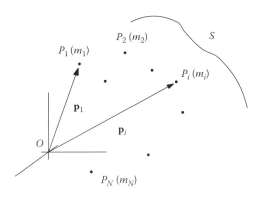

FIGURE 9.4   A set S of N particles.

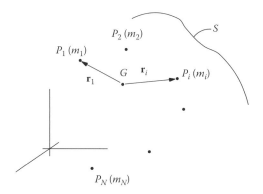

FIGURE 9.5 A set of particles with mass center G.

To illustrate this concept, let G be the mass center of the set S of particles of Figure 9.4, as suggested in Figure 9.5. Let $\mathbf{r}_i$ ($i = 1, \ldots, N$) locate typical particle P_i relative to G. Then an equivalent expression for the definition of Equation 9.8 is

$$\phi^{S/G} = \sum_{i=1}^{N} m_i \mathbf{r}_i = 0 \tag{9.9}$$

Two questions arise in view of the definition of Equation 9.8: Does a point G satisfying Equation 9.8 always exist? If so, is G unique? To answer these questions, consider again the set S of particles of Figure 9.4. Let O be an arbitrary reference point and let G be the sought after mass center as represented in Figure 9.6. Let $\mathbf{p}_G$ locate G relative to O, and as before, let $\mathbf{p}_i$ and $\mathbf{r}_i$ locate typical particle P_i relative to O and G, respectively, as shown in Figure 9.6. Then by simple vector addition, we have

$$\mathbf{p}_i = \mathbf{p}_G + \mathbf{r}_i \quad \text{or} \quad \mathbf{r}_i = \mathbf{p}_i - \mathbf{p}_G \tag{9.10}$$

Then from the definitions of Equation 9.9, we have

$$\sum_{i=1}^{N} m_i \mathbf{r}_i = \sum_{i=1}^{N} m_i (\mathbf{p}_i - \mathbf{p}_G) = \sum_{i=1}^{N} m_i \mathbf{p}_i - \sum_{i=1}^{N} m_i \mathbf{p}_G = 0 \tag{9.11}$$

or

$$\left(\sum_{i=1}^{N} m_i \right) \mathbf{p}_G = \sum_{i=1}^{N} m_i \mathbf{p}_i \tag{9.12}$$

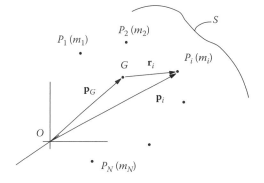

FIGURE 9.6 A set of particles with reference point O and mass center G.

or

$$\mathbf{P}_G = \frac{\displaystyle\sum_{i=1}^{N} m_i \mathbf{p}_i}{\displaystyle\sum_{i=1}^{N} m_i = (1/M)\sum_{i=1}^{N} m_i \mathbf{p}_i = (1/M)\boldsymbol{\phi}^{S/O}} \tag{9.13}$$

where M is the total mass $\left(\displaystyle\sum_{i=1}^{N} m_i\right)$ of the particles of S.

Equation 9.13 determines the existence of G by locating it relative to an arbitrarily chosen reference point O.

Regarding the uniqueness of G, let $\hat{G}$ be an alternative mass center, satisfying the definitions of Equations 9.8 and 9.9 as represented in Figure 9.7 where now $\mathbf{r}_i$ locates typical particle P_i relative to $\hat{G}$. Then from Figure 9.7, we have

$$\mathbf{p}_i = \mathbf{p}_{\hat{G}} + \hat{\mathbf{r}}_i \quad \text{or} \quad \hat{\mathbf{r}}_i = \mathbf{p}_i - \mathbf{p}_{\hat{G}} \tag{9.14}$$

From the definition of Equation 9.9, we have

$$\sum_{i=1}^{N} m_i \hat{\mathbf{r}}_i = 0 = \sum_{i=1}^{N} m_i \mathbf{p}_i - \left(\sum_{i=1}^{N} m_i\right)\mathbf{p}_{\hat{G}} \tag{9.15}$$

or

$$\mathbf{p}_{\hat{G}} = \left(\frac{1}{M}\right)\sum_{i=1}^{N} m_i \mathbf{p}_i \tag{9.16}$$

where, as before, M is the total mass $\left(\displaystyle\sum_{i=1}^{N} m_i\right)$ of S. Since the result of Equation 9.16 is the same as the result of Equation 9.13, we see that

$$\mathbf{p}_{\hat{G}} = \mathbf{p}_G \tag{9.17}$$

Therefore, $\hat{G}$ is at G and thus the mass center is unique.

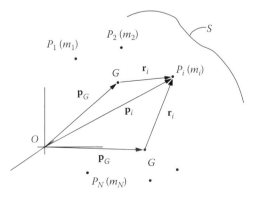

FIGURE 9.7 Set S with distinct mass centers.

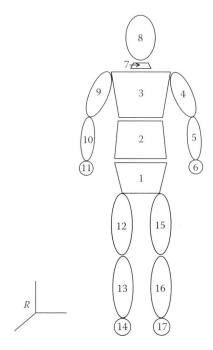

FIGURE 9.8 Human body model (17 bodies).

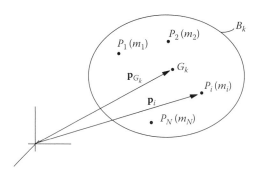

FIGURE 9.9 Typical body B_k of a human body model.

Finally, consider again a 17-member human body model as in Figure 9.8. We can locate the mass center of the model by using Equation 9.12. Consider a typical body B_k of the model as in Figure 9.9, where O is an arbitrary reference point and G_k is the mass center of B_k. Let B_k itself be modeled by a set of particles P_i with masses m_i $(i = 1, \ldots, N)$. Let $\mathbf{p}_{G_k}$ and $\mathbf{p}_i$ locate G_k and P_i relative to O. Then from Equation 9.11, we have

$$\sum_{i=1}^{N} m_i \mathbf{p}_i = \left(\sum_{i=1}^{N} m_i \right) \mathbf{p}_{G_k} = m_k \mathbf{p}_{G_k} \qquad (9.18)$$

where m_k is the mass of B_k.

Equation 9.18 shows that the sum of the first moments of all the particles of B_k relative to O is simply the first moment of a particle at G_k with the total mass m_k of B_k. Thus, for the purpose of determining the first moment of a body B_k (represented as a set of particles) relative to a reference

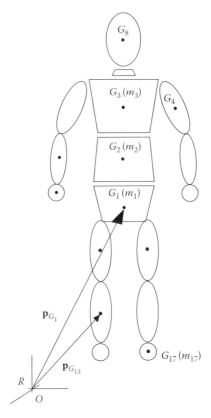

FIGURE 9.10   Particles representing the bodies of the human body model of Figure 9.8.

point O, we need to simply determine the first moment of the mass center G_k, with associated mass m_k, relative to O. That is, G_k with mass m_k can represent the entire body B_k for the purpose of finding the mass center of the human body model.

Returning now to the 17-body human body model of Figure 9.8, we can find the mass center of the model, whatever its configuration, by representing the model by a set of 17 particles of the mass centers of the bodies and with masses equal to the masses of the respective bodies, as in Figure 9.10. Then the mass center G of the human body model is positioned relative to reference point O by position vector $\mathbf{p}_G$ given by

$$\mathbf{p}_G = \left(\frac{1}{M}\right)\sum_{k=1}^{17} m_k \mathbf{p}_{G_k} \tag{9.19}$$

where M is the total mass of the model.

9.3   EQUIVALENT INERTIA FORCE SYSTEMS

In Chapter 4, we saw that if we are given any force system, no matter how large, there exists an equivalent force system consisting of a single force passing through an arbitrary point, together with a couple (see Section 4.5.3). In this regard, consider the inertia force system acting on a typical body of a human body model. Let B be such a body. Then, as B moves, each particle (or cell) of B will experience an inertia force proportional to the cell mass and the acceleration

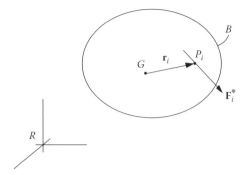

FIGURE 9.11   Inertia force on a typical particle P_i of a typical body B of a human body model.

of the cell in an inertial reference frame R. Let P_i be a typical cell or particle of B, and let m_i be the mass of P_i. Then the inertia force $\mathbf{F}_i^*$ on P_i is simply

$$\mathbf{F}_i^* = -m_i \mathbf{a}^{P_i} \tag{9.20}$$

where, as before, $\mathbf{a}^{P_i}$ is the acceleration of P_i in R. Then from Equation 9.5, $\mathbf{F}_i^*$ may be expressed as

$$\mathbf{F}_i^* = -m_i \left[\mathbf{a}^G + \boldsymbol{\alpha} \times \mathbf{r}_i + \boldsymbol{\omega} \times (\boldsymbol{\omega} \times \mathbf{r}_i) \right] \tag{9.21}$$

where

    G is the mass center of B

    $\boldsymbol{\omega}$ and $\boldsymbol{\alpha}$ are the angular velocity and angular acceleration of B

    $\mathbf{r}_i$ is a position vector locating P_i relative to G, as represented in Figure 9.11

Observe that if B contains N particles (or cells), then B will be subjected to N inertia forces. Consider now the task of obtaining an equivalent inertia force system: Specifically, let this equivalent system consist of a single force $\mathbf{F}^*$ passing through mass center G together with a couple with torque $\mathbf{T}^*$. Then, from Equations 4.4 and 4.5, $\mathbf{F}^*$ and $\mathbf{T}^*$ are

$$\mathbf{F}^* = \sum_{i=1}^{N} \mathbf{F}_i^* \tag{9.22}$$

and

$$\mathbf{T}^* = \sum_{i=1}^{N} \mathbf{r}_i \times \mathbf{F}_i^* \tag{9.23}$$

Then by substituting from Equation 9.21, $\mathbf{F}^*$ becomes

$$\mathbf{F}^* = \sum_{i=1}^{N} -m_i \mathbf{a}^{P_i} = \sum_{i=1}^{N} -m_i \left[\mathbf{a}^G + \boldsymbol{\alpha} \times \mathbf{r}_i + \boldsymbol{\omega} \times (\boldsymbol{\omega} \times \mathbf{r}_i) \right]$$

$$= -\left(\sum_{i=1}^{N} m_i \right) \mathbf{a}^G - \boldsymbol{\alpha} \times \left(\sum_{i=1}^{N} m_i \mathbf{r}_i \right) - \boldsymbol{\omega} \times \left[\boldsymbol{\omega} \times \left(\sum_{i=1}^{N} m_i \mathbf{r}_i \right) \right]$$

or

$$\mathbf{F}^* = -m \mathbf{a}^G \tag{9.24}$$

where m is the total mass of the particles of B and where, in view of Equation 9.9, the last two terms in the penultimate line are zero since G is the mass center.

Similarly, by substituting from Equation 9.21, $\mathbf{T}^*$ becomes

$$\mathbf{T}^* = \sum_{i=1}^{N} \mathbf{r}_i \times (-m_i \mathbf{a}^{P_i}) = \sum_{i=1}^{N} \mathbf{r}_i \times (-m_i) \left[\mathbf{a}^G + \boldsymbol{\alpha} \times \mathbf{r}_i + \boldsymbol{\omega} \times (\boldsymbol{\omega} \times \mathbf{r}_i) \right]$$

$$= -\left(\sum_{i=1}^{N} \overset{0}{m_i \mathbf{r}_i} \right) \mathbf{a}^G - \sum_{i=1}^{N} m_i \mathbf{r}_i \times (\boldsymbol{\alpha} \times \mathbf{r}_i) - \sum_{i=1}^{N} m_i \mathbf{r}_i \times \left[\boldsymbol{\omega} \times (\boldsymbol{\omega} \times \mathbf{r}_i) \right]$$

$$= -\sum_{i=1}^{N} m_i \mathbf{r}_i \times (\boldsymbol{\alpha} \times \mathbf{r}_i) - \sum_{i=1}^{N} m_i \mathbf{r}_i \times \left[\boldsymbol{\omega} \times (\boldsymbol{\omega} \times \mathbf{r}_i) \right] \tag{9.25}$$

where, as before, $\sum_{i=1}^{N} m_i \mathbf{r}_i$ is zero since G is the mass center of B. By using the properties of the vector triple products, we see that

$$\mathbf{r}_i \times \left[\boldsymbol{\omega} \times (\boldsymbol{\omega} \times \mathbf{r}_i) \right] = \boldsymbol{\omega} \times \left[\mathbf{r}_i \times (\boldsymbol{\omega} \times \mathbf{r}_i) \right] \tag{9.26}$$

Hence, $\mathbf{T}^*$ becomes

$$\mathbf{T}^* = -\sum_{i=1}^{N} m_i \mathbf{r}_i \times (\boldsymbol{\alpha} \times \mathbf{r}_i) - \boldsymbol{\omega} \times \sum_{i=1}^{N} m_i \mathbf{r}_i \times (\boldsymbol{\omega} \times \mathbf{r}_i) \tag{9.27}$$

Observe that this expression for $\mathbf{T}^*$ is not nearly as simple as the expression for $\mathbf{F}^*$ (Equation 9.24). Indeed, $\mathbf{T}^*$ has two terms, each involving large sums of vector triple products. These sums, however, have similar forms: Their only difference is that one has $\boldsymbol{\alpha}$ and the other $\boldsymbol{\omega}$. They can be cast into the same form by observing that if $\mathbf{n}_\alpha$ and $\mathbf{n}_\omega$ are unit vectors parallel to $\boldsymbol{\alpha}$ and $\boldsymbol{\omega}$, then $\boldsymbol{\alpha}$ and $\boldsymbol{\omega}$ may be expressed as

$$\boldsymbol{\alpha} = \alpha \mathbf{n}_\alpha \quad \text{and} \quad \boldsymbol{\omega} = \omega \mathbf{n}_\omega \tag{9.28}$$

Then $\mathbf{T}^*$ in turn may be expressed as

$$\mathbf{T}^* = -\alpha \sum_{i=1}^{N} m_i \mathbf{r}_i \times (\mathbf{n}_\alpha \times \mathbf{r}_i) - \boldsymbol{\omega} \times \omega \sum_{i=1}^{N} m_i \mathbf{r}_i \times (\mathbf{n}_\omega \times \mathbf{r}_i) \tag{9.29}$$

We will discuss and develop these triple product sums in Chapter 10.

PROBLEMS

Section 9.1

P9.1.1   A car is going at a constant speed of 45 mph around a curve with radius 250 ft. A passenger in the car weighs 150 lb. By modeling the passenger as a point mass, determine the inertia force experienced by the passenger.

P9.1.2   See Problem P9.1.1. Suppose the vehicle operator begins to slow the car with a braking rate of 10 mph/s. What is now the inertia force experienced by the passenger? What is the magnitude of the inertia force?

P9.1.3 Suppose a diner is sitting in a booth in a revolving restaurant, which has a turning rate of one revolution per hour. Suppose further that the booth is approximately 50 ft away from the axis of rotation. What is the inertia force on a 16 ounce bottle of fruit juice?

P9.1.4 See Problems P9.1.1 and P9.1.3. Suppose a 175 lb man (modeled as a point mass) is standing at the rim of a 30 ft. diametric turntable similar to that of a carousel. Suppose further that the turntable is rotating at a constant rate of one revolution every 10 s. Determine the inertia force on the man.

P9.1.5 See Problem P9.1.4. Suppose the man on the turntable is walking along a radial line toward the axis of rotation (the turntable center) at a rate of 4 ft/s. What is the inertia force experienced by the man when he is (a) 10 ft from the center and (b) when he reaches the center?

Section 9.2

P9.2.1 Three particles P_1, P_2, and P_3 having masses 4, 5, and 2 kg, respectively, are placed along an X-axis at coordinates 1, 3, and 7 m as indicated in the following figure. What is the X-coordinate of the mass center G of the particles?

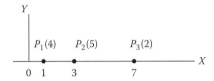

P9.2.2 Consider the set of six particles in an X–Y plane as represented in the following figure, where the numbers in parentheses are the X,Y coordinates (in m) and where the masses of the particles are listed in the following table (which also lists the coordinates).

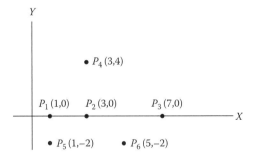

Particle	Mass (kg)	X	Y
P_1	4	1	0
P_2	5	3	0
P_3	2	7	0
P_4	8	3	4
P_5	3	1	−2
P_6	3	5	−2

Find the coordinates of the mass center of the six particles.

P9.2.3 See Problems P9.2.1 and P9.2.2. Repeat Problem P9.2.2 using the results of Problem P9.2.1. Specifically, let particles P_1, P_2, and P_3 be represented by a single particle P having mass (4 + 5 + 2 = 11 kg) located at the mass center of P_1, P_2, and P_3. Then find the mass center of the four particles: P, P_4, P_5, and P_6. Compare the results and the level of effort.

P9.2.4 Repeat Problem P9.2.3 by also letting particles P_5 and P_6 be represented by a single particle.

P9.2.5 Consider a set of 10 particles with masses and Cartesian coordinates as listed in the following table where the masses are in slugs and the coordinates are in feet.

Observe further, in the following table, the last three columns have selected products of the masses and coordinate values.

a. Calculate and fill in the missing entries in the last three columns.
b. Add the entries in columns 3, 7, 8, and 9, and label the sums as Σm_i, $\Sigma m_i x_i$, $\Sigma m_i y_i$, and $\Sigma m_i z_i$.
c. Evaluate the quotients $\Sigma m_i x_i / \Sigma m_i$, $\Sigma m_i y_i / \Sigma m_i$, and $\Sigma m_i z_i / \Sigma m_i$.
d. If the quotients of (c) are called $\bar{x}$, $\bar{y}$, and $\bar{z}$, respectively, show that $(\bar{x}, \bar{y}, \bar{z})$ are the X, Y, Z coordinates of the mass center of the set of particles.

I	Particle	m_i	x_i	y_i	z_i	$m_i x_i$	$m_i y_i$	$m_i z_i$
1	P_1	8	-7	7	8	-56		
2	P_2	2	8	-4	4		-8	
3	P_3	4	4	8	-7	16		-28
4	P_4	6	0	9	6	0		36
5	P_5	5	-5	-6	5		-30	
6	P_6	5	-6	9	-4			
7	P_7	7	9	-3	3			21
8	P_8	1	0	3	3			3
9	P_9	12	8	5	-6		60	
10	P_{10}	3	2	-2	1	6		
Total		53						

P9.2.6 Suppose a plane figure F contains particles with masses and locations such that there is symmetry about a line L as in the following figure. That is, for a typical particle P with mass m on one side of L, there is a particle $\hat{P}$, also with mass m, on the other side of L where P and $\hat{P}$ are equidistant from L.

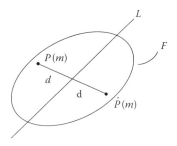

Show that the mass center of the particles of F lies on L.

P9.2.7 See Problem P9.2.6. Show that analogous to a line of symmetry for the particles of a plane figure, there is a plane of symmetry Π for the particles of a body B. Show that the mass center of B lies on Π.

P9.2.8 Find the mass center G of an arm where the upper arm is inclined at 30° to the vertical, the lower arm is horizontal, and the hand is inclined at 45° to the vertical as modeled in the following figure. Specifically, locate G by its horizontal and vertical distance from the elbow joint. Use the data for both a 50 percentile male and a 50 percentile female from the Appendix tables.

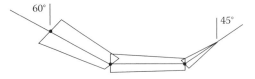

Section 9.3

P9.3.1 Consider a body B, having an irregular shape, tossed tumbling into the air as represented in the following figure. Let G be the mass center of B and let R be an inertial reference frame with Cartesian axes X, Y, Z and unit vectors $\mathbf{N}_X$, $\mathbf{N}_Y$, $\mathbf{N}_Z$ parallel to X, Y, and Z as shown.

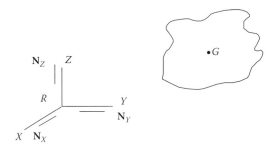

a. Construct a free-body diagram for B with the gravity (weight) forces on B being represented by a single vertical (downward) force W, with magnitude mg where m is the mass of B and g is the gravity acceleration. Let the inertia forces on B be represented by a single force **F*** together with a couple having torque **T***.

b. Let F* be expressed as $-m{}^R\mathbf{a}^G$ (see Equation 9.20) where ${}^R\mathbf{a}^G$ in turn is expressed as

$$^R\mathbf{a}^G = \ddot{x}\mathbf{N}_X + \ddot{y}\mathbf{N}_Y + \ddot{z}\mathbf{N}_Z$$

where (x, y, z) are the X, Y, Z coordinates of G. From the free-body diagram of (a), show that the equations governing the motion of G are

$$\ddot{x} = 0, \quad \ddot{y} = 0, \quad \ddot{z} = -g$$

c. Identify the governing equations of (b) as the differential equations for the motion of a projectile, and thereby show that G moves on a parabolic path.

d. Observe that the angular motion (the rotation or spinning) of B does not affect the path of motion of G. What does this imply about the perceived instantaneous axis of rotation of B?

P9.3.2 See Problem P9.3.1. Consider a multibody system S projected into the air as represented in the following figure. Let G be the mass center of S.

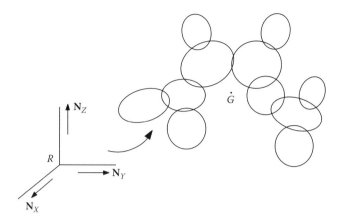

Conduct an analysis for S and G, which is similar to steps (a) through (d) of Problem P9.3.1.

P9.3.3 See the results of Problem P9.3.2. What do the results imply about the perceived instantaneous rotation axis of a gymnast and/or a diver?

10 Human Body Inertia Properties

Consider a representation or model of the human body as in Figure 9.8 and as shown again in Figure 10.1. As noted earlier, this is a finite-segment (or lumped-mass) model with the segments representing the major limbs of the human frame. Consider a typical segment or body, B_k of the model as in Figure 10.2. In our analysis in Chapter 9, we discovered that the inertia forces on B_k are equivalent to a single force $\mathbf{F}_k^*$ passing through the mass center G_k of B_k together with a couple with torque $\mathbf{T}_k^*$ where (Equations 9.24 and 9.27)

$$\mathbf{F}_k^* = -M_k{}^R\mathbf{a}^{G_k} \tag{10.1}$$

and

$$\mathbf{T}_k^* = -\sum_{i=1}^{N} m_i \mathbf{r}_i \times \left({}^R\boldsymbol{\alpha}^{B_k} \times \mathbf{r}_i \right) - {}^R\boldsymbol{\omega}^B \times \sum_{i=1}^{N} m_i \mathbf{r}_i \times \left({}^R\boldsymbol{\omega}^B \times \mathbf{r}_i \right) \tag{10.2}$$

where
  M_k is the mass of B_k
  m_i is the mass of a typical particle, P_i of B_k
  $\mathbf{r}_i$ is a position vector locating P_i relative to G_k
  N is the number of particles P_i of B_k
  ${}^R\boldsymbol{\omega}^{B_k}$ and ${}^R\boldsymbol{\alpha}^{B_k}$ are the respective angular velocity and angular accelerations of B_k in an inertial (Newtonian) reference frame R_i

In this chapter, we will explore ways of simplifying the expressions for the inertia torque $\mathbf{T}_k^*$. Specifically, we will investigate the properties of the triple vector product of Equation 10.2 and show that they may be expressed in terms of the inertia dyadic of B_k relative to the mass center G_k.

10.1 SECOND MOMENT VECTORS, MOMENTS, AND PRODUCTS OF INERTIA

Let P be a particle with mass m. Let R be a reference frame with a Cartesian axis system XYZ with origin O as in Figure 10.3. Let (x, y, z) be the coordinates of P relative to O of the XYZ axis system; let $\mathbf{p}$ be the position vector locating P relative to O. Let $\mathbf{n}_1$, $\mathbf{n}_2$, and $\mathbf{n}_3$ be unit vectors parallel to X, Y, and Z, and let $\mathbf{n}_a$ and $\mathbf{n}_b$ be arbitrarily directed unit vectors, as represented in Figure 10.3.

Given these definitions and notations, the second moment of P relative to O for the direction $\mathbf{n}_a$, written as $\mathbf{I}_a^{P/O}$, is defined as

$$\mathbf{I}_a^{P/O} = m\mathbf{p} \times (\mathbf{n}_a \times \mathbf{p}) \tag{10.3}$$

The second moment vector is sometimes called the inertia vector. The second moment vector is primarily useful in defining other inertia quantities that are useful in dynamic analyses. Specifically, the product of inertia of P relative to O for the directions of $\mathbf{n}_a$ and $\mathbf{n}_b$, written as $I_{ab}^{P/O}$, is defined as the projection of the second moment vector along $\mathbf{n}_b$. That is,

$$I_{ab}^{P/O} = \mathbf{I}_a^{P/O} \cdot \mathbf{n}_b \tag{10.4}$$

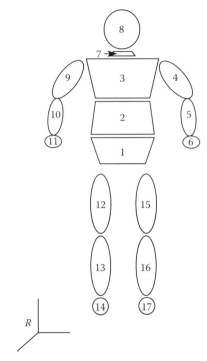

FIGURE 10.1 Human body model (17 bodies).

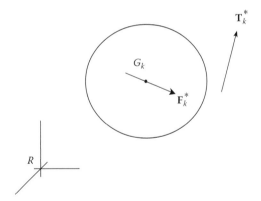

FIGURE 10.2 Typical body B_k of a human body model.

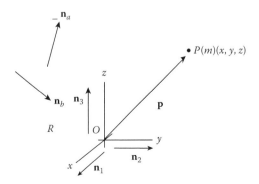

FIGURE 10.3 Particle P with mass m in a reference frame R with Cartesian axes XYZ.

Then by substituting Equation 10.3 into 10.4, we have

$$I_{ab}^{P/O} = m\left[\mathbf{p}\times(\mathbf{n}_a\times\mathbf{p})\right]\cdot\mathbf{n}_b = m(\mathbf{n}_b\times\mathbf{p})\cdot(\mathbf{n}_a\times\mathbf{p})$$

$$= m(\mathbf{n}_a\times\mathbf{p})\cdot(\mathbf{n}_b\times\mathbf{p}) = I_{ba}^{P/O} \tag{10.5}$$

where the second and third equalities follow from properties of the triple scalar product and the scalar (dot) product of vectors (Equation 3.68).

Similarly, the moment of inertia of P relative to O for the direction $\mathbf{n}_a$, written as $I_{aa}^{P/O}$, is defined as the projection of the second moment vector along $\mathbf{n}_a$. That is,

$$I_{aa}^{P/O} = \mathbf{I}_{a}^{P/O}\cdot\mathbf{n}_a \tag{10.6}$$

By substituting Equation 10.3 into Equation 10.6, we have

$$I_{aa}^{P/O} = m\left[\mathbf{p}\times(\mathbf{n}_a\times\mathbf{p})\right]\cdot\mathbf{n}_a = m(\mathbf{n}_a\times\mathbf{p})\cdot(\mathbf{n}_a\times\mathbf{p}) = m(\mathbf{n}_a\times\mathbf{p})^2 \tag{10.7}$$

Examination of Equations 10.5 and 10.6 shows that the product of inertia can have positive, negative, or zero values, whereas the moment of inertia has only positive or possibly zero values.

There are simple geometric interpretations of the products and moments of inertia: Consider again particle P with coordinates (x, y, z) and position vector $\mathbf{p}$ relative to origin O. Then in terms of the unit vectors $\mathbf{n}_x$, $\mathbf{n}_y$, and $\mathbf{n}_z$, $\mathbf{p}$ may be written as

$$\mathbf{p} = x\mathbf{n}_x + y\mathbf{n}_y + z\mathbf{n}_z \tag{10.8}$$

Next, in the definition of Equation 10.3, let $\mathbf{n}_a$ be $\mathbf{n}_x$. Then by substituting Equation 10.8 into 10.3, $I_x^{P/O}$ is

$$\mathbf{I}_x^{P/O} = m\mathbf{p}\times(\mathbf{n}_x\times\mathbf{p})$$

$$= m(x\mathbf{n}_x + y\mathbf{n}_y + z\mathbf{n}_z)\times[\mathbf{n}_x\times(x\mathbf{n}_x + y\mathbf{n}_y + z\mathbf{n}_z)]$$

$$= m(x\mathbf{n}_x + y\mathbf{n}_y + z\mathbf{n}_z)\times(y\mathbf{n}_z - z\mathbf{n}_y) \tag{10.9}$$

or

$$\mathbf{I}_x^{P/O} = m(y^2 + z^2)\mathbf{n}_x - mxy\mathbf{n}_y - mxz\mathbf{n}_z$$

Similarly, $\mathbf{I}_y^{P/O}$ and $\mathbf{I}_z^{P/O}$ are

$$\mathbf{I}_y^{P/O} = -mxy\mathbf{n}_x + m(x^2 + z^2)\mathbf{n}_y - myz\mathbf{n}_z \tag{10.10}$$

and

$$\mathbf{I}_z^{P/O} = -mxz\mathbf{n}_x - myz\mathbf{n}_y + m(x^2 + y^2)\mathbf{n}_z \tag{10.11}$$

The moments and products of inertia of P relative to O for the directions of $\mathbf{n}_x$, $\mathbf{n}_y$, and $\mathbf{n}_z$ are then simply the $\mathbf{n}_x$, $\mathbf{n}_y$, and $\mathbf{n}_z$ components of $\mathbf{I}_x^{P/O}$, $\mathbf{I}_y^{P/O}$, and $\mathbf{I}_z^{P/O}$. That is,

$$
\begin{aligned}
&I_{xx}^{P/O} = m(y^2 + z^2) \qquad I_{xy}^{P/O} = -mxy \qquad\qquad I_{xz}^{P/O} = -mxz \\
&I_{yx}^{P/O} = -myx \qquad\quad I_{yy}^{P/O} = m(x^2 + z^2) \qquad I_{yz}^{P/O} = -myz \\
&I_{zx}^{P/O} = -mzx \qquad\quad I_{zy}^{P/O} = -mzy \qquad\quad I_{zz}^{P/O} = m(x^2)+ y
\end{aligned}
\qquad (10.12)
$$

These moments and products of inertia are conveniently arranged into a matrix $I^{P/O}$ with elements I_{ij} $(i, j = x, y, z)$ as

$$
I^{P/O} = [I_{ij}] = m \begin{bmatrix} (y^2 + z^2) & -xy & -xz \\ -xy & (x^2 + z^2) & -yz \\ -xz & -yz & (x^2 + y^2) \end{bmatrix}
\qquad (10.13)
$$

Observe that in this representation, the moments of inertia are along the diagonal of the matrix and the products of inertia are the off-diagonal elements. Observe further that the matrix is symmetric, that is, $I_{ij} = I_{ji}$. Also observe that the moments of inertia may be interpreted as products of the particle mass and the square of the distance from the particle P to a coordinate axis. For example, I_{xx} is

$$
I_{xx} = m(y^2 + z^2) = md_x^2
\qquad (10.14)
$$

where d_x is the distance from P to the X-axis. Finally, observe that the products of inertia may be interpreted as the negative of the product of the particle mass with the product of the distances from the particle P to coordinate planes. For example, I_{xy} is

$$
I_{xy} = -mxy = -md_{yz}d_{xz}
\qquad (10.15)
$$

where d_{yz} and d_{xz} are the distances from P to the Y–Z plane and to the X–Z plane, respectively.

10.2 INERTIA DYADICS

We can use an inertia matrix as in Equation 10.13, together with its associated unit vectors, to form an inertia dyadic.* Specifically, let the inertia dyadic of a particle P relative to reference point O be defined as

$$
\mathbf{I}^{P/O} = \mathbf{n}_i I_{ij} \mathbf{n}_j \quad \text{(sum over repeated indices)}
\qquad (10.16)
$$

Since from Equation 10.12 we see that the moments and products of inertia $\mathbf{I}_{ij}$ are components of the second moment vector, or inertia vector, we can express the inertia dyadic as

$$
\mathbf{I}^{P/O} = \mathbf{n}_i \mathbf{I}_i^{P/O} = \mathbf{I}_j^{P/O} \mathbf{n}_j
\qquad (10.17)
$$

Consequently, the second moment vector $\mathbf{I}^{P/O}$ may be expressed in terms of the inertia dyadic as

$$
\mathbf{I}_j^{P/O} = \mathbf{I}^{P/O} \cdot \mathbf{n}_j = \mathbf{n}_j \cdot \mathbf{I}^{P/O}
\qquad (10.18)
$$

---

* See Sections 3.3 and 3.4 for a discussion about dyadics.

The inertia dyadic is thus a convenient item for determining moments and products of inertia. That is,

$$I_{ij}^{P/O} = \mathbf{n}_i \cdot \mathbf{I}^{P/O} \cdot \mathbf{n}_j \tag{10.19}$$

Since dyadics (or vector–vectors) can be expressed in terms of any unit vector system, we can conveniently use the inertia dyadic to determine moments and products of inertia for any desired directions. If, for example, $\mathbf{n}_a$ and $\mathbf{n}_b$ are arbitrarily directed unit vectors, and $\mathbf{I}^{P/O}$ is the inertia dyadic of a particle P for reference point O, expressed in terms of any convenient unit vector system, say $\mathbf{n}_i$ ($i = 1, 2, 3$), then the product of inertia of P relative to O for the directions of $\mathbf{n}_a$ and $\mathbf{n}_b$ is

$$I_{ab}^{P/O} = \mathbf{n}_a \cdot \mathbf{I}^{P/O} \cdot \mathbf{n}_b \tag{10.20}$$

Similarly, the moment of inertia of P relative to O for $\mathbf{n}_a$ is

$$I_{aa}^{P/O} = \mathbf{n}_a \cdot \mathbf{I}^{P/O} \cdot \mathbf{n}_a \tag{10.21}$$

Next, let $\mathbf{n}_a$ and $\mathbf{n}_b$ be expressed in terms of the $\mathbf{n}_i$ as

$$\mathbf{n}_a = a_k\mathbf{n}_k \quad \text{and} \quad \mathbf{n}_b = b_l\mathbf{n}_l \tag{10.22}$$

Then by substituting into Equation 10.20, we have

$$I_{ab}^{P/O} = a_k\mathbf{n}_k \cdot \mathbf{n}_i \cdot \mathbf{I}_{ij}^{P/O} \cdot \mathbf{n}_j \cdot b_l\mathbf{n}_l = a_k\delta_{ki} \cdot \mathbf{I}_{ij}^{P/O} \cdot \delta_{jk}b_l$$

or

$$I_{ab}^{P/O} = a_k b_l I_{kl}^{P/O} \tag{10.23}$$

Finally, let $\hat{\mathbf{n}}_{ij}$ ($i = 1, 2, 3$) form a mutually perpendicular unit vector set, and by using the notation of Equation 3.165, let $\hat{\mathbf{n}}_j$ be expressed as

$$\hat{\mathbf{n}}_j = S_{ij}\mathbf{n}_i \quad \text{so that } \mathbf{n}_i = S_{ij}\hat{\mathbf{n}}_i \tag{10.24}$$

where the transformation matrix elements are (Equation 3.160)

$$S_{ij} = \mathbf{n}_i \cdot \mathbf{n}_j \tag{10.25}$$

Thus by comparing Equation 10.24 with 10.22, we obtain the following results:

$$\hat{I}_{kl}^{P/O} = S_{ik}S_{jl}I_{ij}^{P/O} \tag{10.26}$$

Since the S_{ij} are elements of an orthogonal matrix, we can solve Equation 10.26 for the $I_{ij}^{P/O}$ in terms of the $\hat{I}_{kl}^{P/O}$ as

$$\hat{I}_{ij}^{P/O} = S_{ik}S_{jl}\hat{I}_{kl}^{P/O} \tag{10.27}$$

10.3 SETS OF PARTICLES

Consider now a set S of N particles P_i, having masses m_i ($i = 1, ..., N$) as in Figure 10.4. Let O be a reference point, and let $\mathbf{n}_a$ and $\mathbf{n}_b$ be arbitrarily directed unit vectors. Then the second moment of S relative to O for the direction of $\mathbf{n}_a$ is simply the sum of the second moments of the individual particles of S relative to O for the direction of $\mathbf{n}_a$. That is,

$$\mathbf{I}_a^{S/O} = \sum_{i=1}^{N} \mathbf{I}_a^{P/O} \tag{10.28}$$

Once the second moment of S is known, the moments and products of inertia and the inertia dyadic of S are immediately determined by using the definitions of the foregoing section. Specifically

$$I_{ab}^{S/O} = \mathbf{I}_a^{S/O} \cdot \mathbf{n}_b = \mathbf{I}_b^{S/O} \cdot \mathbf{n}_a = I_{ba}^{S/O} \tag{10.29}$$

$$I_{aa}^{S/O} = \mathbf{I}_a^{S/O} \cdot \mathbf{n}_a \tag{10.30}$$

and

$$\mathbf{I}^{S/O} = \mathbf{n}_i I_{ij}^{S/O} \mathbf{n}_j = \mathbf{I}_j^{S/O} \mathbf{n}_j \tag{10.31}$$

where as usual, $\mathbf{n}_1$, $\mathbf{n}_2$, and $\mathbf{n}_3$ are mutually perpendicular dextral unit vectors.

By inspection of these equations, we can also express the moments and products of inertia of S and the second moment vectors of S in terms of the inertia dyadic of S as

$$I_{aa}^{S/O} = \mathbf{n}_a \cdot \mathbf{I}^{S/O} \cdot \mathbf{n}_a \tag{10.32}$$

$$I_{ab}^{S/O} = \mathbf{n}_a \cdot \mathbf{I}^{S/O} \cdot \mathbf{n}_b \tag{10.33}$$

$$\mathbf{I}_a^{S/O} = \mathbf{n}_a \cdot \mathbf{I}^{S/O} = \mathbf{I}^{S/O} \cdot \mathbf{n}_a \tag{10.34}$$

From Equation 10.28, we can readily see that the moments and products of inertia, the second moment vector, and the inertia dyadic can be expressed as the sum of those quantities for the individual particles. That is,

$$\mathbf{I}_{ab}^{S/O} = \sum_{i=1}^{N} \mathbf{I}_{ab}^{P_i/O} \tag{10.35}$$

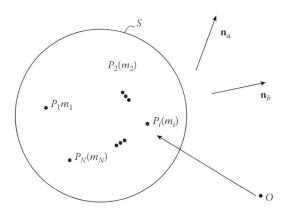

FIGURE 10.4 A set S of N particles.

$$\mathbf{I}_{aa}^{S/O} = \sum_{i=1}^{N} \mathbf{I}_{aa}^{P_i/O} \qquad (10.36)$$

$$\mathbf{I}_{a}^{S/O} = \sum_{i=1}^{N} \mathbf{I}_{a}^{P_i/O} \qquad (10.37)$$

$$\mathbf{I}^{S/O} = \sum_{i=1}^{N} \mathbf{I}_{a}^{P_i/O} \qquad (10.38)$$

10.4 BODY SEGMENTS

Consider a body segment B in a reference frame R as represented in Figure 10.5. If we regard B as being composed of particles P_i ($i = 1, \ldots, N$), we can immediately apply Equations 10.28 through 10.34 to B by simply replacing S with B. We can thus think of a body, or body segment, as being equivalent to a large set of particles. If further, the particles remain at fixed distances relative to each other, the body is rigid.

We can now express the inertia torque $\mathbf{T}^*$ of Equations 9.27 and 10.2 in terms of the inertia dyadic: Let G be the mass center of B (Section 9.2), and let $\mathbf{r}_i$ locate a typical particle of B relative to G as in Figure 10.6.

Then from Equation 10.3, the second moment vector of P_i relative to G for the direction $\mathbf{n}_a$ is

$$\mathbf{I}_{a}^{P_i/G} = m_i \mathbf{r}_i \times (\mathbf{n}_a \times \mathbf{r}_i) \qquad (10.39)$$

Therefore, from Equation 10.28, the second moment vector of B relative to G for the direction $\mathbf{n}_a$ is

$$\mathbf{I}_{a}^{B/G} = \sum_{i=1}^{N} \mathbf{I}_{a}^{P_i/G} = \sum_{i=1}^{N} m_i \mathbf{r}_i \times (\mathbf{n}_a \times \mathbf{r}_i) \qquad (10.40)$$

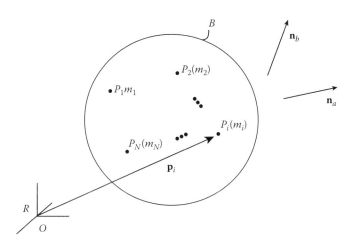

FIGURE 10.5 A body segment B composed of particles P_i ($i = 1, \ldots, N$).

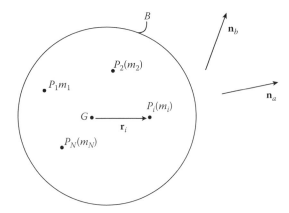

FIGURE 10.6 Body segment composed of particles P_i with mass center G.

Finally, observe from Equation 10.34 that $\mathbf{I}_a^{B/G}$ may also be expressed in terms of the inertia dyadic of B relative to G as

$$\mathbf{I}_a^{B/G} = \mathbf{I}^{B/G} \cdot \mathbf{n}_a = \mathbf{n}_a \cdot \mathbf{I}_a^{B/G} \tag{10.41}$$

Returning now to the inertia torque $\mathbf{T}^*$, from Equation 9.27, we have

$$\mathbf{T}^* = -\sum_{i=1}^{N} m_i \mathbf{r}_i \times (\boldsymbol{\alpha} \times \mathbf{r}_i) - \boldsymbol{\omega} \times \sum_{i=1}^{N} m_i \mathbf{r}_i \times (\boldsymbol{\omega} \times \mathbf{r}_i) \tag{10.42}$$

where $\boldsymbol{\alpha}$ and $\boldsymbol{\omega}$ are the angular accelerations and angular velocity of B in R. Let $\mathbf{n}_\alpha$ and $\mathbf{n}_\omega$ be unit vectors parallel to $\boldsymbol{\alpha}$ and $\boldsymbol{\omega}$. That is, let $\mathbf{n}_\alpha$ and $\mathbf{n}_\omega$ be

$$\mathbf{n}_\alpha = \frac{\boldsymbol{\alpha}}{|\boldsymbol{\alpha}|} \quad \text{and} \quad \mathbf{n}_\omega = \frac{\boldsymbol{\omega}}{|\boldsymbol{\omega}|} \tag{10.43}$$

Then, $\boldsymbol{\alpha}$ and $\boldsymbol{\omega}$ may be expressed as

$$\boldsymbol{\alpha} = \mathbf{n}_\alpha |\boldsymbol{\alpha}| = \alpha \mathbf{n}_\alpha \quad \text{and} \quad \boldsymbol{\omega} = \mathbf{n}_\omega |\boldsymbol{\omega}| = \omega \mathbf{n}_\omega \tag{10.44}$$

where α and ω are the magnitudes of $\boldsymbol{\alpha}$ and $\boldsymbol{\omega}$.

By substituting Equation 10.44 into 10.42, we see that $\mathbf{T}^*$ may be expressed as

$$\begin{aligned}
\mathbf{T}^* &= -\sum_{i=1}^{N} m_i \mathbf{r}_i \times (\alpha \mathbf{n}_\alpha \times \mathbf{r}_i) - \boldsymbol{\omega} \times \sum_{i=1}^{N} m_i \mathbf{r}_i \times (\omega \mathbf{n}_\omega \times \mathbf{r}_i) \\
&= -\alpha \sum_{i=1}^{N} m_i \mathbf{r}_i \times (\mathbf{n}_\alpha \times \mathbf{r}_i) - \boldsymbol{\omega} \times \omega \sum_{i=1}^{N} m_i \mathbf{r}_i \times (\mathbf{n}_\omega \times \mathbf{r}_i) \\
&= -\alpha \mathbf{I}_\alpha^{B/G} - \boldsymbol{\omega} \times \left(\omega \mathbf{I}_\omega^{B/G}\right) \\
&= -\alpha \mathbf{n}_\alpha \cdot \mathbf{I}_\alpha^{B/G} - \boldsymbol{\omega} \times \left(\omega \mathbf{n}_\omega \cdot \mathbf{I}_\omega^{B/G}\right)
\end{aligned}$$

or

$$\mathbf{T}^* = -\alpha \mathbf{I}^{B/G} - \boldsymbol{\omega} \times (\boldsymbol{\omega} \cdot \mathbf{I}^{B/G})$$

$$= -\mathbf{I}^{B/G} \cdot \boldsymbol{\omega} - \boldsymbol{\omega} \times (\mathbf{I}^{B/G} \cdot \boldsymbol{\omega}) \qquad (10.45)$$

Biological bodies are generally not homogeneous (nor are they isotropic). Also, they do not have simple geometric shapes. Nevertheless, for gross dynamic modeling, it is often reasonable to represent the limbs of a model (a human body model) by homogeneous bodies with simple shapes (structures of cones and ellipsoids). In such cases, it is occasionally convenient to use the concept of radius of gyration defined simply as the square root of the moment of inertia/mass ratio. Specifically, for a given moment of inertia say I_{aa} of a body having mass m, the radius of gyration k_a (for the direction $\mathbf{n}_a$) is defined by the expression

$$k_a^2 = \frac{I_{aa}}{m} \quad \text{or} \quad I_{aa} = mk_a^2 \qquad (10.46)$$

The radius of gyration is thus simply a geometric property.

10.5 PARALLEL AXIS THEOREM

Observe that the second moment of a body B for a point O for the direction of a unit vector $\mathbf{n}_a$, $\mathbf{I}_a^{B/O}$, is dependent both upon the unit vector direction and the point O. The same may be said for the moments and products of inertia. By using Equations 10.22, 10.26, and 10.32, we can change the directions of the unit vectors. In this section, we present a procedure for changing the position of the reference point. Specifically, we obtain an expression relating the second moment vector between an arbitrary reference point O and the mass center G.

Consider again a body B modeled as a set of N particles P_i (with masses m_i) as in Figure 10.7. Let G be the mass center of B, and let $\mathbf{p}_i$ locate a typical particle P_i relative to G. Let O be an arbitrary reference point, and let $\mathbf{p}_i$ locate P_i relative to O. Let $\mathbf{p}_G$ locate G relative to O. Finally, let $\mathbf{n}_a$ and $\mathbf{n}_b$ be arbitrarily directed unit vectors.

From Equations 10.3 and 10.28, the second moment vector of B relative to O for $\mathbf{n}_a$ is

$$\mathbf{I}_a^{B/O} = \sum_{i=1}^{N} m_i \mathbf{p}_i \times (\mathbf{n}_a \times \mathbf{p}_i) \qquad (10.47)$$

But from Figure 10.7, we see that $\mathbf{p}_i$ is

$$\mathbf{p}_i = \mathbf{p}_G + \mathbf{r}_i \qquad (10.48)$$

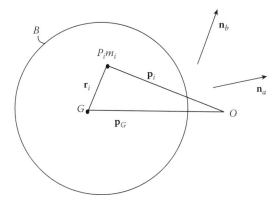

FIGURE 10.7 Body segment B, with mass center G, modeled as a set of particles P_i, with reference point O, and unit vectors $\mathbf{n}_a$ and $\mathbf{n}_b$.

Then by substituting for $\mathbf{p}_i$ in Equation 10.47, we have

$$\mathbf{I}_a^{B/O} = \sum_{i=1}^{N} m_i(\mathbf{p}_G + \mathbf{r}_i) \times \left[\mathbf{n}_a \times (\mathbf{p}_G + \mathbf{r}_i)\right]$$

$$= \sum_{i=1}^{N} m_i\mathbf{p}_G \times (\mathbf{n}_a \times \mathbf{p}_G) + \sum_{i=1}^{N} m_i\mathbf{p}_G \times (\mathbf{n}_a \times \mathbf{r}_i)$$

$$+ \sum_{i=1}^{N} m_i\mathbf{r}_i \times (\mathbf{n}_a \times \mathbf{p}_G) + \sum_{i=1}^{N} m_i\mathbf{r}_i \times (\mathbf{n}_a \times \mathbf{r}_i)$$

$$= \left(\sum_{i=1}^{N} m_i\right)\mathbf{p}_G \times (\mathbf{n}_a \times \mathbf{p}_G) + \mathbf{p}_G \times \mathbf{n}_a \times \left(\sum_{i=1}^{N} m_i\mathbf{r}_i\right)$$

$$+ \left(\sum_{i=1}^{N} m_i\mathbf{r}_i\right) \times (\mathbf{n}_a \times \mathbf{p}_G) + \sum_{i=1}^{N} m_i\mathbf{r}_i \times (\mathbf{n}_a \times \mathbf{r}_i)$$

$$= M\mathbf{p}_G \times (\mathbf{n}_a \times \mathbf{p}_G) + \mathbf{I}_a^{B/G}$$

or

$$\mathbf{I}_a^{B/O} = \mathbf{I}_a^{B/G} + \mathbf{I}_a^{G/O} \tag{10.49}$$

where
   M is the total mass of B
   $\mathbf{I}_a^{G/O}$ is the second moment of a particle with mass M at G relative to O for the direction of $\mathbf{n}_a$
   $\sum_{i=1}^{N} m_i\mathbf{r}_i = 0$ since G is the mass center of B

   Equation 10.49 is commonly called the parallel axis theorem. It can be used to develop a series of relations concerning inertia between O and G. Specifically,

$$\mathbf{I}_{aa}^{B/O} = \mathbf{I}_{aa}^{B/G} + \mathbf{I}_{aa}^{G/O} \tag{10.50}$$

$$\mathbf{I}_{ab}^{B/O} = \mathbf{I}_{ab}^{B/G} + \mathbf{I}_{ab}^{G/O} \tag{10.51}$$

$$\mathbf{I}^{B/O} = \mathbf{I}^{B/G} + \mathbf{I}^{G/O} \tag{10.52}$$

   In Equation 10.52, $\mathbf{I}^{B/G}$ is sometimes called the central inertia dyadic of B.
   The term parallel axis arises from the properties of $\mathbf{I}_{aa}^{G/O}$. That is,

$$\mathbf{I}_{aa}^{G/O} = M\mathbf{p}_G \times (\mathbf{n}_a \times \mathbf{p}_G) = M(\mathbf{p}_G \times \mathbf{n}_a)^2 \tag{10.53}$$

Consider the points G and O with lines parallel to $\mathbf{n}_a$ passing through them as in Figure 10.8.
   Let d be the distance between the lines. Then d is simply

$$d = |\mathbf{p}_G|\sin(\theta) = |\mathbf{p}_G||\mathbf{n}_a|\sin(\theta) = |\mathbf{p}_G \times \mathbf{n}_a| \tag{10.54}$$

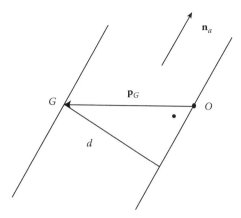

FIGURE 10.8   Parallel lines passing through points O and G (Figure 10.7).

Since $(\mathbf{p}_G \times \mathbf{n}_a)^2$ is the same as $|\mathbf{p}_G \times \mathbf{n}_a|^2$, we have from Equation 10.53

$$\mathbf{I}_{aa}^{G/O} = M(\mathbf{p}_G \times \mathbf{n}_a)^2 = Md^2 \tag{10.55}$$

That is, the moment of inertia $\mathbf{I}_{aa}^{G/O}$ is simply the product of the body mass M and the square of the distance between parallel lines passing through G and O and parallel to $\mathbf{n}_a$.

10.6   EIGENVALUES OF INERTIA: PRINCIPAL DIRECTIONS

In Chapter 3, we saw that symmetric dyadics may have their matrices cast into diagonal form by appropriately choosing their unit vector bases (Section 3.9). Since inertia dyadics are symmetric, their associated inertia matrices can thus be diagonalized by an appropriate choice of unit vectors. That is, by choosing the unit vectors along principal directions, the products of inertia are zero, and the inertia matrix then consists only of moments of inertia on the diagonal.

Also, by using these principal unit vectors (unit eigenvectors), these moments of inertia, along the inertia matrix diagonal, contain the maximum and minimum values of the moments of inertia for all possible directions (Section 3.10). The moments of inertia of the diagonalized inertia matrix are sometimes called the eigenvalues of inertia.

Briefly, the procedure for finding the eigenvalues of inertia and the associated unit eigenvectors (or principal vectors) is as follows (Section 3.9).

Let the inertia matrix of a body B for a reference point O (typically the mass center of B) be found in the usual manner relative to any convenient set of unit vectors, $\mathbf{n}_i$. Let the matrix elements $I_{ij}^{B/O}$ be represented simply as I_{ij}. Next, form the following equations:

$$(I_{11} - \lambda)a_1 + I_{12}a_2 + I_{13}a_3 = 0 \tag{10.56}$$

$$I_{21}a_1 + (I_{22} - \lambda)a_2 + I_{23}a_3 = 0 \tag{10.57}$$

$$I_{31}a_1 + I_{32}a_2 + (I_{33} - \lambda)a_3 = 0 \tag{10.58}$$

where
   λ is the eigenvalue (to be determined)
   a_i are the components of the principal unit vector or unit eigenvector $\mathbf{n}_a$ associated with λ and given by

$$\mathbf{n}_a = a_i\mathbf{n}_i \tag{10.59}$$

Equations 10.56 through 10.58 form a set of three simultaneous linear algebraic equations for a_1, a_2, and a_3. Since the equations are homogeneous, that is, with the right-hand sides all zero, the only solution is the trivial solution, with all a_i being 0, unless the determinant of the coefficients is zero. Thus, there is a nontrivial solution only if

$$\begin{vmatrix} (I_{11} - \lambda) & I_{12} & I_{13} \\ I_{21} & (I_{22} - \lambda) & I_{23} \\ I_{31} & I_{32} & (I_{33} - \lambda) \end{vmatrix} = 0 \tag{10.60}$$

When Equation 10.60 holds, Equations 10.56 through 10.58 are no longer independent. Then at most two of the three equations are independent. But then a_i cannot be uniquely determined unless we have an additional equation. However, since $\mathbf{n}_a$ is a unit vector, its magnitude is unity so that the coefficients a_i satisfy

$$a_1^2 + a_2^2 + a_3^2 = 1 \tag{10.61}$$

Finally, when Equation 10.60 is expanded, we obtain the equation

$$\lambda^3 - I_{\mathrm{I}}\lambda^2 + I_{\mathrm{II}}\lambda - I_{\mathrm{III}} = 0 \tag{10.62}$$

where I_{I}, I_{II}, and I_{III} are (Equation 3.140)

$$I_{\mathrm{I}} = I_{\mathrm{II}} + I_{22} + I_{33} \tag{10.63}$$

$$I_{\mathrm{II}} = I_{22}I_{33} - I_{32}I_{23} + I_{11}I_{33} - I_{31}I_{13} + I_{11}I_{22} - I_{12}I_{21} \tag{10.64}$$

$$I_{\mathrm{III}} = I_{11}I_{22}I_{33} - I_{11}I_{32}I_{23} + I_{12}I_{31}I_{23} - I_{12}I_{21}I_{33} + I_{21}I_{32}I_{13} - I_{31}I_{13}I_{22} \tag{10.65}$$

The solution steps may now be listed as follows:

1. From Equations 10.63 through 10.65, calculate I_{I}, I_{II}, and I_{III}.
2. Using the results of Step 1, form Equation 10.62.
3. Solve Equation 10.62 for $\lambda = \lambda_a$, $\lambda = \lambda_b$, and $\lambda = \lambda_c$.
4. Select one of these roots, say $\lambda = \lambda_a$, and substitute for λ into Equations 10.56 through 10.58.
5. Select two of the equations in Step 4, and combine them with Equation 10.61 to obtain three independent equations for a_1, a_2, and a_3.
6. Solve the equations of Step 5 for a_1, a_2, and a_3.
7. Repeat Steps 4 through 6 for $\lambda = \lambda_b$ and $\lambda = \lambda_c$.

Since the inertia matrix is symmetric, it can be shown [1,2] that the roots of Equation 10.62 are real. Then in the absence of repeated roots, the aforementioned procedure will lead to three distinct eigenvalues of inertia together with three associated unit eigenvectors of inertia. It is also seen that these unit vectors are mutually perpendicular [1,2].

If two of the roots of Equation 10.62 are real, then there are an infinite number of unit eigenvectors. These are in all directions perpendicular to the unit vector of the distinct eigenvalue [1,2]. When all three roots of Equation 10.62 are equal, a unit vector in any direction is a unit eigenvector of inertia [1,2], or equivalently, all unit vectors are then unit eigenvectors.

Section 3.9 presents a numerical example illustrating the determination of eigenvalues and unit eigenvectors. Here, for convenience as a quick reference, we present another numerical illustration. Suppose the inertia matrix I has the simple form as follows:

$$\mathbf{I} = [I_{ij}] = \begin{bmatrix} 30 & -\sqrt{6} & -\sqrt{6} \\ -\sqrt{6} & 41 & -15 \\ -\sqrt{6} & -15 & 41 \end{bmatrix} \tag{10.66}$$

where the units ml^2 are mass–(length)^2, typically kg m^2 or slug ft^2.

Following the foregoing steps, we see that from Equations 10.63 through 10.65, I_I, I_II, and I_III are

$$I_\mathrm{I} = 112, \quad I_\mathrm{II} = 3,904, \quad \text{and} \quad I_\mathrm{III} = 43,008 \tag{10.67}$$

Then Equation 10.62 becomes

$$\lambda^3 - 112\lambda^2 + 3,904\lambda - 43,008 = 0 \tag{10.68}$$

The roots are then found to be

$$\lambda_a = 24, \quad \lambda_b = 32, \quad \text{and} \quad \lambda_c = 56 \tag{10.69}$$

where, as in Equation 10.66, the units for the roots are ml^2.

Next, if $\lambda = \lambda_a = 24$, Equations 10.56 through 10.58 become

$$(30 - 24)a_1 - \sqrt{6}a_2 - \sqrt{6}a_3 = 0 \tag{10.70}$$

$$-\sqrt{6}a_1 + (41 - 24)a_2 - 15a_3 = 0 \tag{10.71}$$

$$-\sqrt{6}a_1 - 15a_2 + (41 - 24)a_3 = 0 \tag{10.72}$$

These equations are seen to be dependent. By multiplying Equations 10.71 and 10.72 each by $-\sqrt{6}$ and then by adding and dividing by 2, we have Equation 10.70.

We can obtain an independent set of equations by appending Equation 10.61 to Equations 10.70 through 10.72. Specifically, Equation 10.61 is

$$a_1^2 + a_2^2 + a_3^2 = 1 \tag{10.73}$$

Then by using Equations 10.70 and 10.71 together with 10.73, we find a_1, a_2, and a_3 to be

$$a_1 = \frac{1}{2}, \quad a_2 = \frac{\sqrt{6}}{4}, \quad a_3 = \frac{\sqrt{6}}{4} \tag{10.74}$$

And then from Equation 10.59, unit eigenvector $\mathbf{n}_a$ is

$$\mathbf{n}_a = \frac{1}{2}\mathbf{n}_1 + \frac{\sqrt{6}}{4}\mathbf{n}_2 + \frac{\sqrt{6}}{4}\mathbf{n}_3 \tag{10.75}$$

Similarly, if $\lambda = \lambda_b = 32$, we obtain unit eigenvector $\mathbf{n}_b$ as

$$\mathbf{n}_b = \frac{-\sqrt{3}}{2}\mathbf{n}_1 + \frac{\sqrt{2}}{4}\mathbf{n}_2 + \frac{\sqrt{2}}{4}\mathbf{n}_3 \tag{10.76}$$

Finally, if $\lambda = \lambda_c = 56$, we obtain unit eigenvector $\mathbf{n}_c$ to be

$$\mathbf{n}_c = 0\mathbf{n}_1 - \frac{\sqrt{2}}{2}\mathbf{n}_2 + \frac{\sqrt{2}}{2}\mathbf{n}_3 \tag{10.77}$$

By observing Equations 10.75 through 10.77, we see that $\mathbf{n}_a$, $\mathbf{n}_b$, and $\mathbf{n}_i$ are mutually perpendicular and that $\mathbf{n}_a \times \mathbf{n}_a = \mathbf{n}_i$. Thus, $\mathbf{n}_a$, $\mathbf{n}_b$, and $\mathbf{n}_i$ form a unit vector basis, and from their components relative to $\mathbf{n}_1$, $\mathbf{n}_2$, and $\mathbf{n}_3$, we can obtain a transformation matrix S relating the unit vector sets (Section 3.9). That is,

$$\{\mathbf{n}_i\} = S\{\mathbf{n}_\alpha\} \quad \text{and} \quad \{\mathbf{n}_\alpha\}S^T\{\mathbf{n}_\alpha\} \tag{10.78}$$

where $\{\mathbf{n}_i\}$ and $\{\mathbf{n}_\alpha\}$ are column arrays of the unit vectors $\mathbf{n}_i$ $(i = 1, 2, 3)$ and $\{\mathbf{n}_\alpha\}$ $(\alpha = a, b, c)$. Specifically, from Equations 10.75 through 10.77, S and S^T are

$$S = \begin{bmatrix} \dfrac{1}{2} & -\dfrac{\sqrt{3}}{2} & 0 \\[2mm] \dfrac{\sqrt{6}}{4} & \dfrac{\sqrt{2}}{4} & -\dfrac{\sqrt{2}}{2} \\[2mm] \dfrac{\sqrt{6}}{4} & \dfrac{\sqrt{2}}{4} & \dfrac{\sqrt{2}}{2} \end{bmatrix} \quad \text{and} \quad S^T = \begin{bmatrix} \dfrac{1}{2} & \dfrac{\sqrt{6}}{4} & \dfrac{\sqrt{6}}{4} \\[2mm] -\dfrac{\sqrt{3}}{2} & \dfrac{\sqrt{2}}{4} & \dfrac{\sqrt{2}}{4} \\[2mm] 0 & -\dfrac{\sqrt{2}}{2} & \dfrac{\sqrt{2}}{2} \end{bmatrix} \tag{10.79}$$

Finally, analogous to Equation 3.171, we can express the inertia matrix (Equation 10.66) relative to the unit eigenvector $\{\mathbf{n}_\alpha\}$ as

$$\begin{bmatrix} \dfrac{1}{2} & -\dfrac{\sqrt{6}}{4} & \dfrac{\sqrt{6}}{4} \\[2mm] -\dfrac{\sqrt{3}}{2} & \dfrac{\sqrt{2}}{4} & -\dfrac{\sqrt{2}}{4} \\[2mm] 0 & -\dfrac{\sqrt{2}}{2} & \dfrac{\sqrt{2}}{2} \end{bmatrix} \begin{bmatrix} 30 & -\sqrt{6} & -\sqrt{6} \\ -\sqrt{6} & 41 & -15 \\ -\sqrt{6} & -15 & 41 \end{bmatrix} \begin{bmatrix} \dfrac{1}{2} & -\dfrac{\sqrt{3}}{2} & 0 \\[2mm] \dfrac{\sqrt{6}}{4} & \dfrac{\sqrt{2}}{4} & -\dfrac{\sqrt{2}}{2} \\[2mm] \dfrac{\sqrt{6}}{4} & \dfrac{\sqrt{2}}{4} & \dfrac{\sqrt{2}}{2} \end{bmatrix} = \begin{bmatrix} 24 & 0 & 0 \\ 0 & 32 & 0 \\ 0 & 0 & 56 \end{bmatrix}$$

$$\tag{10.80}$$

10.7 EIGENVALUES OF INERTIA: SYMMETRICAL BODIES

On some occasions, it is possible to determine unit eigenvectors by inspection and thus avoid the foregoing analysis. This occurs if there are planes of symmetry. A normal plane of symmetry is parallel to the direction of a unit eigenvector. To see this, let Π be a plane of symmetry of a body B, as represented in Figure 10.9.

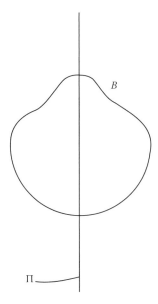

FIGURE 10.9   Edge view of a plane of symmetry of a body B.

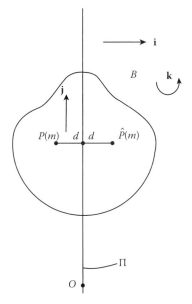

FIGURE 10.10   Equal mass particles P and $\hat{P}$ on either side of a plane of symmetry.

Let O be an arbitrary reference point in Π, let $\mathbf{i}$ be a unit vector normal to Π, let d be the distance from Π to P and $\hat{P}$, and let Q be the midpoint of the line connecting P and $\hat{P}$, as shown in Figure 10.10. Then the second movement vector $\mathbf{I}_n$ of P and $\hat{P}$ for O for the direction $\mathbf{i}$ is

$$\mathbf{I}_i = m\mathbf{OP}\times(\mathbf{i}\times\mathbf{OP})+m\mathbf{O\hat{P}}\times(\mathbf{i}_i\times\mathbf{O\hat{P}})$$

$$= m(\mathbf{OQ}-d\mathbf{i})\times\left[\mathbf{i}\times(\mathbf{OQ}+d\mathbf{i})\right]+m(\mathbf{OQ}+d\mathbf{i})\times\left[\mathbf{i}\times(\mathbf{OQ}+d\mathbf{i})\right]$$

or

$$\mathbf{I}_i = 2m\mathbf{OQ}\times(\mathbf{i}\times\mathbf{OQ})\qquad\qquad(10.81)$$

Next, in Figure 10.10, let $\mathbf{j}$ be a vertical vector parallel to Π as shown, and let $\mathbf{k}$ be a unit vector also parallel to Π such that $\mathbf{k} = \mathbf{i} \times \mathbf{j}$. Then since O and Q are both in Π, $\mathbf{OQ}$ may be written as

$$\mathbf{OQ} = y\mathbf{j} + z\mathbf{k} \tag{10.82}$$

By substituting into Equation 10.81, $\mathbf{I}_i$ becomes

$$\mathbf{I}_i = 2m(y\mathbf{j} + z\mathbf{k}) \times \left[\mathbf{i} \times (y\mathbf{j} + z\mathbf{k}) \right] = \left[2m(y\mathbf{j} + z\mathbf{k}) \times (-z\mathbf{j} + y\mathbf{k}) \right]$$

or

$$\mathbf{I}_i = 2m(y^2 + z^2)\mathbf{i} \tag{10.83}$$

Finally, if $\mathbf{I}$ is the inertia dyadic of the particle, we see from Equations 3.135 and 3.143 that $\mathbf{i}$ is a unit eigenvector if

$$\mathbf{I} \cdot \mathbf{i} = \lambda \mathbf{i} \tag{10.84}$$

But also, from Equation 10.17, it is clear that the second movement vector is relative to the inertia dyadic as

$$\mathbf{I}_i = \mathbf{I} \cdot \mathbf{i} \tag{10.85}$$

Thus, by comparing Equations 10.84 and 10.83, we see that $\mathbf{i}$ is a unit eigenvector if

$$\mathbf{I}_i = \lambda \mathbf{i} \tag{10.86}$$

From Equation 10.83, it is obvious that Equation 10.86 is satisfied with λ being $2m(y^2 + z^2)$. Since P and $\hat{P}$ are typical of pairs of particles of B divided by Π, we see that Π identifies a unit eigenvector by its normal.

10.8   APPLICATION WITH HUMAN BODY MODELS

Consider again the human body model of Figure 10.11. We observed earlier (Chapter 6) that this model is useful for gross motion similarities. From a purely mechanical perspective, however, the human frame is of course considerably more complex (Figure 10.11). But still, for large overall motion, the model can provide reasonable simulations. Such simulations require data for the inertia properties of the body segments. Here again, the use of common geometric shapes is but a gross representation of the human limbs.

Moreover, if the model segments are assumed to be homogeneous, the representation is even more approximate. Nevertheless, for gross motion simulations, the approximations are generally reasonable. Mindful of these limitations in the modeling, we present, in the following paragraph and tables, inertia data for various human body models. These data are expected to be useful for a large class of simulations.

To categorize the data, it is helpful to reintroduce and expand some notational conventions; consider again a system of connective bodies (a multibody system) as in Figure 6.2 and as shown again in Figure 10.12. Let the bodies be constructed with spherical joints.

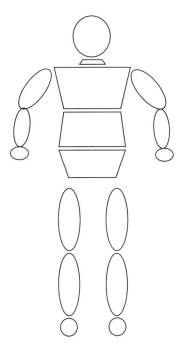

FIGURE 10.11 Human body model.

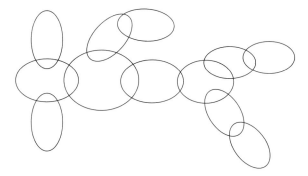

FIGURE 10.12 An open-chain multibody system.

As before, let the system be numbered or labeled as in Figure 10.13. Recall that in this numbering, we arbitrarily selected a body, say one of the larger bodies, as a reference body and call it B or simply 1. Then we number the bodies in ascending progression away from B, toward the extremities as in the figure.

Observe again that this is not a unique numbering system (Section 6.2), but once the numbering is set, each body has a unique adjoining lower-numbered body. Specifically, if we assign the number 0 to an inertial reference frame R, we can form a lower-body array listing the unique lower-body numbers for each of the bodies, as in Table 6.1 and as listed again in Table 10.1. Next by following the procedures of Section 6.2, we can form higher-order lower-body array for the system, leading to Table 10.2.

Consider now a typical pair of bodies say, B_j and B_k, of the system in Figure 10.13, as shown in Figure 10.14. Let B_j be the adjacent lower-numbered body of B_k. Let O_k be at the center of the spherical joint connecting B_j and B_k. Similarly, let O_k be at the center of the joint connecting B_k to its adjacent lower body. Let G_j and G_k be the mass centers of B_j and B_k, and let $\mathbf{r}_j$ and $\mathbf{r}_k$ locate G_j and G_k relative to O_j and O_k as shown. Finally, let $\boldsymbol{\xi}_j$ locate O_k relative to O_j.

272

Fundamentals of Biomechanics

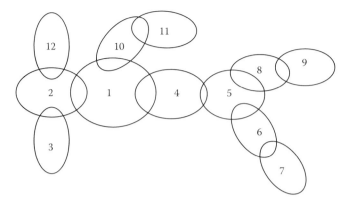

FIGURE 10.13 Numbered (labeled) multibody system.

TABLE 10.1
Lower-Body Average for the System in Figure 10.13

K	1	2	3	4	5	6	7	8	9	10	11	12
$L(K)$	0	1	2	1	4	5	6	5	8	1	10	2

TABLE 10.2
Higher-Order Lower-Body Arrays for the System of Figure 10.13

K	1	2	3	4	5	6	7	8	9	10	11	12
$L(K)$	0	1	2	1	4	5	6	5	8	1	10	2
$L^2(K)$	0	0	1	0	1	4	5	4	5	0	1	1
$L^3(K)$	0	0	0	0	0	1	4	1	4	0	0	0
$L^4(K)$	0	0	0	0	0	0	1	0	1	0	0	0
$L^5(K)$	0	0	0	0	0	0	0	0	0	0	0	0

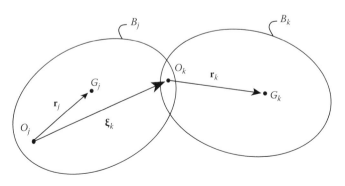

FIGURE 10.14 Two typical adjoining bodies.

Observe that with this notation and nomenclature, O_j and O_k are fixed in B_j and B_k, respectively. Thus, O_j and O_k may be considered as reference points or origins of B_j and B_k. Also observe that $\boldsymbol{\xi}_j$ is fixed in B_j and that $\mathbf{r}_j$ and $\mathbf{r}_k$ are fixed in B_j and B_k, respectively.

As an illustration of the use of their labeling and notational procedure, consider the multibody system of Figure 10.13, showing many body origins and mass centers together with many of their position vectors.

Specifically, O_k is the origin or reference point of body B_k ($k = 1, \ldots, 12$). O_k is also at the center of the spherical joint connecting B_k with its adjacent lower-numbered body except for O_1, which is simply an arbitrarily chosen reference point for B_1. (Observe that R plays the role of the adjacent lower-numbered body of B_1, but B_1 does not in general have a fixed point in R.)

G_k is the mass center of body B_k, and $\mathbf{r}_k$ locates G_k relative to O_k ($k = 1, \ldots, N$). Finally, $\boldsymbol{\xi}_k$ locates O_k relative to O_j where O_j is the origin of B_j, the adjacent lower-numbered body of B_k. Observe that aside from $\boldsymbol{\xi}_1$, $\boldsymbol{\xi}_k$ is fixed in B_j.

Consider now, for example, the mass center of an extremity of the system, say B_9. From Figure 10.15, we see that the position vector $\mathbf{p}_9$ locating G_9 relative to the origin O of R is

$$\mathbf{p}_9 = \boldsymbol{\xi}_1 + \boldsymbol{\xi}_4 + \boldsymbol{\xi}_5 + \boldsymbol{\xi}_8 + \boldsymbol{\xi}_9 + \mathbf{r}_9 \tag{10.87}$$

Observe that the subscripts on the $\boldsymbol{\xi}$ vectors (1, 4, 5, 8, and 9) are the same as the entries in the column for B_q in Table 10.2. Observe further, that as noted previously, each of the positive vectors on the right-hand side of Equation 10.87 is a fixed vector in one of the bodies in the branch containing B_q, except for $\boldsymbol{\xi}_1$.

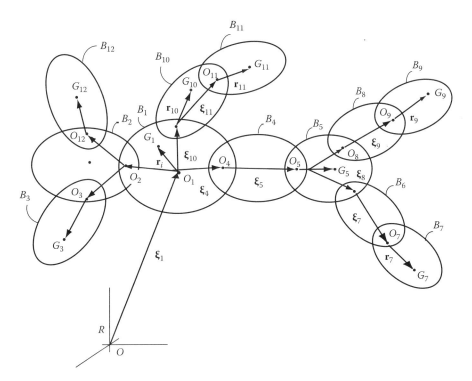

FIGURE 10.15   Position vectors for the multibody system of Figure 10.13.

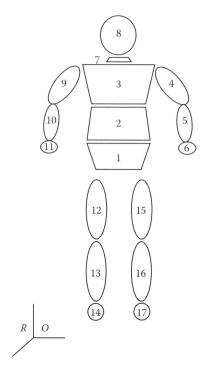

FIGURE 10.16 Numbering and labeling of the model of Figure 10.11.

TABLE 10.3
Higher-Order Lower-Body Arrays for the Model of Figure 10.16

K	1	2	3	4	5	6	7	8	9	10	11	12	13	14	15	16	17
$L(K)$	0	1	2	3	4	5	3	7	3	9	10	1	12	13	1	15	16
$L^2(K)$	0	0	1	2	3	4	2	3	2	3	9	0	1	12	0	1	15
$L^3(K)$	0	0	0	1	2	3	1	2	1	2	3	0	0	1	0	0	1
$L^4(K)$	0	0	0	0	1	2	0	1	0	1	2	0	0	0	0	0	0
$L^5(K)$	0	0	0	0	0	1	0	0	0	0	1	0	0	0	0	0	0
$L^6(K)$	0	0	0	0	0	0	0	0	0	0	0	0	0	0	0	0	0

Consider now the human body model of Figure 10.11 and as shown again in Figure 10.16, where we have assigned numbers to the bodies, following the numbering of Figure 6.9. Table 10.3 provides lists of the associated lower-body arrays (from Table 6.3). Let O be at the origin of a reference frame R fixed in an inertial space in which the model moves.

Let $\boldsymbol{\xi}_1$ be a vector locating a reference point (or origin) O, of body B_1 (the lower torso) relative to O (O_1 could be any convenient point on B_1, perhaps the mass center G). Next, using the notation of Figure 10.14, let O_k ($k = 1, \ldots, 17$) be the origins of bodies B_k ($k = 2, \ldots, 17$), and as in the foregoing illustrations, let O_k be at the centers of the spherical connecting joints of the B_k with their adjoining lower-numbered bodies.

Let G_k ($k = 1, \ldots, 17$) be the mass centers of the bodies B_k. Finally, referring again to the notations of Figure 10.14, let $\mathbf{r}_k$ ($k = 1, \ldots, 17$) be vectors locating the G_k relative to the O_k and let $\boldsymbol{\xi}_k$ ($k = 2, \ldots, 17$) be vectors locating the O_j in the adjacent lower-numbered body.

Let $\mathbf{p}_k$ ($k = 1, \ldots, 17$) be position vectors locating the mass center G_k relative to the origin O in R. Then from Figure 10.16 and Table 10.3, the $\mathbf{p}_k$ are seen to be

$$\mathbf{p}_1 = \xi_1 + \mathbf{r}_1$$
$$\mathbf{p}_2 = \xi_1 + \xi_2 + \mathbf{r}_2$$
$$\mathbf{p}_3 = \xi_1 + \xi_2 + \xi_3 + \mathbf{r}_3$$
$$\mathbf{p}_4 = \xi_1 + \xi_2 + \xi_3 + \xi_4 + \mathbf{r}_4$$
$$\mathbf{p}_5 = \xi_1 + \xi_2 + \xi_3 + \xi_4 + \xi_5 + \mathbf{r}_5$$
$$\mathbf{p}_6 = \xi_1 + \xi_2 + \xi_3 + \xi_4 + \xi_5 + \xi_6 + \mathbf{r}_6$$
$$\mathbf{p}_7 = \xi_1 + \xi_2 + \xi_5 + \xi_7 + \mathbf{r}_7$$
$$\mathbf{p}_8 = \xi_1 + \xi_2 + \xi_3 + \xi_7 + \xi_8 + \mathbf{r}_8$$
$$\mathbf{p}_9 = \xi_1 + \xi_2 + \xi_3 + \xi_9 + \mathbf{r}_9 \qquad (10.88)$$
$$\mathbf{p}_{10} = \xi_1 + \xi_2 + \xi_3 + \xi_9 + \xi_{10} + \mathbf{r}_{10}$$
$$\mathbf{p}_{11} = \xi_1 + \xi_2 + \xi_3 + \xi_9 + \xi_{10} + \xi_{11} + \mathbf{r}_{11}$$
$$\mathbf{p}_{12} = \xi_1 + \xi_{12} + \mathbf{r}_{12}$$
$$\mathbf{p}_{13} = \xi_1 + \xi_{12} + \xi_{13} + \mathbf{r}_{13}$$
$$\mathbf{p}_{14} = \xi_1 + \xi_{12} + \xi_{13} + \xi_{14} + \mathbf{r}_{14}$$
$$\mathbf{p}_{15} = \xi_1 + \xi_{15} + \mathbf{r}_{15}$$
$$\mathbf{p}_{16} = \xi_1 + \xi_{15} + \xi_{16} + \mathbf{r}_{16}$$
$$\mathbf{p}_{17} = \xi_1 + \xi_{15} + \xi_{16} + \xi_{17} + \mathbf{r}_{17}$$

Observe that each of the vectors on the right-hand side of Equation 10.88, except for ξ_1, is fixed in one of the bodies of the model. Specifically, the $\mathbf{r}_k$ are fixed in the B_k, and the ξ_k are fixed in the $L(B_k)$, the adjoining lower-numbered bodies of B_k, respectively. This means that the components of these vectors are constant when referred to basic unit vectors of the bodies in which they are fixed. Thus, if $\mathbf{n}_{k1}$, $\mathbf{n}_{k2}$, and $\mathbf{n}_{k3}$ are mutually perpendicular unit vectors fixed in B_k, and if $\mathbf{r}_k$ is expressed as

$$\mathbf{r}_k = r_{k1}\mathbf{n}_{k1} + r_{k2}\mathbf{n}_{k2} + r_{k3}\mathbf{n}_{k3} \qquad (10.89)$$

then the r_{ki} ($i = 1, 2, 3$) are constants. Similarly, if $\mathbf{n}_{j1}$, $\mathbf{n}_{j2}$, and $\mathbf{n}_{j3}$ are mutually perpendicular unit vectors fixed in B_j, where $B_j = L(B_k)$ and if ξ_k is expressed as

$$\xi_k = \xi_{k1}\mathbf{n}_{j1} + \xi_{k2}\mathbf{n}_{j2} + \xi_{k3}\mathbf{n}_{j3} \qquad (10.90)$$

then the ξ_{ki} ($i = 1, 2, 3$) are constants. Therefore, if we know the orientation of the unit vector sets, we can use anthropometric data, as in Section 2.7, to determine the value of the components of the ξ_k and $\mathbf{r}_k$.

Following the conventions of Chapter 2, let X represent the forward direction, let Z be vertically up, and then in the dextral sense, let Y be to the left. Let each body segment have an X, Y, Z coordinate system imbedded in it, with the origin of the coordinate system located at the body reference points. Let the axes of all the coordinate systems be aligned when the model is in a standing reference configuration. Figure 10.17 illustrates the reference configuration and aligned coordinate axes of several of the bodies of the model.

This notation and convention allows us to list data for the masses' reference point locations, mass center locations, and inertia dyadic components for men and women in the reference configuration [4,5]. Tables 10.4 through 10.11 provide lists for 50 percentile male and female average bodies. Appendix A provides lists for 5, 50, and 95 percentile male and female.

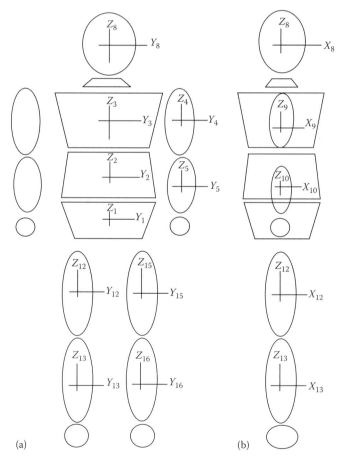

FIGURE 10.17 Model reference configuration and aligned coordinate axes of several of the bodies. (a) Front view and (b) side view.

TABLE 10.4
Inertia Data for a 50 Percentile Male—Body Segment Masses

Body Segment Number	Name	Mass lb. (Weight)	Mass Slug	Mass kg
1	Lower torso (pelvis)	22.05	0.685	10.00
2	Middle torso (lumbar)	24.14	0.750	10.95
3	Upper torso (chest)	40.97	1.270	18.58
4	Upper left arm	4.92	0.153	2.23
5	Lower left arm	3.06	0.095	1.39
6	Left hand	1.15	0.036	0.52
7	Neck	3.97	0.123	1.80
8	Head	10.91	0.339	4.95
9	Upper right arm	4.92	0.153	2.23
10	Lower right arm	3.06	0.095	1.39
11	Right hand	1.15	0.036	0.52
12	Upper right leg	18.63	0.578	8.45
13	Lower right leg	7.61	0.236	3.45
14	Right foot	2.27	0.070	1.03
15	Upper left leg	18.63	0.578	8.45
16	Lower left leg	7.61	0.236	3.45
17	Left foot	2.27	0.070	1.03
	Total	177.32	5.51	80.42

Source: Churchill, E. et al., in *Anthropometric Source Book, Volume I: Anthropometry for Designers,* Webb Associates (Ed.), NASA Reference Publications, Yellow Springs, OH, 1978, III-84 to III-97, and IV-37.

TABLE 10.5
Inertia Data for a 50 Percentile Male—Reference Point (Body Origin) Location

Body Segment Number	Name	Component (ft) (Relative to Adjoining Lower-Numbered Body Frame) X	Y	Z
1	Lower torso (pelvis)	0.0	0.0	0.0
2	Middle torso (lumbar)	0.0	0.0	0.3375
3	Upper torso (chest)	0.0	0.0	0.675
4	Upper left arm	0.0	0.696	0.483
5	Lower left arm	0.0	0.0	−0.975
6	Left hand	0.0	0.0	−0.975
7	Neck	0.0	0.0	0.658
8	Head	0.0	0.0	0.392
9	Upper right arm	0.0	−0.696	0.483
10	Lower right arm	0.0	0.0	−0.975
11	Right hand	0.0	0.0	−0.975
12	Upper right leg	0.0	−0.256	−0.054
13	Lower right leg	0.0	0.0	−1.53
14	Right foot	0.0	0.0	−1.391
15	Upper left leg	0.0	0.256	−0.054
16	Lower left leg	0.0	0.0	−1.55
17	Left foot	0.0	0.0	−1.391

TABLE 10.6
Inertia Data for a 50 Percentile Male—Mass Center Location

Body Segment	Number	Name Component (ft) (Relative to Body Frame)		
		X	Y	Z
1	Lower torso (pelvis)	0.0	0.0	0.0
2	Middle torso (lumbar)	0.0	0.0	0.3375
3	Upper torso (chest)	0.0	0.0	0.329
4	Upper left arm	0.0	0.0	−0.372
5	Lower left arm	0.0	0.0	−0.483
6	Left hand	0.0	0.0	−0.283
7	Neck	0.0	0.0	0.196
8	Head	0.0	0.0	0.333
9	Upper right arm	0.0	0.0	−0.372
10	Lower right arm	0.0	0.0	−0.483
11	Right hand	0.0	0.0	−0.283
12	Upper right leg	0.0	0.0	−0.8225
13	Lower right leg	0.0	0.0	−0.692
14	Right foot	0.333	0.0	−0.167
15	Upper left leg	0.0	0.0	−0.8223
16	Lower left leg	0.0	0.0	−0.692
17	Left foot	0.333	0.0	−0.167

TABLE 10.7

Inertia Data for a 50 Percentile Male—Inertia Dyadic Matrices

Body Segment Number	Name	Inertia Matrix (Slug ft²) (Relative to Body Frame Principal Direction)		
1	Lower torso (pelvis)	0.1090	0.0	0.0
		0.0	0.0666	0.0
		0.0	0.0	0.1060
2	Middle torso (lumbar)	0.1090	0.0	0.0
		0.09	0.0666	0.0
		0.0	0.0	0.1060
3	Upper torso (chest)	0.0775	0.0	0.0
		0.0	0.05375	0.0
		0.0	0.0	0.0775
4	Upper left arm	0.0196	0.0	0.0
		0.0	0.0196	0.0
		0.0	0.0	0.0021
5	Lower left arm	0.0154	0.0	0.0
		0.0	0.0154	0.0
		0.0	0.0	0.0010
6	Left hand	0.0025	0.0	0.0
		0.0	0.0013	0.0
		0.0	0.0	0.0013
7	Neck	0.0114	0.0	0.0
		0.0	0.0114	0.0
		0.0	0.0	0.0021
8	Head	0.0278	0.0	0.0
		0.0	0.0278	0.0
		0.0	0.0	0.0139
9	Upper right arm	0.0196	0.0	0.0
		0.0	0.0196	0.0
		0.0	0.0	0.0021
10	Lower right arm	0.0154	0.0	0.0
		0.0	0.0154	0.0
		0.0	0.0	0.0010
11	Right hand	0.0025	0.0	0.6
		0.0	0.0013	0.0
		0.0	0.0	0.0013
12	Upper right leg	0.0706	0.0	0.0
		0.0	0.0706	0.0
		0.0	0.0	0.0180
13	Lower right leg	0.00569	0.0	0.0
		0.0	0.0059	0.0
		0.0	0.0	0.0008
14	Right foot	0.0008	0.0	0.0
		0.0	0.0046	0.0
		0.0	0.0	0.0050
15	Upper left leg	0.0706	0.0	0.0
		0.0	0.0706	0.0
		0.0	0.0	0.0180
16	Lower left leg	0.0059	0.0	0.0
		0.0	0.0059	0.0
		0.0	0.0	0.0008
17	Left foot	0.0008	0.0	0.0
		0.0	0.0046	0.0
		0.0	0.0	0.0054

TABLE 10.8

Inertia Data for a 50 Percentile Female—Body Segment Masses

Body Segment Number	Name	Mass		
		lb (Weight)	Slug	kg
1	Lower torso (pelvis)	22.05	0.685	10.00
2	Middle torso (lumbar)	14.53	0.457	6.59
3	Upper torso (chest)	20.5	0.636	9.30
4	Upper left arm	3.77	0.1117	1.02
5	Lower left arm	2.25	0.070	1.02
6	Left hand	0.922	0.0286	0.418
7	Neck	3.2	0.099	1.45
8	Head	8.84	0.274	4.01
9	Upper right arm	3.77	0.117	1.71
10	Lower right arm	2.25	0.070	1.02
11	Right hand	0.922	0.0286	0.418
12	Upper right leg	16.6	0.516	7.53
13	Lower right leg	5.97	0.185	2.71
14	Right foot	1.89	0.0587	0.857
15	Upper left leg	16.6	0.516	7.53
16	Lower left leg	5.97	0.185	2.71
17	Left foot	1.89	0.0587	0.857
	Total	131.9	4.0956	59.84

TABLE 10.9

Inertia Data for a 50 Percentile Female—Reference Point (Body Origin) Location

Body Segment Number	Name	Component (ft) (Relative to Adjoining Lower-Numbered Body Frame)		
		X	Y	Z
1	Lower torso (pelvis)	0.0	0.0	0.0
2	Middle torso (lumbar)	0.0	0.0	0.308
3	Upper torso (chest)	0.0	0.0	0.617
4	Upper left arm	0.0	0.635	0.442
5	Lower left arm	0.0	0.0	−0.892
6	Left hand	0.0	0.0	−0.892
7	Neck	0.0	0.0	0.600
8	Head	0.0	0.0	0.357
9	Upper right arm	0.0	−0.635	0.458
10	Lower right arm	0.0	0.0	−0.882
11	Right hand	0.0	0.0	−0.882
12	Upper right leg	0.0	−0.235	−0.049
13	Lower right leg	0.0	0.0	−1.417
14	Right foot	0.0	0.0	−1.285
15	Upper left leg	0.0	0.233	−0.049
16	Lower left leg	0.0	0.0	−1.417
17	Left foot	0.0	0.0	−1.2851

TABLE 10.10

Inertia Data for a 50 Percentile Female 132 lb—Mass Center Location

Body Segment Number	Name	Component (ft) (Relative to Body Frame)		
		X	Y	Z
1	Lower torso (pelvis)	0.0	0.0	0.0
2	Middle torso (lumbar)	0.0	0.0	0.308
3	Upper torso (chest)	0.0	0.0	0.300
4	Upper left arm	0.0	0.0	−0.339
5	Lower left arm	0.0	0.0	−0.442
6	Left hand	0.0	0.0	−0.258
7	Neck	0.0	0.0	0.179
8	Head	0.0	0.0	0.304
9	Upper right arm	0.0	0.0	−0.339
10	Lower right arm	0.0	0.0	−0.442
11	Right hand	0.0	0.0	−0.258
12	Upper right leg	0.0	0.0	−0.750
13	Lower right leg	0.0	0.0	−0.632
14	Right foot	0.304	0.0	−0.153
15	Upper left leg	0.0	0.0	−0.750
16	Lower left leg	0.0	0.0	−0.632
17	Left foot	0.304	0.0	−0.153

TABLE 10.11

Inertia Data for a 50 Percentile Female—Inertia Dyadic Matrices

Body Segment Number	Name	Inertia Matrix (Slug ft²) (Relative to Body Frame Principal Direction)		
1	Lower torso (pelvis)	0.0910	0.0	0.0
		0.0	0.0555	0.0
		0.0	0.0	0.0889
2	Middle torso (lumbar)	0.0546	0.0	0.0
		0.09	0.0334	0.0
		0.0	0.0	0.534
3	Upper torso (chest)	0.0326	0.0	0.0
		0.0	0.05375	0.0
		0.0	0.0	0.0775
4	Upper left arm	0.0124	0.0	0.0
		0.0	0.0124	0.0
		0.0	0.0	0.0014
5	Lower left arm	0.0095	0.0	0.0
		0.0	0.0095	0.0
		0.0	0.0	0.0006
6	Left hand	0.0017	0.0	0.0
		0.0	0.0008	0.0
		0.0	0.0	0.0008
7	Neck	0.0077	0.0	0.0
		0.0	0.0077	0.0
		0.0	0.0	0.0014
8	Head	0.0187	0.0	0.0
		0.0	0.0187	0.0
		0.0	0.0	0.0094
9	Upper right arm	0.0124	0.0	0.0
		0.0	0.0124	0.0
		0.0	0.0	0.0014
10	Lower right arm	0.0095	0.0	0.0
		0.0	0.0095	0.0
		0.0	0.0	0.0006
11	Right hand	0.0017	0.0	0.6
		0.0	0.0008	0.0
		0.0	0.0	0.0008
12	Upper right leg	0.0525	0.0	0.0
		0.0	0.0525	0.0
		0.0	0.0	0.0134
13	Lower right leg	0.0039	0.0	0.0
		0.0	0.0039	0.0
		0.0	0.0	0.0005
14	Right foot	0.0005	0.0	0.0
		0.0	0.0032	0.0
		0.0	0.0	0.0038
15	Upper left leg	0.0525	0.0	0.0
		0.0	0.0525	0.0
		0.0	0.0	0.0134
16	Lower left leg	0.0039	0.0	0.0
		0.0	0.0039	0.0
		0.0	0.0	0.0005
17	Left foot	0.0005	0.0	0.0
		0.0	0.0032	0.0
		0.0	0.0	0.0038

PROBLEMS

Section 10.1

P10.1.1 Let a particle P have mass 3 kg and X, Y, Z Cartesian coordinates: (3, −2, 4 m). Evaluate and compute the moments and products of inertia of P relative to the origin O for the directions of X, Y, and Z.

P10.1.2 See Problem P10.1.1. Using the results of Problem P10.1.1, determine the associated inertia matrix $I^{P/O}$.

P10.1.3 See Problems P10.1.1 and P10.1.2. Let A be a reference point with coordinates: (1, 2, 3 m). Repeat Problems P10.1.1 and P10.1.2 with the origin O replaced by A.

P10.1.4 Repeat Problems P10.1.1, P10.1.2, and P10.1.3 for a particle Q having a mass of 0.5 slug and X,Y,Z coordinates: (−1, 3, 5 ft).

Section 10.2

P10.2.1 See Problem P10.1.1. As before, let a particle P have mass 3 kg and X, Y, Z Cartesian coordinates: (3, −2, 4 m). Let $\mathbf{n}_1$, $\mathbf{n}_2$, $\mathbf{n}_3$ be unit vectors parallel to X, Y, Z, respectively. Determine the inertia dyadic $\mathbf{I}^{P/O}$.

P10.2.2 Suppose an inertia dyadic $\mathbf{I}$ is expressed in terms of mutually perpendicular unit vectors $\mathbf{n}_i$ ($i = 1, 2, 3$) as follows:

$$\mathbf{I} = I_{ij}\mathbf{n}_i\mathbf{n}_j$$

where the inertia matrix elements are

$$I_{ij} = \begin{bmatrix} 8 & \sqrt{3} & 1 \\ \sqrt{3} & 7 & \sqrt{3} \\ 1 & \sqrt{3} & 5 \end{bmatrix}$$

Let $\hat{\mathbf{n}}_i$ ($i = 1, 2, 3$) be a set of mutually perpendicular unit vectors with components relative to the $\mathbf{n}_i$ given as

$$\hat{\mathbf{n}}_1 = -\left(\frac{1}{2}\right)\mathbf{n}_2 + \left(\frac{\sqrt{3}}{2}\right)\mathbf{n}_3$$

$$\hat{\mathbf{n}}_2 = \left(\frac{-\sqrt{2}}{2}\right)\mathbf{n}_1 + \left(\frac{\sqrt{6}}{4}\right)\mathbf{n}_2 + \left(\frac{\sqrt{2}}{4}\right)\mathbf{n}_3$$

$$\hat{\mathbf{n}}_3 = \left(\frac{-\sqrt{2}}{2}\right)\mathbf{n}_1 + \left(\frac{-\sqrt{6}}{4}\right)\mathbf{n}_2 + \left(\frac{-\sqrt{2}}{4}\right)\mathbf{n}_3$$

If now $\mathbf{I}$ is expressed in terms of the $\hat{\mathbf{n}}_i$ as

$$\mathbf{I} = \hat{I}_{ij}\hat{\mathbf{n}}_i\hat{\mathbf{n}}_j$$

determine the elements: $\hat{I}_{ij}$.

Section 10.3

P10.3.1 Consider a set S of four particles P_i ($i = 1, \ldots, 4$) with masses m_i and X, Y, Z (Cartesian) coordinates (x_i, y_i, z_i) as listed in the following table, where the units are arbitrary (e.g., kg, slug, g and m, ft, cm).

Particle	Mass	X	Y	Z
P_1	3	0	3	-2
P_2	1	4	-3	6
P_3	4	-4	1	5
P_4	5	7	-4	1

Determine the elements of the inertia matrix $I^{S/O}$ where O is the origin of the coordinate system.

P10.3.2 See Problem P10.3.1. Find the coordinates of the mass center G of S.

P10.3.3 See Problems P10.3.1 and P10.3.2. Determine the elements of the inertia matrix $I^{S/G}$.

P10.3.4 See Problems P10.3.1, P10.3.2, and P10.3.3. Let the total mass M of S be associated with the mass center G. Determine the inertia matrix $I^{G/O}$.

Section 10.4

P10.4.1 Verify the steps used in the development of $\mathbf{T}^*$ of Equation 10.45.

Section 10.5

P10.5.1 See Problems P10.3.1 to P10.3.4. Using the results of those problems, verify that

$$\mathbf{I}^{S/O} = \mathbf{I}^{S/G} + \mathbf{I}^{G/O}$$

Section 10.6

P10.6.1 Consider the inertia dyadic $\mathbf{I}$ given by the expression

$$\mathbf{I} = I_{ij}\mathbf{n}_i\mathbf{n}_j$$

with, as is customary, the $\mathbf{n}_i$ ($i = 1,2,3$) are mutually perpendicular unit vectors and the I_{ij} are elements of the inertia matrix I, with values given by

$$I = [I_{ij}] = \begin{bmatrix} 27 & 5 & -12 \\ 5 & 16 & 4 \\ -12 & 4 & 13 \end{bmatrix}$$

where the units are arbitrary (e.g., slug ft^2 or kg m^2).

Following the procedures of Section 10.6, determine the eigenvalues and the unit eigenvectors of $\mathbf{I}$.

Section 10.8

P10.8.1 Outline an algorithm for numerically determining the location of the center of mass of a 17-body human body model as in Figure 10.16.

P10.8.2 See Problem P10.8.1. Outline an algorithm for numerically determining the inertia matrix I_{ij}, of an inertia dyadic $\mathbf{I}_O$, a 17-body human body model, where

$$\mathbf{I}_O = I_{ij}\mathbf{N}_i\mathbf{N}_j$$

where

$\mathbf{N}_i$ are mutually perpendicular unit vectors of a fixed Cartesian frame R
O is the origin of the frame.

P10.8.3 See Problems P10.8.1 and P10.8.2. Repeat Problem P10.8.2, for the inertia dyadic $\mathbf{I}_G$, with G being the mass center of the model.

REFERENCES

1. R. L. Huston, *Multibody Dynamics*, Butterworth Heinemann, Boston, MA, 1990, pp. 153–213.
2. R. L. Huston and C. Q. Liu, *Formulas for Dynamic Analysis*, Mercer Dekker, New York, 2001, pp. 328–336.
3. E. Churchill et al., *Anthropometric Source Book, Volume I: Anthropometry for Designers*, Webb Associates (Ed.), III–84 to III–97, and IV–37, NASA Reference Publications, Yellow Springs, OH, 1978.
4. E. P. Hanavan, A mathematical model of the human body, Report No. AMRL-TR-64–102. Aerospace Medical Research Laboratory, WPAFB, OH, 1964.
5. H. Hatze, A model for the computational determination of parameter values of anthropometric segments, National Research Institute for Mathematical Sciences. Technical Report TWISK 79, Pretoria, South Africa, 1979.

11 Kinematics of Human Body Models

Consider the 17-member human body model of Chapters 6 and 8 and as shown again in Figure 11.1.

As noted earlier, we may regard this model as an open-chain or open-tree collection of rigid bodies. As such, we can number or label the bodies as before and as shown in the figure. Recall that with this numbering system, each body of the model has a unique adjacent lower-numbered body. Specifically, as in Chapters 6 and 8, we may list the adjacent lower-numbered bodies $L(K)$ of each of the bodies (K) as in Equation 6.11 and as listed again here:

K	1	2	3	4	5	6	7	8	9	10	11	12	13	14	15	16	17
$L(K)$	0	1	2	3	4	5	3	7	3	9	10	1	12	13	1	15	16

$$(11.1)$$

Observe further that with this modeling, the bodies are connected to each other by spherical (ball-and-socket) joints. Then each body (except for body B_1) has three rotational degrees of freedom relative to its adjacent lower-numbered body. Body B_1, whose adjacent lower-numbered body is the fixed (inertial) frame R, has six degrees of freedom in R (three rotations and three translations). The entire model thus has a total of $(17 \times 3) + 3$ or 54 degrees of freedom.

In this chapter, we will develop the kinematics of this model as well as procedures for determining explicit expressions for the kinematics of human body models in general.

11.1 NOTATION, DEGREES OF FREEDOM, AND COORDINATES

As we just observed, the human body model of Figure 11.1, with 17 spherical joint connecting bodies, has 54 degrees of freedom (3 rotational degrees of freedom for each of the 17 bodies and, in addition, three translational degrees of freedom for body B_1. As noted earlier, the model is an open-chain (or open-tree) multibody system (Section 6.1). To develop the kinematics of the model, it is helpful to simultaneously consider a simpler but yet more general, generic multibody system, as for example, the system shown in Figure 11.2. We will use this particular system throughout the chapter as a precursor for studying the human body model of Figure 11.1 and for studying more elaborate or more specific human body models.

In this generic system, the bodies are also connected to one another by spherical joints (but without closed loops). The entire system is free to move in an inertial frame R. As such, the system has $3N + 3$ degrees of freedom where N is the number of bodies—10 in the system of Figure 11.2. Each of the degrees of freedom, except for three, is a rotation, or orientation change, of one of the bodies. The other three represent translation in R of a body, the reference body B_1, of the system.

As in Section 6.3, we have numbered and labeled the bodies of the system as shown in Figure 11.2, where we have selected a major body (a large body) of the system for our reference, or first, body and called it 1 or B_1. Thus, we have numbered the other bodies of the system in an ascending progression array from B_1, through the branches of the system. Recall and observe that with this

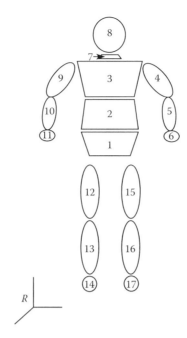

FIGURE 11.1   A 17-member human body model.

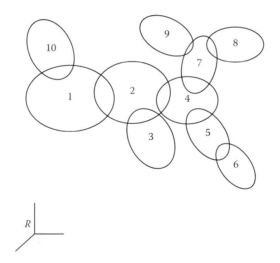

FIGURE 11.2   Generic multibody system.

numbering procedure, each body has a unique adjoining lower-numbered body. The listing of these lower-numbered bodies for the system of Figure 11.2 is

Body number, (K)	1	2	3	4	5	6	7	8	9	10	
Adjacent lower body number, $L(K)$	0	1	2	2	4	5	4	7	7	1	(11.2)

Consider a typical pair of adjoining bodies, say B_j and B_k, as shown in Figure 11.3. Let $\mathbf{n}_{ji}$ and $\mathbf{n}_{ki}$ ($i = 1, 2, 3$) be mutually perpendicular unit vector sets fixed in B_j and B_k.

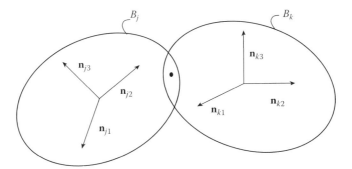

FIGURE 11.3   Typical pair of adjoining bodies (B_j is the lower-numbered body).

Let B_j be the lower-numbered body. Then the orientation of B_k relative to B_j may be defined in terms of the orientation of the $\mathbf{n}_{ki}$ relative to the $\mathbf{n}_{ji}$, which in turn may be defined in terms of orientation angles α_k, β_k, and γ_k as in Section 8.8. Specifically, let B_k initially be oriented so that the $\mathbf{n}_{ki}$ are mutually aligned with the $\mathbf{n}_{ji}$. Then let B_k be brought into a general orientation relative to B_j by three successive rotations of B_k about $\mathbf{n}_{k1}$, $\mathbf{n}_{k2}$, and $\mathbf{n}_{k3}$ through the angles α_k, β_k, and γ_k, respectively. When these rotations obey the right-hand rule, they are known as dextral or Bryant or orientation angles (Section 8.8). Observe that B_k may also be oriented relative to B_j by different sequences of rotations of the $\mathbf{n}_{ki}$, such as with Euler angles. Reference [1] provides a complete listing of these alternative orientation angles.

If we use dextral orientation angles to describe the body orientation with respect to their adjoining lower-numbered bodies, we can list the variables describing the degrees of freedom as in Table 11.1. For this purpose, it is convenient to list the variables in sets of three.

Observe that there are a total of 33 variables corresponding to the 33 degrees of freedom.

We can use the same procedure to define the connection configurations and the degrees of freedom of a human body model. Specifically, let the human body model be the 17-member model of Figure 11.1.

TABLE 11.1
Variable Names for the Multibody
System of Figure 11.2

Variables	Description
x, y, z	Translations of B_1 in R
$\alpha_1, \beta_1, \gamma_1$	Rotation of B_1 in R
$\alpha_2, \beta_2, \gamma_2$	Rotation of B_2 in B_1
$\alpha_3, \beta_3, \gamma_3$	Rotation of B_3 in B_2
$\alpha_4, \beta_4, \gamma_4$	Rotation of B_4 in B_2
$\alpha_5, \beta_5, \gamma_5$	Rotation of B_5 in B_4
$\alpha_6, \beta_6, \gamma_6$	Rotation of B_6 in B_5
$\alpha_7, \beta_7, \gamma_7$	Rotation of B_7 in B_4
$\alpha_8, \beta_8, \gamma_8$	Rotation of B_8 in B_7
$\alpha_9, \beta_9, \gamma_9$	Rotation of B_9 in B_7
$\alpha_{10}, \beta_{10}, \gamma_{10}$	Rotation of B_{10} in B_1

TABLE 11.2

Variable Names for the Human Body
Model of Figure 11.1

Variables	Description
x, y, z	Translations of B_1 in R
$\alpha_1, \beta_1, \gamma_1$	Rotation of B_1 in R
$\alpha_2, \beta_2, \gamma_2$	Rotation of B_2 in B_1
$\alpha_3, \beta_3, \gamma_3$	Rotation of B_3 in B_2
$\alpha_4, \beta_4, \gamma_4$	Rotation of B_4 in B_3
$\alpha_5, \beta_5, \gamma_5$	Rotation of B_5 in B_4
$\alpha_6, \beta_6, \gamma_6$	Rotation of B_6 in B_5
$\alpha_7, \beta_7, \gamma_7$	Rotation of B_7 in B_3
$\alpha_8, \beta_8, \gamma_8$	Rotation of B_8 in B_7
$\alpha_9, \beta_9, \gamma_9$	Rotation of B_9 in B_3
$\alpha_{10}, \beta_{10}, \gamma_{10}$	Rotation of B_{10} in B_9
$\alpha_{11}, \beta_{11}, \gamma_{11}$	Rotation of B_{11} in B_{10}
$\alpha_{12}, \beta_{12}, \gamma_{12}$	Rotation of B_{12} in B_1
$\alpha_{13}, \beta_{13}, \gamma_{13}$	Rotation of B_{13} in B_{12}
$\alpha_{14}, \beta_{14}, \gamma_{14}$	Rotation of B_{14} in B_{13}
$\alpha_{15}, \beta_{15}, \gamma_{15}$	Rotation of B_{15} in B_1
$\alpha_{16}, \beta_{16}, \gamma_{16}$	Rotation of B_{16} in B_{15}
$\alpha_{17}, \beta_{17}, \gamma_{17}$	Rotation of B_{17} in B_{16}

With 17 bodies connected by spherical joints, there are $(17 \times 3) + 3$ or 54 degrees of freedom for the system (three rotations for each body and three translations for body 1 in R).

With the bodies of the model numbered and labeled as in Figure 11.1, we can identify the lower-body array $L(K)$ as in Equation 11.1 and as repeated here:

Body number, (K)	1	2	3	4	5	6	7	8	9	10	
Adjacent lower body number, $L(K)$	0	1	2	2	4	5	4	7	7	1	(11.3)

With 17 bodies, connected by spherical joints, the model has $(17 \times 3) + 3$ or 54 degrees of freedom (three for rotation of each body and three for translation of body 1). As with the generic system of Figure 11.2, we can describe the rotational degrees of freedom by dextral orientation angles. Table 11.2 then presents a list (in sets of three) of variables describing the degrees of freedom of the model.

Observe how the lower-body array $L(K)$ of Equation 11.3 may be used in the variable descriptions in Table 11.2.

11.2   ANGULAR VELOCITIES

Consider the labeled multibody system of Figure 11.2.

We can use the addition theorem for angular velocity (Equation 8.51) to obtain expressions for the angular velocity for each of the bodies of the system relative to a fixed (inertial) frame R. To establish a convenient notation and terminology, consider two typical adjoining bodies of the

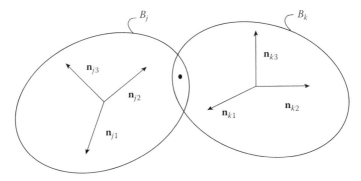

FIGURE 11.4 Two typical adjoining bodies.

system, say B_j and B_k, as in Figure 11.4 (see also Figure 8.34). As before, let $\mathbf{n}_{ji}$ and $\mathbf{n}_{ki}$ be mutually perpendicular unit vectors fixed in B_j and B_k.

Consider first the angular velocity of B_k relative to B_j (notationally $^{B_j}\boldsymbol{\omega}^{B_k}$). From Sections 8.8 and 8.10 by using configuration graphs, and from Equation 8.99, it is clear that $^{B_j}\boldsymbol{\omega}^{B_k}$ may be expressed in terms of the dextral orientation angles as

$$^{B_j}\boldsymbol{\omega}^{B_k} = (\dot{\alpha}_k + \dot{\gamma}_k s_{\beta_k})\mathbf{n}_{j1} + (\dot{\beta}_k c_{\alpha_k} - \dot{\gamma}_k s_{\alpha_k} c_{\beta_k})\mathbf{n}_{j2} + (\dot{\beta}_k s_{\alpha_k} \dot{\gamma}_k c_{\alpha_k} c_{\beta_k})\mathbf{n}_{j3} \tag{11.4}$$

where, as before, s and c are abbreviations for sine and cosine. Since with the numbering system of Figure 11.2, each body has a unique adjacent lower-numbered body, there is no ambiguity in writing $^{B_j}\boldsymbol{\omega}^{B_k}$ as (Equation 8.101)

$$^{B_j}\boldsymbol{\omega}^{B_k} = \hat{\boldsymbol{\omega}}_k \tag{11.5}$$

where, the overhat signifies relative angular velocity. Then in terms of the $\mathbf{n}_{ji}$, $\hat{\boldsymbol{\omega}}_k$ may be written as

$$\hat{\boldsymbol{\omega}}_k = \hat{\omega}_{k1}\mathbf{n}_{j1} + \hat{\omega}_{k2}\mathbf{n}_{j2} + \hat{\omega}_{k3}\mathbf{n}_{j3} = \hat{\omega}_{km}\mathbf{n}_{jm} \tag{11.6}$$

where from Equation 11.4, the $\hat{\omega}_{km}$ ($m = 1, 2, 3$) are

$$\hat{\omega}_{k1} = \dot{\alpha}_k + \dot{\gamma}_k s_{\beta_k}$$

$$\hat{\omega}_{k2} = \dot{\beta}_k c_{\alpha_k} - \dot{\gamma}_k s_{\alpha_k} c_{\beta_k} \tag{11.7}$$

$$\hat{\omega}_{k3} = \dot{\beta}_k s_{\alpha_k} + \dot{\gamma}_k c_{\alpha_k} c_{\beta_k}$$

Angular velocity symbols ($\boldsymbol{\omega}$) without the overhat designate absolute angular velocity, that is, angular velocity relative to the fixed or inertial frame R. Specifically,

$$\boldsymbol{\omega}_k = {}^R\boldsymbol{\omega}^{B_k} \tag{11.8}$$

Using this notation together with the addition theorem for angular velocity, we can express the angular velocities of the bodies of the generic multibody system of Figure 11.2 as

$$\boldsymbol{\omega}_1 = \hat{\boldsymbol{\omega}}_1$$

$$\boldsymbol{\omega}_2 = \hat{\boldsymbol{\omega}}_2 + \hat{\boldsymbol{\omega}}_1$$

$$\boldsymbol{\omega}_3 = \hat{\boldsymbol{\omega}}_3 + \hat{\boldsymbol{\omega}}_2 + \hat{\boldsymbol{\omega}}_1$$

$$\boldsymbol{\omega}_4 = \hat{\boldsymbol{\omega}}_4 + \hat{\boldsymbol{\omega}}_2 + \hat{\boldsymbol{\omega}}_1$$

$$\boldsymbol{\omega}_5 = \boldsymbol{\omega}_5 + \hat{\boldsymbol{\omega}}_4 + \hat{\boldsymbol{\omega}}_2 + \hat{\boldsymbol{\omega}}_1 \qquad (11.9)$$

$$\boldsymbol{\omega}_6 = \hat{\boldsymbol{\omega}}_6 + \hat{\boldsymbol{\omega}}_5 + \hat{\boldsymbol{\omega}}_4 + \hat{\boldsymbol{\omega}}_2 + \hat{\boldsymbol{\omega}}_1$$

$$\boldsymbol{\omega}_7 = \hat{\boldsymbol{\omega}}_7 + \hat{\boldsymbol{\omega}}_4 + \hat{\boldsymbol{\omega}}_2 + \hat{\boldsymbol{\omega}}_1$$

$$\boldsymbol{\omega}_8 = \hat{\boldsymbol{\omega}}_8 + \hat{\boldsymbol{\omega}}_7 + \hat{\boldsymbol{\omega}}_4 + \hat{\boldsymbol{\omega}}_2 + \hat{\boldsymbol{\omega}}_1$$

$$\boldsymbol{\omega}_9 = \hat{\boldsymbol{\omega}}_9 + \hat{\boldsymbol{\omega}}_7 + \hat{\boldsymbol{\omega}}_4 + \hat{\boldsymbol{\omega}}_2 + \hat{\boldsymbol{\omega}}_1$$

$$\boldsymbol{\omega}_{10} = \hat{\boldsymbol{\omega}}_{10} + \hat{\boldsymbol{\omega}}_1$$

Recall from Equation 11.2 that the lower-body array $L(K)$ for the generic system is

$$L(K): \quad 0 \quad 1 \quad 2 \quad 2 \quad 4 \quad 5 \quad 4 \quad 7 \quad 7 \quad 1 \qquad (11.10)$$

Then from the procedures of Section 6.2, we can form a table of lower-body arrays as in Table 11.3.

Observe that the subscript indices in Equation 11.9 are identical to the nonzero entries in the columns in Table 11.3. This relationship may be exploited to develop an algorithm for the computation of the angular velocities. Specifically, from Section 8.10, we see that the entire set of Equations 11.9 is contained within the compact expression (Equation 8.104):

$$\boldsymbol{\omega}_k = \sum_{p=0}^{r} \hat{\boldsymbol{\omega}}_q \quad q = L^p(K) \quad (k = 1, \ldots, 10) \qquad (11.11)$$

where r is the index such that $L^r(K) = 1$ and where $k = k$.

TABLE 11.3

Higher-Order Lower-Body Arrays for the Generic System of Figure 11.2

K	1	2	3	4	5	6	7	8	9	10
$L(k)$	0	1	2	2	4	5	4	7	7	1
$L^2(k)$	0	0	1	1	2	4	2	4	4	0
$L^3(k)$	0	0	0	0	1	2	1	2	2	0
$L^4(k)$	0	0	0	0	0	1	0	1	1	0
$L^5(k)$	0	0	0	0	0	0	0	0	0	0

Next, consider the human body model of Figure 11.1.

Analogous to Table 11.3, the lower-body arrays for the human body model are listed in Table 11.4 (see Table 8.1 and Equation 11.3).

Analogous to Equation 11.9, the angular velocities, in inertia frame R of the bodies of the model, are

$$\omega_1 = \hat{\omega}_1$$
$$\omega_2 = \hat{\omega}_2 + \hat{\omega}_1$$
$$\omega_3 = \hat{\omega}_3 + \hat{\omega}_2 + \hat{\omega}_1$$
$$\omega_4 = \hat{\omega}_4 + \hat{\omega}_3 + \hat{\omega}_2 + \hat{\omega}_1$$
$$\omega_5 = \omega_5 + \hat{\omega}_4 + \hat{\omega}_3 + \hat{\omega}_2 + \hat{\omega}_1$$
$$\omega_6 = \hat{\omega}_6 + \hat{\omega}_5 + \hat{\omega}_4 + \hat{\omega}_3 + \hat{\omega}_2 + \hat{\omega}_1$$
$$\omega_7 = \hat{\omega}_7 + \hat{\omega}_3 + \hat{\omega}_2 + \hat{\omega}_1$$
$$\omega_8 = \hat{\omega}_8 + \hat{\omega}_7 + \hat{\omega}_3 + \hat{\omega}_2 + \hat{\omega}_1$$
$$\omega_9 = \hat{\omega}_9 + \hat{\omega}_3 + \hat{\omega}_2 + \hat{\omega}_1 \tag{11.12}$$
$$\omega_{10} = \hat{\omega}_{10} + \hat{\omega}_9 + \hat{\omega}_3 + \hat{\omega}_2 + \hat{\omega}_1$$
$$\omega_{11} = \hat{\omega}_{11} + \hat{\omega}_{10} + \hat{\omega}_9 + \hat{\omega}_3 + \hat{\omega}_2 + \hat{\omega}_1$$
$$\omega_{12} = \hat{\omega}_{12} + \hat{\omega}_1$$
$$\omega_{13} = \hat{\omega}_{13} + \hat{\omega}_{12} + \hat{\omega}_1$$
$$\omega_{14} = \hat{\omega}_{14} + \hat{\omega}_{13} + \hat{\omega}_{12} + \hat{\omega}_1$$
$$\omega_{15} = \hat{\omega}_{15} + \hat{\omega}_1$$
$$\omega_{16} = \hat{\omega}_{16} + \hat{\omega}_{15} + \hat{\omega}_1$$
$$\omega_{17} = \hat{\omega}_{17} + \hat{\omega}_{16} + \hat{\omega}_{15} + \hat{\omega}_1$$

Observe again that the subscript indices in Equation 11.12 are the same as the entries in the columns in Table 11.4.

TABLE 11.4
Higher-Order Lower-Body Arrays for the System of Figure 11.1

K	1	2	3	4	5	6	7	8	9	10	11	12	13	14	15	16	17
$L(K)$	0	1	2	3	4	5	3	7	3	9	10	1	12	13	1	15	16
$L^2(K)$	0	0	1	2	3	4	2	3	2	3	9	0	1	12	0	1	15
$L^3(K)$	0	0	0	1	2	3	1	2	1	2	3	0	0	1	0	0	1
$L^4(K)$	0	0	0	0	1	2	0	1	0	1	2	0	0	0	0	0	0
$L^5(K)$	0	0	0	0	0	1	0	0	0	0	1	0	0	0	0	0	0
$L^6(K)$	0	0	0	0	0	0	0	0	0	0	0	0	0	0	0	0	0

Finally, observe that Equation 11.12 may be embodied in Equation 11.11 as

$$\boldsymbol{\omega}_k = \sum_{p=0}^{r} \boldsymbol{\omega}_q \quad q = L^p(K) \quad (k = 1, \ldots, 17) \tag{11.13}$$

where r is the index such that $L^r(K) = 1$ and where $k = K$.

11.3 GENERALIZED COORDINATES

When a multibody system such as a human body model has, say, n degrees of freedom, the system may be viewed as moving in an n-dimensional space. The variables describing the configuration and movement of the system are then viewed as coordinates of the system. There is thus one variable (or coordinate) for each degree of freedom. Consequently, the number of variables needed to define the configuration and movement of the system is then equal to the number of degrees of freedom of the system. The choice of variables, however, is not unique. (e.g., in 3D space, a point representing a particle may be treated using various coordinate systems such as Cartesian, cylindrical, and spherical coordinates.)

In general, or in an abstract manner, it is convenient to name or express the variables defining the configuration and movement of a multibody system as q_r, $r = 1, \ldots, n$ where, as before, n is the number of degrees of freedom. When this is done, the q_r are conventionally called generalized coordinates.

To illustrate and develop these concepts, consider the generic multibody system of Figure 11.2.

Table 11.1 provides a variable listing for the system. The system has 33 degrees of freedom. From the foregoing discussion, we may identify 22 generalized coordinates q_r, $(r = 1, \ldots, 33)$ defined as

$$
\begin{aligned}
x, y, z &\rightarrow q_1, q_2, q_3 \\
\alpha_1, \beta_1, \gamma_1 &\rightarrow q_4, q_5, q_6 \\
\alpha_2, \beta_2, \gamma_2 &\rightarrow q_7, q_8, q_9 \\
\alpha_3, \beta_3, \gamma_3 &\rightarrow q_{10}, q_{11}, q_{12} \\
\alpha_4, \beta_4, \gamma_4 &\rightarrow q_{13}, q_{14}, q_{15} \\
\alpha_5, \beta_5, \gamma_5 &\rightarrow q_{16}, q_{17}, q_{18} \\
\alpha_6, \beta_5, \gamma_5 &\rightarrow q_{19}, q_{20}, q_{21} \\
\alpha_7, \beta_7, \gamma_7 &\rightarrow q_{22}, q_{23}, q_{24} \\
\alpha_8, \beta_8, \gamma_8 &\rightarrow q_{25}, q_{26}, q_{27} \\
\alpha_9, \beta_9, \gamma_9 &\rightarrow q_{28}, q_{29}, q_{30} \\
\alpha_{10}, \beta_{10}, \gamma_{10} &\rightarrow q_{31}, q_{32}, q_{33}
\end{aligned} \tag{11.14}
$$

Consider next the human body model as shown in Figure 11.1.

With 17 bodies, this system has 54 degrees of freedom, leading to 54 generalized coordinates $q_r (r = 1, \ldots, 54)$, defined as

$$x, y, z \rightarrow q_1, q_2, q_3$$

$$\alpha_1, \beta_1, \gamma_1 \rightarrow q_4, q_5, q_6$$

$$\alpha_2, \beta_2, \gamma_2 \rightarrow q_7, q_8, q_9$$

$$\alpha_3, \beta_3, \gamma_3 \rightarrow q_{10}, q_{11}, q_{12}$$

$$\alpha_4, \beta_4, \gamma_4 \rightarrow q_{13}, q_{14}, q_{15}$$

$$\alpha_5, \beta_5, \gamma_5 \rightarrow q_{16}, q_{17}, q_{18}$$

$$\alpha_6, \beta_5, \gamma_5 \rightarrow q_{19}, q_{20}, q_{21}$$

$$\alpha_7, \beta_7, \gamma_7 \rightarrow q_{22}, q_{23}, q_{24}$$

$$\alpha_8, \beta_8, \gamma_8 \rightarrow q_{25}, q_{26}, q_{27}$$

$$\alpha_9, \beta_9, \gamma_9 \rightarrow q_{28}, q_{29}, q_{30}$$ (11.15)

$$\alpha_{10}, \beta_{10}, \gamma_{10} \rightarrow q_{31}, q_{32}, q_{33}$$

$$\alpha_{11}, \beta_{11}, \gamma_{11} \rightarrow q_{34}, q_{35}, q_{36}$$

$$\alpha_{12}, \beta_{12}, \gamma_{12} \rightarrow q_{37}, q_{38}, q_{39}$$

$$\alpha_{13}, \beta_{13}, \gamma_{13} \rightarrow q_{40}, q_{41}, q_{42}$$

$$\alpha_{14}, \beta_{14}, \gamma_{14} \rightarrow q_{43}, q_{44}, q_{45}$$

$$\alpha_{15}, \beta_{15}, \gamma_{15} \rightarrow q_{46}, q_{47}, q_{48}$$

$$\alpha_{16}, \beta_{16}, \gamma_{16} \rightarrow q_{49}, q_{50}, q_{51}$$

$$\alpha_{17}, \beta_{17}, \gamma_{17} \rightarrow q_{52}, q_{53}, q_{54}$$

11.4 PARTIAL ANGULAR VELOCITIES

Next, observe from Equations 11.4, 11.5, 11.9, and 11.11 that the angular velocities $\boldsymbol{\omega}_k (k = 1, \ldots, 10)$ are linear functions of the orientation angle derivatives, and thus, considering Equation 11.14, they are then linear functions of the generalized coordinate derivatives $\dot{q}_r (r = 1, \ldots, 33)$. Since the angular velocities are vectors, they may be expressed in terms of the unit vectors $\mathbf{n}_{om} (m = 1, 2, 3)$ fixed in the inertial frame R. Therefore, the angular velocities may be expressed as

$$\boldsymbol{\omega}_k = \omega_{klm} \dot{q}_l \mathbf{n}_{om} \quad (k = 1, \ldots, 10; l = 1, \ldots, 33; m = 1, 2, 3) \tag{11.16}$$

where the coefficients ω_{klm} are called partial angular velocity components. By inspection of Equations 11.4, 11.5, and 11.9, we see that the majority of the ω_{klm} are zero.

It happens that the ω_{klm} plays a central role in multibody dynamic analyses (see Refs [2,3]). To obtain explicit expressions for these, recall from Equations 11.6 and 11.7 that the relative angular velocities $\hat{\boldsymbol{\omega}}_k$ are

$$\hat{\boldsymbol{\omega}}_k = \hat{\omega}_{k1} \mathbf{n}_{f1} + \hat{\omega}_{k2} \mathbf{n}_{j2} + \hat{\omega}_{k3} \mathbf{n}_{j3} \tag{11.17}$$

where ω_{klm} $(m = 1, 2, 3)$ are

$$\hat{\omega}_{k1} = \dot{\alpha}_k + \dot{\gamma}_k s_{\beta_k}$$

$$\hat{\omega}_{k2} = \dot{\beta}_k c_{\alpha_k} - \dot{\gamma}_k s_{\alpha_k} c_{\beta_k} \qquad (11.18)$$

$$\hat{\omega}_{k3} = \dot{\beta}_k s_{\alpha_k} + \dot{\gamma}_k c_{\alpha_k} c_{\beta_k}$$

Observe from Equation 11.14 that the $\dot{\alpha}_k$, $\dot{\beta}_k$, and $\dot{\gamma}_k$ may be identified with the $\dot{q}_r (r = 3k + 1, 3k + 2, 3b + 3)$. By substituting from Equations 11.18 into 11.17, we see that the coefficients of the $\dot{q}_r$ are simply the partial derivatives of $\hat{\omega}_k$ with respect to $\dot{\alpha}_k$, $\dot{\beta}_k$, and $\dot{\gamma}_k$. That is,

$$\frac{\partial \hat{\omega}_k}{\partial \dot{\alpha}_k} = \mathbf{n}_{j1} \qquad (11.19)$$

$$\frac{\partial \hat{\omega}_k}{\partial \dot{\beta}_k} = c_{\alpha_k} \mathbf{n}_{j2} + s_{\alpha_k} \mathbf{n}_{j3} \qquad (11.20)$$

$$\frac{\partial \hat{\omega}_k}{\partial \dot{\gamma}_k} = s_{\beta_k} \mathbf{n}_{j1} + s_{\alpha_k} c_{\beta_k} \mathbf{n}_{j2} + c_{\alpha_k} c_{\beta_k} \mathbf{n}_{j3} \qquad (11.21)$$

where, keeping in tune with the previous terminology, these derivatives might be called partial relative angular velocities.

Next, observe that the angular velocity vector of Equation 11.9 as well as the partial angular velocity vectors of Equations 11.19, 11.21, and 11.23 is expressed in terms of unit vectors $\mathbf{n}_{jm}$ fixed in body B_j as opposed to unit vectors $\mathbf{n}_{om}$ fixed in the inertial frame R. Thus, to use Equation 11.16 to determine the partial angular velocity components (the ω_{klm}), it is necessary to express the relative angular velocities in terms of the $\mathbf{n}_{om}$. We can easily make this transformation by using orthogonal transformation matrices S whose elements are defined by Equations 3.160 through 3.166. Specifically, let the transformation matrix SOJ be defined by its elements SOJ_{mn} given by

$$SOJ_{mn} = \mathbf{n}_{om} \cdot \mathbf{n}_{jn} \qquad (11.22)$$

Then in view of Equation 3.166, $\mathbf{n}_{jm}$ are

$$\mathbf{n}_{jn} = SOJ_{mn} \cdot \mathbf{n}_{om} \qquad (11.23)$$

By substituting Equation 11.23 into Equations 11.17, 11.19, 11.20, and 11.21, we have

$$\hat{\omega}_k = \hat{\omega}_{kn} \mathbf{n}_{jn} = \hat{\omega}_{kn} SOJ_{mn} \mathbf{n}_{om} \qquad (11.24)$$

and

$$\frac{\partial \hat{\omega}_k}{\partial \dot{\alpha}_k} = SOJ_{m1} \mathbf{n}_{om} \qquad (11.25)$$

$$\frac{\partial \hat{\boldsymbol{\omega}}_k}{\partial \dot{\beta}_k} = c_{\alpha_k} SOJ_{m2} \cdot \mathbf{n}_{om} + s_{\alpha_k} SOJ_{m3} \cdot \mathbf{n}_{om} \tag{11.26}$$

$$\frac{\partial \hat{\boldsymbol{\omega}}_k}{\partial \dot{\gamma}_k} = \left(s_{\beta_k} SOJ_{m1} - s_{\alpha_k} c_{\beta_k} SOJ_{m2} + c_{\alpha_k} c_{\beta_k} SOJ_{m3} \right) \cdot \mathbf{n}_{om} \tag{11.27}$$

11.5   TRANSFORMATION MATRICES: RECURSIVE FORMULATION

Observe that by repeated use of Equation 11.22 together with the use of configuration graphs of Section 8.8, we can readily obtain explicit expressions for the *SOJ* matrices. To see this, consider three typical adjoining bodies of the system say B_i, B_j, and B_k as in Figure 11.5, where the $\mathbf{n}_{im}$, $\mathbf{n}_{jm}$, and $\mathbf{n}_{km}$ are mutually perpendicular unit vectors fixed in the respective bodies.

From the analysis of Section 8.8 and specifically from Figure 8.29, the configuration graph with dextral angles relating the n_{im} and the $\mathbf{n}_{jm}$ ($m = 1, 2, 3$) is shown in Figure 11.6.

If these unit vector sets are arranged in column arrays as

$$\mathbf{n}_i = \begin{bmatrix} \mathbf{n}_{i1} \\ \mathbf{n}_{i2} \\ \mathbf{n}_{i3} \end{bmatrix}_j \quad \text{and} \quad \mathbf{n}_j = \begin{bmatrix} \mathbf{n}_{j1} \\ \mathbf{n}_{j2} \\ \mathbf{n}_{j3} \end{bmatrix}_j \tag{11.28}$$

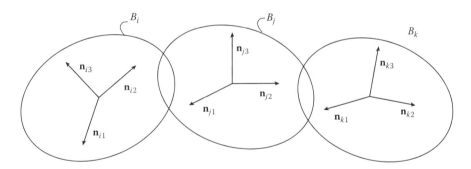

FIGURE 11.5   Three typical adjoining bodies with embedded unit vector sets.

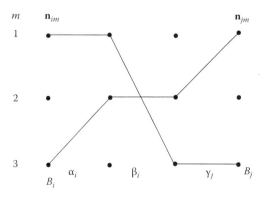

FIGURE 11.6   Dextral angle configuration graph for the $\mathbf{n}_{jm}$ ($m = 1, 2, 3$).

then from Equation 8.73, these arrays are related as

$$\mathbf{n}_i = SIJ\mathbf{n}_j \tag{11.29}$$

where from Equation 8.74, the SIJ transformation array is

$$SIJ = \begin{bmatrix} c_{\beta_j}c_{\gamma_j} & -c_{\beta_j}s_{\gamma_j} & s_{\beta_j} \\ (c_{\alpha_j}s_{\gamma_j} + s_{\alpha_j}s_{\beta_j}c_{\gamma_j}) & (c_{\alpha_j}c_{\gamma_j} - s_{\alpha_j}s_{\beta_j}s_{\gamma_j}) & -s_{\alpha_j}c_{\beta_j} \\ (s_{\alpha_j}s_{\gamma_j} - c_{\alpha_j}s_{\beta_j}c_{\gamma_j}) & (s_{\alpha_j}c_{\gamma_j} + c_{\alpha_j}s_{\beta_j}s_{\gamma_j}) & c_{\alpha_j}c_{\beta_j} \end{bmatrix} \tag{11.30}$$

Similarly, a dextral angle configuration graph relating the $\mathbf{n}_{jm}$ and the $\mathbf{n}_{km}$ is shown in Figure 11.7. Again, if these unit vector sets are arranged in column arrays as

$$\mathbf{n}_j = \begin{bmatrix} \mathbf{n}_{j1} \\ \mathbf{n}_{j2} \\ \mathbf{n}_{j3} \end{bmatrix}_j \quad \text{and} \quad \mathbf{n}_k = \begin{bmatrix} \mathbf{n}_{k1} \\ \mathbf{n}_{k2} \\ \mathbf{n}_{k3} \end{bmatrix}_j \tag{11.31}$$

then from Equation 8.73, these arrays are related as

$$\mathbf{n}_j = SIK\mathbf{n}_k \tag{11.32}$$

where from Equation 8.74, the SJK transformation array is

$$SIJ = \begin{bmatrix} c_{\beta_k}c_{\gamma_k} & -c_{\beta_k}s_{\gamma_k} & s_{\beta_k} \\ (c_{\alpha_k}s_{\gamma_k} + s_{\alpha_k}s_{\beta_k}c_{\gamma_k}) & (c_{\alpha_k}c_{\gamma_k} - s_{\alpha_k}s_{\beta_k}s_{\gamma_k}) & -s_{\alpha_k}c_{\beta_k} \\ (s_{\alpha_k}s_{\gamma_k} - c_{\alpha_k}s_{\beta_k}c_{\gamma_k}) & (s_{\alpha_k}c_{\gamma_k} + c_{\alpha_k}s_{\beta_k}s_{\gamma_k}) & c_{\alpha_k}c_{\beta_k} \end{bmatrix} \tag{11.33}$$

By substituting Equation 11.32 into 11.29, we have

$$\mathbf{n}_i = SIJ \cdot SJK\mathbf{n}_k \tag{11.34}$$

Let SIK be the dextral angle transformation matrix relating the $\mathbf{n}_{im}$ and the $\mathbf{n}_{km}$, as

$$\mathbf{n}_i = SIK\mathbf{n}_k \tag{11.35}$$

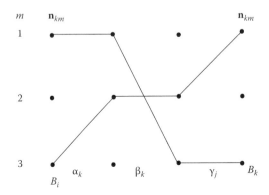

FIGURE 11.7   Dextral angle configuration graph for the $\mathbf{n}_{jm}$ and $\mathbf{n}_{km}$ ($m = 1, 2, 3$).

Thus, from Equation 11.34, we have the transitive result

$$SIK = SIJ \cdot SJK \tag{11.36}$$

Equation 11.36 may now be used, together with the connection configurations as defined by the lower-body array, to obtain the transformation matrices for each of the bodies. Specifically, from Table 11.3, we have the following relations for the generic system of Figure 11.2:

$$S01 = S01$$

$$S02 = S01S12$$

$$S03 = S01S12S23$$

$$S04 = S01S12S24$$

$$S05 = S01S12S24S45 \tag{11.37}$$

$$S06 = S01S12S24S45S56$$

$$S07 = S01S12S24S47$$

$$S08 = S01S12S24S47S78$$

$$S09 = S01S12S24S47S79$$

$$S10 = S01S1,10$$

Finally, consider the 17-member human body model shown in Figure 11.1.

By inspection of the figure, or by use of the lower-body arrays of Table 11.4, we have the following expressions for the transformation matrices of the bodies of the model:

$$S01 = S01$$

$$S02 = S01S12$$

$$S03 = S01S12S23$$

$$S04 = S01S12S23S34$$

$$S05 = S01S12S23S34S45$$

$$S06 = S01S12S23S34S45S46$$

$$S07 = S01S12S23S37$$

$$S08 = S01S12S23S37S78 \tag{11.38}$$

$$S09 = S01S12S23S39$$

$$S10 = S01S12S23S39S9,10$$

$$S11 = S01S12S23S39S9,10S10,11$$

$$S12 = S01S1,12$$

$$S13 = S01S1,12S12,13$$

$$S14 = S01S1,12,S12,13S13,14$$

$$S15 = S01S1,15$$

$$S16 = S01S1,15S15,16$$

$$S17 = S01S1,15S15,16S16,17$$

11.6 GENERALIZED SPEEDS

Recall that in Equation 11.16, we have the following compact expressions for the angular velocities of the bodies of a multibody system in general and of our human body model in particular:

$$\boldsymbol{\omega}_k = \omega_{klm}\dot{q}_l \mathbf{n}_{om} \quad (k = 1, \ldots, N) \tag{11.39}$$

Observe that in this expression, the ω_{klm} essentially determine the angular velocities. Alternatively, if we know the angular velocities, we can determine the ω_{klm} by inspection.

Taken altogether, the ω_{klm} form a block array with dimensions $N \times (3N + 3) \times 3$.

The ω_{klm} are called partial angular velocity components since they are the components of partial angular velocity vectors defined as $\partial \boldsymbol{\omega}_k / \partial \dot{q}_l$. Then from Equation 11.39, we have

$$\frac{\partial \boldsymbol{\omega}_k}{\partial \dot{q}_l} = \omega_{klm}\mathbf{n}_{om} \tag{11.40}$$

Next, recall from Equations 11.25 through 11.27 that the partial derivatives of the relative angular velocities $\hat{\boldsymbol{\omega}}_k$ (angular velocity of B_k relative to B_j) with respect to $\dot{\alpha}_k$, $\dot{\beta}_k$, and $\dot{\gamma}_k$ are

$$\frac{\partial \hat{\boldsymbol{\omega}}_k}{\partial \dot{\alpha}_k} = SOJ_{m1}\mathbf{n}_{om} \tag{11.41}$$

$$\frac{\partial \hat{\boldsymbol{\omega}}_k}{\partial \dot{\beta}_k} = \left(c_{\alpha_k}SOJ_{m2} + s_{\alpha_k}SOJ_{m1}\right) \cdot \mathbf{n}_{om} \tag{11.42}$$

and

$$\frac{\partial \hat{\boldsymbol{\omega}}_k}{\partial \dot{\gamma}_k} = \left(s_{\beta_k}SOJ_{m1} - s_{\alpha_k}c_{\beta_k}SOJ_{m2} + c_{\alpha_k}c_{\beta_k}SOJ_{m3}\right) \cdot \mathbf{n}_{om} \tag{11.43}$$

By comparing Equation 11.40 with Equations 11.41 through 11.43, we see that the nonzero ω_{klm} are cumbersome expressions of sums and multiples of sines and cosines of the relative orientation angles. Also note that the absolute angular velocity $\boldsymbol{\omega}_k$ are composed of sums of relative angular velocities $\hat{\boldsymbol{\omega}}_k$ as demonstrated in Equation 11.9 and in general by Equation 11.11 as

$$\boldsymbol{\omega}_k = \sum_{p=0}^{r} \hat{\boldsymbol{\omega}}_q \quad q = L^p(k) \quad \text{with} \quad L^r(k) = 1 \tag{11.44}$$

Although the expressions for the ω_{klm} are not intractable, they are nevertheless more complex than one might expect. It happens that we can obtain a considerably simpler analysis through the introduction of generalized speeds, which are linear combinations of orientation angle derivatives or more specifically—relative angular velocity components.

To develop this, observe from Equation 11.39 that the angular velocities are linear combinations of generalized coordinate derivations, the $\dot{q}_r$. If the system has, say, n degrees of freedom (so that $r = 1, \ldots, n$), let there be n parameters y_s (called generalized speeds) defined as

$$y_s = \sum_{r=1}^{n} c_{sr}\dot{q}_r = \dot{c}_{rs}\dot{q}_r \tag{11.45}$$

where we continue to use the repeated index summation convention and where the c_{sr} are arbitrary elements of an $n \times n$ matrix C, provided only that the equations may be solved for the $\dot{q}_r$ in terms of the y_s (i.e., C is nonsingular). Then by solving for $\dot{q}_r$, we have

$$\dot{q}_r = C_{rs}^{-1} y_s \tag{11.46}$$

where the C_{rs}^{-1} are the elements of C^{-1}.

By substituting from Equation 11.46 into 11.39, the angular velocities become

$$\omega_k = \omega_{klm} c_{ls}^{-1} y_s \mathbf{n}_{om} = \tilde{\omega}_{ksm} y_s \mathbf{n}_{om} \tag{11.47}$$

where $\tilde{\omega}_{ksm}$ are defined by inspection.

Since y_s (the generalized speeds) are arbitrarily defined, we can define them so that the resulting analysis is simplified. To this end, as observed previously, it is particularly advantageous to define the generalized speeds as components of the relative angular velocity components. These assertions and the selection procedure are perhaps best understood via a simple illustration example, as with our generic example system of Figure 11.2, as outlined in the following sections.

Recall that this system has $(10 \times 3) + 3$ or 33 degrees of freedom. Let these degrees of freedom be represented by variables $x_r (r = 1, \ldots, 33)$ whose derivatives $\dot{x}_r$ are the generalized speeds y_r. That is,

$$y_s = \dot{x}_r \quad (r = 1, \ldots, 33) \tag{11.48}$$

Next, let y_r be identified with the kinematical quantities of the system and arranged in triplets as in Table 11.5, where x, y, and z are the Cartesian coordinates of the relative points of body B_1 in R and $\hat{\omega}_{ki}$ are the angular velocity components of B_k, relative to the adjacent lower-numbered body B_j.

TABLE 11.5
Generalized Speeds for the Generic Multibody System of Figure 11.2

Generalized Speeds	Kinematic Variables
y_1, y_2, y_3	x, y, z
y_4, y_5, y_6	$\hat{\omega}_{11}, \hat{\omega}_{12}, \hat{\omega}_{13}$
y_7, y_8, y_9	$\hat{\omega}_{21}, \hat{\omega}_{22}, \hat{\omega}_{23}$
y_{10}, y_{11}, y_{12}	$\hat{\omega}_{31}, \hat{\omega}_{32}, \hat{\omega}_{33}$
y_{13}, y_{14}, y_{15}	$\hat{\omega}_{41}, \hat{\omega}_{42}, \hat{\omega}_{43}$
y_{16}, y_{17}, y_{18}	$\hat{\omega}_{51}, \hat{\omega}_{52}, \hat{\omega}_{53}$
y_{19}, y_{20}, y_{21}	$\hat{\omega}_{61}, \hat{\omega}_{62}, \hat{\omega}_{63}$
y_{22}, y_{23}, y_{24}	$\hat{\omega}_{71}, \hat{\omega}_{72}, \hat{\omega}_{73}$
y_{25}, y_{26}, y_{27}	$\hat{\omega}_{81}, \hat{\omega}_{82}, \hat{\omega}_{83}$
y_{28}, y_{29}, y_{30}	$\hat{\omega}_{91}, \hat{\omega}_{92}, \hat{\omega}_{93}$
y_{31}, y_{32}, y_{33}	$\hat{\omega}_{10,1}, \hat{\omega}_{10,2}, \hat{\omega}_{10,3}$

Observe in Table 11.5 that the relative angular velocity components are referred to unit vectors fixed in the adjoining lower-numbered bodies. Then in view of Equation 11.23 if we wish to express all the vectors in terms of unit vectors $\mathbf{n}_{om}$ fixed in the inertia frame R, we can readily use the transformation matrix to obtain the R-frame expressions. That is, for typical body B_k, we have

$$\hat{\boldsymbol{\omega}}_k = {}^{B_j}\boldsymbol{\omega}^{B_k} = \hat{\omega}_{k1}\mathbf{n}_{j1} + \hat{\omega}_{k2}\mathbf{n}_{j2} + \hat{\omega}_{k3}\mathbf{n}_{j3} = \hat{\omega}_{km}\mathbf{n}_{jm} = SOJ_{nm}\hat{\omega}_{km}\mathbf{n}_{on}$$

$$= SOJ_{nm}y_{3k+m}\mathbf{n}_{on} \tag{11.49}$$

where the last equality follows Table 11.5.

Recall that angular velocity components are not, in general, integrable in terms of elementary functions—except in the case of simple rotation about a fixed line (so-called simple angular velocity [see Section 8.6]). Therefore, with the generalized speeds chosen as relative angular velocity components, the variables x_r do not in general exist as elementary functions. Thus, x_r are sometimes called quasicoordinates.

11.7   ANGULAR VELOCITIES AND GENERALIZED SPEEDS

Suppose that in view of Equation 11.47, we may express the angular velocities in terms of the generalized speeds as

$$\boldsymbol{\omega}_k = \omega_{ksm}y_s\mathbf{n}_{om} \tag{11.50}$$

where for simplicity we have omitted the overhat. Considering (11.9), (11.12), and (11.48), we see that the nonzero value of the ω_{klm} is simply elements of the transformation matrices.

To illustrate this, consider the angular velocity of, say, body B_8 of the generic system of Figure 11.2. From Equation 11.9, $\boldsymbol{\omega}_8$ is

$$\boldsymbol{\omega}_8 = \hat{\boldsymbol{\omega}}_8 + \hat{\boldsymbol{\omega}}_7 + \hat{\boldsymbol{\omega}}_4 + \hat{\boldsymbol{\omega}}_2 + \hat{\boldsymbol{\omega}}_1 \tag{11.51}$$

or equivalently

$$\boldsymbol{\omega}_8 = \hat{\boldsymbol{\omega}}_1 + \hat{\boldsymbol{\omega}}_2 + \hat{\boldsymbol{\omega}}_4 + \hat{\boldsymbol{\omega}}_7 + \hat{\boldsymbol{\omega}}_8 \tag{11.52}$$

Thus, using Equation 11.48, $\boldsymbol{\omega}_8$ may be expressed as

$$\boldsymbol{\omega}_8 = [y_{3+n}\delta_{mn} + SO1_{mn}y_{6+m} + SO2_{mn}y_{12+n} + SO4_{mn}y_{21+n} + SO7_{mn}y_{24+n}]\mathbf{n}_{om} \tag{11.53}$$

From Equation 11.50, it is clear that the ω_{8lm} are

$$\omega_{8lm} = \frac{\partial\boldsymbol{\omega}_8}{\partial y_l}\cdot\mathbf{n}_{om} \tag{11.54}$$

Then by comparing Equations 11.53 and 11.54, we see that the nonzero ω_{8lm} are

$$\omega_{8lm} \begin{cases} \delta_{m(l-3)}l = 4,5,6 \\ S01_{m(l-6)}l = 7,8,9 \\ S02_{m(l-12)}l = 13,14,15 \\ S04_{m(l-21)}l = 22,23,24 \\ S07_{m(l-24)}l = 25,26,27 \end{cases} \quad m = 1, 2, 3 \tag{11.55}$$

Observing the patterns of the results of Equation 11.55, we can list all the ω_{klm} in a relatively compact form as in Table 11.6.

Note that the vast majority of the ω_{klm} are zero. Observe also that here with the generalized speeds selected as relative angular velocity components, the nonzero ω_{klm} are simply transformation matrix elements.

Finally, observe the pattern of the nonzero entries in Table 11.6. Recall from Equation 11.37 that the global transfer matrices SOK are

$$\begin{aligned} S01 &= S01 \\ S02 &= S01S12 \\ S03 &= S01S12S23 \\ S04 &= S01S12S24 \\ S05 &= S01S12S24S45 \\ S06 &= S01S12S24S45S56 \\ S07 &= S01S12S24S47 \\ S08 &= S01S12S24S47S78 \\ S09 &= S01S12S24S47S79 \\ S10 &= S01S1,10 \end{aligned} \tag{11.56}$$

TABLE 11.6
Partial Angular Velocity Components ω_{klm} with Generalized Speeds as Relative Angular Velocity Components for the Generic Multibody System of Figure 11.2

	y_1										
	1	4	7	10	13	16	19	22	25	28	31
	2	5	8	11	14	17	20	23	26	29	32
Body	3	6	9	12	15	18	21	24	27	30	33
1	0	0	0	0	0	0	0	0	0	0	0
2	0	I	$SO1$	0	0	0	0	0	0	0	0
3	0	I	$SO1$	$SO2$	0	0	0	0	0	0	0
4	0	I	$SO1$	0	$SO2$	0	0	0	0	0	0
5	0	I	$SO1$	0	$SO2$	$SO4$	0	0	0	0	0
6	0	I	$SO1$	0	$SO2$	$SO4$	$SO5$	0	0	0	0
7	0	I	$SO1$	0	$SO2$	0	0	$SO4$	0	0	0
8	0	I	$SO1$	0	$SO2$	0	0	$SO4$	$SO7$	0	0
9	0	I	$SO1$	0	$SO2$	0	0	$SO4$	0	$SO7$	0
10	0	I	0	0	0	0	0	0	0	0	$SO1$

Recall further that these equations are obtained directly from Table 11.3, which is a listing of the higher-order lower-body arrays. By comparing the pattern of the entries of Table 11.3 with the recursive pattern in Equation 11.56, we see that they are the same except for the spacing, and the numbering in the table is absolute (relative to R) and one less than that in the equation. This means that we may readily prepare an algorithm, based on the lower-body array $L(K)$, to generate the entries in Table 11.6. This in turn shows that the partial angular velocity components, and hence also the angular velocities themselves, may be determined once the lower-body array $L(K)$ is known.

To apply these concepts with our human body model shown in Figure 11.1, consider Equation 11.38, providing the recursive relations for the transformation matrices:

$$S01 = S01$$
$$S02 = S01S12$$
$$S03 = S01S12S23$$
$$S04 = S01S23S23S34$$
$$S05 = S01S12S23S34S45$$
$$S06 = S01S12S23S34S45S56$$
$$S07 = S01S12S23S37$$
$$S08 = S01S12S23S37S78$$
$$S09 = S01S12S23S39$$
$$S10 = S01S12S23S39S9,10$$
$$S11 = S01S12S23S39S9,10S10,11$$
$$S12 = S01S1,12$$
$$S13 = S01S1,12S12,13$$
$$S14 = S01S1,12S12,13S13,14$$
$$S15 = S01S1,15$$
$$S16 = S01S1,15S15,16$$
$$S17 = S01S1,15S15,16S16,17$$

(11.57)

By examining the recursive pattern in the equation, we immediately obtain a table for the partial angular velocity components. Table 11.7 displays the results.

11.8 ANGULAR ACCELERATION

When generalized speeds are used as generalized coordinate derivatives, the angular velocities of the bodies of a multibody system and, hence, of a human body model may be expressed in the compact form of Equation 11.50 or as

$$\omega_k = \omega_{klm}y_l\mathbf{n}_{om}$$

(11.58)

where the ω_{klm} coefficients (partial angular velocity components) are either zero or they are elements of transformation matrices as seen in Tables 11.6 and 11.7.

TABLE 11.7

Partial Angular Velocity Components ω_{klm} with Generalized Speeds as Relative Angular Velocity Components for the Human Body Model of Figure 11.1

y_1

Body	1/2/3	4/5/6	7/8/9	10/11/12	13/14/15	16/17/18	19/20/21	22/23/24	25/26/27	28/29/30	31/32/33	34/35/36	37/38/39	40/41/42	43/44/45	46/47/48	49/50/51	52/53/54
1	0	*I*	0	0	0	0	0	0	0	0	0	0	0	0	0	0	0	0
2	0	*I*	S01	0	0	0	0	0	0	0	0	0	0	0	0	0	0	0
3	0	*I*	S01	S02	0	0	0	0	0	0	0	0	0	0	0	0	0	0
4	0	*I*	S01	S02	S03	0	0	0	0	0	0	0	0	0	0	0	0	0
5	0	*I*	S01	S02	S03	S04	0	0	0	0	0	0	0	0	0	0	0	0
6	0	*I*	S01	S02	S03	S04	S05	0	0	0	0	0	0	0	0	0	0	0
7	0	*I*	S01	S02	0	0	0	S06	0	0	0	0	0	0	0	0	0	0
8	0	*I*	S01	S02	0	0	0	S06	S07	0	0	0	0	0	0	0	0	0
9	0	*I*	S01	S02	0	0	0	0	0	S08	0	0	0	0	0	0	0	0
10	0	*I*	S01	S02	0	0	0	0	0	S08	S09	0	0	0	0	0	0	0
11	0	*I*	S01	S02	0	0	0	0	0	S08	S09	S010	0	0	0	0	0	0
12	0	*I*	0	0	0	0	0	0	0	0	0	0	S011	0	0	0	0	0
13	0	*I*	0	0	0	0	0	0	0	0	0	0	S011	S012	0	0	0	0
14	0	*I*	0	0	0	0	0	0	0	0	0	0	S011	S012	S013	0	0	0
15	0	*I*	0	0	0	0	0	0	0	0	0	0	0	0	0	S014	0	0
16	0	*I*	0	0	0	0	0	0	0	0	0	0	0	0	0	S014	S015	0
17	0	*I*	0	0	0	0	0	0	0	0	0	0	0	0	0	S014	S015	S016

Considering Equation 11.58, the angular accelerations $\boldsymbol{\alpha}_k$ are obtained by differentiation as

$$\boldsymbol{\alpha}_k = (\omega_{klm}\dot{y} + \dot{\omega}_{klm}y_l)\mathbf{n}_{om} \tag{11.59}$$

where the nonzero $\dot{\omega}_{klm}$ are elements of the transformation matrix derivatives.

Recall from Equation 8.123 that if S is a transformation matrix between unit vectors of the body B and those of a reference frame R, the time derivative of S may be expressed simply as

$$\dot{S} = WS \tag{11.60}$$

where W (the angular velocity matrix) is (see Equation 8.122)

$$W = \begin{bmatrix} 0 & -\Omega_3 & \Omega_2 \\ \Omega & 0 & -\Omega_1 \\ -\Omega_2 & \Omega_1 & 0 \end{bmatrix} \tag{11.61}$$

where Ω_m $(m = 1, 2, 3)$ are the components of the angular velocity of B in R referred to unit vectors fixed in R.

We can generalize Equation 11.60 so that it is applicable to the multibody systems. Specifically, for a body B_k of a multibody system, we have

$$\dot{SOK} = WK\,SOK \tag{11.62}$$

where now the WK matrix is

$$WK = \begin{bmatrix} 0 & -\omega_{k3} & \omega_{k2} \\ \omega_{k3} & 0 & -\omega_{k1} \\ -\omega_{k2} & \omega_{k1} & 0 \end{bmatrix} \tag{11.63}$$

where the ω_{km} are the $\mathbf{n}_{om}$ components of $\boldsymbol{\omega}_k$, the angular velocity of B_k in R. From Equation 11.58, ω_{km} are

$$\omega_{km} = \omega_{klm}y_l \tag{11.64}$$

Alternately, if we denote the elements of matrix WK as WK_{ij}, we see from Equation 8.121 that the WK_{ij} may be expressed as

$$WK_{ij} = -e_{ijm}\omega_{km} \tag{11.65}$$

Thus, by substituting from Equation 11.64, the WK_{ij} becomes

$$WK_{ij} = -e_{ijm}\omega_{klm}y_i \tag{11.66}$$

(Observe the presence of the ω_{klm} in the expression.)

TABLE 11.8

Derivatives of Partial Angular Velocity Components ω_{klm} of Table 11.6 for the Generic Multibody System of Figure 11.2

Body	y_1 1 2 3	4 5 6	7 8 9	10 11 12	13 14 15	16 17 18	19 20 21	22 23 24	25 26 27	28 29 30	31 32 33
1	0	0	0	0	0	0	0	0	0	0	0
2	0	0	$S\dot{O}1$	0	0	0	0	0	0	0	0
3	0	0	$S\dot{O}1$	$S\dot{O}2$	0	0	0	0	0	0	0
4	0	0	$S\dot{O}1$	0	$S\dot{O}2$	0	0	0	0	0	0
5	0	0	$S\dot{O}1$	0	$S\dot{O}2$	$S\dot{O}4$	0	0	0	0	0
6	0	0	$S\dot{O}1$	0	$S\dot{O}2$	$S\dot{O}4$	$S\dot{O}5$	0	0	0	0
7	0	0	$S\dot{O}1$	0	$S\dot{O}2$	0	0	$S\dot{O}4$	0	0	0
8	0	0	$S\dot{O}1$	0	$S\dot{O}2$	0	0	$S\dot{O}4$	$S\dot{O}7$	0	0
9	0	0	$S\dot{O}1$	0	$S\dot{O}2$	0	0	$S\dot{O}4$	0	$S\dot{O}7$	0
10	0	0	0	0	0	0	0	0	0	0	$S\dot{O}1$

Finally, Equation 11.59 provides the angular accelerations of the bodies, once ω_{klm} and $\dot{\omega}_{klm}$ are known. Tables 11.6 and 11.7 provide lists of the ω_{klm} for the generic multibody system and for our human body model. Similarly, Tables 11.8 and 11.9 provide lists of the ω_{klm} for the generic multibody system and for our human body model. In Tables 11.8 and 11.9, $\dot{S}OK$ are obtained via Equations 11.62 through 11.66.

11.9 MASS CENTER POSITIONS

Consider the positions of the mass center of the bodies of the human body model. Let these positions be defined by position vectors $\mathbf{p}_k$ ($k = 1, \ldots, 17$) locating mass centers G_k relative to a fixed point O in the fixed or inertial frame R, as illustrated in Figure 11.8 for the right hand, the lower right leg, and the left foot.

To develop expressions for these mass centers, it is helpful, as before, to use a generic multibody system to establish the notation and procedure. To this end, consider the generic system shown in Figure 11.2, and consider three typical adjoining bodies, such as bodies B_4, B_7, and B_9, as in Figure 11.9, where the bodies are called B_i, B_j, and B_k.

Let the bodies be connected by the spherical joints, and let the centers of these joints be the origins of reference frames fixed in the respective bodies.

Let ξ be the position vector locating connecting joints relative to each other within a given body. Similarly, let $\mathbf{r}$ be the name of the position vector locating the mass center of a body relative to the connecting joint origins, as illustrated in Figure 11.10. Specifically, let the connecting joint origin be O_i, O_j, and O_k, and let the body mass centers be G_i, G_j, and G_k. Thus, ξ_j locates O_j relative to O_i, and ξ_k locates O_k relative to O_j; and $\mathbf{r}_i$, $\mathbf{r}_j$, and $\mathbf{r}_k$ locate G_i, G_j, and G_k relative to O_i, O_j, and O_k, respectively.

Observe that with these definitions, ξ_i is fixed in B_i and ξ_k is fixed in B_j and $\mathbf{r}_i$, $\mathbf{r}_j$, and $\mathbf{r}_k$ are fixed in B_i, B_j, and B_k, respectively. Note further that $i = L(j)$ and $j = L(k)$.

Consider the generic multibody system of Figure 11.2. Figure 11.11 illustrates the relations of the foregoing paragraph. Observe, however, that ξ_1, which locates the origin O_1 of B_1 (which may be

TABLE 11.9
Derivatives of Partial Angular Velocity Components ω_klm of Table 11.9 for the Human Body Model of Figure 11.1

Y_1

	1	4	7	10	13	16	19	22	25	28	31	34	37	40	43	46	49	52
	2	5	8	11	14	17	20	23	26	29	32	35	38	41	44	47	50	53
Body	3	6	9	12	15	18	21	24	27	30	33	36	39	42	45	48	51	54
1	0	0	0	0	0	0	0	0	0	0	0	0	0	0	0	0	0	0
2	0	0	SȮ1	0	0	0	0	0	0	0	0	0	0	0	0	0	0	0
3	0	0	SȮ1	SȮ2	0	0	0	0	0	0	0	0	0	0	0	0	0	0
4	0	0	SȮ1	SȮ2	SȮ3	0	0	0	0	0	0	0	0	0	0	0	0	0
5	0	0	SȮ1	SȮ2	SȮ3	SȮ4	0	0	0	0	0	0	0	0	0	0	0	0
6	0	0	SȮ1	SȮ2	SȮ3	SȮ4	SȮ5	0	0	0	0	0	0	0	0	0	0	0
7	0	0	SȮ1	SȮ2	0	0	0	SȮ3	0	0	0	0	0	0	0	0	0	0
8	0	0	SȮ1	SȮ2	0	0	0	SȮ3	SȮ7	0	0	0	0	0	0	0	0	0
9	0	0	SȮ1	SȮ2	0	0	0	0	0	SȮ3	0	0	0	0	0	0	0	0
10	0	0	SȮ1	SȮ2	0	0	0	0	0	SȮ3	SȮ9	0	0	0	0	0	0	0
11	0	0	SȮ1	SȮ2	0	0	0	0	0	SȮ3	SȮ9	SȮ10	0	0	0	0	0	0
12	0	0	0	0	0	0	0	0	0	0	0	0	SȮ1	0	0	0	0	0
13	0	0	0	0	0	0	0	0	0	0	0	0	SȮ1	SȮ12	0	0	0	0
14	0	0	0	0	0	0	0	0	0	0	0	0	SȮ1	SȮ12	SȮ13	0	0	0
15	0	0	0	0	0	0	0	0	0	0	0	0	0	0	0	SȮ1	0	0
16	0	0	0	0	0	0	0	0	0	0	0	0	0	0	0	SȮ1	SȮ15	0
17	0	0	0	0	0	0	0	0	0	0	0	0	0	0	0	SȮ1	SȮ15	SȮ16

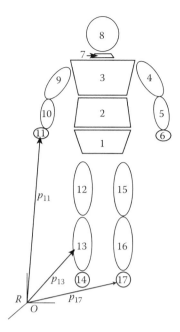

FIGURE 11.8 Sample mass center position vectors for the human body model.

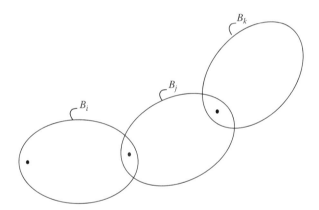

FIGURE 11.9 Three typical adjoining bodies.

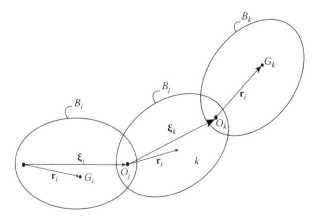

FIGURE 11.10 Body origins, mass centers, and locating position vectors.

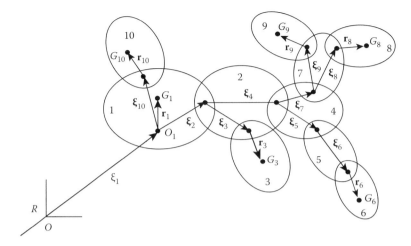

FIGURE 11.11 Selected position vectors, origins, and mass center of the generic multibody system of Figure 11.9.

chosen arbitrarily) relative to the origin O of R, is not fixed in either R or B_1. Instead, $\boldsymbol{\xi}_1$ defines the translation of B_1, and it may be expressed as

$$\boldsymbol{\xi}_1 = x\mathbf{n}_{o1} + y\mathbf{n}_{o2} + z\mathbf{n}_{o3} \tag{11.67}$$

From Figure 11.11, it is clear that the position vectors $\mathbf{p}_k$ locating the mass centers G_k relative to O for the 10 bodies of the generic system are

$$
\begin{aligned}
\mathbf{p}_1 &= \boldsymbol{\xi}_1 + \mathbf{r}_1 \\
\mathbf{p}_2 &= \boldsymbol{\xi}_1 + \boldsymbol{\xi}_2 + \mathbf{r}_2 \\
\mathbf{p}_3 &= \boldsymbol{\xi}_1 + \boldsymbol{\xi}_2 + \boldsymbol{\xi}_3 + \mathbf{r}_3 \\
\mathbf{p}_4 &= \boldsymbol{\xi}_1 + \boldsymbol{\xi}_2 + \boldsymbol{\xi}_3 + \mathbf{r}_4 \\
\mathbf{p}_5 &= \boldsymbol{\xi}_1 + \boldsymbol{\xi}_2 + \boldsymbol{\xi}_4 + \boldsymbol{\xi}_5 + \mathbf{r}_5 \\
\mathbf{p}_6 &= \boldsymbol{\xi}_1 + \boldsymbol{\xi}_2 + \boldsymbol{\xi}_4 + \boldsymbol{\xi}_5 + \boldsymbol{\xi}_6 + \mathbf{r}_6 \\
\mathbf{p}_7 &= \boldsymbol{\xi}_1 + \boldsymbol{\xi}_2 + \boldsymbol{\xi}_4 + \boldsymbol{\xi}_7 + \mathbf{r}_7 \\
\mathbf{p}_8 &= \boldsymbol{\xi}_1 + \boldsymbol{\xi}_2 + \boldsymbol{\xi}_4 + \boldsymbol{\xi}_7 + \boldsymbol{\xi}_8 + \mathbf{r}_8 \\
\mathbf{p}_9 &= \boldsymbol{\xi}_1 + \boldsymbol{\xi}_2 + \boldsymbol{\xi}_4 + \boldsymbol{\xi}_7 + \boldsymbol{\xi}_9 + \mathbf{r}_9 \\
\mathbf{p}_{10} &= \boldsymbol{\xi}_1 + \boldsymbol{\xi}_{10} + \mathbf{r}_{10}
\end{aligned}
\tag{11.68}
$$

These vectors may be expressed in terms of the $\mathbf{n}_{om}\,(m = 1, 2, 3)$ as

$$
\begin{aligned}
\mathbf{p}_1 &= [\boldsymbol{\xi}_1 + S01_{mn}r_{1m}]\mathbf{n}_{om} \\
\mathbf{p}_2 &= [\xi_{1m} + S01_{mn}\xi_{2n} + S02_{mn}r_{2n}]\mathbf{n}_{om} \\
\mathbf{p}_3 &= [\xi_{1m} + S01_{mn}\xi_{2n} + S02_{mn}\xi_{3n} + S03r_{3n}]\mathbf{n}_{om} \\
\mathbf{p}_4 &= [\xi_{1m} + S01_{mn}\xi_{2n} + S02_{mn}\xi_{4n} + S04r_{4n}]\mathbf{n}_{om} \\
\mathbf{p}_5 &= [\xi_{1m} + S01_{mn}\xi_{2n} + S02_{mn}\xi_{4n} + S04_{mn}\xi_{5n} + S05r_{5n}]\mathbf{n}_{om} \\
\mathbf{p}_6 &= [\xi_{1m} + S01_{mn}\xi_{2n} + S02_{mn}\xi_{4n} + S04_{mn}\xi_{5n} + S05_{mn}\xi_{6n} + S06r_{6n}]\mathbf{n}_{om} \\
\mathbf{p}_7 &= [\xi_{1m} + S01_{mn}\xi_{2n} + S02_{mn}\xi_{4n} + S04_{mn}\xi_{7n} + S07r_{7n}]\mathbf{n}_{om} \\
\mathbf{p}_8 &= [\xi_{1m} + S01_{mn}\xi_{2n} + S02_{mn}\xi_{4n} + S04_{mn}\xi_{7n} + S07_{mn}\xi_{8n} + S08r_{8n}]\mathbf{n}_{om} \\
\mathbf{p}_9 &= [\xi_{1m} + S01_{mn}\xi_{2n} + S02_{mn}\xi_{4n} + S04_{mn}\xi_{7n} + S07_{mn}\xi_{9n} + S09r_{9n}]\mathbf{n}_{om} \\
\mathbf{p}_{10} &= [\xi_{1m} + S01_{mn}\xi_{10n} + S010_{mn}r_{10n}]\mathbf{n}_{om}
\end{aligned}
\tag{11.69}
$$

By observing the pattern of the integers in Equation 11.69, we see that it is the same as the pattern of the integers in the columns of Table 11.3 for the higher-order lower-body array of the generic multibody systems. Thus, by using the lower-body arrays, we can express Equation 11.69 in the compact form:

$$\mathbf{p}_k = \left[\xi_{1m} + \sum_{s=0}^{r-1} SOP_{mn}\xi_{qn} + SOK_{mn}r_{kn} \right]\mathbf{n}_{0m} \quad (k = 1,\ldots,10) \tag{11.70}$$

where q, P, and r are given by

$$q = L^s(k), \quad p = L(q) = L^{s+1}(K), \quad L^r(K) = 1 \tag{11.71}$$

where $k = K$.

Consider the human body model of Figure 11.8 and of Figure 11.12 where selected position vectors are shown. Then analogous to Equations 11.68 and 11.69, we have the following mass center position vectors relative to the origin O of k:

$$\mathbf{p}_1 = \xi_1 + \mathbf{r}_1 = [\xi_1 + SO1_{mn}r_{1m}]\mathbf{n}_{om}$$

$$\mathbf{p}_2 = \xi_1 + \xi_2 + \mathbf{r}_2 = [\xi_{1m} + SO1_{mn}\xi_{2n} + SO2_{mn}r_{2n}]\mathbf{n}_{om}$$

$$\mathbf{p}_3 = \xi_1 + \xi_2 + \xi_3 + \mathbf{r}_3 = [\xi_{1m} + SO1_{mn}\xi_{2n} + SO2_{mn}\xi_{3n} + SO3r_{3n}]\mathbf{n}_{om}$$

$$\mathbf{p}_4 = \xi_1 + \xi_2 + \xi_3 + \xi_4 + \mathbf{r}_4 = [\xi_{1m} + SO1_{mn}\xi_{2n} + SO2_{mn}\xi_{4n} + SO4r_{4n}]\mathbf{n}_{om}$$

$$\mathbf{p}_5 = \xi_1 + \xi_2 + \xi_3 + \xi_4 + \xi_5 + \mathbf{r}_5 = [\xi_{1m} + SO1_{mn}\xi_{2n} + SO2_{mn}\xi_{3n} + SO3_{mn}\xi_{4n} + SO4_{mn}\xi_{5n} + SO5r_{5n}]\mathbf{n}_{om}$$

$$\mathbf{p}_6 = \xi_1 + \xi_2 + \xi_3 + \xi_4 + \xi_5 + \xi_6 + \mathbf{r}_6 = [\xi_{1m} + SO1_{mn}\xi_{2n} + SO2_{mn}\xi_{3n} + SO3_{mn}\xi_{4n} + SO4_{mn}\xi_{5n}$$
$$+ SO5_{mn}\xi_{6n} + SO6r_{6n}]\mathbf{n}_{om}$$

$$\mathbf{p}_7 = \xi_1 + \xi_2 + \xi_3 + \xi_7 + \mathbf{r}_7 = [\xi_{1m} + SO1_{mn}\xi_{2n} + SO2_{mn}\xi_{3n} + SO3_{mn}\xi_{4n} + SO7r_{7n}]\mathbf{n}_{om}$$

$$\mathbf{p}_8 = \xi_1 + \xi_2 + \xi_3 + \xi_7 + \xi_8 + \mathbf{r}_8 = [\xi_{1m} + SO1_{mn}\xi_{2n} + SO2_{mn}\xi_{3n} + SO3_{mn}\xi_{7n}$$
$$+ SO7_{mn}\xi_{8n} + SO8r_{8n}]\mathbf{n}_{om}$$

$$\mathbf{p}_9 = \xi_1 + \xi_2 + \xi_3 + \xi_9 + \mathbf{r}_9 = [\xi_{1m} + SO1_{mn}\xi_{2n} + SO2_{mn}\xi_{4n} + SO3_{mn}\xi_{9n} + SO9_{9n}]\mathbf{n}_{om}$$

$$\mathbf{p}_{10} = \xi_1 + \xi_2 + \xi_3 + \xi_9 + \xi_{10} + \mathbf{r}_{10} = [\xi_{1m} + SO1_{mn}\xi_{2n} + SO2_{mn}\xi_{3n} + SO3_{mn}\xi_{9n} + SO9_{mn}\xi_{10n}$$
$$+ SO10_{mn}r_{10n}]\mathbf{n}_{om}$$

$$\mathbf{p}_{11} = \xi_1 + \xi_2 + \xi_3 + \xi_9 + \xi_{10} + \xi_{11} + \mathbf{r}_{11} = [\xi_{1m} + SO1_{mn}\xi_{2n} + SO2_{mn}\xi_{3n} + SO3_{mn}\xi_{9n} + SO9_{mn}\xi_{10n}$$
$$+ SO10\xi_{11n} + SO11_{mn}r_{11n}]\mathbf{n}_{om}$$

$$\mathbf{p}_{12} = \xi_1 + \xi_{12} + \mathbf{r}_{12} = [\xi_{1m} + SO1_{mn}\xi_{12n} + SO12_{mn}r_{12n}]\mathbf{n}_{om}$$

$$\mathbf{p}_{13} = \xi_1 + \xi_{12} + \xi_{13} + \mathbf{r}_{13} = [\xi_{1m} + SO1_{mn}\xi_{12n} + SO12_{mn}\xi_{13n} + SO13_{mn}r_{13n}]\mathbf{n}_{om}$$

$$\mathbf{p}_{14} = \xi_1 + \xi_{12} + \xi_{13} + \xi_{14} + \mathbf{r}_{14} = [\xi_{1m} + SO1_{mn}\xi_{12n} + SO12_{mn}\xi_{13n} + SO13_{mn}\xi_{14n} + SO14_{mn}r_{14n}]\mathbf{n}_{om}$$

$$\mathbf{p}_{15} = \xi_1 + \xi_{15} + \mathbf{r}_{15} = [\xi_{1m} + SO1_{mn}\xi_{15n} + SO15_{mn}r_{15n}]\mathbf{n}_{om}$$

$$\mathbf{p}_{16} = \xi_1 + \xi_{15} + \xi_{16} + \mathbf{r}_{16} = [\xi_{1m} + SO1_{mn}\xi_{15n} + SO15_{mn}\xi_{16n} + SO16_{mn}r_{16n}]\mathbf{n}_{om}$$

$$\mathbf{p}_{17} = \xi_1 + \xi_{15} + \xi_{16} + \xi_{17} + \mathbf{r}_{17} = [\xi_{1m} + SO1_{mn}\xi_{15n} + SO15_{mn}\xi_{16n} + SO16_{mn}\xi_{17n} + SO17r_{17n}]\mathbf{n}_{om}$$

$$\tag{11.72}$$

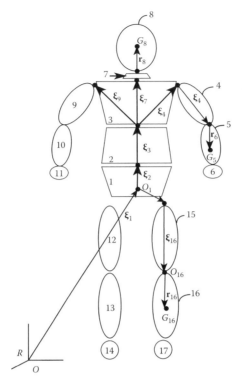

FIGURE 11.12   Selected position vectors, origins, and mass centers of the human body model.

Here again, Equations 11.70 and 11.71 provide compact expressions for this equation. That is,

$$\mathbf{p}_k = \left[\xi_{1m} + \sum_{s=0}^{r-1} SOP_{mn}\xi_{qn} + SOK_{mn}r_{kn} \right]\mathbf{n}_{om} \quad (k = 1, \dots, 17) \tag{11.73}$$

where q, P, and r are given by

$$q = L^s(k), \quad P = L(q) = L^{s+1}(K), \quad L^r(K) = 1 \tag{11.74}$$

where $k = K$.

11.10   MASS CENTER VELOCITIES

The mass center velocities $\mathbf{v}_k$ ($k = 1, \dots, 17$) of the human body model may now be obtained by differentiation of the mass center position vectors. Specifically, by differentiating in Equation 11.73, we obtain

$$\mathbf{v}_k = \frac{d\mathbf{p}_k}{dt} = \left[\dot{\xi}_{1m} + \sum_{s=0}^{r-1} S\dot{O}P_{mn}\xi_{qn} + S\dot{O}K_{mn}r_{kn} \right]\mathbf{n}_{om} \tag{11.75}$$

Where, as before,

$$q = L^s(k), \quad P = L(q) = L^{s+1}(K), \quad L^r(K) = 1 \tag{11.76}$$

where $k = K$. Recall that the ξ_{kn}, $k > 1$ are constants.

From Equations 11.62 through 11.66, we see that the transformation matrix derivatives SOK_{mn} may be expressed as

$$\dot{SOK}_{mn} = W_{ms}SOK_{sn} \tag{11.77}$$

where

$$W_{ms} = -e_{rms}\omega_{klr}y_l \tag{11.78}$$

so that

$$\dot{SOK}_{mn} = -e_{rms}\omega_{klr}y_l SOK_{sn} \tag{11.79}$$

To simplify the foregoing expressions, let a new set of parameters UK_{mln} be defined as

$$UK_{mln} \overset{D}{=} -e_{rms}\omega_{klr}SOK_{sn} \tag{11.80}$$

Then the $\dot{SOK}_{mn}$ take the simplified form:

$$\dot{SOK}_{mn} = UK_{mln}y_l \tag{11.81}$$

Thus, from Equation 11.75, the mass center velocities are

$$\mathbf{v}_k = \left[\dot{\xi}_{1m} + \sum_{s=0}^{r-1} UP_{mln}y_l\xi_{qn} + UK_{mln}y_l r_{kn}\right]\mathbf{n}_{om} \tag{11.82}$$

where from Equation 11.76, q, P, and r are given by

$$q = L^s(k), \quad P = L(q) = L^{s+1}(K), \quad L^r(K) = 1, \quad k = K \tag{11.83}$$

In a more compact form, the $\mathbf{v}_k$ may be expressed as

$$\mathbf{v}_k = v_{klm}y_l\mathbf{n}_{0m} \quad (k = 1, \ldots, 17) \tag{11.84}$$

where the v_{klm}, called partial velocity components, are

$$v_{klm} = \delta_{km} \quad (\text{for } l = 1, 2, 3; m = 1, 2, 3) \tag{11.85}$$

and

$$v_{klm} = \sum_{s=0}^{r-1} UP_{mln}\xi_{qm} + UK_{mln}r_{km} \quad (\text{for } l = 4, \ldots, 54; m = 1, 2, 3) \tag{11.86}$$

and as before

$$q = L^s(k), \quad p = L(q) = L^{s+1}(K), \quad L^r(K) = 1 \tag{11.87}$$

Observe the similarity of Equation 11.84 with Equation 11.50, for angular velocities. Observe further that the v_{klm} depend upon the ω_{klm} through the U_{mln} of Equation 11.80.

As noted previously, the v_{klm} are components of the partial velocity vectors defined as $\partial \mathbf{v}_k / \partial y_l$. As with the partial angular vectors, the partial velocity vectors are useful in determining generalized forces—as discussed in Chapter 12.

11.11   MASS CENTER ACCELERATIONS

The mass center acceleration a_k ($k = 1, \ldots, 17$) of the human body model may now be obtained by differentiation of the mass center velocities. Specifically, by differentiating in Equation 11.84, we have

$$a_k = (v_{klm} \dot{y}_l + \dot{v}_{klm} y_l) \mathbf{n}_{om} \tag{11.88}$$

where from Equations 11.85, 11.86, and 11.87, the v_{klm} are

$$v_{klm} = \delta_{km} \quad \text{(for } l = 1, 2, 3; m = 1, 2, 3) \tag{11.89}$$

$$v_{klm} = \sum_{s=0}^{r-1} UP_{mln} \xi_{qm} + UK_{mln} r_{km} \quad \text{(for } l = 4, \ldots, 54; m = 1, 2, 3) \tag{11.90}$$

where q, P, and r are given by

$$q = L^s(k), \quad p = L(q) = L^{s+1}(K), \quad L^r(K) = 1, \quad k = K \tag{11.91}$$

Equation 11.80 defines the UK_{mln} as

$$UK_{mln} \overset{D}{=} -e_{rms} \omega_{klr} SOK_{sm} \tag{11.92}$$

Then by differentiating in Equations 11.89 and 11.90, the $\dot{v}_{klm}$ are

$$\dot{v}_{klm} = 0 \quad \text{(for } l = 1, 2, 3; m = 1, 2, 3) \tag{11.93}$$

and

$$\dot{v}_{klm} = \sum_{s=0}^{r-1} \dot{U}P_{mln} \xi_{qn} + \dot{U}K_{mln} r_{kn} \quad \text{(for } l = 4, \ldots, 54; m = 1, 2, 3) \tag{11.94}$$

where from Equation 11.92, $\dot{U}K_{mln}$ are

$$\dot{U}K_{mln} \overset{D}{=} -e_{rms} \dot{\omega}_{klr} SOK_{sn} - e_{rms} \omega_{klr} \dot{SOK}_{sn} \tag{11.95}$$

Recall from Equations 11.79 and 11.81 that the $\dot{SOK}_{sm}$ are

$$\dot{SOK}_{sn} = -e_{rsi} \omega_{klr} y_e SOK_{sn} = UK_{sln} y_e \tag{11.96}$$

11.12   SUMMARY: HUMAN BODY MODEL KINEMATICS

Since this has been a rather lengthy chapter, it may be helpful to summarize the results, particularly since they will be useful for our analysis of human body kinetics and dynamics. Also the results are useful for algorithmic-based software development.

Figure 11.1 shows our 17-member human body model. By modeling the human body as a system of body segments as shown (i.e., as a multibody system), we can describe the entire system kinematics by determining the kinematics of the individual body segments. To this end, we model these segments as rigid bodies. Thus, for each of the bodies, we know the entire kinematics of the body where we know the velocity and acceleration of a point (say the mass center) of the body, and the angular velocity and angular acceleration of the body itself, all relative to the fixed or inertial frame R. That is, for each body B_k ($k = 11, \ldots, 17$), we seek four kinematic quantities:

1. The velocity $\mathbf{v}_k$ of the mass center G_k in R
2. The acceleration $\boldsymbol{\alpha}_k$ of the mass center G_k in R
3. The angular velocity $\boldsymbol{\omega}_k$ of B_k in R
4. The angular acceleration $\boldsymbol{\alpha}_k$ of B_k in R

Knowing these four quantities for each of the bodies enables us to determine the velocity and acceleration of every point of the human body model.

Of these four kinematic quantities, the angular velocity is the most fundamental in that it is used to determine the other three quantities. In this regard, it is convenient to use generalized speeds y_1 as the fundamental variables for the model where the first three of these are simply Cartesian coordinate derivatives of a reference point of B_1 in R, and the remaining y_r are 17 triplets of angular velocity components of the bodies relative to their adjacent lower-numbered bodies. Then by examining the body angular velocities in terms of unit vectors $\mathbf{n}_{om}$ fixed in R, the coefficients of the y_1 and $\mathbf{n}_{om}$ (i.e., the ω_{klm}) form the components of the partial angular velocities of the bodies, which in turn are elements of a block array, which are fundamental in determining the components of the other kinematical quantities.

To summarize the results, these four kinematic quantities for the bodies are

$$\text{Masscenter velocities: } \mathbf{v}_k = v_{klm}y_l\mathbf{n}_{om} \tag{11.97}$$

$$\text{Mass center acceleration: } \mathbf{a}_k = (v_{klm}\dot{y}_l + \dot{v}_{klm}y_l)\mathbf{n}_{om} \tag{11.98}$$

$$\text{Angular velocities: } \boldsymbol{\omega}_k = \omega_{klm}y_l\mathbf{n}_{om} \tag{11.99}$$

$$\text{Angular acceleration: } \boldsymbol{\alpha}_k = (\omega_{klm}\dot{y} + \dot{\omega}_{klm}y_l)\mathbf{n}_{om} \tag{11.100}$$

where $k = 1, \ldots, 17$ and v_{klm}, $\dot{v}_{klm}$, ω_{klm}, and $\dot{\omega}_{klm}$ are

$$v_{klm} = \delta_{km} \quad \text{(for } l = 1, 2, 3; m = 1, 2, 3) \tag{11.101}$$

$$v_{klm} = \sum_{s=0}^{r-1} UP_{mln}\xi_{qn} + UK_{mln}r_{kn} \quad \text{(for } l = 4, \ldots, 54; m = 1, 2, 3) \tag{11.102}$$

(from Equations 11.85 and 11.86) where UK_{mln} are

$$UK_{mln} = -e_{rms}\omega_{klr}SOK_{sn} \tag{11.103}$$

(from Equation 11.92) and

$$\dot{v}_{klm} = 0 \quad \text{(for } l = 1, 2, 3; m = 1, 2, 3)} \tag{11.104}$$

$$\dot{v}_{klm} = \sum_{s=0}^{r-1} \dot{U}P_{mln}\xi_{qn} + \dot{U}K_{mln}r_{kn} \quad \text{(for } l = 4, \ldots, 54; m = 1, 2, 3)} \tag{11.105}$$

(from Equations 11.93 and 11.94) where $\dot{U}K_{mln}$ are

$$\dot{U}K_{mln} = -e_{rms}\dot{\omega}_{klr}SOK_{sn} - e_{rms}\omega_{klr}\dot{SOK}_{sn} \tag{11.106}$$

(from Equation 11.95), where $\dot{SOK}_{sm}$ are

$$\dot{SOK}_{sn} = -e_{rsi}\omega_{klr}y_e SOK_{sn} = UK_{sln}y_e \tag{11.107}$$

(from Equation 11.96) and ω_{klm} and $\dot{\omega}_{klm}$ are listed in Tables 11.7 and 11.9.

In the summations of Equations 11.102 and 11.105, q, P, and r are given by

$$q = L^s(k), \quad p = L(q) = L^{s+1}(K), \quad L^r(K) = 1, \quad k = K \tag{11.108}$$

(from Equation 11.76). Also recall that the ξ_{kn} and r_{kn} are position vector components, referred to body-fixed unit vectors, locating higher-numbered body origins and mass centers (see Figure 11.12).

Finally, observe the central role played by the ω_{klm} in these kinematic expressions. Observe further from Table 11.7 that most of the ω_{klm} are zero and that the nonzero ω_{klm} are transformation matrix elements.

PROBLEMS

INTRODUCTION

The concepts of this chapter are easy to understand by simply working through examples similar to the illustrative examples of the chapter itself. To this end, consider the 10-body system of the following figure. Let the system be numbered/labeled as in the figures for Problems P11.1.1 and P11.1.2.

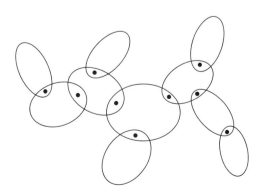

Section 11.1

P11.1.1 Consider the multibody system of the following figure. Develop a table analogous to Table 11.1 for the system.

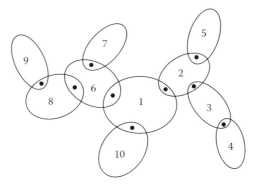

P11.1.2 Repeat Problem P11.1.1 for the multibody system of the following figure.

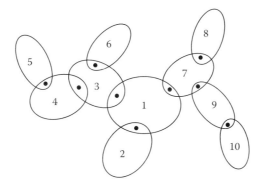

Section 11.2

P11.2.1 Consider again the multibody system of the figure for Problem P11.1.1. Develop a table of higher-order lower-body arrays analogous to Table 11.3.

P11.2.2 Repeat Problem P11.2.1 for the multibody system of the figure for Problem P11.1.2.

P11.2.3 Develop angular velocity expressions for the system of the figure for Problem P11.1.1 analogous to Equation 11.9.

P11.2.4 Repeat Problem P11.2.3 for the multibody system of the figure for Problem P11.1.2.

Section 11.3

P11.3.1 Develop equations analogous to those of Equation 11.14 for the systems of the figures for Problems P11.1.1 and P11.1.2. Compare the forms of expressions.

Section 11.4

P11.4.1 Verify the terms, the indices, and the signs in Equations 11.18 through 11.21.

Section 11.5

P11.5.1 Develop equations analogous to those of Equation 11.37 for the systems of the figures for Problems P11.1.1 and P11.1.2.

Section 11.6

P11.6.1   Verify the terms, the indices, and the signs in Equations 11.41 through 11.43.

Section 11.7

P11.7.1   Consider the multibody system of the following figure. Develop a table of partial angular velocity components where the generalized speeds are relative angular velocity components as in Table 11.6.

Section 11.8

P11.8.1   See Problem P11.7.1. Develop a table analogous to Table 11.8 for the derivatives of the partial angular velocity components of the table developed in Problem P11.7.1.

Section 11.9

P11.9.1   Continuing the analysis of the system of the figure for Problem P11.7.1, construct expressions analogous to those of Equation 11.68 for the mass center position vectors of the 10 bodies.

P11.9.2   See Problem P11.9.1. Express the mass center position vectors in terms of unit vectors $\mathbf{n}_{om}$ ($m = 1, 2, 3$) fixed in an inertia frame R, as in Equation 11.6.9.

Section 11.10

P11.10.1   Verify the terms and the indices of Equations 11.82 through 11.87.

Section 11.11

P11.11.1   Verify the terms and the indices of Equations 11.94 through 11.96.

REFERENCES

1. R. L. Huston and C. Q. Liu, *Formulas for Dynamic Analysis*, Marcel Dekker, New York, 2001, pp. 202–215.
2. H. Josephs and R. L. Huston, *Dynamics of Mechanical Systems*, CRC Press, Boca Raton, FL, 2002.
3. R. L. Huston, *Multibody Dynamics*, Butterworth Heinemann, Boston, MA, 1990.

12 Kinetics of Human Body Models

Consider again the 17-member human body model of Chapters 6, 8, and 11 and as shown again in Figure 12.1. Suppose this model is intended to represent a person in a force field. For example, we are all subject to gravity forces—at least on the Earth. In addition, a person may be engaged in a sport such as swimming where the water will exert forces on the limbs and torso. Alternatively, a person may be a motor vehicle occupant, wearing a seatbelt, in an accident where high accelerations are occurring. Or still another example, a person may be carrying a bag or parcel while descending a stair and grasping a railing.

In each of these cases, the person will experience gravity forces, contact forces, and inertia forces. In this chapter, we will consider procedures for efficiently accounting for these various force systems. We will do this primarily through the use of equivalent force systems (see Chapter 4) and generalized forces (developed in this chapter). We will develop expressions for use with Kane's dynamical equations and then for the development of numerical algorithms.

12.1 APPLIED (ACTIVE) AND INERTIA (PASSIVE) FORCES

We often think of a force as a push or a pull. This is a good description of contact forces on a particle or body. People on Earth also experience gravity (or weight) forces. Contact forces, weight forces, as well as other externally applied forces are simply called applied or active forces.

As a contrast, in Chapter 9, we discussed how forces may also arise due to motion—the so-called inertia or passive forces. These forces occur when a body is accelerated in an inertial or fixed frame R. Specifically, for a particle P with mass m having an acceleration $\mathbf{a}$ in R, the inertia force on P is (see Equation 9.3)

$$\mathbf{F}^* = -m\mathbf{a} \tag{12.1}$$

For a body B modeled as a set of particles, we found (see Sections 9.3 and 10.4) that the inertia forces on B are equivalent to a single force $\mathbf{F}^*$ passing through the mass center G of B together with a couple with torque $\mathbf{T}^*$ where $\mathbf{F}^*$ and $\mathbf{T}^*$ are (see Equations 9.24 and 10.45) [1,2]

$$\mathbf{F}^* = -M\mathbf{a}^G \quad \text{and} \quad \mathbf{T}^* = -\mathbf{I} \cdot \boldsymbol{\alpha}^B - \boldsymbol{\omega}^B \times (\mathbf{I} \cdot \boldsymbol{\omega}^B) \tag{12.2}$$

where
  M is the total mass of B
  $\mathbf{I}$ is the central inertia dyadic of B (see Section 10.6)
  $\mathbf{a}^G$ is the acceleration of G in R
  $\boldsymbol{\alpha}^B$ is the angular acceleration of G in R
  $\boldsymbol{\omega}^B$ is the angular velocity of B in R

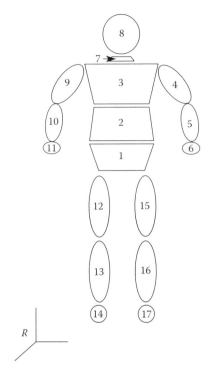

FIGURE 12.1 Human body model.

In this manner, it is convenient to express the applied forces on a body in terms of an equivalent force system consisting of a single force **F** passing through G together with a couple with torque **T**, where **F** and **T** are

$$\mathbf{F} = \sum_{i=1}^{N} \mathbf{F}_i \quad \text{and} \quad \mathbf{T} = \sum_{i=1}^{N} \mathbf{p}_i \times \mathbf{F}_i \tag{12.3}$$

where $\mathbf{F}_i$ is the force at a point P_i relative to G.

12.2 GENERALIZED FORCES

For large multibody systems, such as our human body model, it is useful not only to organize the geometry and the kinematic descriptions (as in Chapter 11), but also to organize and efficiently account for the kinetics—that is, the various forces exerted on the bodies of the system. We can do this by using the concept of generalized forces.

Generalized forces provide the desired efficiency by automatically eliminating the so-called nonworking forces such as interactive forces at joints, which do not ultimately contribute to the governing dynamical equations. This elimination is accomplished by projecting the forces along the partial velocity vectors (see Section 11.10). Partial velocity vectors, which are velocity vector coefficients of the generalized speeds, may be interpreted as base vectors in the n-dimensional space corresponding to the degrees of freedom of the system. The projection of forces along these base vectors may be interpreted as a generalized work.

The procedure for computing the generalized forces is remarkably simple: Consider again a multibody system S as in Figure 12.2.

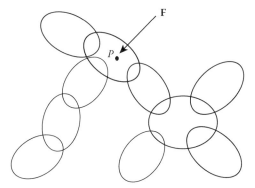

FIGURE 12.2   A force exerted at a point of a multibody system.

Let P be a point of a body of the system and let $\mathbf{F}$ be a force applied at P_1. Let S have n degrees of freedom represented by generalized speeds y_s ($s = 1, \ldots, n$). Let $\mathbf{v}$ be the velocity of P and let $\partial \mathbf{v}/\partial y_s$ be the partial velocity of P for y_s. That is,

$$\mathbf{v}_{y_s} = \frac{\partial \mathbf{v}}{\partial y_s} \tag{12.4}$$

Then, the contribution $F_{y_s}^P$ to the generalized active force F_{y_s} due to $\mathbf{F}$ is simply

$$F_{y_s}^P = \mathbf{F} \cdot \mathbf{v}_{y_s} \quad s = 1, \ldots, n \tag{12.5}$$

Observe that $\mathbf{F}$ will potentially contribute to each of the n generalized active forces. Observe further, however, that if $\mathbf{F}$ is perpendicular to $\mathbf{v}_{y_s}$, or if $\mathbf{v}_{y_s}$ is zero, then $F_{y_s}^P$ will be zero. Observe further that depending upon the units of the partial velocity vector, the units of $F_{y_s}^P$ are not necessarily the same as those of force $\mathbf{F}$.

Next, consider a typical body B_k of the multibody system S where there are several forces $\mathbf{F}_j$ ($j = 1, \ldots, N$) exerted at points P_j of B_k as in Figure 12.3. Then the contribution $F_{y_s}^{B_k}$ to the generalized active force for y_s from the $\mathbf{F}_{y_s}^P$ is simply the sum of the contribution from the individual forces. That is,

$$F_{y_s}^{B_k} = \sum_{j=1}^{N} F_{y_s}^{P_j} = \sum_{j=1}^{N} \mathbf{v}_{y_s}^{P_j} \cdot \mathbf{F}_j \tag{12.6}$$

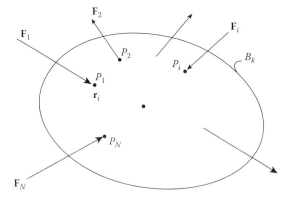

FIGURE 12.3   A typical body subjected to several forces.

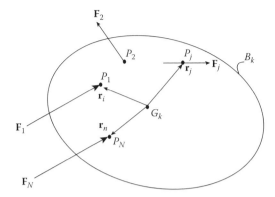

FIGURE 12.4 Forces exerted on typical body B_k.

If a body B_k has many forces exerted on it, it is convenient to represent these forces by a single force $\mathbf{F}_k$ passing through the mass center G_k of B_k together with a couple having a torque $\mathbf{T}_k$.

To see this, consider a body B_k with forces $\mathbf{F}_j$ acting through points P_j ($j = 1, ..., N$) of B_k as in Figure 12.4. Let G_k be the mass center of B_k and let $\mathbf{r}_j$ locate P_j relative to G_k.

Recall that since P_j and G_k are both fixed in B_k, their velocities in the inertial frame R are related by the expression (see Equation 8.126)

$$^R\mathbf{v}^{P_j} = {}^R\mathbf{v}^G + \boldsymbol{\omega}_k \times \mathbf{r}_j \tag{12.7}$$

where, as before, $\boldsymbol{\omega}_k$ is the angular velocity of B_k in R. Then by differentiating with respect to y_s, we have

$$\mathbf{v}_{y_s}^{P_i} = \mathbf{v}_{y_s}^{G_k} + \left(\frac{\partial\boldsymbol{\omega}_k}{\partial y_s}\right) \times \mathbf{r}_j \tag{12.8}$$

By substituting into Equation 12.6, we see that the contribution to the generalized forces for y_s by the set of forces $\mathbf{F}_j$ ($j = 1, ..., N$) on B_k is

$$F_{y_s}^{B_k} = \sum_{j=1}^{N} \mathbf{v}_{y_s}^{P_j} \cdot \mathbf{F}_j = \sum_{j=1}^{N} \mathbf{v}_{y_s}^{G_k} \cdot \mathbf{F}_j = \sum_{j=1}^{N}\left[\left(\frac{\partial\boldsymbol{\omega}_k}{\partial y_s}\right) \times \mathbf{r}_j\right] \cdot \mathbf{F}_j = \mathbf{v}_{y_s}^{G_k} \cdot \left(\sum_{j=1}^{N}\mathbf{F}_j\right) + \left(\frac{\partial\boldsymbol{\omega}_k}{\partial y_s}\right) \cdot \left[\sum_{j=1}^{N}\mathbf{r}_j \times \mathbf{F}_j\right] \tag{12.9}$$

where the last term is obtained by recalling that in triple scalar products of vectors, the vector and scalar operation ($\times$) and ($\cdot$) may be interchangeable. Finally, $\sum_{j=1}^{N}\mathbf{F}_j$ is $\mathbf{F}_k$ and $\sum_{j=1}^{N}\mathbf{r}_j \times \mathbf{F}_k$ is $\mathbf{T}_k$; $F_{y_s}^{B_k}$ takes the simple form

$$F_{y_s}^{B_k} = \mathbf{N}_{y_s}^{B_k} \cdot \mathbf{F}_k + \boldsymbol{\omega}_{y_s}^{B_k} \cdot \mathbf{T}_k \tag{12.10}$$

where $\boldsymbol{\omega}_{y_s}^{B_k}$ is $\partial\boldsymbol{\omega}_k/\partial y_s$.

12.3 GENERALIZED APPLIED (ACTIVE) FORCES ON A HUMAN BODY MODEL

Consider again the human body model as shown in Figure 12.1. Let the model be placed in a gravity field so that the bodies of the model experience weight forces. Let there also be contact forces on the bodies such as that could occur in daily activities, in the workplace, in sports, or in traumatic accidents. On each body B_k of the model, let these forces be represented by a single force $\mathbf{F}_k$ passing through the mass center G_k together with a couple with torque $\mathbf{M}_k$.

Let $F_{y_s}^{(k)}$ be the contribution to the generalized force F_{y_s} due to the applied forces on B_k. Then $F_{y_s}^{(k)}$ is

$$F_{y_s}^{(k)} = \mathbf{F}_k \cdot \mathbf{v}_{y_s}^{G_k} + \mathbf{M}_k \cdot \boldsymbol{\omega}_{y_s}^{B_k} \tag{12.11}$$

Let the vectors of Equation 12.11 be expressed in terms of unit vectors $\mathbf{n}_{om}$ ($m = 1, 2, 3$) fixed in R as

$$\mathbf{F}_k = F_{km}\mathbf{n}_{om}, \quad \mathbf{M}_k = M_{km}\mathbf{n}_{om}, \quad \mathbf{v}_{y_s}^{G_k} = v_{kom}\mathbf{n}_{om}, \quad \boldsymbol{\omega}_{y_s}^{B_k} = \omega_{ksm}\mathbf{n}_{om} \tag{12.12}$$

Then by resubstituting into Equation 12.11, $F_{y_s}^{(k)}$ becomes

$$F_{y_s}^{(k)} = F_{km}v_{ksm} + M_{km} \cdot \omega_{ksm} \quad \text{(no sum on } k\text{)} \tag{12.13}$$

12.4 FORCES EXERTED ACROSS ARTICULATING JOINTS

Next consider internally applied forces such as those transmitted across joints by contact and by ligaments and tendons, due to muscle activity.

12.4.1 Contact Forces across Joints

Consider the contact forces on an articulating joint such as a shoulder, an elbow, a wrist, or a similar leg joint. Healthy joint surfaces are quite smooth. Therefore, joint contact forces are approximately normal to the joint surface.

To model this, let the surfaces be represented as shown in Figure 12.5, where B_j and B_k are mating bodies or bones at the joint. Let P_j and P_k be points of contact between the surfaces, as shown.

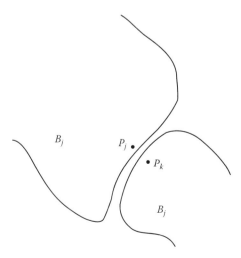

FIGURE 12.5 A schematic representation of an articulating joint.

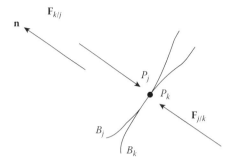

FIGURE 12.6 Forces exerted across contact points P_j and P_k.

Let $\mathbf{F}_{j/k}$ be the force exerted on B_j by B_k at P_j, and similarly, let $\mathbf{F}_{k/i}$ be the force exerted on B_j by B_k at P_k, as in Figure 12.6, where $\mathbf{n}$ is a unit vector normal to the surfaces at the contact points P_j and P_k.

Observe again that since the surfaces are nearly smooth, $\mathbf{F}_{j/k}$ and $\mathbf{F}_{k/i}$ are essentially parallel to $\mathbf{n}$.

Let $\mathbf{v}^{P_j}$ and $\mathbf{v}^{P_k}$ be the velocities of P_j and P_k in R, and as before, let $\mathbf{v}_{y_s}^{P_i}$ be the partial velocities of P_j and P_k for the generalized speeds y_s ($s = 1, \ldots, n$), with n being the number of degrees of freedom. Then the contributions $F_{y_s}^C$ to the generalized forces from $\mathbf{F}_{j/k}$ and $\mathbf{F}_{k/i}$ are

$$F_{y_s}^C = \mathbf{v}_{y_s}^{P_j} \cdot \mathbf{F}_{k/j} + \mathbf{v}_{y_s}^{P_k} \cdot \mathbf{F}_{j/k} \tag{12.14}$$

Since B_j and B_k are in contact at P_j and P_k without penetration, we have

$$\mathbf{v}^{P_i/P_k} \cdot \mathbf{n} = 0 \quad \text{or} \quad (\mathbf{v}^{P_i} - \mathbf{v}^{P_k}) \cdot \mathbf{n} = 0 \quad \text{or} \quad \mathbf{v}^{P_i} \cdot \mathbf{n} = \mathbf{v}^{P_k} \cdot \mathbf{n} \tag{12.15}$$

Consequently, we have

$$\mathbf{v}_{y_s}^{P_i} \cdot \mathbf{n} = \mathbf{v}_{y_s}^{P_k} \cdot \mathbf{n} \tag{12.16}$$

Recall from the law of action–reaction that the interactive contact forces are equal in magnitude but oppositely directed. That is,

$$\mathbf{F}_{j/k} = -\mathbf{F}_{k/j} = F\mathbf{n} \tag{12.17}$$

where F is the magnitude of the mutually interactive force. By substituting from Equations 12.16 and 12.17 into Equation 12.14, we find the generalized force contribution to be

$$F_{y_s}^C = \mathbf{v}_{y_s}^{P_j} \cdot \left(-F\mathbf{n}\right) + \mathbf{v}_{y_s}^{P_k} \cdot F\mathbf{n}$$

$$= \left(\mathbf{v}_{y_s}^{P_k} - \mathbf{v}_{y_s}^{P_j}\right) \cdot F\mathbf{n}$$

$$= F\left(\mathbf{v}_{y_s}^{P_k} \cdot \mathbf{n} - \mathbf{v}_{y_s}^{P_j} \cdot \mathbf{n}\right) = 0 \tag{12.18}$$

Equation 12.18 shows that interactive contact forces exerted across smooth surfaces do not make any contribution to the generalized forces. Therefore, these forces may be ignored in generalized force calculations.

12.4.2 LIGAMENT AND TENDON FORCES

Consider the contributions of ligament and tendon forces to the generalized active forces. Recall that ligaments connect bones to bones, whereas tendons connect muscles to bones. To determine the force contribution, consider again typical adjoining bodies B_j and B_k as in Figure 12.7, where P_j and P_k are attachment points for a ligament connecting B_j and B_k.

Let l be the natural length of the ligament. If forces are exerted on the bones tending to separate them, the ligament will be stretched to a length, say, $l + x$ where x is a measure of the extension. When the ligament is stretched, it will be in tension, tending to keep the bones together.

To quantify the forces exerted by the ligament on the bones and their contribution to the generalized forces, let $\mathbf{F}_{j/k}$ be the force exerted on B_j at P_j as represented in Figure 12.8, where $\mathbf{n}$ is a unit vector parallel to the ligament.

By considering the equilibrium of the ligament itself, we immediately see that the forces at P_j and P_k are of equal magnitude and opposite direction. That is,

$$\mathbf{F}_{j/k} = -\mathbf{F}_{k/j} = F\mathbf{n} \tag{12.19}$$

where F is the force magnitude. Since ligaments are both elastic and viscoelastic, F depends upon both the extensions x and the extensions rate $\dot{x}$. That is,

$$F = F(x, \dot{x}) \tag{12.20}$$

As before, let $\mathbf{v}_{y_s}^{P_j}$ and $\mathbf{v}_{y_s}^{P_k}$ be the partial velocities of P_j and P_k for the generalized speed y_s. Then the contribution $\hat{F}_{y_s}$ to the generalized speed F_{y_s} due to $\mathbf{F}_{j/k}$ and $\mathbf{F}_{k/j}$ is

$$\hat{F}_{y_s} = \mathbf{F}_{j/k} \cdot \mathbf{v}_{y_s}^{P_j} + \mathbf{F}_{k/j} \cdot \mathbf{v}_{y_s}^{P_k} = F\mathbf{n} \cdot \left(\mathbf{v}_{y_s}^{P_j} - \mathbf{v}_{y_s}^{P_k} \right)$$

$$= F(x, \dot{x})\mathbf{n} \cdot \left(\mathbf{v}_{y_s}^{P_j} - \mathbf{v}_{y_s}^{P_k} \right) \tag{12.21}$$

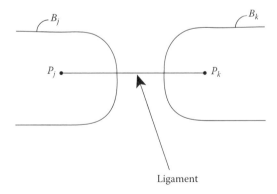

FIGURE 12.7 Schematic of a ligament connecting adjoining bones: B_j and B_k.

FIGURE 12.8 Ligament forces between B_j and B_k.

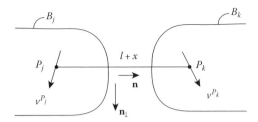

FIGURE 12.9 Schematic of ligament connecting bones: B_j and B_k.

Consider the schematic of the ligament connecting the bones as in Figure 12.7 and as shown again in Figure 12.9 where we have shown the velocities of P_j and P_k and the distance $l + x$. Then by inspection of the figure, the velocities are related as

$$v^{P_k} = v^{P_j} + \dot{x}\mathbf{n} + ()\mathbf{n}_\perp \tag{12.22}$$

where
   $\mathbf{n}_1$ represents a unit vector perpendicular to $\mathbf{n}$ (not necessarily in the place of the figure)
   () is an unknown scalar quantity

Then the partial velocities are related as

$$\mathbf{v}_{y_s}^{P_k} = \mathbf{v}_{y_s}^{P_j} + \left(\frac{\partial \dot{x}}{\partial y_s}\right)\mathbf{n} + \left[\frac{\partial ()}{\partial y_s}\right]\mathbf{n}_\perp \tag{12.23}$$

Hence, the term $\mathbf{n}\cdot\left(\mathbf{v}_{y_s}^{P_j} - \mathbf{v}_{y_s}^{P_k}\right)$ of Equation 12.21 is

$$\mathbf{n}\cdot\left(\mathbf{v}_{y_s}^{P_j} - \mathbf{v}_{y_s}^{P_k}\right) = -\frac{\partial \dot{x}}{\partial y_s} \tag{12.24}$$

Finally, the contribution to the generalized force is

$$\hat{F}_{y_s} = -F(x, \dot{x})\frac{\partial \dot{x}}{\partial y_s} \tag{12.25}$$

To obtain insight into the meaning of this last expression, suppose that $F(x, \dot{x})$ is simply k with k being a constant as with a linear spring. Further, suppose that the generalized speed y_s is itself $\dot{x}$. Then $\hat{F}_{y_s}$ becomes

$$\hat{F}_{y_s} = -kx \tag{12.26}$$

Next, regarding tendons connecting muscles to bones, the force in the tendon is due to the muscle shortening. Thus, tendon forces are due to an activation of the muscle, whereas ligament forces are passive, occurring as bones tend to separate. Tendon/muscle forces across skeletal joints lead to skeletal movement and limb articulation. In effect, these forces create moments at the joints. In the next section, we examine the contribution of these joint moments to the generalized forces.

12.4.3 Joint Articulation Moments

Consider two adjoining bones at an articulation joint of an extremity (say, a hip, knee, ankle, shoulder, elbow, or wrist joint), as represented by B_j and B_k in Figure 12.10 where O_k is the joint center.

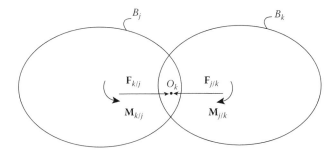

FIGURE 12.10 Forces/moments at an articulation joint due to muscle/tendon activation.

Let the forces exerted on B_k by a muscle/tendon system between B_j and B_k be represented on B_k by a single force $\mathbf{F}_{k/j}$ passing through O_k together with a couple with torque $\mathbf{M}_{k/j}$. Similarly, let the forces exerted on B_j by B_k by the muscle/tendon system be represented on B_j by a single force $\mathbf{F}_{j/k}$ passing through O_k together with a couple with torque $\mathbf{M}_{j/k}$. Then by the action–reaction principle, we have

$$\mathbf{F}_{k/j} = -\mathbf{F}_{j/k} \quad \text{and} \quad \mathbf{M}_{j/k} = -\mathbf{M}_{k/j} \tag{12.27}$$

Let $\boldsymbol{\omega}_j$ and $\boldsymbol{\omega}_k$ be the angular velocities of B_j and B_k in the inertia frame R_j and let $\boldsymbol{\omega}_k$ be the angular velocity of B_k relative to B_j. That is,

$$\boldsymbol{\omega}_k = \boldsymbol{\omega}_j + \hat{\boldsymbol{\omega}}_k \tag{12.28}$$

Then the partial angular velocities for generalized speed y_s are

$$\frac{\partial \boldsymbol{\omega}_k}{\partial y_s} = \frac{\partial \boldsymbol{\omega}_j}{\partial y_s} + \frac{\partial \hat{\boldsymbol{\omega}}_k}{\partial y_s} \tag{12.29}$$

The contributions of $\mathbf{F}_{j/k}$ and $\mathbf{F}_{k/j}$ to the generalized forces are the same as those discussed in the previous section. Specifically, if there is no translational separation between the bodies at O_k (and the joint is smooth surfaced), then the contribution of $\mathbf{F}_{j/k}$ and $\mathbf{F}_{k/j}$ to the generalized forces is zero (see Section 12.4.1). If, however, there is separation between the bodies at O_k, then the tendon behaves as a ligament with the generalized force contribution being the same as that in Section 12.4.2.

The contribution $\hat{F}_{y_s}$ of $\mathbf{M}_{j/k}$ and $\mathbf{M}_{k/j}$ to the generalized forces is

$$\hat{F}_{y_s} = \mathbf{M}_{j/k} \cdot \frac{\partial \boldsymbol{\omega}_j}{\partial y_s} + \mathbf{M}_{k/j} \cdot \frac{\partial \boldsymbol{\omega}_k}{\partial y_s}$$

$$= -\mathbf{M}_{k/j} \cdot \frac{\partial \boldsymbol{\omega}_j}{\partial y_s} + \mathbf{M}_{k/j} \cdot \frac{\partial \boldsymbol{\omega}_k}{\partial y_s}$$

$$= \mathbf{M}_{k/j} \cdot \left[\frac{\partial \boldsymbol{\omega}_k}{\partial y_s} - \frac{\partial \boldsymbol{\omega}_j}{\partial y_s} \right]$$

$$= \mathbf{M}_{k/j} \cdot \frac{\partial \hat{\boldsymbol{\omega}}_k}{\partial y_s} \tag{12.30}$$

For simplicity, let $\mathbf{M}_{k/j}$ be expressed in terms of unit vectors fixed in B_j as

$$\mathbf{M}_{k/j} = M_1 \mathbf{n}_{j1} + M_2 \mathbf{n}_{j2} + M_3 \mathbf{n}_{j3} \tag{12.31}$$

Also, let $\hat{\boldsymbol{\omega}}_k$ be expressed as

$$\hat{\boldsymbol{\omega}}_k = \hat{\omega}_{k1} \mathbf{n}_{j1} + \hat{\omega}_{k2} \mathbf{n}_{j2} + \hat{\omega}_{k3} \mathbf{n}_{j3} \tag{12.32}$$

Further, observe that the $\hat{\boldsymbol{\omega}}_{ki}$ are generalized speeds. That is,

$$\hat{\omega}_{k1} = y_{3k+1} \qquad \hat{\omega}_{k2} = y_{3k+2} \qquad \hat{\omega}_{k3} = y_{3k+3} \tag{12.33}$$

Then in Equation 12.30, the $\partial \hat{\boldsymbol{\omega}}_j / \partial y_s$ are simply $\mathbf{n}_{j1}$, $\mathbf{n}_{j2}$, and $\mathbf{n}_{j3}$.

   With these observations, Equations 12.30 and 12.31 show that the contribution to the generalized forces by the joint moments (due to the muscles) is

$$\hat{\mathbf{F}}_{y_s} = 0 \qquad s \neq 3k + i \quad (i = 1, 2, 3)$$

$$\hat{\mathbf{F}}_{y_s} = M_i \qquad s = 3k + i \quad (i = 1, 2, 3) \tag{12.34}$$

12.5   CONTRIBUTION OF GRAVITY (WEIGHT) FORCES TO THE GENERALIZED ACTIVE FORCES

Consider again a typical body, say, B_k of a human body model as in Figure 12.11. Let G_k be the mass center of B_k and let m_k be the mass of B_k.

   Then the weight force $\mathbf{w}_k$ on B_k may be expressed as

$$\mathbf{w}_k = -m_k g \mathbf{k} \tag{12.35}$$

where
   $\mathbf{k}$ is a vertical unit vector
   g is the gravity acceleration

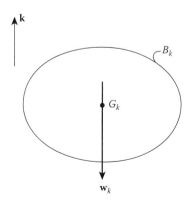

FIGURE 12.11   Weight force on a typical body of a human body model.

Let the velocity of G_k in the inertia frame R be expressed as

$$^R\mathbf{v}^{G_k} = \mathbf{v_k} = v_{ksm}y_s\mathbf{n}_{om} \tag{12.36}$$

where, as before, the partial velocity of G_k with respect to y_s is

$$\frac{\partial \mathbf{v}_k}{\partial y_s} = v_{ksm}\mathbf{n}_{om} \tag{12.37}$$

Using Equations 12.35 and 12.37, the contribution $\hat{F}_{y_s}^{(g)}$ to the generalized active force by the weight force is

$$\hat{F}_{y_s}^{(g)} = \left(\frac{\partial \mathbf{v}_k}{\partial y_s}\right) \cdot \mathbf{w}_k = v_{ksm}\mathbf{n}_{om} \cdot (-m_k g\mathbf{k})$$

$$= -v_{ksm}m_k g \tag{12.38}$$

where we have identified $\mathbf{k}$ with $\mathbf{n}_{03}$.

12.6   GENERALIZED INERTIA FORCES

Recall in Chapter 9 that by using d'Alembert's principle and the concept of inertia forces (forces due to motion), we found that for a body B moving in an inertia frame R, the inertia force on the particles of B may be represented by a single force $\mathbf{F}^*$ passing through the mass center G of B together with a couple with torque $\mathbf{T}^*$, as in Figure 12.12, where $\mathbf{F}^*$ and $\mathbf{T}^*$ are (see Equations 9.24, 9.27, and 10.45)

$$\mathbf{F}^* = -m\mathbf{a}^G \tag{12.39}$$

and

$$\mathbf{T}^* = -\mathbf{T} \cdot \boldsymbol{\alpha} - \boldsymbol{\omega} \times (\mathbf{I} \cdot \boldsymbol{\omega}) \tag{12.40}$$

where
   m is the mass of B
   $\mathbf{I}$ is the central inertia dyadic
   $\boldsymbol{\omega}$ and $\boldsymbol{\alpha}$ are the respective angular velocity and angular acceleration of B in R
   $\mathbf{a}^G$ is the acceleration of G in R

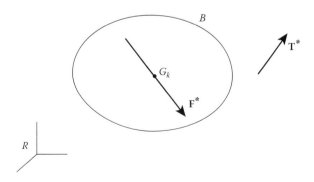

FIGURE 12.12   Equivalent inertia force system on a body B moving in an inertia frame R.

In the context of a multibody system and specifically a human body model, we can envision an equivalent inertia force system on each body of the model. Thus, for a typical body B_k of the model, the inertia force system is equivalent to a force $\mathbf{F}_k$ passing through the mass center G_k together with a couple with torque $\mathbf{T}_k$, where, considering Equations 12.39 and 12.40, $\mathbf{F}_k$ and $\mathbf{T}_k$ are

$$\mathbf{F}_k^* = -m_k \mathbf{a}_k \quad \text{(no sum on } k) \tag{12.41}$$

and

$$\mathbf{T}_k^* = -\mathbf{T}_k \cdot \boldsymbol{\alpha}_k - \boldsymbol{\omega}_k \times (\mathbf{I}_k \cdot \boldsymbol{\omega}_k) \quad \text{(no sum on } k) \tag{12.42}$$

where
  m_k is the mass of B_k
  $\mathbf{I}_k$ is the central inertia dyadic
  $\boldsymbol{\omega}_k$ and $\boldsymbol{\alpha}_k$ are the respective angular velocity and angular acceleration of B_k in the inertia frame R
  $\boldsymbol{\alpha}_k$ is the acceleration of the mass center G_k in R

With this notation, we can readily use the results of Chapter 11 to obtain compact expressions for $\mathbf{F}_k$ and $\mathbf{T}_k$. From Equations 11.98 through 11.100, we can express $\mathbf{a}_k$, $\boldsymbol{\omega}_k$, and $\boldsymbol{\alpha}_k$ as

$$\mathbf{a}_k = (v_{klm}\dot{y}_l + \dot{v}_{klm}y_l)\mathbf{n}_{om} \tag{12.43}$$

$$\boldsymbol{\omega}_k = \omega_{klm}y_l\mathbf{n}_{om} \tag{12.44}$$

$$\boldsymbol{\alpha}_k = (\omega_{klm}\dot{y} + \dot{\omega}_{klm}y_l)\mathbf{n}_{om} \tag{12.45}$$

Also, we can express the inertia dyadic $\mathbf{I}_k$ in terms of the moments and products of inertia as (see Equation 10.16)

$$\mathbf{I}_k = I_{klm}\mathbf{n}_{om}\mathbf{n}_{om} \tag{12.46}$$

By substituting from Equations 12.43 through 12.46, we can express $\mathbf{F}_k^*$ and $\mathbf{T}_k^*$ as

$$\mathbf{F}_k^* = F_{km}^*\mathbf{n}_{om} \quad \text{and} \quad \mathbf{T}_k^* = T_{km}^*\mathbf{n}_{om} \tag{12.47}$$

where F_{km}^* and T_{km}^* are

$$F_{km}^* = -m_k(v_{klm}\dot{y}_l + \dot{v}_{klm}y_l)\mathbf{n}_{om} \tag{12.48}$$

and

$$T_{km}^* = -\left[I_{kmn}(\omega_{kln}\dot{y}_l + \dot{\omega}_{kln}y_l) + e_{rsm}\omega_{klr}\omega_{kpm}I_{ksn}y_ly_k \right] \tag{12.49}$$

With these results, we can obtain the generalized inertia force $F_{y_s}^*$ for the generalized speeds y_s as (see Equations 12.10 and 12.13)

$$F_{y_s}^* = v_{ksm} F_{km}^* + \omega_{ksm} T_{km}^* \tag{12.50}$$

and thus by substituting from Equations 12.48 and 12.49, F_l^* becomes

$$F_{y_s}^* = -m_k v_{ksm}(v_{kpm}\dot{y}_p + \dot{v}_{kpm}y_p) - I_{kmn}\omega_{ksm}(\omega_{kpn}\dot{y}_p + \dot{\omega}_{kpn}y_p) - I_{ktn}\omega_{ksm}e_{rtn}\omega_{kqr}\omega_{kpn}y_q y_p \tag{12.51}$$

PROBLEMS

Section 12.1

P12.1.1 Consider a free-body diagram of a body B where the applied forces and the inertia forces on B are represented as in Equations 12.2 and 12.3. That is, both the applied and inertia force systems on B are represented by single forces $\mathbf{F}$ and $\mathbf{F}^*$ passing through the mass center G of B together with couples with torques $\mathbf{T}$ and $\mathbf{T}^*$, respectively.

Determine the vector equations of motion for B.

Section 12.2

P12.2.1 Verify Equation 12.9 by checking the validity of the terms and their indices.

Section 12.3

P12.3.1 Observe in Equation 12.13 the parenthetical comment: "no sum on k." Suppose that we want to obtain the generalized force F_{y_ℓ} for the entire system by deleting the "no sum" restrictions and by applying the summation convention for all repeated indices. How then should Equation 12.13 be written and how should the repeated indices be interpreted?

Section 12.4

P12.4.1 Review the analysis preceding and up to Equation 12.34. How would this analysis be different if orientation angle derivatives were used (instead of generalized speeds) for obtaining the contributions of the joint moments to the generalized forces?

Section 12.5

P12.5.1 Discuss the relative advantages and disadvantages of representing the weight forces on a human body model by (a) a single total weight force W passing through the mass center G of the entire mode or (b) individual weight forces W_i ($i = 1, ..., N$) passing through the N bodies of the model.

Section 12.6

P12.6.1 Verify the terms and the indices occurring in Equation 12.51.

REFERENCES

1. R. L. Huston and C. Q. Liu, *Formulas for Dynamic Analysis*, Marcel Dekker, New York, Chapter 14, 2001, pp. 202–215.
2. H. Josephs and R. L. Huston, *Dynamics of Mechanical Systems*, CRC Press, Boca Raton, FL, Chapter 18, 2002.

13 Dynamics of Human Body Models

By having explicit expressions for the generalized forces on a human body model, we can readily obtain expressions for the governing dynamical equations. This can be accomplished by using Kane's equations, which are ideally suited for obtaining governing equations for large systems.

If, in addition to the applied (active) and inertia (passive) forces, there are constraints and constraint forces imposed on the model, we can append constraint equations to the dynamical equations and constraint forces to the equations themselves.

In this chapter, we explore and document these concepts. We then go on to discuss solution procedures.

13.1 KANE'S EQUATIONS

Kane's equations were originally introduced by Kane in 1961 [1] to study nonholonomic systems. Their full potential and use, however, were not known until many years later. In recent years, they have become the principle of choice for studying multibody systems such as the human body model of Figure 15.1.

Kane's equations simply state that the sum of the generalized active (applied) and the passive (inertia) forces is zero for each generalized coordinate or generalized speed. In the notation described in Chapter 12, Kane's equations are

$$F_{y_s} + F_{y_s}^* = 0, \quad s = 1, \dots, n \tag{13.1}$$

where
  y_s are the generalized speeds
  n is the number of degrees of freedom

For large multibody systems, Kane's equations have distinct advantages over other dynamics principles such as Newton's laws or Lagrange's equations. Kane's equations provide the exact same number of equations as there are degrees of freedom, without the introduction and subsequent elimination of nonworking interval constraint forces (as in Newton's laws) and without the tedious and often unwieldy differentiation of energy functions (as with Lagrange's equations).

13.2 GENERALIZED FORCES FOR A HUMAN BODY MODEL

The human body model of Figure 13.1 has 17 bodies and 54 degrees of freedom. Table 11.2 lists the variables describing these degrees of freedom. Equation 11.15 lists the generalized coordinates, and Section 11.12 lists the generalized speeds for the model.

From Equation 12.13, the corresponding generalized (applied) forces have the form

$$F_{y_s} = F_{km} v_{ksm} + M_{km} \omega_{ksm} \tag{13.2}$$

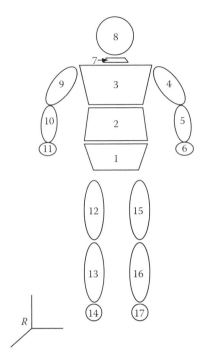

FIGURE 13.1 Human body model.

where in terms of unit vectors $\mathbf{n}_{om}$ ($m = 1, 2, 3$) fixed in the inertial frame R for a typical body B_k, F_{km} and M_{km} are the $\mathbf{n}_{om}$ components of a force $\mathbf{F}_k$ passing through the mass center G_k of B_k and a couple with torque $\mathbf{M}_k$, which, taken together, provide an equivalent representation of the force system applied to B_k; and as before, v_{klm} and ω_{klm} are the $\mathbf{n}_{om}$ components of the partial velocity of G_k and the partial angular velocity of B_k for the generalized speed y_s.

The form of the contribution to the F_{y_s} for joint forces, moments, and ligament/tendon compliance is given by Equations 12.18, 12.25, and 12.34, as discussed in Section 12.4. The contribution of the weight (gravity) forces is given by Equation 12.38, as discussed in Section 12.5.

From Equation 12.51, the corresponding generalized passive (inertia) forces are

$$F_{y_s}^* = -m_k v_{ksm}(v_{ksm}\dot{y}_p + \dot{v}_{ksm}y_p) - NI_{kmn}\omega_{ksm}(\omega_{kpm}\dot{y}_p + \dot{\omega}_{kpm}y_p)$$

$$- I_{ktm}\omega_{ksm}e_{rtm}\omega_{kqr}\omega_{kpn}y_q y_p \tag{13.3}$$

where
m_k is the mass of B_k
I_{ktm} are the $\mathbf{n}_{ot}$, $\mathbf{n}_{om}$ components of the central inertia dyadic I_k of B_k

13.3 DYNAMICAL EQUATIONS

With expressions for the generalized forces in Equations 13.2 and 13.3, we can immediately obtain governing dynamical equations by direct substitution into Kane's equations, as expressed in Equation 13.1. Specifically, they are

$$F_{km}v_{ksm} + M_{km}\omega_{ksm} - m_k v_{ksm}v_{kpm}\dot{y}_p - m_k v_{ksm}\dot{v}_{kpm}y_p$$

$$- I_{kmn}\omega_{ksm}\omega_{kpn}\dot{y}_p - I_{kmn}\omega_{ksm}\dot{\omega}_{kpm}y_p - I_{ktm}\omega_{ksm}e_{rtm}\omega_{kpm}y_q y_p = 0 \tag{13.4}$$

where

$$k = 1, \ldots, N; \quad s, p = 1, \ldots, n; \quad m, n, t, r = 1, 2, 3 \tag{13.5}$$

where

    N is the number of bodies (17 for the model of Figure 13.1)
    n is the number of degrees of freedom (54 for the model of Figure 13.1)

Equation 13.4 may be expressed in a more compact form as

$$a_{sp} = \dot{y}_p = f_s, \quad (s = 1, \ldots, n) \tag{13.6}$$

where the a_{sp} and f_s are

$$a_{sp} = m_k v_{ksm} v_{kpm} + I_{kmn} \omega_{ksm} \omega_{kpm} \tag{13.7}$$

and

$$f_s = F_{ys} - m_k v_{ksm} \dot{v}_{kpm} y_p - I_{kmn} \omega_{ksm} \dot{\omega}_{kpm} y_p - e_{rtm} I_{ktn} \omega_{ksm} \omega_{kqr} \omega_{kpn} y_q y_p \tag{13.8}$$

Observe in Equations 13.4 and 13.6 that y_s (the generalized speeds) are the dependent variables. Thus, Equation 13.6 forms a set of n first-order ordinary differential equations for the n_{ys}. Observe further, in Equations 13.4 and 13.6, that the equations are linear in the $\dot{y}_s$ but nonlinear in the y_s. Moreover, the nonlinear terms also involve orientation variables (Euler parameters), which are related to the generalized speeds through auxiliary first-order ordinary differential equations as developed in Chapter 8 (we discuss this further in the following section).

    Finally, observe that Equation 13.6 may be expressed in matrix form as

$$A\dot{y} = f \tag{13.9}$$

where
    A is a symmetric $n \times n$ array whose elements are the A_{sp}
    y and f are the $n \times 1$ column arrays whose elements are y_s and f_s

13.4 FORMULATION FOR NUMERICAL SOLUTIONS

In the governing equations, as expressed in Equation 13.9, the coefficient array A may be thought of as a generalized mass array. In addition to being symmetric, A is nonsingular. Therefore, its inverse A^{-1} exists, and thus the solution for the array of generalized speed derivatives $\dot{y}$ is given as

$$\dot{y} = A^{-1} f \tag{13.10}$$

Equation 13.10 is in a form that is ideally suited for numerical integration. The generalized speeds, however, are not geometric parameters, which describe the configuration of the system. Instead, except for the first three, they are relative angular velocity components, and as such, they do not immediately determine the orientations of the bodies. But, as noted in Chapter 8, we can use Euler parameters to define these orientations. Moreover, the relation between the Euler parameters and the relative angular velocity components (the generalized speeds) is also first-order ordinary differential equations and ideally suited for numerical integration.

To comprehend this in more detail, recall from Equations 8.62 to 8.213 that Euler parameter derivatives are related to angular velocity components by the expressions

$$\dot{\varepsilon}_1 = \left(\frac{1}{2}\right)(\varepsilon_4\Omega_1 + \varepsilon_3\Omega_2 - \varepsilon_2\Omega_3) \tag{13.11}$$

$$\dot{\varepsilon}_2 = \left(\frac{1}{2}\right)(-\varepsilon_3\Omega_1 + \varepsilon_4\Omega_2 + \varepsilon_1\Omega_3) \tag{13.12}$$

$$\dot{\varepsilon}_3 = \left(\frac{1}{2}\right)(\varepsilon_2\Omega_1 - \varepsilon_1\Omega_2 + \varepsilon_4\Omega_3) \tag{13.13}$$

$$\dot{\varepsilon}_4 = \left(\frac{1}{2}\right)(-\varepsilon_1\Omega_1 + \varepsilon_2\Omega_2 - \varepsilon_3\Omega_3) \tag{13.14}$$

In these equations, the Euler parameters define the orientation of a body in a reference frame R, and the Ω are the angular velocity components of B referred to as unit vectors fixed in R.

We can readily extend these equations so that they are applicable with multibody systems and, specifically, human body models. Recall that for convenience in formulation, we describe the orientations of the bodies relative to their adjacent lower-numbered bodies. Consider, for example, a typical pair of bodies B_j and B_k as in Figure 13.2 where $\mathbf{n}_{jm}$ and $\mathbf{n}_{km}$ are mutually perpendicular unit vectors fixed in B_j and B_k as shown.

Then, the orientation of B_k relative to B_j and the angular velocity of B_k relative to B_j may be defined as the orientation and angular velocity of the $\mathbf{n}_{km}$ relative to the $\mathbf{n}_{jm}$. Therefore, let ε_{k1}, ε_{k2}, ε_{k3}, and ε_{k4} be Euler parameters describing the orientation of the $\mathbf{n}_{km}$ relative to the $\mathbf{n}_{jm}$ and thus also of B_k relative to B_j. Recall from Chapter 8 that these parameters may be defined in terms of a rotation of B_k through an angle θ_k about line L_k fixed in both B_j and B_k (see Section 8.16). That is, the ε_{kn} ($n = 1, \ldots, 4$) may be defined as

$$\varepsilon_{k1} = \gamma_{k1}\sin\left(\frac{\theta_k}{2}\right) \tag{13.15}$$

$$\varepsilon_{k2} = \gamma_{k2}\sin\left(\frac{\theta_k}{2}\right) \tag{13.16}$$

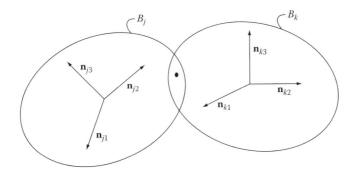

FIGURE 13.2   Two typical adjoining bodies.

$$\varepsilon_{k3} = \gamma_{k3} \sin\left(\frac{\theta_k}{2}\right) \tag{13.17}$$

$$\varepsilon_{k4} = \cos\left(\frac{\theta_k}{2}\right) \tag{13.18}$$

where y_{km} ($m = 1, ..., 4$) are components referring the $\mathbf{n}_{jm}$ of a unit vector $\mathbf{y}_k$ parallel to the rotation axis L_k.

As before, let $\hat{\boldsymbol{\omega}}_k$ be the angular velocity of B_k relative to B_j expressed in terms of the $\mathbf{n}_{jm}$ as

$$\hat{\boldsymbol{\omega}}_k = \hat{\omega}_{k1}\mathbf{n}_{j1} + \hat{\omega}_{k2}\mathbf{n}_{j2} + \hat{\omega}_{k3}\mathbf{n}_{j3} \tag{13.19}$$

Then, analogous to Equations 13.15 through 13.18, the ε_{kn} may be expressed in terms of the $\hat{\omega}_{km}$ as

$$\dot{\varepsilon}_{k1} = \left(\frac{1}{2}\right)(\varepsilon_{k4}\hat{\omega}_{k1} + \varepsilon_{k3}\hat{\omega}_{k2} - \varepsilon_{k2}\hat{\omega}_{k3}) \tag{13.20}$$

$$\dot{\varepsilon}_{k2} = \left(\frac{1}{2}\right)(-\varepsilon_{k3}\hat{\omega}_{k1} + \varepsilon_{k4}\hat{\omega}_{k2} + \varepsilon_{k1}\hat{\omega}_{k3}) \tag{13.21}$$

$$\dot{\varepsilon}_{k3} = \left(\frac{1}{2}\right)(\varepsilon_{k2}\hat{\omega}_{k1} + \varepsilon_{k1}\hat{\omega}_{k2} + \varepsilon_{k4}\hat{\omega}_{k3}) \tag{13.22}$$

$$\dot{\varepsilon}_{k4} = \left(\frac{1}{2}\right)(-\varepsilon_{k1}\hat{\omega}_{k1} - \varepsilon_{k2}\hat{\omega}_{k2} - \varepsilon_{k3}\hat{\omega}_{k3}) \tag{13.23}$$

As noted, the relative angular velocity components are identified with the generalized speeds. From Equation 11.39 and Table 11.5, we see that

$$\hat{\omega}_{km} = y_{3k+m}, \quad (k = 1, ..., N; m = 1, 2, 3) \tag{13.24}$$

and y_1, y_2, and y_3 are translation variable derivatives: If (x, y, z) are the Cartesian coordinates of a reference point (say, the origin O or, alternatively, the mass center G_1) of body B_1 relative to a Cartesian frame fixed in R, then y_1, y_2, and y_3 are

$$y_1 = \dot{x}, \quad y_2 = \dot{y}, \quad y_3 = \dot{z} \tag{13.25}$$

Observe again in Equation 13.10 that $\dot{y}$ is an $n \times 1$ array of generalized speeds where n is the number of degrees of freedom. For the 17-body model of Figure 13.1, there are 54 degrees of freedom ($n = 54$). Then, Equation 13.10 is equivalent to 54 scalar differential equations. Next, observe that Equations 13.20 to 13.23 provide four differentiated equations for each body and Equation 13.25 provides an additional three equations. Therefore, for the model, there are a total of $54 + (17 \times 4) + 3$ or 125 first-order ordinary differential equations. Correspondingly, there are 125 dependent variables, 54 generalized speeds, 68 Euler parameters, and 3 displacement variables.

13.5   CONSTRAINT EQUATIONS

When using a human body model to study actual physical activity, there will need to be constraints placed on the model to accurately simulate the activity. For example, suppose a person is simply standing with hands on hips as represented in Figure 13.3 at the beginning of and during some activity such as twisting, knee bending, or head turning.

In this instance, the movements of the feet are constrained, and the arms form closed loops restricting this movement. Consequently, the joints are not all freely moving. Analytically, such constraints may be expressed as

$$\phi_j(\varepsilon_{ki}) = 0, \quad j = 1, \ldots, m_i \tag{13.26}$$

where m_i is the number of constraints.

In addition to these position, or geometric, constraints, there may be motion, or kinematic, constraints. That is, the movement of a body segment and/or joint may be specified, such as when a person throws a ball. These constraints may be expressed in terms of the generalized speeds as

$$\psi_j(y_s) = g_j(t), \quad j = 1, \ldots, m_2 \tag{13.27}$$

where
  g_j are given functions of time
  m_2 is the number of constraints

Equations 13.26 and 13.27 represent constraints of different kinds and different forms. Interestingly, they may be cast into the same form by an observation and a simple analysis. Specifically, for the

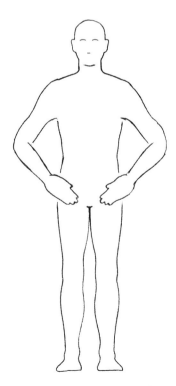

FIGURE 13.3   Standing human body model with hands on hips (line drawing needed).

kinematic constraints of Equation 13.27, it happens that the vast majority of these are specifications of velocities. This in turn means that the constraints may be written in the form

$$\psi_j(y_s) = \sum_{s=1}^{n} c_{js} y_s = g_j(t) \tag{13.28}$$

where
  coefficient c_{js} may be functions of the joint angles
  n is the number of generalized speeds

That is, the constraints are linear in the generalized speeds.

Next, recall that the Euler parameter derivative may be expressed in terms of the generalized speeds through Equations 13.20 through 13.24. Thus, by differentiating Equation 13.26, we have

$$\frac{d\phi_j}{dt} = \sum_{i=1}^{4} \frac{\partial \phi_j}{\partial \varepsilon_{ki}} \dot{\varepsilon}_{ki} = 0 \tag{13.29}$$

Then, by substituting for $\dot{\varepsilon}_{ki}$, we can express these equations in the form

$$\sum_{s=1}^{n} d_{js} y_s = 0 \tag{13.30}$$

where the coefficients d_{js} are determined by observation. (Recall that in Equations 13.20 to 13.24, $\dot{\varepsilon}_{ki}$ are linear functions of the generalized speeds.)

By combining Equations 13.28 and 13.30, we can express the totality of the constraints in the simple form:

$$b_{js} y_s = g_j, \quad j = 1, \ldots, m \tag{13.31}$$

where the b_{js} are obtained by superposing the c_{js} of Equation 13.28 with the d_{js} of Equation 13.30, and m is the sum: $m_1 + m_2$.

Equation 13.31 may be written in the compact matrix form:

$$By = g \tag{13.32}$$

where
  B is an $m \times n$ $(m < n)$ constraint array
  y is an $n \times 1$ column array of generalized speeds
  g is an $n \times 1$ column array of specified functions of time

13.6   CONSTRAINT FORCES

For constraints to be imposed upon the movement of a person or the model of a person, there need to be "controlling forces" or "constraint reactions" exerted on the person or model. Such forces (and moments) give rise to generalized constraint forces analogous to the generalized applied and inertia forces.

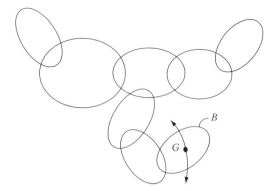

FIGURE 13.4   A multibody system with a body having a specified motion constraint.

To develop these concepts, consider a multibody system where a typical body B of the system has a specified motion and/or constraint as represented in Figure 13.4.

(A system with a body having a specified motion could model a person throwing a ball where B would be the throwing hand.)

To analytically describe the movement of B, let the velocity of the mass center G of B and the angular velocity of B in an inertial frame R be expressed as

$$\mathbf{v}^G = V_i(t)\mathbf{N}_i \quad \text{and} \quad \boldsymbol{\omega}^B = \Omega_i(t)\mathbf{N}_i \tag{13.33}$$

where
   $\mathbf{N}_i$ are mutually perpendicular unit vectors fixed in R
   the components $V_i(t)$ and $\Omega_i(t)$ are known (or specified) functions of time

Next, suppose that the unconstrained system has n degrees of freedom characterized by the generalized speeds y_s ($s = 1, \ldots, n$). Then, in terms of these y_s, $\mathbf{v}^G$ and ω^B may be expressed as (see Equations 11.50 and 11.84)

$$\mathbf{v}^G = V_{is}^G y_s \mathbf{N}_i \quad \text{and} \quad \boldsymbol{\omega}^B = \Omega_{is}^B y_s \mathbf{N}_i \tag{13.34}$$

Then, by comparing Equations 13.33 and 13.34, we see that

$$V_{is}^G y_s = V_i(t) \quad \text{and} \quad \Omega_{is}^B = \Omega_i(t) \tag{13.35}$$

These expressions are of the form of Equation 13.31, which in turn have the matrix form of Equation 13.32 as $By = g$. Indeed, by combining Equation 13.35 and matching the expressions with Equations 13.31 and 13.32, we see that the constraint matrix B is a $6 \times n$ array with elements:

$$B = 6\left[b_{ij}^n\right] = \begin{bmatrix} V_{11}^G & V_{12}^G & V_{13}^G & \cdots & V_{1n}^G \\ V_{21}^G & V_{22}^G & V_{23}^G & \cdots & V_{2n}^G \\ V_{31}^G & V_{32}^G & V_{33}^G & \cdots & V_{3n}^G \\ \Omega_{11}^B & \Omega_{12}^B & \Omega_{13}^B & \cdots & \Omega_{1n}^B \\ \Omega_{21}^B & \Omega_{22}^B & \Omega_{23}^B & \cdots & \Omega_{2n}^B \\ \Omega_{31}^B & \Omega_{32}^B & \Omega_{33}^B & \cdots & \Omega_{3n}^B \end{bmatrix} \tag{13.36}$$

Similarly, the array g is seen to be

$$g = \begin{bmatrix} V_1(t) \\ V_2(t) \\ V_3(t) \\ \Omega_1(t) \\ \Omega_2(t) \\ \Omega_3(t) \end{bmatrix} \tag{13.37}$$

Let the constraining forces needed to give B its specified motion be represented by an equivalent force system consisting of a single force $\mathbf{F}_G'$ passing through G together with a couple with torque $\mathbf{T}_G'$. Then, from Equation 12.10, the contribution F_{y_s}' to the generalized force on B, for the generalized speed y_s, is

$$F_{y_s}' = \left(\frac{\partial \mathbf{V}^G}{\partial_{y_s}} \right) \cdot \mathbf{F}_G' + \left(\frac{\partial \boldsymbol{\omega}^B}{\partial_{y_s}} \right) \cdot \mathbf{T}' \tag{13.38}$$

Let $\mathbf{F}_G'$ and $\mathbf{T}'$ be expressed in terms of the R-based unit vectors $\mathbf{N}_i$ as

$$\mathbf{F}_G' = F_{G_i}' \mathbf{N}_i \tag{13.39}$$

Then, by substituting into Equation 13.38 and using Equation 13.34, we have

$$F_{y_s}' = V_{is}^G F_{G_i} + \Omega_{is}^B T_i' \tag{13.40}$$

We can express Equation 13.40 as a matrix of a row and column arrays as

$$F_{y_s}' = \begin{bmatrix} V_{1s}^G V_{2s}^G V_{3s}^G \Omega_{1s}^B \Omega_{2s}^B \Omega_{3s}^B \end{bmatrix} \begin{bmatrix} F_{G1}' \\ F_{G2}' \\ F_{G3}' \\ T_1' \\ T_2' \\ T_3' \end{bmatrix} = B_S^T \lambda \tag{13.41}$$

where B_s^T and λ are row and column arrays defined by inspection. Observe that the B_s^T array is a column of the constraint matrix B of Equation 13.36. Therefore, if we form an array F^1 of generalized constraint forces as

$$F' = \begin{bmatrix} F_{y_1}' \\ F_{y_2}' \\ \vdots \\ F_{y_s}' \\ \vdots \\ F_{y_n}' \end{bmatrix} \tag{13.42}$$

then in view of Equations 13.35 and 13.41, F' may be written in the compact form:

$$F' = B^T \lambda \tag{13.43}$$

The column array λ defined by Equation 13.41 is sometimes called the "constraint force array." Observe that Equation 13.43 provides a relation between the constraint matrix of Section 13.5 and the constraint force array, thus providing a relation between the kinematic and kinetic representations of constraints.

13.7 CONSTRAINED SYSTEM DYNAMICS

We can develop the dynamical equations for constrained multibody systems, and specifically for constrained human body models, by initially regarding the system as being unconstrained. The constraints are then imposed upon the system via constraint forces, through the generalized constraint force array as defined in Equations 13.42 and 13.43. That is, we can divide the forces on the system into three categories: (1) externally applied ("active") forces, (2) inertia ("passive") forces, and (3) constraint forces. In this context, Kane's equations take the form

$$F_{y_s} + F^*_{y_s} + F'_{y_s}, \quad (s = 1, \ldots, n) \tag{13.44}$$

or in matrix form

$$F + F^* + F' = 0 \tag{13.45}$$

Geometrically, we can represent Equation 13.45 as a force triangle as in Figure 13.5, where the sides of the triangle are the force arrays.

By substituting from Equation 13.43, we can express Equation 13.45 as

$$F + F^* + B^T \lambda = 0 \tag{13.46}$$

From Equation 13.3, we may express the generalized inertia force array as

$$F^* = -A\dot{y} + f^* \tag{13.47}$$

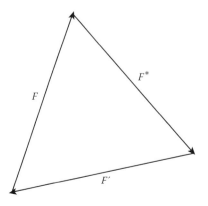

FIGURE 13.5 Generalized force triangle.

where the elements a_{sp} and f_s^* of the A and f^* arrays are given by Equations 13.7 and 13.8 as*

$$a_{sp} = m_k v_{ksm} v_{kpm} + I_{kmn} \omega_{ksm} \omega_{kpn} \tag{13.48}$$

and

$$f_s^* = -m_k v_{ksm} \dot{v}_{kpm} y_p - I_{kmn} \omega_{ksm} \dot{\omega}_{kpn} - e_{rtm} I_{ktm} \omega_{ksm} \omega_{kpr} \omega_{kpn} y_q y_p \tag{13.49}$$

Equation 13.46 may then be written as

$$A\dot{y} = F + B^T \lambda + f^* = f + B^T \lambda \tag{13.50}$$

The scalar form of Equation 13.50 is

$$a_{sp} \dot{y}_p = F_{y_s} + b_{js} \lambda_j + f_s^*, \quad (s, p = 1, \dots, n; \ j = 1, \dots, n) \tag{13.51}$$

Observe in Equation 13.51 that there are n equations for ny_s and $m\lambda_r$, that is, n equations for $n + m$ unknowns. We can obtain an additional m equation by appending the constraint equations (Equations 13.31):

$$B_{js} y_s = g_i, \quad (j = 1, \dots, m) \tag{13.52}$$

In many instances, we are primarily interested in the movement of the system and not particularly concerned with determining the elements λ_j of the constraint force array. In such cases, we can eliminate the λ_j from the analysis by premultiplying Equation 13.46 by the transpose of an orthogonal complement C of B. That is, let C be an $n \times (n - m)$ array such that

$$BC = 0 \quad \text{or} \quad C^T B^T = 0 \tag{13.53}$$

Then, by premultiplying Equation 13.46 by C^T, we have

$$C^T F + C^T F^* = 0 \tag{13.54}$$

Consequently, Equation 13.50 becomes

$$C^T A\dot{y} + C^T F + C^T f^* \tag{13.55}$$

Equation 13.55 is equivalent to $n - m$ scalar equations for the ny_s. Then, with constraint equations of Equation 13.52, or alternatively $By = g$, we have n equations for the ny_s.

We may view Equation 13.53 as the projection of Equation 13.45 onto the direction normal to the constraint matrix B. That is, in view of Figure 13.5, we have the geometric interpretation of Figure 13.6.

---

* Note that the difference between f_s^* and f_s of Equations 13.49 and 13.8 is F_{y_s}. That is, $f_s = F_{y_s} + f_s^*$.

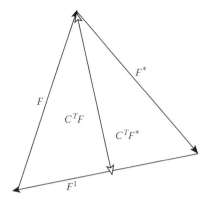

FIGURE 13.6   Projection of generalized forces along the direction of the orthogonal complement array.

13.8   DETERMINATION OF ORTHOGONAL COMPLEMENT ARRAYS

In view of the analysis of the foregoing section and particularly in view of Equation 13.53, the question arising is as follows: How can we find the orthogonal complement C of B? An answer is to use an ingenious zero eigenvalues theorem documented by Walton and Steeves [2].

Observe that if the rank of B is m, then the rank of $B^T B$ is an $n \times n$ symmetric array. Therefore, $B^T B$ will have $n - m$ zero eigenvalues and consequently $n - m$ eigenvectors associated with these zero eigenvalues. If n is an eigenvalue of $B^T B$ and if x is the associated eigenvector, we have

$$B^T B X = \mu X \tag{13.56}$$

Thus, if n is zero, we have

$$B^T B X = 0 \tag{13.57}$$

If we arrange the $n - m$ eigenvectors associated with the eigenvalues into an array C, we have

$$B^T B C = 0 \tag{13.58}$$

Finally, by multiplying by C^T, we have

$$C^T B^T B C = 0 \quad \text{or} \quad (BC)^T (BC) = 0 \quad \text{or} \quad BC = 0 \tag{13.58}$$

Therefore, C as the assemblage of eigenvectors of the zero eigenvalues is the desired orthogonal complement of B.

13.9   SUMMARY

From Equation 13.9, the governing dynamical equations for a human body model are expressed in matrix form as

$$A\dot{y} = f \tag{13.59}$$

In scalar form, these equations are (Equation 13.6)

$$a_{sp}\dot{y}_p = f_s, \quad (s, p = 1, \ldots, n) \tag{13.60}$$

where n is the number of degrees of freedom, and from Equations 13.7 and 13.8, a_{sp} and f_s are

$$a_{sp} = m_k v_{ksm} v_{kpm} + I_{kmn}\omega_{ksm}\omega_{kpn} \tag{13.61}$$

and

$$f_s = F_{y_s} - m_k v_{ksm}\dot{v}_{kpm}y_p - I_{kmn}\omega_{ksm}\dot{\omega}_{kpn}y_p - e_{rtm}I_{ktn}\omega_{ksm}\omega_{kqr}\omega_{kpn}y_q y_p \tag{13.62}$$

Except for movement in free space, however, virtually all real-life simulations of human motion will require constraints on the model. Most of these constraints can be represented analytically as (see Equation 13.32)

$$By = g \tag{13.63}$$

where
  B is an $m \times n(m < n)$ constraint array
  g is an $m \times 1$ array of time functions for constraints with specified motion

Constraints on the movement of the model give rise to constraint forces, which in turn produce generalized constraint forces. These generalized constraint forces may be assembled into an array F' given by (see Equation 13.43)

$$F' = B^T \lambda \tag{13.64}$$

where λ is an array of constraint force and moment components.

When constraints are imposed, the governing dynamical equations take the form (see Equation 13.50)

$$A\dot{y} = f + B^T y \tag{13.65}$$

Taken together, Equations 13.64 and 13.65 are equivalent to a set of $m + n$ scalar equations for the n kinematic variables (the generalized speeds) and the m constraint force and moment component in the array λ. This set of equations can be reduced in number by eliminating the constraint force array λ. This can be accomplished using an orthogonal complement array C of B, that is, an array C such that $BC = 0$ and $C^T B^T = 0$. Then, by multiplying Equation 13.65 by C^T, λ is eliminated, resulting in

$$C^T A y = C^T f \tag{13.66}$$

Since B is an $m \times n$ array, C is an $n \times (n - m)$ array, and, thus, C^T is an $(n - m) \times n$ array. Therefore, Equation 13.66 is equivalent to $(n - m)$ scalar equations. Then by appending Equation 13.63, which is equivalent to m scalar equations, we have a net of $(n - m) + m$ or n equations for the ny_s.

Given suitable auxiliary conditions (initial conditions), Equations 13.66 and 13.63 can be integrated (numerically) for the generalized speeds. With the use of Euler parameters and generalized speeds as dependent variables, the equations may all be cast into first-order form and thus are ideally suited for numerical integration.

By differentiating Equation 13.63, we have

$$B\dot{y} = \dot{g} - \dot{B}y \tag{13.67}$$

Then the combination of Equations 13.66 and 13.67 may be written in the matrix form:

$$\hat{A}\dot{y} = \hat{f} \tag{13.68}$$

where $\hat{A}$ is an $n \times m$ nonsingular array defined in partitioned form as

$$\hat{A} = \left[\frac{B}{C^T A} \right] \tag{13.69}$$

and $\hat{f}$ is an $n \times 1$ column array defined as

$$\hat{f} = \left[\frac{g - B'y}{C^T f} \right] \tag{13.70}$$

Chapter 14 provides an outline of algorithms for numerically developing and solving these equations.

PROBLEMS

Section 13.1

P13.1.1 Under what conditions might differentiation of an energy function, say, kinetic energy, be unwieldy?

P13.1.2 How do Kane's equations avoid the tedious differentiation of energy functions? Hint: Consider Equation 8.18 and the text comments that follow that expression.

Section 13.2

P13.2.1 What are the ranges for the sums of the repeated and free indices in Equations 13.2 and 13.3?

Section 13.3

P13.3.1 Verify the terms and their indices in Equations 13.6 through13.8.

P13.3.2 Discuss the pattern of the indices in Equation 13.6. Specifically, identify the similarities with kinetic energy.

Section 13.4

P13.4.1 In the common terminology used for classifying differential equations, describe the expressions of Equations 13.10 through 13.14.

P13.4.2 Consider a six-body planar model of the human body consisting of one body for the head/neck system, one body for the torso, two bodies for the legs, and two bodies for the arms, as represented in the following figure.
  a. How many degrees of freedom does this six-body model have?
  b. Prepare a list of variables for the degrees of freedom of the model.
  c. Using the notation of Equations 13.10 through 13.14, what are the forms of the governing differential equations of the model?

Section 13.5

P13.5.1 Describe the constraints on a model of a person who is
  a. Walking (one to two feet on the ground)
  b. Running (none to one foot on the ground)

P13.5.2 See Problem P13.5.1. Describe the constraints for
  a. An automobile operator
  b. An automobile passenger

Section 13.6

P13.6.1 Verify the terms and their indices in Equation 13.34.

P13.6.2 Verify the terms and their indices in Equations 13.40 through 13.43.

Section 13.7

P13.7.1 An optional problem: intended for those familiar with the principle of virtual work commonly used in structural problems: Compare and contrast the procedures of Section 13.7 with the procedures of virtual work.

P13.7.2 Discuss the relative advantages and disadvantages of using Equation 13.54 instead of Equation 13.44 for dynamic analyses of human body models.

Section 13.8

P13.8.1 An orthogonal complement array C for a matrix B might be viewed as a matrix that is "perpendicular" to B. In this regard, suppose a vector $\mathbf{b}$ is given by

$$\mathbf{b} = 8\mathbf{n}_1 - 2\mathbf{n}_2 + 6\mathbf{n}_3$$

where $\mathbf{n}_1$, $\mathbf{n}_2$, and $\mathbf{n}_3$ are mutually perpendicular unit vectors.

    a. Find a nonzero vector $\mathbf{c}$ such that $\mathbf{c} \cdot \mathbf{b} = 0$.

    b. Find a second nonzero vector $\hat{\mathbf{c}}$ such that $\hat{\mathbf{c}} \cdot \mathbf{b} = 0$ and thereby show that a vector perpendicular to $\mathbf{b}$ is not unique.

    c. See the results of (b). Find a third nonzero vector $\hat{\hat{\mathbf{c}}}$ such that $\hat{\hat{\mathbf{c}}} \cdot \mathbf{b} = 0$ and $\hat{\hat{\mathbf{c}}} \cdot \hat{\mathbf{c}} = 0$.

P13.8.2 Repeat Problem P13.8.1 for the generic vector:

$$\mathbf{b} = b_1 \mathbf{n}_1 + b_2 \mathbf{n}_2 + b_3 \mathbf{n}_3$$

P13.8.3 See Problems P13.8.1 and P13.8.2. By extending the concepts from vectors to matrices, show that orthogonal complement arrays are not unique.

REFERENCES

1. T. R. Kane, Dynamics of nonholonomic systems, *Journal of Applied Mechanics*, 28, 1961, 574–578.
2. W. C. Walton, Jr. and E. C. Steeves, A new matrix theory and its application for establishing independent coordinates for complex dynamical systems with constraints, NASA Technical Report TR-326, 1969.

14 Numerical Methods

Two tasks must be performed to obtain numerical simulations of human body motion: First, the governing equation must be developed, and second, the equation must be solved.

The form of the governing equations depends to the same extent upon the particular application of interest. Since applications vary considerably, the preferred approach is to develop general dynamics equations and then apply constraints appropriate to the application.

The numerical solution of the equations is facilitated by expressing the governing differential equations in the first-order form as developed in the previous chapter. Then, with suitable initial conditions, a numerical integration (solution) may be used to obtain a time history of the movement.

In this chapter, we consider an outline (or sketch) of procedures for both the numerical development and the solution of the governing equations. Details of the procedures depend upon the software employed. In 1983, Kamman and coworkers [1] wrote a Fortran program for this purpose. We will follow Kamman's procedure for the numerical development of the equations.

14.1 GOVERNING EQUATIONS

To develop a numerical methodology of human body dynamics, which is applicable to a general class of problems, it is useful to start with a many-bodied human body model whose limbs are constrained. The governing dynamical equations are then easily developed, as discussed in the foregoing chapters, culminating in Chapter 13. Next, constraint equations for a large class of problems of interest need to be developed and appended to the dynamical equations, as in Chapter 13. The resulting reduced equations may then be solved numerically.

In matrix form, the dynamical equations may be written as (see Equation 13.9)

$$A\dot{y} = f \tag{14.1}$$

where, if the unconstrained model has n degrees of freedom, y is an $n \times 1$ column array of generalized speeds corresponding to these degrees of freedom; A is an $n \times n$ symmetric "generalized mass" array; and f is an $n \times 1$ column array of generalized forces and inertia force terms. The elements of A and f are given by Equations 13.7 and 13.8 as

$$a_{sp} = m_k v_{ksm} v_{kpm} + I_{kmn} \omega_{ksm} \omega_{kpn} \tag{14.2}$$

and

$$f_s = F_{ys} - m_k v_{ksm} \dot{v}_{kpm} y_p - I_{kmn} \omega_{ksm} \dot{\omega}_{kpm} y_p - e_{rtm} I_{ktn} \omega_{ksm} \omega_{kqr} \omega_{kpn} y_q y_p \tag{14.3}$$

where the notation is as described in the previous chapters.

Now, if there are constraints applied to the model, whether they are geometric or kinematic, they may often be written in the matrix form

$$By = g \tag{14.4}$$

349

where B is an $m \times n$ array of constraint equation coefficients and g is an $m \times 1$ column array of given applied force components as discussed in Section 13.5 and where m is the number of imposed constraints ($m < n$).

When movement constraints are imposed on the model, these constraints will cause forces and moments to be applied to the model. Specifically, if there are m movement constraints, there will be m force and/or moment components applied to the model. If these forces and moment components are assembled into an $m \times 1$ column array λ, then the dynamics equations take the form

$$A = f + B^T \lambda \tag{14.5}$$

As discussed in Sections 13.7 and 13.8, we can reduce the number of equations needed to study constrained systems by eliminating the constraint force/moment array λ. We can do this by multiplying Equation 14.5 by the transpose of an orthogonal complement C of the constraint matrix B. That is, with $BC = 0$, we have $C^T A y \doteq 0$, and then by multiplying in Equation 14.4 by C^T, we have

$$C^T A y = \dot{C}^T f \tag{14.6}$$

Next, by differentiating Equation 14.4, we obtain

$$B\dot{y} = g - \dot{B}y \tag{14.7}$$

Finally, by combining Equations 14.5 and 14.6, we can express the governing equations as

$$\hat{A}\dot{y} = \hat{f} \tag{14.8}$$

where $\hat{A}$ and $\hat{f}$ are the arrays

$$\hat{A} = \left[\frac{B}{C^T A}\right] \quad \text{and} \quad \hat{f} = \left[\frac{\dot{g} - By}{C^T f}\right] \tag{14.9}$$

where
  $\hat{A}$ is an $n \times n$ square array
  $\hat{f}$ is an $n \times 1$ column array

14.2 NUMERICAL DEVELOPMENT OF THE GOVERNING EQUATIONS

Observe in Equations 14.2 and 14.3 the dominant role played by the partial velocity and the partial angular velocity array elements (v_{klm} and ω_{klm}) and their derivatives (v_{klm} and ω_{klm}). Recall from Chapter 11 that the values of the ω_{klm} may be obtained from the elements of the transformation matrices SOK (see Sections 11.7 and 11.12). The $\dot{\omega}_{klm}$ are thus identified with the elements of the transformation matrix derivatives. Recall further, from Chapter 11, that the v_{klm} and the $\dot{v}_{klm}$ may be evaluated in terms of the ω_{klm} and the $\dot{\omega}_{klm}$ together with geometric parameters locating the mass centers and connecting joints (see Sections 11.9 through 11.11). Also, recall that the generalized speeds y_l of Equations 14.1 and 14.4 through 14.9 are components of the relative angular velocities $\hat{\omega}_k$, except for the first three, which are displacement variables (see Section 11.6). Therefore, if we know the transformation matrix elements SOK_{mn} and their derivatives, $\dot{SOK}_{mn}$, we can determine the ω_{klm} and the $\dot{\omega}_{klm}$ and then in turn, the v_{klm} and the $\dot{v}_{klm}$ and the y_l. That is, aside from the three translation variables, the various terms of the governing equations are directly dependent upon the transformation matrix elements and their derivatives.

Recall also in Chapter 8 that the transformation matrix elements may be expressed in terms of the Euler parameters (see Equations 8.176 and 8.210 through 8.213) and that the Euler parameter derivatives are linearly related to the relative angular velocity components—the generalized speeds themselves. Specifically, these equations are (see Equation 8.176)

$$
SOK = \begin{bmatrix}
\left(\varepsilon_{k1}^2 - \varepsilon_{k2}^2 - \varepsilon_{k3}^2 + \varepsilon_{k4}^2\right) & 2(\varepsilon_{k1}\varepsilon_{\varepsilon k2} + \varepsilon_{k3}\varepsilon_{k4}) & 2(\varepsilon_{k1}\varepsilon_{k2} + \varepsilon_{k2}\varepsilon_{k4}) \\
2(\varepsilon_{k1}\varepsilon_{k2} + \varepsilon_{k3}\varepsilon_{k4}) & \left(-\varepsilon_{k1}^2 - \varepsilon_{k2}^2 - \varepsilon_{k3}^2 + \varepsilon_{k4}^2\right) & 2(\varepsilon_{k2}\varepsilon_{k3} + \varepsilon_{k1}\varepsilon_{k4}) \\
2(\varepsilon_{k1}\varepsilon_{k3} + \varepsilon_{k2}\varepsilon_{k4}) & 2(\varepsilon_{k2}\varepsilon_{k3} + \varepsilon_{k1}\varepsilon_{k4}) & \left(-\varepsilon_{k1}^2 - \varepsilon_{k2}^2 - \varepsilon_{k3}^2 + \varepsilon_{k4}^2\right)
\end{bmatrix} \quad (14.10)
$$

and Equations 8.210 through 8.213

$$
\dot{\varepsilon}_{k1} = \left(\frac{1}{2}\right)(\varepsilon_{k4}\hat{\omega}_{k1} + \varepsilon_{k3}\hat{\omega}_{k2} - \varepsilon_{k2}\hat{\omega}_{k3})
$$

$$
\dot{\varepsilon}_{k1} = \left(\frac{1}{2}\right)(-\varepsilon_{k3}\hat{\omega}_{k1} + \varepsilon_{k4}\hat{\omega}_{k2} - \varepsilon_{k1}\hat{\omega}_{k3})
$$

$$
\dot{\varepsilon}_{k1} = \left(\frac{1}{2}\right)(\varepsilon_{k2}\hat{\omega}_{k1} + \varepsilon_{k1}\hat{\omega}_{k2} - \varepsilon_{k4}\hat{\omega}_{k3})
$$

$$
\dot{\varepsilon}_{k1} = \left(\frac{1}{2}\right)(\varepsilon_{k1}\hat{\omega}_{k1} + \varepsilon_{k2}\hat{\omega}_{k2} - \varepsilon_{k2}\hat{\omega}_{k3})
$$

$$(14.11)$$

These equations show that by integrating Equations 14.11, we can obtain the Euler parameters. Then, by Equation 14.10, we obtain the transformation matrix elements of each integrated time step. This in turn gives us the partial velocity and partial angular velocity arrays and their derivatives (v_{klm}, $\dot{v}_{klm}$, ω_{klm}, and $\dot{\omega}_{klm}$). Using these data, we obtain the dynamical equations (Equations 14.1) by using Equations 14.2 and 14.3. Constraint equations (Equation 14.7) may then be developed for applications of interest.

The following section outlines the details of this procedure.

14.3 OUTLINE OF NUMERICAL PROCEDURES

From the observations of the foregoing section, we can outline an algorithmic procedure for numerically generating and solving the governing equations. Tables 14.1 through 14.3 provide "flowcharts" of such algorithms for input, computation, and output of data, respectively.

TABLE 14.1
Input Data for an Algorithmic Human Body Model

1. Number of bodies in the model
2. Masses of the bodies
3. Inertia matrices of the bodies
4. Mass center location (coordinates) of the bodies
5. Connection joint locations (coordinates) of the bodies
6. Constraint descriptions
7. Specified motion descriptions
8. Initial values of the dependent variables
9. Integration parameters (time steps, time duration, accuracy)
10. Specification of the amount and style of output data derived

---

TABLE 14.2
Computation Steps

1. Identify known (specified) and unknown (dependent) variables and place them in separate arrays
2. Knowing the specified variables, the specified motions, and the initial values of the dependent variables, establish arrays of all variables
3. Calculate initial values of the transformation matrices
4. Calculate initial values of the ω_{klm}, $\dot{\omega}_{klm}$, v_{klm}, and $\dot{v}_{klm}$ arrays
5. Calculate initial values of the a_{ij} and f_i arrays (see Equations 13.7 and 13.8)
6. Form governing differential equations
7. Isolate the differential equations associated with the unknown variables
8. Reduce and assemble the equations into a first-order system
9. Integrate the equations to the first time increment
10. Repeat steps 4 through 9 for subsequent time increment

TABLE 14.3
Output Data

1. Documentation of input data (see Table 14.1)
2. For selected integration time steps, list values of the following:
   a. Specified variables
   b. Dependent variables
   c. Derivatives of dependent variables
   d. Second derivatives of dependent variables
   e. Joint force components
   f. Joint moment components
3. Graphical representation of variables of interest
4. Animation of model movement

14.4 ALGORITHM ACCURACY AND EFFICIENCY

Questions arising with virtually every extensive numerical procedure and simulation are as follows: How accurate is the simulation? How efficient is the computation? The answer to these questions depends upon both modeling and the algorithm of the numerical procedure.

Generally, the more comprehensive and detailed the modeling, the more accurate the simulation. However, detailed modeling in body regions not pertinent or significant in the simulation will only increase computation time, and this in turn may possibly adversely affect accuracy by introducing increased numerical error. Thus, accuracy is also dependent upon efficiency to an extent.

By using a multibody systems approach, the modeling may be made as comprehensive as it appears to be appropriate. For gross-motion simulations, we have opted to use the 17-body model of Figure 13.1. For more focused interest upon a particular region of the body, such as the head/neck system or the arm/hand system, we may want to use models such as those of Figures 6.10 and 6.11.

The computational efficiency is dependent upon two items: (1) the efficiency of the simulation algorithm and (2) the efficiency of the numerical integration. The algorithm efficiency of the analyses of the foregoing chapter stems from the use of the differentiation algorithms facilitated by the angular velocities (see Equations 8.18 and 8.123) and from the use of lower-body arrays (see Section 6.2).

Recall that if a vector $\mathbf{c}$ is fixed in a body B, which in turn is moving in a reference frame R, the derivative of $\mathbf{c}$ relative to an observer in R is simply (see Equation 8.18)

$$\frac{{}^R\mathrm{d}\mathbf{c}}{\mathrm{d}t} = \boldsymbol{\omega} \times \mathbf{c} \tag{14.12}$$

where $\boldsymbol{\omega}$ is the angular velocity of B in R.

By applying Equation 14.12 with transformation matrix derivatives, we obtain the result (see Equation 8.123)

$$\frac{\mathrm{d}S}{\mathrm{d}t} = WS \tag{14.13}$$

where

S is the transformation matrix between unit vectors fixed in B and R
W is the angular velocity matrix defined as (see Equations 8.122 and 8.123)

$$W = \begin{bmatrix} 0 & -\Omega_3 & \Omega_2 \\ \Omega_3 & 0 & -\Omega_1 \\ -\Omega_2 & \Omega_1 & 0 \end{bmatrix} \tag{14.14}$$

where the Ω_i ($i = 1, 2, 3$) are the components of the angular velocity ω of B in R relative to unit vectors fixed in R.

Observe in Equations 14.12 and 14.13 that the derivations are calculated by a multiplication—an efficient and accurate procedure for numerical (computer) analysis.

By using lower-body arrays, we can efficiently compute the system kinematics and in the process obtain the all-important partial velocity and partial angular velocity arrays and their derivatives. This leads immediately to the governing dynamical equations.

The development of the governing equations is enabled by using Kane's equations as the dynamics principle. Kane's equations involve the concepts of generalized forces—both applied (or active) and inertial (or passive) forces. Indeed, Kane's equations simply state that the sum of the generalized applied and inertia forces is zero for each degree of freedom (represented by generalized coordinates).

The generalized forces are obtained using the partial velocity and partial angular velocity vectors and their associated array elements (v_{klm} and ω_{klm}) (see Equations 11.50 and 11.84). The use of partial velocity and partial angular velocity vectors produces the automatic elimination of nonworking internal constraint forces (noncontributing "action–reaction" forces).

The use of Kane's equations allows us to numerically obtain the governing differential equations without the tedious and inefficient numerical differentiation of energy functions, as is needed with Lagrangian dynamics principles.

In summary, for the large multibody system human body models, Kane's equations, combined with the use of the angular velocity differentiation algorithms and lower-body arrays, provide an extremely efficient numerical development of the governing equations. For large systems, Kane's equations have the advantages of both Newton–Euler methods and Lagrange's equations without the corresponding disadvantages.

Regarding the numerical solution of the developed governing differential equations, it appears that the most stable solution procedures (integrators) are those based upon power series expansions. Of these, fourth-order Runge–Kutta methods have been found to be effective.

The governing differential equations are nonlinear. This means that the initial conditions can have dramatic effect upon the subsequent motion, depending upon the simulation. In the following chapter, we review a few simulations obtained using the methods described herein.

PROBLEMS

Section 14.1

P14.1.1 Verify that Equations 14.1 through 14.9 are consistent with those developed in Chapter 13. Specifically, review the analysis leading up to the equations, checking the terms and the indices.

Section 14.2

P14.2.1 Verify that the procedure outlined in the text is logical and complete.

Section 14.3

P14.3.1 Verify the logic and completeness of the steps of Tables 14.1 through 14.3.

Section 14.4

P14.4.1 Discuss the relative advantages and disadvantages of using Euler parameters for improving accuracy and efficiency in the numerical solution of the governing dynamical equations.

REFERENCE

1. R. L. Huston, T. P. King, and J. W. Kamman, UCIN-DYNOCOMBS-software for the dynamic analysis of constrained multibody systems, in *Multibody Systems Handbook*, W. Schielen (Ed.), Springer-Verlag, New York, 1990, pp. 103–111.

15 Simulations and Applications

Our goal of human body modeling is the accurate simulation of actual human motion events. Movements of interest range from routine daily activities (walking, sitting, standing, lifting, machine operation) to optimal performance (sport activities, playing musical instruments) to accident victim kinematics (falling, motor vehicle collisions). The objectives of the simulations are to obtain accurate, quantitative analyses as well as to identify the effects of important parameters.

In this chapter, we look at some results obtained using simulation software developed in Chapters 10 through 14. We begin with a brief review. We then look at movements of astronauts in free space. Next we consider simple lifting. We follow this with an analysis of walking. We then look at some simple swimming motions. Two sections are then devoted to crash-victim simulation. For a workplace application, we consider a waitperson carrying a tray. We conclude with a presentation of a series of applications, which are yet to be developed.

15.1 REVIEW OF HUMAN MODELING FOR DYNAMIC SIMULATION

Figure 15.1 shows the basic model discussed in Chapters 10 through 14. Although this model is a gross-motion simulator, it is nevertheless useful for studying a wide variety of human motion varying from routine activity in daily life to optimal motion, as in sport activity, to unintended motion as in accidents.

The model with 17 spherical-joint connected bodies has as many as $(17 \times 3) + 3$, or 54, degrees of freedom. If movements corresponding to these degrees of freedom are represented by generalized speeds, say, y_l ($l = 1, \ldots, 54$), the governing dynamical equations of the model may be written as (see Equation 13.6)

$$a_{sp}\dot{y}_p = f_s \quad (s = 1, \ldots, 54) \tag{15.1}$$

where, from Equation 13.7, the coefficients a_{sp} are elements of the symmetrical matrix A given by

$$a_{sp} = m_k v_{ksm} v_{kpm} + I_{kmn} \omega_{ksm} \omega_{kpn} \tag{15.2}$$

where v_{ksm} and ω_{ksm} ($k = 1, \ldots, 17$; $s = 1, \ldots, 54$; $m = 1, 2, 3$) are the components of the partial velocities and partial angular velocities of the mass centers and the bodies as developed in Sections 11.7 and 11.10, and as before, the m_k and the I_{kmn} are the masses of the bodies and components of the central inertia dyadics (see Section 10.2) relative to eigen unit vectors fixed in the bodies. The terms f_s on the right side of Equation 15.1 are given by Equation 13.8 as

$$f_s = F_{y_s} - m_k v_{ksm} \dot{v}_{kpm} y_p - I_{kmn} \omega_{ksm} \dot{\omega}_{kpm} y_p - e_{rtm} I_{ktn} \omega_{ksm} \omega_{kqr} \omega_{kpn} y_q y_p \tag{15.3}$$

where the F_{y_s} ($s = 1, \ldots, 54$) are the generalized active forces developed in Section 12.2, and as before, the e_{rtm} are elements of the permutation symbol.

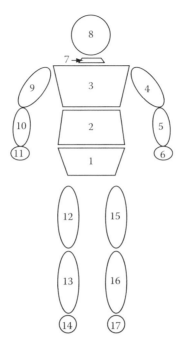

FIGURE 15.1   A human body model.

If there are constraints on the system, the constraints may often be modeled by equations of the form (see Equation 13.31)

$$b_{js}y_s = g_j \quad (j = 1, \ldots, m) \tag{15.4}$$

where
   m is the number of constraints
   b_{js} and g_j are the given functions of time and the geometric parameters (see Section 13.7)

Chapter 14 outlines algorithms for numerically developing and solving Equations 15.1 and 15.4.

15.2 HUMAN BODY IN FREE SPACE: A "SPACEWALK"

In the early years of the U.S. space program, and particularly during the 1960s, as the space agency NASA was preparing for a manned moon landing, astronauts in an orbiting satellite would often get out of the satellite and move about in free space around and about the satellite. This activity was often referred to as a spacewalk. A little documented problem, however, was that the astronauts had difficulty in orienting their bodies in the free-space environment. This problem was unexpected since it is known that a dropped pet house cat will always land on its feet—even if dropped upside down from only a short height. Also, gymnasts and divers jumping off a diving board are able to change their orientation at will.

   The questions arising then are as follows: Why did the astronauts have such difficulty? The answer appears to be simply a matter of improper training or lack of experience. In this section, we explore this assertion and offer some elementary maneuvers, which allow a person in free space to arbitrarily change orientation. Analysis, results, and discussion are based upon the research of Passerello and Huston as documented in Ref. [1].

   Consider again the human body model of Figure 15.1, shown again in Figure 15.2 where, as before, the X-axis is forward, the Y-axis is to the left, and the Z-axis is up. For a person to arbitrarily

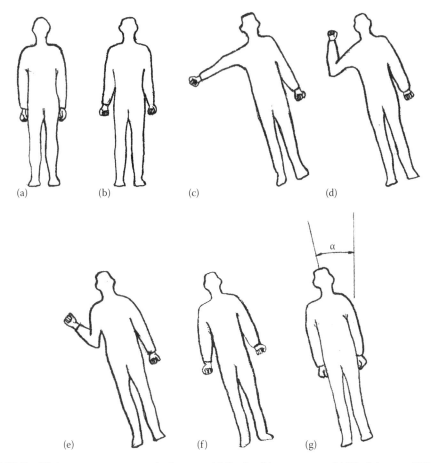

FIGURE 15.2 Right arm maneuver producing yaw. (a) Begin, (b) rotate arms, (c) lift right arm, (d) bend right elbow, (e) lower right arm, (f) straighten right elbow, and (g) reverse arm rotation.

reorient himself or herself in space, it is sufficient for the person to be able to arbitrarily turn about the X, Y, and Z axes, respectively.

Many maneuvers exist, which enable each of these rotations. Since the overall system inertia is smallest about the Z-axis, that rotation is the easiest. Correspondingly, with the inertia largest about the X-axis, that rotation is the most difficult. The Y-axis rotation is of intermediate difficulty. In the following sections, we present maneuvers for each of these axes.

15.2.1 X-Axis (Yaw) Rotation

In looking for sample maneuvers, an immediate issue is as follows: What should be the form of an angular function defining a limb, or body, motion relative to an adjacent body? Smith and Kane have proposed the following function [2]:

$$\theta(t) = \theta_1 + (\theta_2 - \theta_1)\left\{\left[\frac{t}{(t_2 - t_1)}\right] - \left(\frac{1}{2\pi}\right)\sin\left[\frac{2\pi t}{(t_2 - t_1)}\right]\right\} \qquad (15.5)$$

where θ_1 and θ_2 are the values of θ at times t_1 and t_2. This function has the property of having zero, first and second derivatives at times $t = t_1$ and $t = t_2$, respectively.

Figure 15.2a–g demonstrates a right-arm maneuver, which provides for an overall yaw of the person. In the beginning, the person is in reference standing position. Next, the arms are rotated so

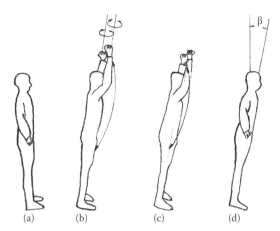

FIGURE 15.3 Arm maneuver to produce pitch. (a) Begin, (b) raise arms to the side, (c) lower arms to the front, and (d) finish.

that the hands are facing away from the body (see Figure 15.2a and b). The right arm is then rotated up approximately 90°, keeping the elbow straight. From this position, the forearm is flexed relative to the upper arm (see Figure 15.2c and d). Finally, the upper arm is brought back to the chest, and then the forearm is rotated back to the reference position, and the arms rotated to the reference configuration (see Figure 15.2e and f).

With the low mass of the arm relative to the remainder of the body, the maneuver will produce only a small yaw rotation. However, the maneuver may be repeated as many times as needed to attain any desired yaw rotation angle.

Observe that the axial rotations of the arms in the beginning and end of the maneuver do not change the orientation of the body. That is, the inertia forces between the left and the right arms are equal and opposite.

15.2.2   *Y*-AXIS (PITCH) ROTATION

Figure 15.3a–d demonstrates arm maneuvers producing overall pitch rotation of the body. In the beginning, the person is in the reference standing position. Next, the arms are raised up over the head in a forward arc as shown in Figure 15.3b. In this position, the arms are rotated about their axes as in Figure 15.3c, and then they are brought down to the sides in an arc in the frontal (*Y*–*Z*) plane. This maneuver will produce a reasonable forward pitch.

15.2.3   *Z*-AXIS (ROLL) ROTATION

Figure 15.4a–e demonstrates arm and leg rotations producing overall roll rotation of the body. In the beginning, the person is in the reference standing position. The arms are then rotated forward (right) and rearward (left) as in Figure 15.4b. In this position, the legs are spread (Figure 15.4e), and the arms are returned to the sides (Figure 15.4d). Finally, the legs are closed (adduction), bringing the body back to a reference configuration, but now rotated through a small angle δ as in Figure 15.4e.

15.3   SIMPLE WEIGHT LIFT

For a second example, consider a simple maneuver described by Huston and Passerello [3]. Although it is an unlikely actual physical movement, it nevertheless illustrates the interactive effects of gravity and inertia forces.

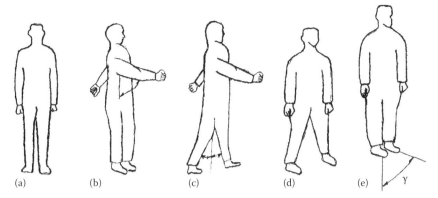

FIGURE 15.4 Arm and leg maneuvers producing roll. (a) Begin, (b) move arms forward and rearward, (c) spread legs, (d) return arms to the sides, and (e) close legs.

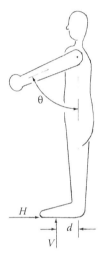

FIGURE 15.5 A lifting simulation.

Consider a person (176 lb male) lifting a weight by simply swinging his arms forward as represented in Figure 15.5. That is, a person keeps his body erect and vertical and then rotates his arms forward through an angle θ while keeping his elbows straight as shown. A weight is held in the hands. To simulate the movement, let θ be described by the function

$$\theta = \theta(t) = \theta_0 + (\theta_T - \theta_0)\left\{\left[\frac{t}{T}\right] - \left(\frac{1}{2\pi}\right)\sin\left[\frac{2\pi t}{T}\right]\right\} \tag{15.6}$$

where θ_0 and θ_T are the values of θ at both $t = 0$ and $t = T$ (see also Equation 15.5). Figure 15.6 illustrates the character of the function.

Suppose the weight is lifted and then lowered in the same fashion so that the lifting/lowering function is as in Figure 15.7 [3].

Let the resulting reaction forces on the feet be represented by a single force **F** passing through a point at distance d in front of the shoulder axis as represented in Figure 15.5, where H and V are the horizontal and vertical components of **F**.

Consider a lift on both the Earth and the moon. Then by integrating the equations of motion, the horizontal and vertical force components, H and V, are found as represented by the graphs of

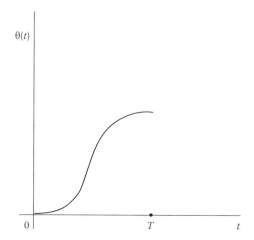

FIGURE 15.6   Rise function for a simple lift.

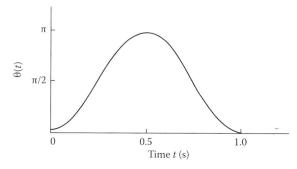

FIGURE 15.7   Lifting/lowering function.

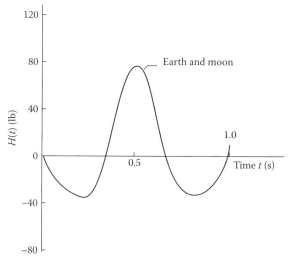

FIGURE 15.8   Horizontal foot reaction force.

Figures 15.8 and 15.9 [3]. Observe in the results that H is the same on both the Earth and the moon, whereas V depends upon the venue. On the moon, the inertia forces from the lifted weight and the arms would cause the man to lift himself off the surface.

Also, on the moon, and even on the Earth, the distance d becomes greater than the foot length, so that the lifter would lose his balance without leaning backward (see Figure 15.10) [3].

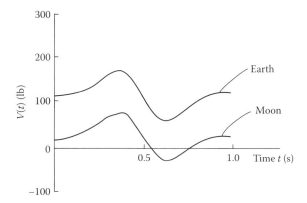

FIGURE 15.9   Vertical foot reaction force.

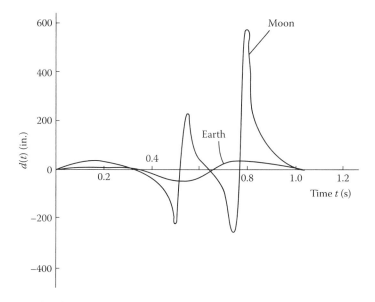

FIGURE 15.10   Reaction force position.

15.4   WALKING

Of all human dynamic activities, walking is perhaps the most fundamental. A parent's milestone is when their child begins to walk.

In one sense, walking analysis (or gait analysis) is relatively simple if we focus upon statistical data (speed, step length, cadence). From a dynamics perspective, however, walking is extremely complex. Walking engages all of the limbs including the arms, upper torso, head, and neck. From the pelvic girdle down, walking involves over 100 individual muscles and 60 bones [4].

Analysts have been studying walking for many years. Abdelnour et al. [4] provide a brief bibliography of some early efforts. But the complexity of the motion makes a comprehensive analysis virtually impossible. Thus, the focus is upon modeling, attempting to obtain a simplified analysis, which can still provide useful information.

In this section, we summarize the modeling and analysis of Abdelnour et al., which in turn is based upon the procedures outlined in Chapters 10 through 14. We begin with a brief review of terminology. We then discuss the modeling and provide results of a simple simulation.

15.4.1 TERMINOLOGY

Walking (as opposed to running) requires that at least one foot must always be in contact with the walking surface or the "ground." Where two feet are on the ground, there is "double support." With only one foot on the ground, it is a "single support," and the free leg is said to be in the "swing phase." Thus, a leg will alternately be in the support phase and the swing phase.

The support phase for a leg begins when the heel first touches the ground (heel strike) and ends when the toe leaves the ground (toe off). The time required to complete both the support phase and the swing phase is called the stride. The distance between two successive footprints of the same foot is called the stride length. The "step length" is the distance (along the direction of walking) of two successive footprints (one from each foot). The number of steps per unit time is called the "cadence." The walking speed is thus the product of the cadence and the foot length.

15.4.2 MODELING/SIMULATION

We use the same model as shown in Figure 15.1. For walking however, we focus upon the movement of the lower extremities. Then to simulate walking, we use data recorded by Lamoreux [5] to create specified movements at the hips, knees, and ankles. That is, with a knowledge of the kinematics, we are able to calculate the foot, ankle, knee, and hip forces.

15.4.3 RESULTS

Figures 15.11 through 15.13 show Z-axis (upward) forces on the foot, knee, and hip, normalized by the weight. Observe that these forces are multiples of the weight.

15.5 SWIMMING

Swimming is a popular team sport, individual sport, and recreational activity. The model used in the foregoing section and in the earlier chapters is ideally suited for studying swimming. We use the same model for swimming as shown in Figure 15.1.

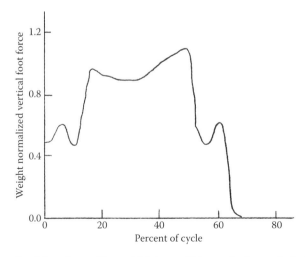

FIGURE 15.11 Normalized foot force. (From Abdelnour, T.A. et al., An analytical analysis of walking, ASME Paper 75-WA/Bio-4, American Society of Mechanical Engineers, New York, 1975.)

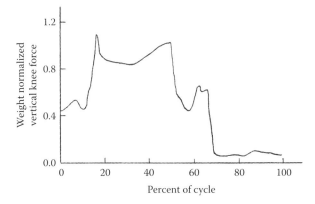

FIGURE 15.12   Normalized knee force. (From Abdelnour, T.A. et al., An analytical analysis of walking, ASME Paper 75-WA/Bio-4, American Society of Mechanical Engineers, New York, 1975. With permission.)

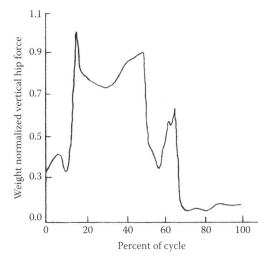

FIGURE 15.13   Normalized hip force. (From Abdelnour, T.A. et al., An analytical analysis of walking, ASME Paper 75-WA/Bio-4, American Society of Mechanical Engineers, New York, 1975. With permission.)

Elementary analyses of swimming are relatively easy to obtain by specifying the limb and body movements and modeling of the ensuing fluid forces on the limbs. In this way, we can obtain a relation between the limb movements (primarily the arm and leg movements) and the overall displacement or progression of the swimmer.

15.5.1   MODELING THE WATER FORCES

The water forces may be modeled using a procedure developed by Gallenstein and coworkers [6–9]. This procedure is based upon the extensive research of Hoerner [10]. For a given limb (say, a hand or a forearm), the water forces may be represented by a single force passing through the centroid* of the limb with the magnitude of the force being proportional to the square of the water velocity past the centroid (i.e., the square of the relative velocity of the centroid and water) and directed perpendicular to the limb axis. The constant of proportionality C depends upon the limb shape and the water properties.

---

* For the model, the geometric center (centroid) is the same as the mass center.

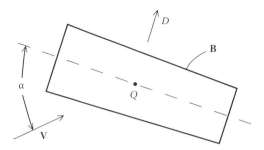

FIGURE 15.14 Modeling of water forces on a swimmer's limb.

Figure 15.14 illustrates the modeling [6] where *B* is a propelling limb (an arm or leg segment), *Q* is the centroid of **B**, **V** is the velocity of the water past the limb, α is the angle between **V** and the limb axis, and **D** is the resulting drag force. Specifically, the magnitude *D* of **D** is expressed as

$$D = C\mathbf{V}^2 \sin^2 \alpha \qquad\qquad (15.7)$$

15.5.2 LIMB MOTION SPECIFICATION

Using the movement modeling suggested by Smith and Kane [2], we can model the limb movements as we did in free-space rotation (see Section 15.2) and in lifting (Section 15.3). Specifically, we model a joint angle movement θ(*t*) as (see Equation 15.6)

$$\theta(t) = \theta_0 + (\theta_T - \theta_0)\left\{\left[\frac{t}{T}\right] - \left(\frac{1}{2\pi}\right)\sin\left[\frac{2\pi t}{T}\right]\right\} \qquad (15.8)$$

where
   θ_0 is the angle at the beginning of the movement
   θ_T is the angle at the end of the movement with *T* being the time of the movement

As noted earlier, this function has the property of having zero first and second derivatives (angular velocity and angular acceleration) at $\theta = B_0$ (*t* = 0) and at $\theta = \theta_T$ (*t* = *T*).

15.5.3 KICK STROKES

In their analyses, Gallenstein and Huston [6] studied several elementary kick strokes. The first of these, called a "simple, symmetrical V-kick," has the swimmer on his (or her) back opening and closing his legs with knees straight—thus forming a "V" pattern. To obtain a forward thrust, the swimmer must close his or her legs faster than he or she opens them. In one simulation, the swimmer opens his (or her) legs to a central angle of 60° in 0.75 s. The swimmer then closes them in 0.25 s. After steady state is obtained, the swimmer's torso advances approximately 10 in./s.

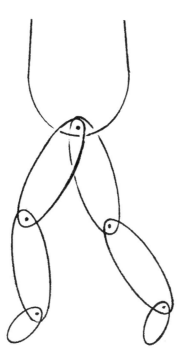

FIGURE 15.15 A flutter kick.

The flutter kick was then studied. In this kick (probably the most common kick stroke), the swimmer is prone (face down) and the legs are alternately kicked up and down as represented in Figure 15.15.

With our finite-segment model, we can conduct a variety of simulations to evaluate the effectiveness of various leg joint movements. For example, we can keep the knees straight. We can even remove the feet, or, alternatively, we can add flippers to the feet.

In their analyses, Gallenstein and Huston [6] considered straight knees, bent knees, and legs without feet. Of these three, the removal of the feet has a dramatic difference in the swimmer's forward (axial) movement. Indeed, the effectiveness of the flutter kick was reduced nearly 90% by the feet removal.

Finally, a breaststroke kick was studied. In this configuration, the swimmer is face down, and the leg joints are bent at the hips, knees, and ankles as represented in Figure 15.16. The swimmer then rapidly returns the legs to the reference configuration, providing thrust to the torso. The induced torso advance depends upon the frequency of the movement and the rapidity of the squeezing of the legs.

15.5.4 BREASTSTROKE

Figure 15.15 also shows arm configurations for the breaststroke: The arm joints are bent at the shoulders, the elbows, and the wrists as shown. The arms are then rapidly returned to a reference configuration, providing thrust to the torso.

Interestingly, the effectiveness of the arms in providing torso thrust was found to be approximately the same as that of the legs.

The breaststroke itself uses both arms and legs. With the combination of arm and leg movement, the torso is advanced approximately twice as far as when only the arms and legs are used.

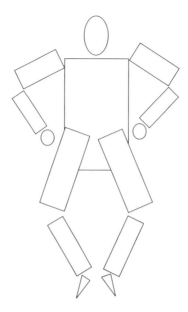

FIGURE 15.16 Breaststroke limb configuration.

15.5.5 Comments

Readers interested in more details about these are encouraged to refer Ref. [6] as well as the references in that article.

15.6 CRASH-VICTIM SIMULATION I: MODELING

There has been no greater application of multibody human body dynamics modeling than with crash-victim simulation. With the ever-increasing number of motor vehicles going at ever-greater speeds, with continuing large numbers of accidents and serious injuries, and with an increasing interest in safety, there is a demanding interest in understanding the movement of vehicle occupants within a vehicle involved in an accident. This is generally referred to as "occupant kinematics." In this section and the next two sections, we briefly review this singularly important application.

The modeling in occupant kinematics is essentially the same as that in Section 15.5 except that here we establish the reference configuration to represent a vehicle occupant in a normal sitting position as in Figure 15.17. As before, the person is modeled by a system of pin-connected rigid bodies representing the human limbs.

The typical model consists of 17 bodies: three for each limb (upper arm/leg, lower arm/leg, hand/foot), three for the torso, one for the neck, and one for the head.

As noted earlier, the objective of the modeling is to be able to simulate occupant movement within a vehicle during a crash. Of particular interest is the movement of the head, neck, and torso since it is with these bodies that the most serious and residual injuries occur. In many cases, the movement of the hands and feet is relatively unimportant. Therefore, the modeling is often simplified by incorporating the hands and feet with the lower arms and legs.

With the model being established, the next step in the simulation is to place the model within a vehicle environment. This in turn necessitates modeling a vehicle interior including the seats, the seat belts, the doors, the roof, the toe pan, the windshield, the dash, the steering wheel, and the air bag. Finally, with the combination of the vehicle occupant and the vehicle model, the simulations

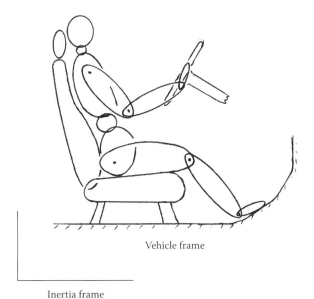

Vehicle frame

Inertia frame

FIGURE 15.17   Vehicle occupant model.

may be obtained by accelerating (or decelerating) the vehicle model to represent vehicle crashes. We discuss this further in Section 15.7.

15.7   CRASH-VICTIM SIMULATION II: VEHICLE ENVIRONMENT MODELING

To simulate the movement of an accident-vehicle occupant relative to the vehicle, it is necessary to have a description of the interior vehicle environment—or cockpit. This description may be relatively simple or quite elaborate. The principal objective of the interior vehicle representation is to model those structures and surfaces, which the occupant is likely to encounter in an accident. These structures include the seat, the seat belt, the doors, the steering wheel, the dash, the windshield, and possibly the air bags.

Of all these structures, undoubtedly the most important are the seat and seat belts. Consequently, accident reconstructionists have given considerable attention to study the effects of seats and seat belts on accident-vehicle occupants. The seats are often modeled by springs and dampers (i.e., as viscoelastic springs) [11–14]. Interestingly, seat belts may also be modeled as viscoelastic springs, but here the forces exerted on the occupant model are "one-way" forces. That is, seat belts can only exert forces in tension.

In violent crashes, seats may lose this integrity. The backrest can bend and collapse. The seat bottom can even come loose from its supports or rail anchors [14]. Also, in violent crashes, the seat belts may slide over the occupant's hips and chest so that the occupant becomes grossly out of position. The seat belt hardware may even fail so that the occupant becomes unbuckled. Numerous studies of seat belt behavior are reported in the literature [15–35].

The behavior of seat belts in successfully restraining an accident-vehicle occupant may be approximately modeled by forces placed on the pelvis (body 1) and on the torso (bodies 2 and 3). More elaborate modeling has been developed by Obergefel [36].

A difficulty with the commonly used three-point (three-anchor) system employed in the majority of current passenger automobiles, sport utility vehicles, and pickup trucks is the asymmetry of the belt system. That is, although the lap belt is symmetric, the shoulder belt comes over either the left or right shoulder for a driver and right front passengers, respectively. That is, only one shoulder is restrained.

This asymmetry can cause occupant spinal twisting, even in low-speed frontal collisions. Also, most current seat belt systems have a continuous webbing so that there may be an interchange of webbing between the lap and shoulder belt. This transfer of webbing in turn allows an occupant to slide under the belt (submarine), or, alternatively, it may allow an occupant to slide forward out of the belt (porpoising) or even to twist about the shoulder belt (barber poling).

A more subtle difficulty with current automobile seat belt systems is that the majority are anchored to the vehicle frame, as opposed to being anchored and integrated into the seat. When an occupant adjusts his or her seat with vehicle-anchored seat belts, the webbing geometry changes relative to the occupant.

Advanced seat belt designs employ pretensioners, which eliminate slack in the webbing just prior to a collision. Other safety devices are webbing arresters (preventing rapid spool out of webbing) and load limiters to reduce harming the occupant by an excessively taut webbing.

With the advent of crash sensors, air bags (supplemental restraint systems [SAS]) have been introduced. Air bags were initially installed on the steering wheel and later in the right dash for right front passenger protection. Door and side curtain air bags are also now being employed to mitigate the hazards in side impact collisions.

Air bag modeling is more difficult than seat belt modeling due to their varied geometries, deployment speeds, and peak pressures. Air bags can deploy in 15 ms at speeds up to 200 miles/h.

During a collision, a vehicle occupant may strike hard interior surfaces, such as the steering wheel hub, the dashboard, or a door. Unrestrained occupants may even collide with the windshield or door windows. Windshields made of laminated glass help to keep occupants from being ejected from the vehicle. Door windows however are made from tempered glass, which crumbles upon impact. Thus, in violent crashes, unrestrained vehicle occupants can be ejected through door windows or even the rear window.

With a modeling of the vehicle occupant space (cockpit) and its surfaces and with knowledge about the movement of the occupant within the vehicle, we can determine the input forces on the occupant during crashes. This in turn can lead to knowledge about injuries occurring during the crash. To accurately simulate a given crash, it is of course necessary to know the vehicle motion during the crash.

The procedure of the analysis is as follows: First, determine the vehicle movement for a given accident (through a reconstruction of the accident); second, determine the movement of the occupant within the vehicle (through numerical analysis as described in Section 15.8); third, determine the impact, or impacts, of the occupant with the interior vehicle surfaces; and finally, determine the forces of the impacts upon the occupant.

15.8   CRASH-VICTIM SIMULATION III: NUMERICAL ANALYSIS

The discussions of Sections 15.6 and 15.7 provide a basis for the development of software for the numerical analysis of occupant movement (kinematics) and impact forces (kinetics) during a crash. These impact forces may then be correlated with the occupant's injuries.

With such software and its generated data, we can evaluate the effect of vehicle impact speed upon injury, and we can also evaluate the effectiveness of the safety systems (seat belts, air bags, collapsing steering wheel, and interior surface padding).

The development of crash-victim simulation software dates back to 1960. In 1963, McHenry [37,38] presented a seven degree of freedom, 2D model for frontal motor vehicle accident victims. Since then, models have become increasingly sophisticated, with greater degrees of freedom and with 3D movement. Currently, there is theoretically no limit on the number of bodies, which may be used in the modeling or on the ranges of motion.

During the development of crash-victim simulation software, there have appeared a number of survey articles documenting the historical developments and also providing critique on the relative advantages and disadvantages of the models [39–43]. The principal issues in these critiques are

(1) accuracy of the simulation, (2) efficiency of the software, (3) range of applicability, (4) ease of use, and (5) means of representing the results.

The issues of accuracy and efficiency depend upon the formulation of the governing differential equations. With improvements in modeling and advances in computer hardware, there is now theoretically no limit to the range of applicability of the models. The issues of ease of use and the representation of results are continually being addressed with the continuing advances in hardware and software.

With the software developed by Huston and colleagues [44,45], the accuracy and efficiency are enabled through the use of Kane's equations and associated procedures [46,47] as outlined in the foregoing chapters. The governing differential equations are solved numerically using a fourth-order Runge–Kutta integration routine (Refs. [48,49] document some of these efforts and results).

15.9 BURDEN BEARING: WAITER/TRAY SIMULATIONS

When people carry objects or burdens, they invariably position the object so that it is the least uncomfortable. Expressed another way, people carry objects in a way that minimizes stress—particularly, muscle stress. For example, when carrying a book, most persons will either cradle the book or use a cupped hand at the end of an extended hanging arm. This is the reason for handles on suitcases and luggage.

On occasion, a person may have to maintain the orientation of an object he or she is carrying. This occurs, for example, when a person is carrying a cup of coffee or a beverage in an open container. Object orientation also needs to be maintained when a waiter, or waitress, is carrying a tray of objects. In this latter case, the waitperson (or waitron) may also need to navigate around tables and other people. To do this, the waitron usually balances the tray while holding it at eye level. But even here, the waitron will configure his/her arm so as to minimize discomfort.

If we equate discomfort with muscle stress, then the discomfort is minimized by appropriate load sharing of the muscles supporting the tray. That is, individual muscles will have the same stress. Interestingly, using this criterion, we can obtain a good representation of the waitron's arm while carrying a tray [50].

15.9.1 HEAVY HANGING CABLE

To develop this, consider the classic strength of materials problem of a heavy, hanging cable with a varying cross-sectional area along the length and varying so that the stress is constant along the length. Specifically, consider the upper end supporting structure of Figure 15.18 where the area increases with the vertical distance so that the stress at any level, due to the weight and end load P, is constant.

The design objective is to determine the cross-sectional area A of the vertical coordinate, so that the stress is constant. (This problem, in essence, is the same problem as the design of a tower with uniform compressive stress at all levels of the tower.)

The problem is solved by considering a finite element of the cable as in Figure 15.19 where Δy is the element height and $A(y)$ is the element cross-sectional area, at the lower end of the cable element.

By Taylor's theorem [51], we can approximate the area $A(y + \Delta y)$ at the upper end of the cable element as

$$A(y + \Delta y) = A(y) + \frac{dA}{dy} \Delta y \tag{15.9}$$

Next, consider a free-body diagram of the element as in Figure 15.20 where σ represents the uniform stress along the length, δ is the weight density of the cable material, and Δv is the volume of the cable element.

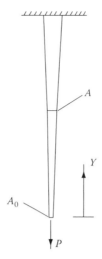

FIGURE 15.18   Heavy hanging cable.

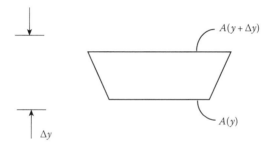

FIGURE 15.19   An element of the hanging cable.

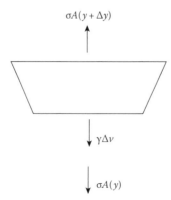

FIGURE 15.20   Free-body diagram of the cable element.

A balancing of forces in Figure 15.20 immediately leads to the expression

$$\sigma A(y + \Delta y) = \gamma \Delta v + \sigma A(y) \tag{15.10}$$

From Figure 15.19 and Equation 15.9, we see that Δv is approximately

$$\Delta v = \frac{[A(y) + A(y + \Delta y)]}{2} \quad \Delta y = \left[A + \frac{1}{2} \frac{dA}{dy} \Delta y \right] \quad \Delta y = A \Delta y \tag{15.11}$$

where in reaching the final equality, all nonlinear terms in Δy are neglected.

By substituting from Equations 15.9 and 15.11 into 15.10, we obtain

$$\sigma \left[A + \frac{dA}{dy} \Delta y \right] = \gamma A \Delta y + \sigma A \tag{15.12}$$

or simply

$$\sigma \frac{dA}{dy} = \gamma A \tag{15.13}$$

Finally, by solving Equation 15.13 for A, we obtain

$$A = A_0 e^{(\gamma/\sigma)y} \tag{15.14}$$

where A_0 is the end area given by

$$\sigma = \frac{P}{A_0} \quad \text{or} \quad A_0 = \frac{P}{\sigma} \tag{15.15}$$

Interestingly, the cross-sectional areas of the legs, arms, and fingers vary in the manner of Equation 15.14.

15.9.2 UNIFORM MUSCLE STRESS CRITERION

To provide rationale for the uniform muscle stress criterion, consider that the strength of a muscle is proportional to the size or cross-sectional area of the muscle. From a dimensional analysis perspective, this means that strength is a length squared parameter.

Similarly, a person's weight is proportional to the person's volume—a length cubed parameter. Thus, if s represents strength and w weight, we have

$$s = \alpha l^2 \quad \text{and} \quad w = \beta l^3 \tag{15.16}$$

where

l is the length parameter
α and β are the constants

By eliminating l between these expressions, we have

$$s = \left(\frac{\alpha}{\beta^{2/3}} \right) W^{2/3} = K W^{2/3} \tag{15.17}$$

where K is the proportion constant defined by inspection.

TABLE 15.1
Weight Lifter Lifts and Lift/Weight 2/3 Ratio for Various Lifting Classes

Weight Lifter Mass (W) (kg(lb))	Winning Lift (S) (kg(lb))	$S/W^{2/3}$
55.8 (123)	318.9 (703.0)	28.4
59.9 (132)	333.2 (734.6)	28.3
67.6 (149)	359.0 (791.4)	28.3
74.8 (165)	388.0 (855.4)	28.4
82.6 (182)	405.7 (894.4)	27.9
89.8 (198)	446.6 (984.6)	28.9
109.8 (242)	463.6 (1022.0)	26.3

Equation 15.17 states that the strength is proportional to the two-thirds root of the weight. Interestingly, this result may be tested. Consider Table 15.1 that lists data for Olympic weight lifting winners over a 40-year period [52]. Specifically, the table provides the winning lift for the various weight classes of the weight lifters. The "lift" is a total for the snatch and the clean and jerk. Table 15.1 also shows the ratio of the lift to the two-thirds root of the weight is approximately constant.

15.9.3 WAITRON/TRAY ANALYSIS

Consider now the analysis for the waitron/tray simulation. Let the waitron's arm and tray be modeled as in Figure 15.21. Let the upper arm, forearm, and hand be numbered or labeled as 1, 2, and 3 as shown and let the orientation of these bodies be defined by angles θ_1, θ_2, and θ_3 as in Figure 15.22. Let the forces acting on the bodies be as those represented in Figure 15.23, where M_1, M_2, and M_3 are joint moments created by the waitron's muscles in supporting the tray.

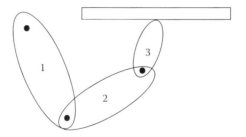

FIGURE 15.21   Schematic representation of a waitron arm and tray.

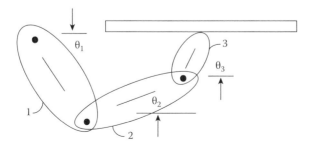

FIGURE 15.22   Orientation angles for the arm segments.

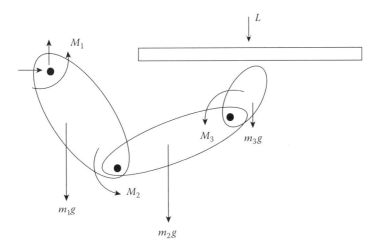

FIGURE 15.23   Forces and moments on the arm segments.

An elementary balance of the forces and moments of Figure 15.23 immediately leads to the equations [50]:

$$M_3 - m_3 g r_3 \cos\theta_3 - L l_3 \cos\theta_3 = 0 \qquad (15.18)$$

$$M_2 - M_3 - m_2 g r_2 \cos\theta_2 - (m_3 g + L) l_2 \cos\theta_2 = 0 \qquad (15.19)$$

$$M_1 - M_2 - m_1 g r_1 \cos\theta_1 - (m_2 g + m_3 g + L) l_1 \cos\theta_1 = 0 \qquad (15.20)$$

where
  l_1, l_2, and l_3 are the lengths of the upper arm, forearm, and hand, respectively
  r_1, r_2, and r_3 are the distances to the mass centers of the upper arm, forearm, and hand, from the shoulder, elbow, and wrist, respectively

For the waitron to keep the tray at shoulder level, the following constraint equation must be satisfied [50]:

$$l_1 \sin\theta_1 + l_2 \sin\theta_2 + l_3 \sin\theta_3 = 0 \qquad (15.21)$$

Next, by applying the uniform muscle stress criterion, the moments in Equations 15.18 through 15.20 may be expressed as

$$M_1 = \kappa A_1, \quad M_2 = \kappa A_2, \quad M_3 = \kappa A_3 \qquad (15.22)$$

where A_1, A_2, and A_3 may be obtained experimentally or analytically using Equation 15.14.

Equations 15.18 through 15.22 form a set of seven algebraic equations for the seven unknowns: M_1, M_2, M_3, θ_1, θ_2, θ_3, and κ. By substituting for M_1, M_2, and M_3 from Equation 15.22 and by solving Equations 15.18 through 15.20 for $\cos\theta_1$, $\cos\theta_2$, and $\cos\theta_3$, we have [50]

$$\cos\theta_1 = \frac{(\kappa A_1 - \kappa A_2)}{[m_1 g r_1 + (m_2 g + m_3 g + L) l_1]} \qquad (15.23)$$

$$\cos\theta_2 = \frac{(\kappa A_2 - \kappa A_3)}{[m_2 g r_2 + (m_3 g + L)l_2]} \tag{15.24}$$

$$\cos\theta_3 = \frac{\kappa A_3}{[(m_3 g r_3 + L)l_3]} \tag{15.25}$$

Equations 15.23 through 15.25 may be solved iteratively. For a given load L, we can select a small value of K and then solve Equations 15.23 through 15.25 for θ_1, θ_2, and θ_3. By substituting the results into Equation 15.21, we can determine how nearly the constraint equation is satisfied. If the terms of Equation 15.21 do not add to zero, we can increase the value of K and repeat the procedure.

Huston and Liu [50] used this method to determine the arm angles for two tray weights: 5 and 8 lb. Figures 15.24 and 15.25 show their results.

The configurations of Figures 15.24 and 15.25 are representative of waitron arm configurations.

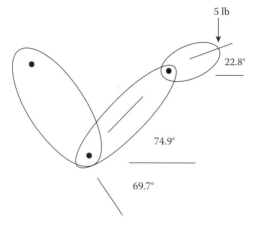

FIGURE 15.24   Waitron arm configuration for holding a 5 lb tray.

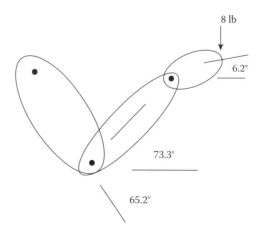

FIGURE 15.25   Waitron arm configuration for holding an 8 lb tray.

15.10   OTHER APPLICATIONS

There is no end to the number of analyses we could make using our dynamic procedures. Indeed, virtually every human movement could be studied. We list here a few of these, which may be of interest but are as yet not fully explored or understood.

15.10.1   LOAD SHARING BETWEEN MUSCLE GROUPS

Many of the major skeletal muscles are actually groupings of parallel muscles. Also, these muscle groups are often aided by adjacent muscle groups. As noted earlier, the major muscles in the upper arm and thigh are called biceps, triceps, and quadriceps. They are assemblages of 2, 3, and 4 muscles, respectively.

When the muscle groups are activated, the resultant tendon force is produced by the tandem/parallel muscles, but the individual muscle contributions to this force are unknown. Also, the effect of adjacent muscle groups is unknown. For example, in weight lifting, the simple biceps curl is difficult to perform without energizing the shoulder and back muscles.

While weight lifting machines are designed to isolate the exercise of individual muscles, there is an ongoing debate about the advantages of these machines compared with lifting free weights. More research and analyses of these issues are needed.

15.10.2   TRANSITION MOVEMENTS

In small, localized movements as with extremity movements (foot tapping, handwriting, or hand tool use), a person will use only a few muscles, and only a few of the bodies of the human frame will be moving. As a greater movement is desired, there is a gradual transition to use a greater number of bodies and muscles.

For example, while eating, a person may use primarily finger, hand, and arm segments. But if the person wants to reach across a table for a food item or an implement, it becomes increasingly uncomfortable to restrict the movement of the arm. Instead, it is more convenient to use the shoulder and upper body as well. The reaching movement thus has a transition from one kind of movement to another. It is analogous to a phase change of a material.

Another common example of a movement transition is a person walking at increasing speed until running becomes more comfortable or more natural.

Dainoff et al. [53] have studied such phenomenon but more analysis is needed—particularly from a dynamic perspective.

The effect of discomfort in causing transition is itself an interesting phenomenon. Consider that in speed walking, the objective is to endure the discomfort of keeping at least one foot on the ground while continuing to increase the speed.

Some sporting events have transition as an integral part of the event. This occurs in the broad jump, high jump, and pole vault, where the athlete must transition from a run to a jump or lift. The skill with which the athlete executes the transition can have a marked effect upon the performance.

As another example, consider a triathlon where the participants must transition between swimming, biking, and running. Some may tend to speed up the end of the bike ride to shorten the time. But such effort may be counterproductive since the leg muscles can become fatigued and ill-prepared for the start of the run. That is, time gained at the end of the bike ride may be more than lost in the beginning of the run.

On occasion, the best athletic strategy may be to avoid transition movements. For example, a basketball player in motion is more likely to be in position to obtain a rebound than a player who is watching where the ball is going before moving toward that position. Interestingly, the moving player has the advantage even if he or she is moving away from the ball when the rebound occurs. Similar reasoning may be made for baseball fielders.

15.10.3   Gyroscopic Effects in Walking

Probably the most familiar manifestation of gyroscopic stabilization is with bicycle and motorcycle riding. Stability occurs while the vehicle is moving. The greater the speed, the greater the stability. At zero speed, however, the vehicle is unstable and will fall over unless it is propped up.

Another familiar manifestation of gyroscopic stability is with spinning projectiles such as a Frisbee or a football. Still another example is a top or a spinning pivoting coin. In each of these cases, the rotary movement helps to maintain the orientation or posture of the spinning body (the wheel, projectile, or top). In stability analyses, the moment of inertia of the spinning body is found to be a multiplying factor producing the stability [54].

In view of this relatively well-understood phenomenon, a question arising is as follows: To what extent do the movements of the arms and legs aid a person in staying erect while walking? It is a commonly known fact that it is easier to remain erect while walking than while standing. Also, it is a natural reaction to rapidly lift one's arm to keep from falling. Quantification of stabilization by limb movement is not yet fully developed.

15.10.4   Neck Injuries in Rollover Motor Vehicle Accidents

Rollovers are the most dramatic of the common motor vehicle accidents. It does not take much thought to imagine the complex dynamics of occupants within a rolling vehicle—even those restrained by seat belts.

Injuries arising from rollover accidents are often fatal or permanently disabling. Among the most disabling are neck injuries with spinal cord damage.

When a vehicle rolls over and strikes its roof on the ground, the roof generally deforms and intrudes into the occupant space. Interestingly, in side-to-side rollovers (barrel rolls), the greater roof damage occurs in the area of the roof opposite to the direction of the roll. For example, for a driver-side leading roll, the greatest roof damage occurs on the passenger side.

When there is extensive roof deformation and where there is neck injury for an occupant sitting under the deformed roof, a question arising is as follows: How did the injury occur? One might initially think that the roof simply came down upon the occupant's head, causing the neck to flex and be broken or be severely injured. Another view however is that during the rolling motion of the vehicle, but before the roof strikes the ground, the occupant is projected onto the roof, causing the neck injury—often characterized as a diving injury. Still a third possibility is that the occupants find themselves with their head near the roof at the instant of roof/ground impact. Then when the roof suddenly deforms, the injury occurs.

While many investigators have firm opinions about which of these scenarios actually occurs, there is no clear agreement or consensus. Therefore, this is still another area where the methodologies developed herein are believed to be ideally suited for application.

PROBLEMS

Section 15.1

P15.1.1   Review Equations 15.1 through 15.4, verifying the terms and the indices.

Section 15.2

P15.2.1   Devise limb motions, which will reverse the overall trunk movements of yaw, pitch, and roll of Sections 15.2.1 through 15.2.3.

P15.2.2   Determine alternative movements to those of Sections 15.2.1 through 15.2.3, which will also produce yaw, pitch, and roll.

Section 15.3

P15.3.1 In Equation 15.6, verify that θ has the values θ_O and θ_T when $t = 0$ and $t = T$, respectively.

P15.3.2 In Equation 15.6, verify that the derivatives $\dot{\theta}$ and $\ddot{\theta}$ are zero at both $t = 0$ and $t = T$.

P15.3.3 Outline a more realistic modeling of a simple lift than that of Section 15.3. Specifically, describe a simple curl—that is, bending at the elbows.

P15.3.4 See Problem P15.3.3. For a simple curl, how much should the torso be leaning backward to keep the equivalent floor reaction force at a fixed point relative to the feet?

P15.3.5 Outline the arm movement of a bench press.

P15.3.6 See Problem P15.3.5. Develop a research proposal for finding the optimal hand grip on the bar for a bench press. Consider such issues as the optimal hand spacing as a function of the arm geometry. Do persons with short arms have an advantage?

P15.3.7 See Problem P15.3.6. Is the speed of the lift a factor in optimal bench press lifting? Propose a method for finding the optimal speed profile—that is, what should be the acceleration profile for optimal lifting?

Section 15.4

P15.4.1 Design an experiment to determine walking speed and stride length for a cross section of the population.

P15.4.2 From an inertia force perspective, describe the roles of the arms and legs in keeping the torso from rotating during walking. How do these roles differ for men and women?

P15.4.3 From a kinematic perspective, explain why the torso moves up and down during the walking cycle.

P15.4.4 See Problem P15.4.3. What are the kinetic and dynamic effects on the torso during the walking cycle?

P15.4.5 See Problems P15.4.3 and P15.4.4. How are the kinetics and dynamics for the torso changed as the walking speed increases?

P15.4.6 See Problems P15.4.3 through P15.4.5. What is the challenge experienced by speed walkers as the speed is increased? How is this challenge typically addressed?

Section 15.5

P15.5.1 Create a model of a fish to simulate its swimming.

P15.5.2 Create a model of a snake to simulate its travel across the land.

Section 15.6

P15.6.1 See Figure 15.17. Develop two convenient reference configurations (initial positioning) for the analysis of a human body model of an automobile operator.

P15.6.2 Repeat Problem P15.6.1 for a front seat passenger.

Section 15.7

P15.7.1 Consider the vehicle environment of an automobile operator: the seat, the seat belts, the door, the door window, the steering wheel, the dash, the windshield, the roof, and the air bags. Envision possible models for each of these vehicle components.

Section 15.8

P15.8.1 List the relative advantages and disadvantages of using human body modeling together with numerical analysis software to simulate vehicle occupant movement during a crash, as opposed to using (a) crash dummies, (b) cadavers, or (c) possibly, human volunteers.

Section 15.9

P15.9.1 In view of the analysis and discussion of the waitron/tray of Section 15.9, consider the arm/leg/torso movements in lifting a barbell from the floor.

Section 15.10

P15.10.1 Outline approaches to answering/resolving the questions/issues posed in Section 15.10.

REFERENCES

1. C. E. Passerello and R. L. Huston, Human altitude control, *Journal of Biomechanics*, 4, 1971, 95–102.
2. P. G. Smith and T. R. Kane, On the dynamics of a human body in free fall, *Journal of Applied Mechanics*, 35, 1968, 167–168.
3. R. L. Huston and C. E. Passerello, On the dynamics of a human body model, *Journal of Biomechanics*, 4, 1971, 369–378.
4. T. A. Abdelnour, C. E. Passerello, and R. L. Huston, An analytical analysis of walking, ASME Paper 75-WA/B10-4, American Society of Mechanical Engineers, New York, 1975.
5. L. W. Lamoreux, Experimental kinematics of human walking, PhD thesis, University of California, Berkeley, CA, 1970.
6. J. Gallenstein and R. L. Huston, Analysis of swimming motions, *Human Factors*, 15(1), 1973, 91–98.
7. J. M. Winget and R. L. Huston, Cable dynamics—A finite segment approach, *Computers and Structures*, 6, 1976, 475–480.
8. R. L. Huston and J. W. Kammon, A representation of fluid forces in finite segment cable models, *Computers and Structures*, 14, 1981, 281–187.
9. J. W. Kammon and R. L. Huston, Modeling of submerged cable dynamics, *Computers and Structures*, 20, 1985, 623–629.
10. S. F. Hoerner, *Fluid-Dynamic Drag*, Hoerner, New York, 1965.
11. R. L. Huston, R. E. Hessel, and C. E. Passerello, A three-dimensional vehicle-man model for collision and high acceleration studies, SAE Paper No. 740275, Society of Automotive Engineers, Warrendale, PA, 1974.
12. R. L. Huston, R. E. Hessel, and J. M. Winget, Dynamics of a crash victim—A finite segment mode, *AIAA Journal*, 14(2), 1976, 173–178.
13. R. L. Huston, C. E. Passerello, and M. W. Harlow, UCIN Vehicle-occupant/crash victim simulation model, *Structural Mechanics Software Series*, University Press of Virginia, Charlottesville, VA, 1977.
14. R. L. Huston, A. M. Genaidy, and W. R. Shapton, Design parameters for comfortable and safe vehicle seats, *Progress with Human Factors in Automated Design, Seating Comfort, Visibility and Safety*, SAE Publication SP-1242, SAE Paper No. 971132, Society of Automotive Engineers, Warrendale, PA, 1997.
15. R. L. Huston, A review of the effectiveness of seat belt systems: Design and safety considerations, *International Journal of Crashworthiness*, 6(2), 2001, 243–252.
16. H. G. Johannessen, Historical perspective on seat belt restraint systems, SAE Paper No. 840392, Society of Automotive Engineers, Warrendale, PA, 1984.
17. J. E. Shanks and J. L. Thompson, Injury mechanisms to fully restrained occupants, SAE Paper No. 791003, *Proceedings of the 23rd Stapp Car Crash Conference*, Society of Automotive Engineers, Warrendale, PA, 1979, pp. 17–38.
18. M. B. James, D. Allsop, T. R. Perl, and D. E. Struble, Inertial seatbelt release, *Frontal Impact Protection Seat Belts and Air Bags*, SAE Publication SP-947, SAE Paper No. 930641, Warrendale, PA, 1993.
19. E. A. Moffatt, T. M. Thomas, and E. R. Cooper, Safety belt buckle inertial responses in laboratory and crash tests, *Advances in Occupant Protection Technologies for the Mid-Nineties*, SAE Publication SP-1077, SAE Paper No. 950887, Society of Automotive Engineers, Warrendale, PA, 1995.
20. D. Andreatta, J. F. Wiechel, T. F. MacLaughlin, and D. A. Guenther, An analytical model of the inertial opening of seat belt latches, SAE Publication SP-1139, SAE Paper No. 960436, Society of Automotive Engineers, Warrendale, PA, 1996.
21. National Transportation Safety Board, Performance of lap belts in 26 frontal crashes, Report No. NTSB/S, 86/03, Government Accession No. PB86917006, Washington, DC, 1986.
22. National Transportation Safety Board, Performance of lap/shoulder belts in 167 motor vehicle crashes, Vols. 1 and 2, Report Nos. N75B/SS-88/02,03, Government Accession Nos. PB88-917002,3, Washington, DC, 1988.

23. D. F. Huelke, M. Ostrom, G. M. Mackay, and A. Morris, Thoracic and lumbar spine injuries and the lap-shoulder belt, *Frontal Impact Protection: Seat Belts and Air Bags*, SAE Publication SP-947, SAE Paper No. 930640, Society of Automotive Engineers, Warrendale, PA, 1993.

24. J. E. Mitzkus and H. Eyrainer, Three-point belt improvements for increased occupant protection, SAE Paper No. 840395, Society of Automotive Engineers, Warrendale, PA, 1984.

25. L. Stacki and R. A. Galganski, Safety performance improvement of production belt system assemblies, SAE Paper No. 870654, Society of Automotive Engineers, Warrendale, PA, 1984.

26. B. J. Campbell, The effectiveness of rear-seat lap-belts in crash injury reduction, SAE Paper No. 870480, Society of Automotive Engineers, Warrendale, PA, 1987.

27. D. J. Dalmotas, Injury mechanisms to occupants restrained by three-point belts in side impacts, SAE Paper No. 830462, Society of Automotive Engineers, Warrendale, PA, 1983.

28. M. Dejeammes, R. Baird, and Y. Derrieu, The three-point belt restraint: Investigation of comfort needs, evaluation of comfort needs, evaluation of efficacy improvements, SAE Paper No. 840333, Society of Automotive Engineers, Warrendale, PA.

29. R. L. Huston, M. W. Harlow, and R. F. Zernicke, Effects of restraining belts in preventing vehicle-occupant/steering-system impact, SAE Paper No. 820471, Society of Automotive Engineers, Warrendale, PA, 1982.

30. R. L. Huston and T. P. King, An analytical assessment of three-point restraints in several accident configurations, SAE Paper No. 880398, Society of Automotive Engineers, Warrendale, PA, 1988.

31. O. H. Jacobson and R. M. Ziernicki, Field investigation of automotive seat belts, *Accident Investigation Quarterly*, 16, 1996, 16–19.

32. C. S. O'Connor and M. K. Rao, Dynamic simulations of belted occupants with submarining, SAE Paper No. 901749, Society of Automotive Engineers, Warrendale, PA, 1990.

33. L. S. Robertson, Shoulder belt use and effectiveness in cars with and without window shade slack devices, *Human Factors*, 32(2), 1990, 235–242.

34. O. Jacobson and R. Ziernicki, Seat belt development and current design features, *Accident Reconstruction Journal*, 6, 1, 1994.

35. J. Marcosky, J. Wheeler, and P. Hight, The development of seat belts and an evaluation of the efficacy of some current designs, *Journal of the National Academy of Forensic Engineers*, 6, 2, 1989.

36. L. Obergefel, *Harness Belt Restraint Modeling*, Doctoral dissertation, University of Cincinnati, Cincinnati, OH, 1992.

37. R. R. McHenry, Analysis of the dynamics of automobile passenger restraint systems, *Proceedings of the 7th Stapp Car Crash Conference*, Society of Automotive Engineers (SAE), Warrendale, PA, 1963, pp. 207–249.

38. R. R. McHenry and K. N. Naab, Computer simulations of the crash victim—A Validation Study, *Proceedings of the 10th Stapp Car Crash Conference*, Warrendale, PA, 1966.

39. A. I. King and C. C. Chou, Mathematical modeling, simulation, and experimental testing of biomechanical system crash response, *Journal of Biomechanics*, 9, 1976, 301–317.

40. R. L. Huston, A summary of three-dimensional gross-motion, crash-victim simulators, *Structural Mechanics Software Series*, Vol. 1, University Press of Virginia, Charlottesville, VA, 1977, pp. 611–622.

41. A. I. King, A review of biomechanical models, *Journal of Biomechanical Engineering*, 106, 1984, 97–104.

42. P. Prasad, An overview of major occupant simulation models, SAE Paper No. 840855, Society of Automotive Engineers, Warrendale, PA, 1984.

43. R. L. Huston, Crash victim simulation: Use of computer models, *International Journal of Industrial Ergonomics*, 1, 1987, 285–291.

44. R. L. Huston, C. E. Passerello, and M. W. Harlow, UCIN vehicle-occupant/crash-victim simulation model, *Structural Mechanics Software Series*, University Press of Virginia, VA, 1977.

45. R. L. Huston, J. W. Kamman, and T. P. King, UCIN-DYNOCOMBS-software for the dynamic analysis of constrained multibody systems, in *Multibody Systems Handbook*, W. Schielen (Ed.), Springer-Verlag, New York, 1990, pp. 103–111.

46. R. L. Huston, C. E. Passerello, and M. W. Harlow, Dynamics of multi-rigid-body systems, *Journal of Applied Mechanics*, 45, 1978, 889–894.

47. T. R. Kane and D. A. Levinson, Formulations of equations of motion for complex spacecraft, *Journal of Guidance and Control*, 3(2), 1980, 99–112.

48. R. L. Huston, Multibody dynamics—Modeling and analysis methods, Feature Article, *Applied Mechanics Reviews*, 44(3), 1991, 109–117.

49. R. L. Huston, Multibody dynamics since 1990, *Applied Mechanics Reviews*, 49(10), 1996, 535–540.

50. R. L. Huston and Y. S. Liu, Optimal human posture—Analysis of a waitperson holding a tray, *Ohio Journal of Science*, 96(4/5), 1996, 93–96.

51. R. E. Johnson and F. L. Kiokemeister, *Calculus with Analytic Geometry*, 3rd edn., Allyn & Bacon, Boston, PA, 1964, p. 449.

52. P. W. Goetz (Ed.), *The Encyclopedia Britannica*, 15th edn., Vol. 8, Chicago, IL, 1974, pp. 935–936.

53. M. J. Dainoff, L. S. Mark, and D. L. Gardner, Scaling problems in the design of workspaces for human use, in *Human Performance and Ergonomics*, P. A. Hancock (Ed.), Academic Press, New York, 1999, pp. 265–296.

54. H. Josephs and R. L. Huston, *Dynamics of Mechanical Systems*, CRC Press, Boca Raton, FL, 2002, Chapter 14.

Appendix: Anthropometric Data Tables

A.1 HUMAN ANTHROPOMETRIC DATA

TABLE A.1
Human Anthropometric Data (in in.)

Name	Figure Dimension	Male 5th%	Male 50th%	Male 95th%	Female 5th%	Female 50th%	Female 95th%
Stature	A	64.9	69.3	73.6	59.8	63.7	67.9
Eye height (standing)	B	60.8	64.7	68.8	56.2	59.8	64.2
Mid shoulder height	C	53.0	56.9	61.6	47.6	51.7	56.7
Waist height	D	39.1	43.4	46.0	35.7	38.8	43.6
Buttocks height	E	30.0	33.0	36.2	27.2	29.2	32.7
Sitting height	F	33.8	36.5	38.4	31.4	33.6	35.9
Eye height (sitting)	G	29.3	31.5	33.7	27.2	29.3	31.1
Upper arm length	H	13.1	14.2	15.3	12.0	13.1	14.1
Lower arm/hand length	I	17.8	19.0	20.4	15.6	16.9	18.0
Upper leg length	J	22.0	23.8	26.0	20.9	22.8	24.7
Lower leg length	K	19.9	21.8	23.6	18.1	19.8	21.5

Note: See Figures A.1 and A.2 and Chapter 2, Refs. [6–10].

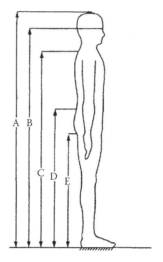

FIGURE A.1   Standing dimensions: (A) Stature, (B) eye height (standing), (C) mid shoulder height, (D) waist height, (E) buttocks height.

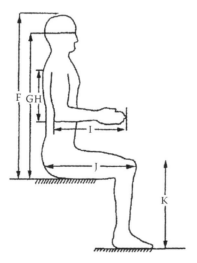

FIGURE A.2   Sitting dimensions: (F) sitting height, (G) eye height (sitting), (H) upper arm length, (I) lower arm/hand length, (J) upper leg length, (K) lower leg length.

TABLE A.2
Human Anthropometric Data (in ft)

Name	Figure Dimension	Male			Female		
		5th%	50th%	95th%	5th%	50th%	95th%
Stature	A	5.41	5.78	6.13	4.98	5.31	5.66
Eye height (standing)	B	5.07	5.39	5.73	4.68	4.98	5.35
Mid shoulder height	C	4.42	4.74	5.13	3.97	4.31	4.73
Waist height	D	3.26	3.62	3.83	2.98	3.23	3.63
Buttocks height	E	2.50	2.75	3.02	2.27	2.43	2.73
Sitting height	F	2.82	3.04	3.20	2.62	2.80	2.99
Eye height (sitting)	G	2.44	2.63	2.81	2.27	2.44	2.59
Upper arm length	H	1.09	1.18	1.28	1.0	1.09	1.18
Lower arm/hand length	I	1.48	1.58	1.70	1.30	1.41	1.50
Upper leg length	J	1.83	1.98	2.17	1.74	1.90	2.06
Lower leg length	K	1.66	1.82	1.97	1.51	1.65	1.79

Note: See Figures A.1 and A.2 and Chapter 2, Refs. [6–10].

TABLE A.3
Human Anthropometric Data (in m)

Name	Figure Dimension	Male			Female		
		5th%	50th%	95th%	5th%	50th%	95th%
Stature	A	1.649	1.759	1.869	1.518	1.618	1.724
Eye height (standing)	B	1.545	1.644	1.748	1.427	1.520	1.630
Mid shoulder height	C	1.346	1.444	1.564	1.210	1.314	1.441
Waist height	D	0.993	1.102	1.168	0.907	0.985	1.107
Buttocks height	E	0.761	0.839	0.919	0.691	0.742	0.832
Sitting height	F	0.859	0.927	0.975	0.797	0.853	0.911
Eye height (sitting)	G	0.743	0.800	0.855	0.692	0.743	0.791
Upper arm length	H	0.333	0.361	0.389	0.306	0.332	0.358
Lower arm/hand length	I	0.451	0.483	0.517	0.396	0.428	0.458
Upper leg length	J	0.558	0.605	0.660	0.531	0.578	0.628
Lower leg length	K	0.506	0.553	0.599	0.461	0.502	0.546

Note: See Figures A.1 and A.2 and Chapter 2, Refs. [6–10].

A.2   HUMAN BODY SEGMENT MASSES

TABLE A.4
Fifth-Percentile (5%) Male Body Segment Masses

Body Segment Number	Name	Mass		
		Weight (lb)	Slug	kg
1	Lower torso (pelvis)	18.17	0.564	8.24
2	Middle torso (lumbar)	19.89	0.618	9.01
3	Upper torso (chest)	33.75	1.048	15.30
4	Upper left arm	4.05	0.126	1.84
5	Lower left arm	2.52	0.078	1.14
6	Left hand	0.95	0.029	0.43
7	Neck	3.27	0.101	1.48
8	Head	8.99	0.279	4.07
9	Upper right arm	4.05	0.126	1.84
10	Lower right arm	2.52	0.078	1.14
11	Right hand	0.95	0.029	0.43
12	Upper right leg	15.35	0.477	6.96
13	Lower right leg	6.27	0.195	2.84
14	Right foot	1.87	0.058	0.85
15	Upper left leg	15.35	0.477	6.96
16	Lower left leg	6.27	0.195	2.84
17	Left foot	1.87	0.058	0.85
	Total	146	4.536	66.21

Note: See Chapter 10, Refs. [3,4].

TABLE A.5
Fiftieth-Percentile (50%) Male Body Segment Masses

Body Segment Number	Name	Mass		
		Weight (lb)	Slug	kg
1	Lower torso (pelvis)	22.05	0.685	10.00
2	Middle torso (lumbar)	24.14	0.750	10.95
3	Upper torso (chest)	40.97	1.272	18.58
4	Upper left arm	4.92	0.153	2.23
5	Lower left arm	3.06	0.095	1.39
6	Left hand	1.15	0.036	0.52
7	Neck	3.97	0.123	1.80
8	Head	10.91	0.339	4.95
9	Upper right arm	4.92	0.153	2.23
10	Lower right arm	3.06	0.095	1.39
11	Right hand	1.15	0.036	0.52
12	Upper right leg	18.63	0.578	8.45
13	Lower right leg	7.61	0.236	3.45
14	Right foot	2.27	0.070	1.03
15	Upper left leg	18.63	0.578	8.45
16	Lower left leg	7.61	0.236	3.45
17	Left foot	2.27	0.070	1.03
	Total	177	5.5	80.5

Note: See Chapter 10, Refs. [3,4].

TABLE A.6
Ninety-Fifth-Percentile (95%) Male Body Segment Masses

Body Segment Number	Name	Mass		
		Weight (lb)	Slug	kg
1	Lower torso (pelvis)	26.45	0.821	11.99
2	Middle torso (lumbar)	28.96	0.8999	13.13
3	Upper torso (chest)	49.15	1.526	22.28
4	Upper left arm	5.90	0.183	2.67
5	Lower left arm	3.67	0.114	1.66
6	Left hand	1.38	0.043	0.63
7	Neck	4.76	0.148	2.16
8	Head	13.09	0.406	5.93
9	Upper right arm	5.90	0.183	2.67
10	Lower right arm	3.67	0.114	1.66
11	Right hand	1.38	0.043	0.63
12	Upper right leg	22.35	0.694	10.13
13	Lower right leg	9.13	0.283	4.14
14	Right foot	2.72	0.084	1.23
15	Upper left leg	22.35	0.694	10.13
16	Lower left leg	9.13	0.283	4.14
17	Left foot	2.72	0.084	1.23
	Total	212.71	6.606	96.41

Note: See Chapter 10, Refs. [3,4].

TABLE A.7
Fifth-Percentile (5%) Female Body Segment Masses

Body Segment Number	Name	Mass		
		Weight (lb)	Slug	kg
1	Lower torso (pelvis)	18.24	0.566	8.27
2	Middle torso (lumbar)	12.02	0.373	5.45
3	Upper torso (chest)	16.96	0.526	7.69
4	Upper left arm	3.12	0.097	1.41
5	Lower left arm	1.86	0.058	0.84
6	Left hand	0.76	0.024	0.34
7	Neck	2.65	0.082	1.20
8	Head	7.31	0.227	3.31
9	Upper right arm	3.12	0.097	1.41
10	Lower right arm	1.86	0.058	0.84
11	Right hand	0.76	0.024	0.34
12	Upper right leg	13.73	0.426	6.22
13	Lower right leg	4.94	0.153	2.24
14	Right foot	1.56	0.048	0.71
15	Upper left leg	13.73	0.426	6.22
16	Lower left leg	4.94	0.153	2.24
17	Left foot	1.56	0.048	0.71
	Total	109.12	3.386	49.44

Note: See Chapter 10, Refs. [3,4].

TABLE A.8
Fiftieth-Percentile (50%) Female Body Segment Masses

Body Segment Number	Name	Mass		
		Weight (lb)	Slug	kg
1	Lower torso (pelvis)	22.05	0.685	10.00
2	Middle torso (lumbar)	14.53	0.451	6.59
3	Upper torso (chest)	20.50	0.636	9.30
4	Upper left arm	3.77	0.117	1.71
5	Lower left arm	2.25	0.070	1.02
6	Left hand	0.92	0.029	0.42
7	Neck	3.20	0.099	1.45
8	Head	8.84	0.274	4.01
9	Upper right arm	3.77	0.117	1.71
10	Lower right arm	2.25	0.070	1.02
11	Right hand	0.92	0.029	0.42
12	Upper right leg	16.6	0.516	7.53
13	Lower right leg	5.97	0.185	2.71
14	Right foot	1.89	0.059	0.86
15	Upper left leg	16.6	0.516	7.53
16	Lower left leg	5.97	0.185	2.71
17	Left foot	1.89	0.059	0.86
	Total	131.92	4.097	59.85

Note: See Chapter 10, Refs. [3,4].

TABLE A.9
Ninety-Fifth-Percentile (95%) Female Body Segment Masses

Body Segment Number	Name	Mass		
		Weight (lb)	Slug	kg
1	Lower torso (pelvis)	26.71	0.829	12.11
2	Middle torso (lumbar)	17.60	0.546	7.98
3	Upper torso (chest)	24.83	0.771	11.25
4	Upper left arm	4.57	0.142	2.07
5	Lower left arm	2.73	0.085	1.24
6	Left hand	1.11	0.034	0.50
7	Neck	3.88	0.120	1.76
8	Head	10.71	0.332	4.85
9	Upper right arm	4.57	0.142	2.07
10	Lower right arm	2.73	0.085	1.24
11	Right hand	1.11	0.034	0.50
12	Upper right leg	20.10	0.624	9.11
13	Lower right leg	7.23	0.224	3.28
14	Right foot	2.29	0.071	1.04
15	Upper left leg	20.10	0.624	9.11
16	Lower left leg	7.23	0.224	3.28
17	Left foot	2.29	0.071	1.04
	Total	159.79	4.958	72.43

Note: See Chapter 10, Refs. [3,4].

A.3   HUMAN BODY SEGMENT ORIGIN COORDINATES

TABLE A.10
Fifth-Percentile (5%) Male Body Segment Origin Coordinates Relative to the Reference Frame of the Adjacent Lower-Numbered Body Segment (in ft)

Body Segment Number	Name	Coordinates (ft)		
		X	Y	Z
1	Lower torso (pelvis)	0.0	0.0	0.0
2	Middle torso (lumbar)	0.0	0.0	0.319
3	Upper torso (chest)	0.0	0.0	0.637
4	Upper left arm	0.0	0.656	0.455
5	Lower left arm	0.0	0.0	−0.919
6	Left hand	0.0	0.0	−0.919
7	Neck	0.0	0.0	0.621
8	Head	0.0	0.0	0.370
9	Upper right arm	0.0	−0.656	0.455
10	Lower right arm	0.0	0.0	−0.919
11	Right hand	0.0	0.0	−0.919
12	Upper right leg	0.0	−0.241	−0.051
13	Lower right leg	0.0	0.0	−1.462
14	Right foot	0.0	0.0	−1.312
15	Upper left leg	0.0	0.241	−0.051
16	Lower left leg	0.0	0.0	−1.462
17	Left foot	0.0	0.0	−1.312

Note: See Section 6.2 and Chapter 10, Refs. [3,4].

TABLE A.11
Fifth-Percentile (5%) Male Body Segment Origin Coordinates Relative to the Reference Frame of the Adjacent Lower-Numbered Body Segment (in m)

Body Segment Number	Name	Coordinates (m)		
		X	Y	Z
1	Lower torso (pelvis)	0.0	0.0	0.0
2	Middle torso (lumbar)	0.0	0.0	0.097
3	Upper torso (chest)	0.0	0.0	0.194
4	Upper left arm	0.0	0.200	0.139
5	Lower left arm	0.0	0.0	−0.280
6	Left hand	0.0	0.0	−0.280
7	Neck	0.0	0.0	0.189
8	Head	0.0	0.0	0.113
9	Upper right arm	0.0	−0.200	0.139
10	Lower right arm	0.0	0.0	−0.280
11	Right hand	0.0	0.0	−0.280
12	Upper right leg	0.0	−0.073	−0.016
13	Lower right leg	0.0	0.0	−1.445
14	Right foot	0.0	0.0	−1.400
15	Upper left leg	0.0	0.073	−0.016
16	Lower left leg	0.0	0.0	−0.446
17	Left foot	0.0	0.0	−0.400

Note: See Section 6.2 and Chapter 10, Refs. [3,4].

TABLE A.12
Fiftieth-Percentile (50%) Male Body Segment Origin Coordinates Relative to the Reference Frame of the Adjacent Lower-Numbered Body Segment (in ft)

Body Segment Number	Name	Coordinates (ft)		
		X	Y	Z
1	Lower torso (pelvis)	0.0	0.0	0.0
2	Middle torso (lumbar)	0.0	0.0	0.338
3	Upper torso (chest)	0.0	0.0	0.675
4	Upper left arm	0.0	0.696	0.483
5	Lower left arm	0.0	0.0	−0.975
6	Left hand	0.0	0.0	−0.975
7	Neck	0.0	0.0	0.658
8	Head	0.0	0.0	0.392
9	Upper right arm	0.0	−0.696	0.483
10	Lower right arm	0.0	0.0	−0.975
11	Right hand	0.0	0.0	−0.975
12	Upper right leg	0.0	−0.256	−0.054
13	Lower right leg	0.0	0.0	−1.55
14	Right foot	0.0	0.0	−1.391
15	Upper left leg	0.0	0.256	−0.054
16	Lower left leg	0.0	0.0	−1.55
17	Left foot	0.0	0.0	−1.391

Note: See Section 6.2 and Chapter 10, Refs. [3,4].

TABLE A.13
Fiftieth-Percentile (50%) Male Body Segment Origin Coordinates Relative to the Reference Frame of the Adjacent Lower-Numbered Body Segment (in m)

Body Segment Number	Name	Coordinates (m)		
		X	Y	Z
1	Lower torso (pelvis)	0.0	0.0	0.0
2	Middle torso (lumbar)	0.0	0.0	0.103
3	Upper torso (chest)	0.0	0.0	0.206
4	Upper left arm	0.0	0.212	0.147
5	Lower left arm	0.0	0.0	−0.297
6	Left hand	0.0	0.0	−0.297
7	Neck	0.0	0.0	0.201
8	Head	0.0	0.0	0.119
9	Upper right arm	0.0	−0.212	0.147
10	Lower right arm	0.0	0.0	−0.297
11	Right hand	0.0	0.0	−0.297
12	Upper right leg	0.0	−0.078	−0.016
13	Lower right leg	0.0	0.0	−1.472
14	Right foot	0.0	0.0	−1.424
15	Upper left leg	0.0	0.078	−0.016
16	Lower left leg	0.0	0.0	−0.472
17	Left foot	0.0	0.0	−0.424

Note: See Section 6.2 and Chapter 10, Refs. [3,4].

TABLE A.14
Ninety-Fifth-Percentile (95%) Male Body Segment Origin Coordinates Relative to the Reference Frame of the Adjacent Lower-Numbered Body Segment (in ft)

Body Segment Number	Name	Coordinates (ft)		
		X	Y	Z
1	Lower torso (pelvis)	0.0	0.0	0.0
2	Middle torso (lumbar)	0.0	0.0	0.358
3	Upper torso (chest)	0.0	0.0	0.714
4	Upper left arm	0.0	0.737	0.511
5	Lower left arm	0.0	0.0	−1.032
6	Left hand	0.0	0.0	−1.032
7	Neck	0.0	0.0	0.696
8	Head	0.0	0.0	0.415
9	Upper right arm	0.0	−0.737	0.511
10	Lower right arm	0.0	0.0	−1.032
11	Right hand	0.0	0.0	−1.032
12	Upper right leg	0.0	−0.271	−0.057
13	Lower right leg	0.0	0.0	−1.641
14	Right foot	0.0	0.0	−1.472
15	Upper left leg	0.0	0.271	−0.057
16	Lower left leg	0.0	0.0	−1.641
17	Left foot	0.0	0.0	−1.472

Note: See Section 6.2 and Chapter 10, Refs. [3,4].

TABLE A.15
Ninety-Fifth-Percentile (95%) Male Body Segment Origin Coordinates Relative to the Reference Frame of the Adjacent Lower-Numbered Body Segment (in m)

Body Segment Number	Name	Coordinates (m)		
		X	Y	Z
1	Lower torso (pelvis)	0.0	0.0	0.0
2	Middle torso (lumbar)	0.0	0.0	0.109
3	Upper torso (chest)	0.0	0.0	0.218
4	Upper left arm	0.0	0.225	0.156
5	Lower left arm	0.0	0.0	−0.315
6	Left hand	0.0	0.0	−0.315
7	Neck	0.0	0.0	0.212
8	Head	0.0	0.0	0.126
9	Upper right arm	0.0	−0.225	0.156
10	Lower right arm	0.0	0.0	−0.315
11	Right hand	0.0	0.0	−0.315
12	Upper right leg	0.0	−0.083	−0.017
13	Lower right leg	0.0	0.0	−1.500
14	Right foot	0.0	0.0	−0.449
15	Upper left leg	0.0	0.083	−0.017
16	Lower left leg	0.0	0.0	−0.500
17	Left foot	0.0	0.0	−0.449

Note: See Section 6.2 and Chapter 10, Refs. [3,4].

TABLE A.16
Fifth-Percentile (5%) Female Body Segment Origin Coordinates Relative to the Reference Frame of the Adjacent Lower-Numbered Body Segment (in ft)

Body Segment Number	Name	Coordinates (ft)		
		X	Y	Z
1	Lower torso (pelvis)	0.0	0.0	0.0
2	Middle torso (lumbar)	0.0	0.0	0.288
3	Upper torso (chest)	0.0	0.0	0.577
4	Upper left arm	0.0	0.594	0.414
5	Lower left arm	0.0	0.0	−0.835
6	Left hand	0.0	0.0	−0.835
7	Neck	0.0	0.0	0.561
8	Head	0.0	0.0	0.334
9	Upper right arm	0.0	−0.594	0.414
10	Lower right arm	0.0	0.0	−0.835
11	Right hand	0.0	0.0	−0.835
12	Upper right leg	0.0	−0.218	−0.046
13	Lower right leg	0.0	0.0	−1.326
14	Right foot	0.0	0.0	−1.201
15	Upper left leg	0.0	0.218	−0.046
16	Lower left leg	0.0	0.0	−1.326
17	Left foot	0.0	0.0	−1.201

Note: See Section 6.2 and Chapter 10, Refs. [3,4].

TABLE A.17

Fifth-Percentile (5%) Female Body Segment Origin Coordinates Relative to the Reference Frame of the Adjacent Lower-Numbered Body Segment (in m)

Body Segment Number	Name	Coordinates (m)		
		X	Y	Z
1	Lower torso (pelvis)	0.0	0.0	0.0
2	Middle torso (lumbar)	0.0	0.0	0.088
3	Upper torso (chest)	0.0	0.0	0.176
4	Upper left arm	0.0	0.181	0.126
5	Lower left arm	0.0	0.0	−0.254
6	Left hand	0.0	0.0	−0.254
7	Neck	0.0	0.0	0.171
8	Head	0.0	0.0	0.102
9	Upper right arm	0.0	−0.181	0.126
10	Lower right arm	0.0	0.0	−0.254
11	Right hand	0.0	0.0	−0.254
12	Upper right leg	0.0	−0.066	−0.014
13	Lower right leg	0.0	0.0	−0.404
14	Right foot	0.0	0.0	−0.366
15	Upper left leg	0.0	0.066	−0.014
16	Lower left leg	0.0	0.0	−0.404
17	Left foot	0.0	0.0	−0.366

Note: See Section 6.2 and Chapter 10, Refs. [3,4].

TABLE A.18
Fiftieth-Percentile (50%) Female Body Segment Origin Coordinates Relative to the Reference Frame of the Adjacent Lower-Numbered Body Segment (in ft)

Body Segment Number	Name	Coordinates (ft)		
		X	Y	Z
1	Lower torso (pelvis)	0.0	0.0	0.0
2	Middle torso (lumbar)	0.0	0.0	0.308
3	Upper torso (chest)	0.0	0.0	0.617
4	Upper left arm	0.0	0.635	0.442
5	Lower left arm	0.0	0.0	−0.892
6	Left hand	0.0	0.0	−0.892
7	Neck	0.0	0.0	0.600
8	Head	0.0	0.0	0.357
9	Upper right arm	0.0	−0.635	0.442
10	Lower right arm	0.0	0.0	−0.892
11	Right hand	0.0	0.0	−0.892
12	Upper right leg	0.0	−0.233	−0.049
13	Lower right leg	0.0	0.0	−1.417
14	Right foot	0.0	0.0	−1.283
15	Upper left leg	0.0	0.233	−0.049
16	Lower left leg	0.0	0.0	−1.417
17	Left foot	0.0	0.0	−1.283

Note: See Section 6.2 and Chapter 10, Refs. [3,4].

TABLE A.19
Ninety-Fifth-Percentile (95%) Female Body Segment Origin Coordinates Relative to the Reference Frame of the Adjacent Lower-Numbered Body Segment (in m)

Body Segment Number	Name	Coordinates (m)		
		X	Y	Z
1	Lower torso (pelvis)	0.0	0.0	0.0
2	Middle torso (lumbar)	0.0	0.0	0.093
3	Upper torso (chest)	0.0	0.0	0.188
4	Upper left arm	0.0	0.194	0.135
5	Lower left arm	0.0	0.0	−0.272
6	Left hand	0.0	0.0	−0.272
7	Neck	0.0	0.0	0.183
8	Head	0.0	0.0	0.109
9	Upper right arm	0.0	−0.194	0.135
10	Lower right arm	0.0	0.0	−0.272
11	Right hand	0.0	0.0	−0.272
12	Upper right leg	0.0	−0.071	−0.015
13	Lower right leg	0.0	0.0	−0.432
14	Right foot	0.0	0.0	−0.391
15	Upper left leg	0.0	0.071	−0.015
16	Lower left leg	0.0	0.0	−0.432
17	Left foot	0.0	0.0	−0.391

Note: See Section 6.2 and Chapter 10, Refs. [3,4].

TABLE A.20
Ninety-Fifth-Percentile (95%) Female Body
Segment Origin Coordinates Relative to the
Reference Frame of the Adjacent Lower-Numbered
Body Segment (in ft)

Body Segment Number	Name	Coordinates (ft)		
		X	Y	Z
1	Lower torso (pelvis)	0.0	0.0	0.0
2	Middle torso (lumbar)	0.0	0.0	0.327
3	Upper torso (chest)	0.0	0.0	0.654
4	Upper left arm	0.0	0.634	0.469
5	Lower left arm	0.0	0.0	−0.946
6	Left hand	0.0	0.0	−0.946
7	Neck	0.0	0.0	0.637
8	Head	0.0	0.0	0.379
9	Upper right arm	0.0	−0.674	0.469
10	Lower right arm	0.0	0.0	−0.946
11	Right hand	0.0	0.0	−0.946
12	Upper right leg	0.0	−0.247	−0.052
13	Lower right leg	0.0	0.0	−1.503
14	Right foot	0.0	0.0	−1.361
15	Upper left leg	0.0	0.247	−0.052
16	Lower left leg	0.0	0.0	−1.503
17	Left foot	0.0	0.0	−1.361

Note: See Section 6.2 and Chapter 10, Refs. [3,4].

TABLE A.21
Ninety-Fifth-Percentile (95%) Female Body Segment Origin Coordinates Relative to the Reference Frame of the Adjacent Lower-Numbered Body Segment (in m)

Body Segment Number	Name	Coordinates (m)		
		X	Y	Z
1	Lower torso (pelvis)	0.0	0.0	0.0
2	Middle torso (lumbar)	0.0	0.0	0.099
3	Upper torso (chest)	0.0	0.0	0.199
4	Upper left arm	0.0	0.206	0.143
5	Lower left arm	0.0	0.0	−0.289
6	Left hand	0.0	0.0	−0.289
7	Neck	0.0	0.0	0.194
8	Head	0.0	0.0	0.116
9	Upper right arm	0.0	−0.206	0.143
10	Lower right arm	0.0	0.0	−0.289
11	Right hand	0.0	0.0	−0.289
12	Upper right leg	0.0	−0.075	−0.016
13	Lower right leg	0.0	0.0	−0.458
14	Right foot	0.0	0.0	−0.415
15	Upper left leg	0.0	0.075	−0.016
16	Lower left leg	0.0	0.0	−0.458
17	Left foot	0.0	0.0	−0.415

Note: See Section 6.2 and Chapter 10, Refs. [3,4].

A.4   HUMAN BODY SEGMENT MASS CENTER COORDINATES

TABLE A.22

Fifth-Percentile (5%) Male Body Segment Mass Center Coordinates Relative to the Reference Frame of the Body Segment (in ft)

Body Segment Number	Name	Coordinates (ft)		
		X	Y	Z
1	Lower torso (pelvis)	0.0	0.0	0.0
2	Middle torso (lumbar)	0.0	0.0	0.319
3	Upper torso (chest)	0.0	0.0	0.310
4	Upper left arm	0.0	0.0	−0.351
5	Lower left arm	0.0	0.0	−0.455
6	Left hand	0.0	0.0	−0.267
7	Neck	0.0	0.0	0.185
8	Head	0.0	0.0	0.314
9	Upper right arm	0.0	−0.674	−0.351
10	Lower right arm	0.0	0.0	−0.455
11	Right hand	0.0	0.0	−0.267
12	Upper right leg	0.0	0.0	−0.776
13	Lower right leg	0.0	0.0	−0.653
14	Right foot	0.314	0.0	−0.157
15	Upper left leg	0.0	0.0	−0.776
16	Lower left leg	0.0	0.0	−0.653
17	Left foot	0.314	0.0	−0.157

TABLE A.23
Fifth-Percentile (5%) Male Body Segment Mass Center Coordinates Relative to the Reference Frame of the Body Segment (in m)

Body Segment Number	Name	Coordinates (m)		
		X	Y	Z
1	Lower torso (pelvis)	0.0	0.0	0.0
2	Middle torso (lumbar)	0.0	0.0	0.097
3	Upper torso (chest)	0.0	0.0	0.094
4	Upper left arm	0.0	0.0	−0.107
5	Lower left arm	0.0	0.0	−0.139
6	Left hand	0.0	0.0	−0.081
7	Neck	0.0	0.0	0.057
8	Head	0.0	0.0	0.095
9	Upper right arm	0.0	0.0	−0.107
10	Lower right arm	0.0	0.0	−0.139
11	Right hand	0.0	0.0	−0.081
12	Upper right leg	0.0	0.0	−0.237
13	Lower right leg	0.0	0.0	−0.199
14	Right foot	0.095	0.0	−0.048
15	Upper left leg	0.0	0.0	−0.237
16	Lower left leg	0.0	0.0	−0.199
17	Left foot	0.095	0.0	−0.048

TABLE A.24

Fiftieth-Percentile (50%) Male Body Segment Mass Center Coordinates Relative to the Reference Frame of the Body Segment (in ft)

Body Segment Number	Name	Coordinates (ft)		
		X	Y	Z
1	Lower torso (pelvis)	0.0	0.0	0.0
2	Middle torso (lumbar)	0.0	0.0	0.338
3	Upper torso (chest)	0.0	0.0	0.329
4	Upper left arm	0.0	0.0	−0.372
5	Lower left arm	0.0	0.0	−0.483
6	Left hand	0.0	0.0	−0.283
7	Neck	0.0	0.0	0.196
8	Head	0.0	0.0	0.333
9	Upper right arm	0.0	0.0	−0.372
10	Lower right arm	0.0	0.0	−0.483
11	Right hand	0.0	0.0	−0.283
12	Upper right leg	0.0	0.0	−0.823
13	Lower right leg	0.0	0.0	−0.692
14	Right foot	0.333	0.0	−0.167
15	Upper left leg	0.0	0.0	−0.823
16	Lower left leg	0.0	0.0	−0.692
17	Left foot	0.333	0.0	−0.167

TABLE A.25

Fiftieth-Percentile (50%) Male Body Segment Mass Center Coordinates Relative to the Reference Frame of the Body Segment (in m)

Body Segment Number	Name	Coordinates (m)		
		X	Y	Z
1	Lower torso (pelvis)	0.0	0.0	0.0
2	Middle torso (lumbar)	0.0	0.0	0.103
3	Upper torso (chest)	0.0	0.0	0.100
4	Upper left arm	0.0	0.0	−0.113
5	Lower left arm	0.0	0.0	−0.147
6	Left hand	0.0	0.0	−0.086
7	Neck	0.0	0.0	0.060
8	Head	0.0	0.0	0.101
9	Upper right arm	0.0	0.0	−0.113
10	Lower right arm	0.0	0.0	−0.147
11	Right hand	0.0	0.0	−0.086
12	Upper right leg	0.0	0.0	−0.251
13	Lower right leg	0.0	0.0	−0.211
14	Right foot	0.095	0.0	−0.057
15	Upper left leg	0.0	0.247	−0.251
16	Lower left leg	0.0	0.0	−0.211
17	Left foot	0.095	0.0	−0.048

TABLE A.26
Ninety-Fifth-Percentile (95%) Male Body Segment Mass Center Coordinates Relative to the Reference Frame of the Body Segment (in ft)

Body Segment Number	Name	Coordinates (ft)		
		X	Y	Z
1	Lower torso (pelvis)	0.0	0.0	0.0
2	Middle torso (lumbar)	0.0	0.0	0.358
3	Upper torso (chest)	0.0	0.0	0.348
4	Upper left arm	0.0	0.0	−0.394
5	Lower left arm	0.0	0.0	−0.511
6	Left hand	0.0	0.0	−0.300
7	Neck	0.0	0.0	0.207
8	Head	0.0	0.0	0.352
9	Upper right arm	0.0	0.0	−0.394
10	Lower right arm	0.0	0.0	−0.511
11	Right hand	0.0	0.0	−0.300
12	Upper right leg	0.0	0.0	−0.871
13	Lower right leg	0.0	0.0	−0.732
14	Right foot	0.352	0.0	−0.177
15	Upper left leg	0.0	0.0	−0.871
16	Lower left leg	0.0	0.0	−0.732
17	Left foot	0.352	0.0	−0.177

TABLE A.27

Ninety-Fifth-Percentile (95%) Male Body Segment Mass Center Coordinates Relative to the Reference Frame of the Body Segment (in m)

Body Segment Number	Name	Coordinates (m)		
		X	Y	Z
1	Lower torso (pelvis)	0.0	0.0	0.0
2	Middle torso (lumbar)	0.0	0.0	0.109
3	Upper torso (chest)	0.0	0.0	0.106
4	Upper left arm	0.0	0.0	−0.120
5	Lower left arm	0.0	0.0	−0.156
6	Left hand	0.0	0.0	−0.091
7	Neck	0.0	0.0	0.064
8	Head	0.0	0.0	0.107
9	Upper right arm	0.0	0.0	−0.120
10	Lower right arm	0.0	0.0	−0.156
11	Right hand	0.0	0.0	−0.091
12	Upper right leg	0.0	0.0	−0.266
13	Lower right leg	0.0	0.0	−0.223
14	Right foot	0.107	0.0	−0.054
15	Upper left leg	0.0	0.0	−0.266
16	Lower left leg	0.0	0.0	−0.223
17	Left foot	0.107	0.0	−0.054

TABLE A.28

Fifth-Percentile (5%) Female Body Segment Mass Center Coordinates Relative to the Reference Frame of the Body Segment (in ft)

Body Segment Number	Name	Coordinates (ft)		
		X	Y	Z
1	Lower torso (pelvis)	0.0	0.0	0.0
2	Middle torso (lumbar)	0.0	0.0	0.288
3	Upper torso (chest)	0.0	0.0	0.281
4	Upper left arm	0.0	0.0	−0.317
5	Lower left arm	0.0	0.0	−0.414
6	Left hand	0.0	0.0	−0.241
7	Neck	0.0	0.0	0.168
8	Head	0.0	0.0	0.285
9	Upper right arm	0.0	0.0	−0.317
10	Lower right arm	0.0	0.0	−0.414
11	Right hand	0.0	0.0	−0.214
12	Upper right leg	0.0	0.0	−0.702
13	Lower right leg	0.0	0.0	−0.592
14	Right foot	0.285	0.0	−0.143
15	Upper left leg	0.0	0.0	−0.702
16	Lower left leg	0.0	0.0	−0.592
17	Left foot	0.285	0.0	−0.143

TABLE A.29

Fifth-Percentile (5%) Female Body Segment Mass Center Coordinates Relative to the Reference Frame of the Body Segment (in m)

Body Segment Number	Name	Coordinates (m)		
		X	Y	Z
1	Lower torso (pelvis)	0.0	0.0	0.0
2	Middle torso (lumbar)	0.0	0.0	0.088
3	Upper torso (chest)	0.0	0.0	0.085
4	Upper left arm	0.0	0.0	−0.096
5	Lower left arm	0.0	0.0	−0.126
6	Left hand	0.0	0.0	−0.074
7	Neck	0.0	0.0	0.051
8	Head	0.0	0.0	0.087
9	Upper right arm	0.0	0.0	−0.096
10	Lower right arm	0.0	0.0	−0.126
11	Right hand	0.0	0.0	−0.074
12	Upper right leg	0.0	0.0	−0.214
13	Lower right leg	0.0	0.0	−0.181
14	Right foot	0.087	0.0	−0.044
15	Upper left leg	0.0	0.0	−0.214
16	Lower left leg	0.0	0.0	−0.181
17	Left foot	0.087	0.0	−0.044

TABLE A.30

Fiftieth-Percentile (50%) Female Body Segment Mass Center Coordinates Relative to the Reference Frame of the Body Segment (in ft)

Body Segment Number	Name	Coordinates (ft)		
		X	Y	Z
1	Lower torso (pelvis)	0.0	0.0	0.0
2	Middle torso (lumbar)	0.0	0.0	0.308
3	Upper torso (chest)	0.0	0.0	0.300
4	Upper left arm	0.0	0.0	−0.339
5	Lower left arm	0.0	0.0	−0.442
6	Left hand	0.0	0.0	−0.258
7	Neck	0.0	0.0	0.179
8	Head	0.0	0.0	0.304
9	Upper right arm	0.0	0.0	−0.339
10	Lower right arm	0.0	0.0	−0.442
11	Right hand	0.0	0.0	−0.258
12	Upper right leg	0.0	0.0	−0.750
13	Lower right leg	0.0	0.0	−0.632
14	Right foot	0.304	0.0	−0.153
15	Upper left leg	0.0	0.0	−0.750
16	Lower left leg	0.0	0.0	−0.632
17	Left foot	0.304	0.0	−0.153

TABLE A.31
Fiftieth-Percentile (50%) Female Body Segment Mass Center Coordinates Relative to the Reference Frame of the Body Segment (in m)

Body Segment Number	Name	Coordinates (m)		
		X	Y	Z
1	Lower torso (pelvis)	0.0	0.0	0.0
2	Middle torso (lumbar)	0.0	0.0	0.094
3	Upper torso (chest)	0.0	0.0	0.091
4	Upper left arm	0.0	0.0	−0.103
5	Lower left arm	0.0	0.0	−0.135
6	Left hand	0.0	0.0	−0.079
7	Neck	0.0	0.0	0.055
8	Head	0.0	0.0	0.093
9	Upper right arm	0.0	0.0	−0.103
10	Lower right arm	0.0	0.0	−0.135
11	Right hand	0.0	0.0	−0.079
12	Upper right leg	0.0	0.0	−0.229
13	Lower right leg	0.0	0.0	−0.193
14	Right foot	0.093	0.0	−0.047
15	Upper left leg	0.0	0.0	−0.229
16	Lower left leg	0.0	0.0	−0.193
17	Left foot	0.093	0.0	−0.047

TABLE A.32

Ninety-Fifth-Percentile (95%) Female Body Segment Mass Center Coordinates Relative to the Reference Frame of the Body Segment (in ft)

Body Segment Number	Name	Coordinates (ft)		
		X	Y	Z
1	Lower torso (pelvis)	0.0	0.0	0.0
2	Middle torso (lumbar)	0.0	0.0	0.327
3	Upper torso (chest)	0.0	0.0	0.318
4	Upper left arm	0.0	0.634	−0.360
5	Lower left arm	0.0	0.0	−0.469
6	Left hand	0.0	0.0	−0.274
7	Neck	0.0	0.0	0.190
8	Head	0.0	0.0	0.322
9	Upper right arm	0.0	0.0	−0.360
10	Lower right arm	0.0	0.0	−0.469
11	Right hand	0.0	0.0	−0.274
12	Upper right leg	0.0	0.0	−0.796
13	Lower right leg	0.0	0.0	−0.670
14	Right foot	0.322	0.0	−0.162
15	Upper left leg	0.0	0.0	−0.796
16	Lower left leg	0.0	0.0	−0.670
17	Left foot	0.322	0.0	−0.162

TABLE A.33
Ninety-Fifth-Percentile (95%) Female Body Segment Mass Center Coordinates Relative to the Reference Frame of the Body Segment (in m)

Body Segment Number	Name	Coordinates (m)		
		X	Y	Z
1	Lower torso (pelvis)	0.0	0.0	0.0
2	Middle torso (lumbar)	0.0	0.0	0.100
3	Upper torso (chest)	0.0	0.0	0.097
4	Upper left arm	0.0	0.0	−0.109
5	Lower left arm	0.0	0.0	−0.143
6	Left hand	0.0	0.0	−0.084
7	Neck	0.0	0.0	0.058
8	Head	0.0	0.0	0.099
9	Upper right arm	0.0	0.0	−0.109
10	Lower right arm	0.0	0.0	−0.143
11	Right hand	0.0	0.0	−0.084
12	Upper right leg	0.0	0.0	−0.243
13	Lower right leg	0.0	0.0	−0.205
14	Right foot	0.099	0.0	−0.050
15	Upper left leg	0.0	0.0	−0.243
16	Lower left leg	0.0	0.0	−0.205
17	Left foot	0.099	0.0	−0.050

A.5 HUMAN BODY SEGMENT PRINCIPAL INERTIA MATRICES

TABLE A.34
Fifth-Percentile (5%) Male Body Segment Principal Inertia Matrices (in Slug ft^2)

Body Segment Number	Name	Inertia Matrix (in Slug ft^2) Relative to Body Frame Principal Directions
1	Lower torso (pelvis)	$\begin{bmatrix} 0.078 & 0.0 & 0.0 \\ 0.0 & 0.048 & 0.0 \\ 0.0 & 0.0 & 0.075 \end{bmatrix}$
2	Middle torso (lumbar)	$\begin{bmatrix} 0.078 & 0.0 & 0.0 \\ 0.0 & 0.048 & 0.0 \\ 0.0 & 0.0 & 0.075 \end{bmatrix}$
3	Upper torso (chest)	$\begin{bmatrix} 0.055 & 0.0 & 0.0 \\ 0.0 & 0.038 & 0.0 \\ 0.0 & 0.0 & 0.055 \end{bmatrix}$
4	Upper left arm	$\begin{bmatrix} 0.014 & 0.0 & 0.0 \\ 0.0 & 0.014 & 0.0 \\ 0.0 & 0.0 & 0.001 \end{bmatrix}$
5	Lower left arm	$\begin{bmatrix} 0.011 & 0.0 & 0.0 \\ 0.0 & 0.011 & 0.0 \\ 0.0 & 0.0 & 0.001 \end{bmatrix}$
6	Left hand	$\begin{bmatrix} 0.002 & 0.0 & 0.0 \\ 0.0 & 0.001 & 0.0 \\ 0.0 & 0.0 & 0.001 \end{bmatrix}$
7	Neck	$\begin{bmatrix} 0.008 & 0.0 & 0.0 \\ 0.0 & 0.008 & 0.0 \\ 0.0 & 0.0 & 0.001 \end{bmatrix}$
8	Head	$\begin{bmatrix} 0.020 & 0.0 & 0.0 \\ 0.0 & 0.020 & 0.0 \\ 0.0 & 0.0 & 0.010 \end{bmatrix}$
9	Upper right arm	$\begin{bmatrix} 0.014 & 0.0 & 0.0 \\ 0.0 & 0.014 & 0.0 \\ 0.0 & 0.0 & 0.001 \end{bmatrix}$
10	Lower right arm	$\begin{bmatrix} 0.011 & 0.0 & 0.0 \\ 0.0 & 0.011 & 0.0 \\ 0.0 & 0.0 & 0.001 \end{bmatrix}$

TABLE A.34 (continued)
Fifth-Percentile (5%) Male Body Segment Principal Inertia Matrices (in Slug ft²)

Body Segment Number	Name	Inertia Matrix (in Slug ft²) Relative to Body Frame Principal Directions
11	Right hand	$\begin{bmatrix} 0.002 & 0.0 & 0.0 \\ 0.0 & 0.001 & 0.0 \\ 0.0 & 0.0 & 0.001 \end{bmatrix}$
12	Upper right leg	$\begin{bmatrix} 0.050 & 0.0 & 0.0 \\ 0.0 & 0.050 & 0.0 \\ 0.0 & 0.0 & 0.013 \end{bmatrix}$
13	Lower right leg	$\begin{bmatrix} 0.004 & 0.0 & 0.0 \\ 0.0 & 0.004 & 0.0 \\ 0.0 & 0.0 & 0.001 \end{bmatrix}$
14	Right foot	$\begin{bmatrix} 0.001 & 0.0 & 0.0 \\ 0.0 & 0.004 & 0.0 \\ 0.0 & 0.0 & 0.004 \end{bmatrix}$
15	Upper left leg	$\begin{bmatrix} 0.050 & 0.0 & 0.0 \\ 0.0 & 0.050 & 0.0 \\ 0.0 & 0.0 & 0.013 \end{bmatrix}$
16	Lower left leg	$\begin{bmatrix} 0.004 & 0.0 & 0.0 \\ 0.0 & 0.004 & 0.0 \\ 0.0 & 0.0 & 0.001 \end{bmatrix}$
17	Left foot	$\begin{bmatrix} 0.001 & 0.0 & 0.0 \\ 0.0 & 0.004 & 0.0 \\ 0.0 & 0.0 & 0.004 \end{bmatrix}$

TABLE A.35

Fifth-Percentile (5%) Male Body Segment Principal Inertia Matrices (in kg m²)

Body Segment Number	Name	Inertia Matrix (in kg m²) Relative to Body Frame Principal Directions
1	Lower torso (pelvis)	$\begin{bmatrix} 0.105 & 0.0 & 0.0 \\ 0.0 & 0.065 & 0.0 \\ 0.0 & 0.0 & 0.105 \end{bmatrix}$
2	Middle torso (lumbar)	$\begin{bmatrix} 0.105 & 0.0 & 0.0 \\ 0.0 & 0.065 & 0.0 \\ 0.0 & 0.0 & 0.105 \end{bmatrix}$
3	Upper torso (chest)	$\begin{bmatrix} 0.075 & 0.0 & 0.0 \\ 0.0 & 0.052 & 0.0 \\ 0.0 & 0.0 & 0.075 \end{bmatrix}$
4	Upper left arm	$\begin{bmatrix} 0.019 & 0.0 & 0.0 \\ 0.0 & 0.019 & 0.0 \\ 0.0 & 0.0 & 0.002 \end{bmatrix}$
5	Lower left arm	$\begin{bmatrix} 0.014 & 0.0 & 0.0 \\ 0.0 & 0.014 & 0.0 \\ 0.0 & 0.0 & 0.001 \end{bmatrix}$
6	Left hand	$\begin{bmatrix} 0.003 & 0.0 & 0.0 \\ 0.0 & 0.001 & 0.0 \\ 0.0 & 0.0 & 0.001 \end{bmatrix}$
7	Neck	$\begin{bmatrix} 0.011 & 0.0 & 0.0 \\ 0.0 & 0.011 & 0.0 \\ 0.0 & 0.0 & 0.002 \end{bmatrix}$
8	Head	$\begin{bmatrix} 0.027 & 0.0 & 0.0 \\ 0.0 & 0.027 & 0.0 \\ 0.0 & 0.0 & 0.014 \end{bmatrix}$
9	Upper right arm	$\begin{bmatrix} 0.019 & 0.0 & 0.0 \\ 0.0 & 0.019 & 0.0 \\ 0.0 & 0.0 & 0.002 \end{bmatrix}$
10	Lower right arm	$\begin{bmatrix} 0.014 & 0.0 & 0.0 \\ 0.0 & 0.014 & 0.0 \\ 0.0 & 0.0 & 0.001 \end{bmatrix}$

TABLE A.35 (continued)
Fifth-Percentile (5%) Male Body Segment Principal Inertia
Matrices (in kg m²)

Body Segment Number	Name	Inertia Matrix (in kg m²) Relative to Body Frame Principal Directions
11	Right hand	$\begin{bmatrix} 0.003 & 0.0 & 0.0 \\ 0.0 & 0.001 & 0.0 \\ 0.0 & 0.0 & 0.001 \end{bmatrix}$
12	Upper right leg	$\begin{bmatrix} 0.069 & 0.0 & 0.0 \\ 0.0 & 0.069 & 0.0 \\ 0.0 & 0.0 & 0.017 \end{bmatrix}$
13	Lower right leg	$\begin{bmatrix} 0.006 & 0.0 & 0.0 \\ 0.0 & 0.006 & 0.0 \\ 0.0 & 0.0 & 0.001 \end{bmatrix}$
14	Right foot	$\begin{bmatrix} 0.001 & 0.0 & 0.0 \\ 0.0 & 0.005 & 0.0 \\ 0.0 & 0.0 & 0.005 \end{bmatrix}$
15	Upper left leg	$\begin{bmatrix} 0.069 & 0.0 & 0.0 \\ 0.0 & 0.069 & 0.0 \\ 0.0 & 0.0 & 0.017 \end{bmatrix}$
16	Lower left leg	$\begin{bmatrix} 0.006 & 0.0 & 0.0 \\ 0.0 & 0.006 & 0.0 \\ 0.0 & 0.0 & 0.001 \end{bmatrix}$
17	Left foot	$\begin{bmatrix} 0.001 & 0.0 & 0.0 \\ 0.0 & 0.005 & 0.0 \\ 0.0 & 0.0 & 0.005 \end{bmatrix}$

TABLE A.36

Fiftieth-Percentile (50%) Male Body Segment Principal Inertia Matrices (in Slug ft²)

Body Segment Number	Name	Inertia Matrix (in Slug ft²) Relative to Body Frame Principal Directions
1	Lower torso (pelvis)	$\begin{bmatrix} 0.109 & 0.0 & 0.0 \\ 0.0 & 0.067 & 0.0 \\ 0.0 & 0.0 & 0.106 \end{bmatrix}$
2	Middle torso (lumbar)	$\begin{bmatrix} 0.109 & 0.0 & 0.0 \\ 0.0 & 0.067 & 0.0 \\ 0.0 & 0.0 & 0.106 \end{bmatrix}$
3	Upper torso (chest)	$\begin{bmatrix} 0.078 & 0.0 & 0.0 \\ 0.0 & 0.054 & 0.0 \\ 0.0 & 0.0 & 0.078 \end{bmatrix}$
4	Upper left arm	$\begin{bmatrix} 0.020 & 0.0 & 0.0 \\ 0.0 & 0.020 & 0.0 \\ 0.0 & 0.0 & 0.002 \end{bmatrix}$
5	Lower left arm	$\begin{bmatrix} 0.015 & 0.0 & 0.0 \\ 0.0 & 0.015 & 0.0 \\ 0.0 & 0.0 & 0.001 \end{bmatrix}$
6	Left hand	$\begin{bmatrix} 0.003 & 0.0 & 0.0 \\ 0.0 & 0.001 & 0.0 \\ 0.0 & 0.0 & 0.001 \end{bmatrix}$
7	Neck	$\begin{bmatrix} 0.011 & 0.0 & 0.0 \\ 0.0 & 0.011 & 0.0 \\ 0.0 & 0.0 & 0.002 \end{bmatrix}$
8	Head	$\begin{bmatrix} 0.028 & 0.0 & 0.0 \\ 0.0 & 0.028 & 0.0 \\ 0.0 & 0.0 & 0.014 \end{bmatrix}$
9	Upper right arm	$\begin{bmatrix} 0.020 & 0.0 & 0.0 \\ 0.0 & 0.020 & 0.0 \\ 0.0 & 0.0 & 0.002 \end{bmatrix}$
10	Lower right arm	$\begin{bmatrix} 0.015 & 0.0 & 0.0 \\ 0.0 & 0.015 & 0.0 \\ 0.0 & 0.0 & 0.001 \end{bmatrix}$

TABLE A.36 (continued)
Fiftieth-Percentile (50%) Male Body Segment Principal Inertia Matrices (in Slug ft²)

Body Segment Number	Name	Inertia Matrix (in Slug ft²) Relative to Body Frame Principal Directions
11	Right hand	$\begin{bmatrix} 0.003 & 0.0 & 0.0 \\ 0.0 & 0.001 & 0.0 \\ 0.0 & 0.0 & 0.001 \end{bmatrix}$
12	Upper right leg	$\begin{bmatrix} 0.071 & 0.0 & 0.0 \\ 0.0 & 0.071 & 0.0 \\ 0.0 & 0.0 & 0.018 \end{bmatrix}$
13	Lower right leg	$\begin{bmatrix} 0.006 & 0.0 & 0.0 \\ 0.0 & 0.006 & 0.0 \\ 0.0 & 0.0 & 0.001 \end{bmatrix}$
14	Right foot	$\begin{bmatrix} 0.001 & 0.0 & 0.0 \\ 0.0 & 0.005 & 0.0 \\ 0.0 & 0.0 & 0.005 \end{bmatrix}$
15	Upper left leg	$\begin{bmatrix} 0.071 & 0.0 & 0.0 \\ 0.0 & 0.071 & 0.0 \\ 0.0 & 0.0 & 0.018 \end{bmatrix}$
16	Lower left leg	$\begin{bmatrix} 0.006 & 0.0 & 0.0 \\ 0.0 & 0.006 & 0.0 \\ 0.0 & 0.0 & 0.001 \end{bmatrix}$
17	Left foot	$\begin{bmatrix} 0.001 & 0.0 & 0.0 \\ 0.0 & 0.005 & 0.0 \\ 0.0 & 0.0 & 0.005 \end{bmatrix}$

TABLE A.37

Fiftieth-Percentile (50%) Male Body Segment Principal Inertia Matrices (in kg m²)

Body Segment Number	Name	Inertia Matrix (in kg m²) Relative to Body Frame Principal Directions
1	Lower torso (pelvis)	$\begin{bmatrix} 0.148 & 0.0 & 0.0 \\ 0.0 & 0.091 & 0.0 \\ 0.0 & 0.0 & 0.148 \end{bmatrix}$
2	Middle torso (lumbar)	$\begin{bmatrix} 0.148 & 0.0 & 0.0 \\ 0.0 & 0.091 & 0.0 \\ 0.0 & 0.0 & 0.148 \end{bmatrix}$
3	Upper torso (chest)	$\begin{bmatrix} 0.106 & 0.0 & 0.0 \\ 0.0 & 0.073 & 0.0 \\ 0.0 & 0.0 & 0.106 \end{bmatrix}$
4	Upper left arm	$\begin{bmatrix} 0.027 & 0.0 & 0.0 \\ 0.0 & 0.027 & 0.0 \\ 0.0 & 0.0 & 0.003 \end{bmatrix}$
5	Lower left arm	$\begin{bmatrix} 0.020 & 0.0 & 0.0 \\ 0.0 & 0.020 & 0.0 \\ 0.0 & 0.0 & 0.001 \end{bmatrix}$
6	Left hand	$\begin{bmatrix} 0.004 & 0.0 & 0.0 \\ 0.0 & 0.001 & 0.0 \\ 0.0 & 0.0 & 0.001 \end{bmatrix}$
7	Neck	$\begin{bmatrix} 0.015 & 0.0 & 0.0 \\ 0.0 & 0.015 & 0.0 \\ 0.0 & 0.0 & 0.003 \end{bmatrix}$
8	Head	$\begin{bmatrix} 0.038 & 0.0 & 0.0 \\ 0.0 & 0.038 & 0.0 \\ 0.0 & 0.0 & 0.019 \end{bmatrix}$
9	Upper right arm	$\begin{bmatrix} 0.027 & 0.0 & 0.0 \\ 0.0 & 0.027 & 0.0 \\ 0.0 & 0.0 & 0.003 \end{bmatrix}$
10	Lower right arm	$\begin{bmatrix} 0.020 & 0.0 & 0.0 \\ 0.0 & 0.020 & 0.0 \\ 0.0 & 0.0 & 0.001 \end{bmatrix}$

TABLE A.37 (continued)

Fiftieth-Percentile (50%) Male Body Segment Principal Inertia Matrices (in kg m²)

Body Segment Number	Name	Inertia Matrix (in kg m²) Relative to Body Frame Principal Directions
11	Right hand	$\begin{bmatrix} 0.004 & 0.0 & 0.0 \\ 0.0 & 0.001 & 0.0 \\ 0.0 & 0.0 & 0.001 \end{bmatrix}$
12	Upper right leg	$\begin{bmatrix} 0.097 & 0.0 & 0.0 \\ 0.0 & 0.097 & 0.0 \\ 0.0 & 0.0 & 0.024 \end{bmatrix}$
13	Lower right leg	$\begin{bmatrix} 0.008 & 0.0 & 0.0 \\ 0.0 & 0.008 & 0.0 \\ 0.0 & 0.0 & 0.001 \end{bmatrix}$
14	Right foot	$\begin{bmatrix} 0.001 & 0.0 & 0.0 \\ 0.0 & 0.007 & 0.0 \\ 0.0 & 0.0 & 0.007 \end{bmatrix}$
15	Upper left leg	$\begin{bmatrix} 0.097 & 0.0 & 0.0 \\ 0.0 & 0.097 & 0.0 \\ 0.0 & 0.0 & 0.024 \end{bmatrix}$
16	Lower left leg	$\begin{bmatrix} 0.008 & 0.0 & 0.0 \\ 0.0 & 0.008 & 0.0 \\ 0.0 & 0.0 & 0.001 \end{bmatrix}$
17	Left foot	$\begin{bmatrix} 0.001 & 0.0 & 0.0 \\ 0.0 & 0.007 & 0.0 \\ 0.0 & 0.0 & 0.007 \end{bmatrix}$

TABLE A.38
Ninety-Fifth-Percentile (95%) Male Body Segment Principal
Inertia Matrices (in Slug ft²)

Body Segment Number	Name	Inertia Matrix (in Slug ft²) Relative to Body Frame Principal Directions
1	Lower torso (pelvis)	$\begin{bmatrix} 0.146 & 0.0 & 0.0 \\ 0.0 & 0.090 & 0.0 \\ 0.0 & 0.0 & 0.142 \end{bmatrix}$
2	Middle torso (lumbar)	$\begin{bmatrix} 0.146 & 0.0 & 0.0 \\ 0.0 & 0.090 & 0.0 \\ 0.0 & 0.0 & 0.142 \end{bmatrix}$
3	Upper torso (chest)	$\begin{bmatrix} 0.105 & 0.0 & 0.0 \\ 0.0 & 0.073 & 0.0 \\ 0.0 & 0.0 & 0.105 \end{bmatrix}$
4	Upper left arm	$\begin{bmatrix} 0.027 & 0.0 & 0.0 \\ 0.0 & 0.027 & 0.0 \\ 0.0 & 0.0 & 0.003 \end{bmatrix}$
5	Lower left arm	$\begin{bmatrix} 0.020 & 0.0 & 0.0 \\ 0.0 & 0.020 & 0.0 \\ 0.0 & 0.0 & 0.001 \end{bmatrix}$
6	Left hand	$\begin{bmatrix} 0.004 & 0.0 & 0.0 \\ 0.0 & 0.001 & 0.0 \\ 0.0 & 0.0 & 0.001 \end{bmatrix}$
7	Neck	$\begin{bmatrix} 0.015 & 0.0 & 0.0 \\ 0.0 & 0.015 & 0.0 \\ 0.0 & 0.0 & 0.003 \end{bmatrix}$
8	Head	$\begin{bmatrix} 0.038 & 0.0 & 0.0 \\ 0.0 & 0.038 & 0.0 \\ 0.0 & 0.0 & 0.019 \end{bmatrix}$
9	Upper right arm	$\begin{bmatrix} 0.027 & 0.0 & 0.0 \\ 0.0 & 0.027 & 0.0 \\ 0.0 & 0.0 & 0.003 \end{bmatrix}$
10	Lower right arm	$\begin{bmatrix} 0.020 & 0.0 & 0.0 \\ 0.0 & 0.020 & 0.0 \\ 0.0 & 0.0 & 0.001 \end{bmatrix}$

TABLE A.38 (continued)
Ninety-Fifth-Percentile (95%) Male Body Segment Principal
Inertia Matrices (in Slug ft²)

Body Segment Number	Name	Inertia Matrix (in Slug ft²) Relative to Body Frame Principal Directions		
11	Right hand	0.004	0.0	0.0
		0.0	0.001	0.0
		0.0	0.0	0.001
12	Upper right leg	0.095	0.0	0.0
		0.0	0.095	0.0
		0.0	0.0	0.024
13	Lower right leg	0.008	0.0	0.0
		0.0	0.008	0.0
		0.0	0.0	0.001
14	Right foot	0.001	0.0	0.0
		0.0	0.007	0.0
		0.0	0.0	0.007
15	Upper left leg	0.095	0.0	0.0
		0.0	0.095	0.0
		0.0	0.0	0.024
16	Lower left leg	0.008	0.0	0.0
		0.0	0.008	0.0
		0.0	0.0	0.001
17	Left foot	0.001	0.0	0.0
		0.0	0.007	0.0
		0.0	0.0	0.007

TABLE A.39
Ninety-Fifth-Percentile (95%) Male Body Segment Principal Inertia Matrices (in kg m²)

Body Segment Number	Name	Inertia Matrix (in kg m²) Relative to Body Frame Principal Directions		
1	Lower torso (pelvis)	0.199	0.0	0.0
		0.0	0.122	0.0
		0.0	0.0	0.199
2	Middle torso (lumbar)	0.199	0.0	0.0
		0.0	0.122	0.0
		0.0	0.0	0.199
3	Upper torso (chest)	0.142	0.0	0.0
		0.0	0.098	0.0
		0.0	0.0	0.142
4	Upper left arm	0.036	0.0	0.0
		0.0	0.036	0.0
		0.0	0.0	0.004
5	Lower left arm	0.027	0.0	0.0
		0.0	0.027	0.0
		0.0	0.0	0.001
6	Left hand	0.005	0.0	0.0
		0.0	0.001	0.0
		0.0	0.0	0.001
7	Neck	0.020	0.0	0.0
		0.0	0.020	0.0
		0.0	0.0	0.004
8	Head	0.051	0.0	0.0
		0.0	0.051	0.0
		0.0	0.0	0.026
9	Upper right arm	0.036	0.0	0.0
		0.0	0.036	0.0
		0.0	0.0	0.004
10	Lower right arm	0.027	0.0	0.0
		0.0	0.027	0.0
		0.0	0.0	0.001

TABLE A.39 (continued)
Ninety-Fifth-Percentile (95%) Male Body Segment Principal
Inertia Matrices (in kg m²)

Body Segment Number	Name	Inertia Matrix (in kg m²) Relative to Body Frame Principal Directions		
11	Right hand	$\begin{bmatrix} 0.005 & 0.0 & 0.0 \\ 0.0 & 0.001 & 0.0 \\ 0.0 & 0.0 & 0.001 \end{bmatrix}$		
12	Upper right leg	$\begin{bmatrix} 0.130 & 0.0 & 0.0 \\ 0.0 & 0.130 & 0.0 \\ 0.0 & 0.0 & 0.032 \end{bmatrix}$		
13	Lower right leg	$\begin{bmatrix} 0.011 & 0.0 & 0.0 \\ 0.0 & 0.011 & 0.0 \\ 0.0 & 0.0 & 0.001 \end{bmatrix}$		
14	Right foot	$\begin{bmatrix} 0.001 & 0.0 & 0.0 \\ 0.0 & 0.009 & 0.0 \\ 0.0 & 0.0 & 0.009 \end{bmatrix}$		
15	Upper left leg	$\begin{bmatrix} 0.130 & 0.0 & 0.0 \\ 0.0 & 0.130 & 0.0 \\ 0.0 & 0.0 & 0.032 \end{bmatrix}$		
16	Lower left leg	$\begin{bmatrix} 0.011 & 0.0 & 0.0 \\ 0.0 & 0.011 & 0.0 \\ 0.0 & 0.0 & 0.001 \end{bmatrix}$		
17	Left foot	$\begin{bmatrix} 0.001 & 0.0 & 0.0 \\ 0.0 & 0.009 & 0.0 \\ 0.0 & 0.0 & 0.009 \end{bmatrix}$		

TABLE A.40

Fifth-Percentile (5%) Female Body Segment Principal Inertia Matrices (in Slug ft²)

Body Segment Number	Name	Inertia Matrix (in Slug ft²) Relative to Body Frame Principal Directions
1	Lower torso (pelvis)	$\begin{bmatrix} 0.063 & 0.0 & 0.0 \\ 0.0 & 0.039 & 0.0 \\ 0.0 & 0.0 & 0.062 \end{bmatrix}$
2	Middle torso (lumbar)	$\begin{bmatrix} 0.038 & 0.0 & 0.0 \\ 0.0 & 0.023 & 0.0 \\ 0.0 & 0.0 & 0.037 \end{bmatrix}$
3	Upper torso (chest)	$\begin{bmatrix} 0.023 & 0.0 & 0.0 \\ 0.0 & 0.015 & 0.0 \\ 0.0 & 0.0 & 0.022 \end{bmatrix}$
4	Upper left arm	$\begin{bmatrix} 0.008 & 0.0 & 0.0 \\ 0.0 & 0.008 & 0.0 \\ 0.0 & 0.0 & 0.001 \end{bmatrix}$
5	Lower left arm	$\begin{bmatrix} 0.007 & 0.0 & 0.0 \\ 0.0 & 0.007 & 0.0 \\ 0.0 & 0.0 & 0.001 \end{bmatrix}$
6	Left hand	$\begin{bmatrix} 0.001 & 0.0 & 0.0 \\ 0.0 & 0.001 & 0.0 \\ 0.0 & 0.0 & 0.001 \end{bmatrix}$
7	Neck	$\begin{bmatrix} 0.001 & 0.0 & 0.0 \\ 0.0 & 0.001 & 0.0 \\ 0.0 & 0.0 & 0.001 \end{bmatrix}$
8	Head	$\begin{bmatrix} 0.013 & 0.0 & 0.0 \\ 0.0 & 0.013 & 0.0 \\ 0.0 & 0.0 & 0.006 \end{bmatrix}$
9	Upper right arm	$\begin{bmatrix} 0.008 & 0.0 & 0.0 \\ 0.0 & 0.008 & 0.0 \\ 0.0 & 0.0 & 0.001 \end{bmatrix}$
10	Lower right arm	$\begin{bmatrix} 0.007 & 0.0 & 0.0 \\ 0.0 & 0.007 & 0.0 \\ 0.0 & 0.0 & 0.001 \end{bmatrix}$

TABLE A.40 (continued)
Fifth-Percentile (5%) Female Body Segment Principal Inertia Matrices (in Slug ft²)

Body Segment Number	Name	Inertia Matrix (in Slug ft²) Relative to Body Frame Principal Directions		
11	Right hand	0.001	0.0	0.0
		0.0	0.001	0.0
		0.0	0.0	0.001
12	Upper right leg	0.037	0.0	0.0
		0.0	0.037	0.0
		0.0	0.0	0.009
13	Lower right leg	0.003	0.0	0.0
		0.0	0.003	0.0
		0.0	0.0	0.001
14	Right foot	0.001	0.0	0.0
		0.0	0.003	0.0
		0.0	0.0	0.003
15	Upper left leg	0.037	0.0	0.0
		0.0	0.037	0.0
		0.0	0.0	0.009
16	Lower left leg	0.003	0.0	0.0
		0.0	0.003	0.0
		0.0	0.0	0.001
17	Left foot	0.001	0.0	0.0
		0.0	0.002	0.0
		0.0	0.0	0.003

TABLE A.41

Fifth-Percentile (5%) Female Body Segment Principal Inertia Matrices (in kg m²)

Body Segment Number	Name	Inertia Matrix (in kg m²) Relative to Body Frame Principal Directions
1	Lower torso (pelvis)	$\begin{bmatrix} 0.086 & 0.0 & 0.0 \\ 0.0 & 0.052 & 0.0 \\ 0.0 & 0.0 & 0.083 \end{bmatrix}$
2	Middle torso (lumbar)	$\begin{bmatrix} 0.051 & 0.0 & 0.0 \\ 0.0 & 0.031 & 0.0 \\ 0.0 & 0.0 & 0.050 \end{bmatrix}$
3	Upper torso (chest)	$\begin{bmatrix} 0.031 & 0.0 & 0.0 \\ 0.0 & 0.021 & 0.0 \\ 0.0 & 0.0 & 0.031 \end{bmatrix}$
4	Upper left arm	$\begin{bmatrix} 0.012 & 0.0 & 0.0 \\ 0.0 & 0.012 & 0.0 \\ 0.0 & 0.0 & 0.001 \end{bmatrix}$
5	Lower left arm	$\begin{bmatrix} 0.009 & 0.0 & 0.0 \\ 0.0 & 0.009 & 0.0 \\ 0.0 & 0.0 & 0.001 \end{bmatrix}$
6	Left hand	$\begin{bmatrix} 0.001 & 0.0 & 0.0 \\ 0.0 & 0.001 & 0.0 \\ 0.0 & 0.0 & 0.001 \end{bmatrix}$
7	Neck	$\begin{bmatrix} 0.007 & 0.0 & 0.0 \\ 0.0 & 0.007 & 0.0 \\ 0.0 & 0.0 & 0.001 \end{bmatrix}$
8	Head	$\begin{bmatrix} 0.017 & 0.0 & 0.0 \\ 0.0 & 0.017 & 0.0 \\ 0.0 & 0.0 & 0.009 \end{bmatrix}$
9	Upper right arm	$\begin{bmatrix} 0.012 & 0.0 & 0.0 \\ 0.0 & 0.012 & 0.0 \\ 0.0 & 0.0 & 0.001 \end{bmatrix}$
10	Lower right arm	$\begin{bmatrix} 0.009 & 0.0 & 0.0 \\ 0.0 & 0.009 & 0.0 \\ 0.0 & 0.0 & 0.001 \end{bmatrix}$

TABLE A.41 (continued)
Fifth-Percentile (5%) Female Body Segment Principal Inertia Matrices (in kg m²)

Body Segment Number	Name	Inertia Matrix (in kg m²) Relative to Body Frame Principal Directions
11	Right hand	$\begin{bmatrix} 0.001 & 0.0 & 0.0 \\ 0.0 & 0.001 & 0.0 \\ 0.0 & 0.0 & 0.001 \end{bmatrix}$
12	Upper right leg	$\begin{bmatrix} 0.049 & 0.0 & 0.0 \\ 0.0 & 0.049 & 0.0 \\ 0.0 & 0.0 & 0.013 \end{bmatrix}$
13	Lower right leg	$\begin{bmatrix} 0.003 & 0.0 & 0.0 \\ 0.0 & 0.003 & 0.0 \\ 0.0 & 0.0 & 0.001 \end{bmatrix}$
14	Right foot	$\begin{bmatrix} 0.001 & 0.0 & 0.0 \\ 0.0 & 0.003 & 0.0 \\ 0.0 & 0.0 & 0.003 \end{bmatrix}$
15	Upper left leg	$\begin{bmatrix} 0.049 & 0.0 & 0.0 \\ 0.0 & 0.049 & 0.0 \\ 0.0 & 0.0 & 0.013 \end{bmatrix}$
16	Lower left leg	$\begin{bmatrix} 0.003 & 0.0 & 0.0 \\ 0.0 & 0.003 & 0.0 \\ 0.0 & 0.0 & 0.001 \end{bmatrix}$
17	Left foot	$\begin{bmatrix} 0.001 & 0.0 & 0.0 \\ 0.0 & 0.003 & 0.0 \\ 0.0 & 0.0 & 0.003 \end{bmatrix}$

TABLE A.42
Fiftieth-Percentile (50%) Female Body Segment Principal
Inertia Matrices (in Slug ft²)

Body Segment Number	Name	Inertia Matrix (in Slug ft²) Relative to Body Frame Principal Directions
1	Lower torso (pelvis)	$\begin{bmatrix} 0.091 & 0.0 & 0.0 \\ 0.0 & 0.056 & 0.0 \\ 0.0 & 0.0 & 0.089 \end{bmatrix}$
2	Middle torso (lumbar)	$\begin{bmatrix} 0.055 & 0.0 & 0.0 \\ 0.0 & 0.033 & 0.0 \\ 0.0 & 0.0 & 0.053 \end{bmatrix}$
3	Upper torso (chest)	$\begin{bmatrix} 0.033 & 0.0 & 0.0 \\ 0.0 & 0.022 & 0.0 \\ 0.0 & 0.0 & 0.032 \end{bmatrix}$
4	Upper left arm	$\begin{bmatrix} 0.012 & 0.0 & 0.0 \\ 0.0 & 0.012 & 0.0 \\ 0.0 & 0.0 & 0.001 \end{bmatrix}$
5	Lower left arm	$\begin{bmatrix} 0.010 & 0.0 & 0.0 \\ 0.0 & 0.010 & 0.0 \\ 0.0 & 0.0 & 0.001 \end{bmatrix}$
6	Left hand	$\begin{bmatrix} 0.002 & 0.0 & 0.0 \\ 0.0 & 0.001 & 0.0 \\ 0.0 & 0.0 & 0.001 \end{bmatrix}$
7	Neck	$\begin{bmatrix} 0.008 & 0.0 & 0.0 \\ 0.0 & 0.008 & 0.0 \\ 0.0 & 0.0 & 0.001 \end{bmatrix}$
8	Head	$\begin{bmatrix} 0.019 & 0.0 & 0.0 \\ 0.0 & 0.019 & 0.0 \\ 0.0 & 0.0 & 0.009 \end{bmatrix}$
9	Upper right arm	$\begin{bmatrix} 0.012 & 0.0 & 0.0 \\ 0.0 & 0.012 & 0.0 \\ 0.0 & 0.0 & 0.001 \end{bmatrix}$
10	Lower right arm	$\begin{bmatrix} 0.010 & 0.0 & 0.0 \\ 0.0 & 0.010 & 0.0 \\ 0.0 & 0.0 & 0.001 \end{bmatrix}$

TABLE A.42 (continued)
Fiftieth-Percentile (50%) Female Body Segment Principal Inertia Matrices (in Slug ft²)

Body Segment Number	Name	Inertia Matrix (in Slug ft²) Relative to Body Frame Principal Directions		
11	Right hand	0.002	0.0	0.0
		0.0	0.001	0.0
		0.0	0.0	0.001
12	Upper right leg	0.053	0.0	0.0
		0.0	0.053	0.0
		0.0	0.0	0.013
13	Lower right leg	0.004	0.0	0.0
		0.0	0.004	0.0
		0.0	0.0	0.001
14	Right foot	0.001	0.0	0.0
		0.0	0.003	0.0
		0.0	0.0	0.004
15	Upper left leg	0.053	0.0	0.0
		0.0	0.053	0.0
		0.0	0.0	0.013
16	Lower left leg	0.004	0.0	0.0
		0.0	0.004	0.0
		0.0	0.0	0.001
17	Left foot	0.001	0.0	0.0
		0.0	0.003	0.0
		0.0	0.0	0.004

TABLE A.43
Fiftieth-Percentile (50%) Female Body Segment Principal Inertia Matrices (in kg m²)

Body Segment Number	Name	Inertia Matrix (in kg m²) Relative to Body Frame Principal Directions		
1	Lower torso (pelvis)	$\begin{bmatrix} 0.123 & 0.0 & 0.0 \\ 0.0 & 0.075 & 0.0 \\ 0.0 & 0.0 & 0.120 \end{bmatrix}$		
2	Middle torso (lumbar)	$\begin{bmatrix} 0.074 & 0.0 & 0.0 \\ 0.0 & 0.045 & 0.0 \\ 0.0 & 0.0 & 0.072 \end{bmatrix}$		
3	Upper torso (chest)	$\begin{bmatrix} 0.044 & 0.0 & 0.0 \\ 0.0 & 0.030 & 0.0 \\ 0.0 & 0.0 & 0.044 \end{bmatrix}$		
4	Upper left arm	$\begin{bmatrix} 0.017 & 0.0 & 0.0 \\ 0.0 & 0.017 & 0.0 \\ 0.0 & 0.0 & 0.002 \end{bmatrix}$		
5	Lower left arm	$\begin{bmatrix} 0.013 & 0.0 & 0.0 \\ 0.0 & 0.013 & 0.0 \\ 0.0 & 0.0 & 0.001 \end{bmatrix}$		
6	Left hand	$\begin{bmatrix} 0.002 & 0.0 & 0.0 \\ 0.0 & 0.001 & 0.0 \\ 0.0 & 0.0 & 0.001 \end{bmatrix}$		
7	Neck	$\begin{bmatrix} 0.010 & 0.0 & 0.0 \\ 0.0 & 0.010 & 0.0 \\ 0.0 & 0.0 & 0.002 \end{bmatrix}$		
8	Head	$\begin{bmatrix} 0.025 & 0.0 & 0.0 \\ 0.0 & 0.025 & 0.0 \\ 0.0 & 0.0 & 0.013 \end{bmatrix}$		
9	Upper right arm	$\begin{bmatrix} 0.017 & 0.0 & 0.0 \\ 0.0 & 0.017 & 0.0 \\ 0.0 & 0.0 & 0.002 \end{bmatrix}$		
10	Lower right arm	$\begin{bmatrix} 0.013 & 0.0 & 0.0 \\ 0.0 & 0.013 & 0.0 \\ 0.0 & 0.0 & 0.001 \end{bmatrix}$		

TABLE A.43 (continued)
Fiftieth-Percentile (50%) Female Body Segment Principal Inertia Matrices (in kg m²)

Body Segment Number	Name	Inertia Matrix (in kg m²) Relative to Body Frame Principal Directions		
11	Right hand	$\begin{bmatrix} 0.002 & 0.0 & 0.0 \\ 0.0 & 0.001 & 0.0 \\ 0.0 & 0.0 & 0.001 \end{bmatrix}$		
12	Upper right leg	$\begin{bmatrix} 0.071 & 0.0 & 0.0 \\ 0.0 & 0.071 & 0.0 \\ 0.0 & 0.0 & 0.018 \end{bmatrix}$		
13	Lower right leg	$\begin{bmatrix} 0.005 & 0.0 & 0.0 \\ 0.0 & 0.005 & 0.0 \\ 0.0 & 0.0 & 0.001 \end{bmatrix}$		
14	Right foot	$\begin{bmatrix} 0.001 & 0.0 & 0.0 \\ 0.0 & 0.004 & 0.0 \\ 0.0 & 0.0 & 0.005 \end{bmatrix}$		
15	Upper left leg	$\begin{bmatrix} 0.071 & 0.0 & 0.0 \\ 0.0 & 0.071 & 0.0 \\ 0.0 & 0.0 & 0.018 \end{bmatrix}$		
16	Lower left leg	$\begin{bmatrix} 0.005 & 0.0 & 0.0 \\ 0.0 & 0.005 & 0.0 \\ 0.0 & 0.0 & 0.001 \end{bmatrix}$		
17	Left foot	$\begin{bmatrix} 0.001 & 0.0 & 0.0 \\ 0.0 & 0.004 & 0.0 \\ 0.0 & 0.0 & 0.005 \end{bmatrix}$		

TABLE A.44
Ninety-Fifth-Percentile (95%) Female Body Segment Principal Inertia Matrices (in Slug ft²)

Body Segment Number	Name	Inertia Matrix (in Slug ft²) Relative to Body Frame Principal Directions
1	Lower torso (pelvis)	$\begin{bmatrix} 0.128 & 0.0 & 0.0 \\ 0.0 & 0.079 & 0.0 \\ 0.0 & 0.0 & 0.126 \end{bmatrix}$
2	Middle torso (lumbar)	$\begin{bmatrix} 0.078 & 0.0 & 0.0 \\ 0.0 & 0.047 & 0.0 \\ 0.0 & 0.0 & 0.075 \end{bmatrix}$
3	Upper torso (chest)	$\begin{bmatrix} 0.047 & 0.0 & 0.0 \\ 0.0 & 0.031 & 0.0 \\ 0.0 & 0.0 & 0.045 \end{bmatrix}$
4	Upper left arm	$\begin{bmatrix} 0.017 & 0.0 & 0.0 \\ 0.0 & 0.017 & 0.0 \\ 0.0 & 0.0 & 0.001 \end{bmatrix}$
5	Lower left arm	$\begin{bmatrix} 0.014 & 0.0 & 0.0 \\ 0.0 & 0.014 & 0.0 \\ 0.0 & 0.0 & 0.001 \end{bmatrix}$
6	Left hand	$\begin{bmatrix} 0.003 & 0.0 & 0.0 \\ 0.0 & 0.001 & 0.0 \\ 0.0 & 0.0 & 0.001 \end{bmatrix}$
7	Neck	$\begin{bmatrix} 0.011 & 0.0 & 0.0 \\ 0.0 & 0.011 & 0.0 \\ 0.0 & 0.0 & 0.001 \end{bmatrix}$
8	Head	$\begin{bmatrix} 0.027 & 0.0 & 0.0 \\ 0.0 & 0.027 & 0.0 \\ 0.0 & 0.0 & 0.013 \end{bmatrix}$
9	Upper right arm	$\begin{bmatrix} 0.017 & 0.0 & 0.0 \\ 0.0 & 0.017 & 0.0 \\ 0.0 & 0.0 & 0.001 \end{bmatrix}$
10	Lower right arm	$\begin{bmatrix} 0.014 & 0.0 & 0.0 \\ 0.0 & 0.014 & 0.0 \\ 0.0 & 0.0 & 0.001 \end{bmatrix}$

TABLE A.44 (continued)
Ninety-Fifth-Percentile (95%) Female Body Segment Principal
Inertia Matrices (in Slug ft²)

Body Segment Number	Name	Inertia Matrix (in Slug ft²) Relative to Body Frame Principal Directions
11	Right hand	$\begin{bmatrix} 0.003 & 0.0 & 0.0 \\ 0.0 & 0.001 & 0.0 \\ 0.0 & 0.0 & 0.001 \end{bmatrix}$
12	Upper right leg	$\begin{bmatrix} 0.075 & 0.0 & 0.0 \\ 0.0 & 0.075 & 0.0 \\ 0.0 & 0.0 & 0.018 \end{bmatrix}$
13	Lower right leg	$\begin{bmatrix} 0.006 & 0.0 & 0.0 \\ 0.0 & 0.006 & 0.0 \\ 0.0 & 0.0 & 0.001 \end{bmatrix}$
14	Right foot	$\begin{bmatrix} 0.001 & 0.0 & 0.0 \\ 0.0 & 0.004 & 0.0 \\ 0.0 & 0.0 & 0.006 \end{bmatrix}$
15	Upper left leg	$\begin{bmatrix} 0.075 & 0.0 & 0.0 \\ 0.0 & 0.075 & 0.0 \\ 0.0 & 0.0 & 0.018 \end{bmatrix}$
16	Lower left leg	$\begin{bmatrix} 0.006 & 0.0 & 0.0 \\ 0.0 & 0.006 & 0.0 \\ 0.0 & 0.0 & 0.001 \end{bmatrix}$
17	Left foot	$\begin{bmatrix} 0.001 & 0.0 & 0.0 \\ 0.0 & 0.004 & 0.0 \\ 0.0 & 0.0 & 0.006 \end{bmatrix}$

TABLE A.45
Ninety-Fifth-Percentile (95%) Female Body Segment Principal Inertia Matrices (in kg m²)

Body Segment Number	Name	Inertia Matrix (in kg m²) Relative to Body Frame Principal Directions
1	Lower torso (pelvis)	$\begin{bmatrix} 0.174 & 0.0 & 0.0 \\ 0.0 & 0.106 & 0.0 \\ 0.0 & 0.0 & 0.169 \end{bmatrix}$
2	Middle torso (lumbar)	$\begin{bmatrix} 0.104 & 0.0 & 0.0 \\ 0.0 & 0.064 & 0.0 \\ 0.0 & 0.0 & 0.102 \end{bmatrix}$
3	Upper torso (chest)	$\begin{bmatrix} 0.062 & 0.0 & 0.0 \\ 0.0 & 0.042 & 0.0 \\ 0.0 & 0.0 & 0.062 \end{bmatrix}$
4	Upper left arm	$\begin{bmatrix} 0.024 & 0.0 & 0.0 \\ 0.0 & 0.024 & 0.0 \\ 0.0 & 0.0 & 0.003 \end{bmatrix}$
5	Lower left arm	$\begin{bmatrix} 0.018 & 0.0 & 0.0 \\ 0.0 & 0.018 & 0.0 \\ 0.0 & 0.0 & 0.001 \end{bmatrix}$
6	Left hand	$\begin{bmatrix} 0.003 & 0.0 & 0.0 \\ 0.0 & 0.001 & 0.0 \\ 0.0 & 0.0 & 0.001 \end{bmatrix}$
7	Neck	$\begin{bmatrix} 0.014 & 0.0 & 0.0 \\ 0.0 & 0.014 & 0.0 \\ 0.0 & 0.0 & 0.003 \end{bmatrix}$
8	Head	$\begin{bmatrix} 0.035 & 0.0 & 0.0 \\ 0.0 & 0.035 & 0.0 \\ 0.0 & 0.0 & 0.018 \end{bmatrix}$
9	Upper right arm	$\begin{bmatrix} 0.024 & 0.0 & 0.0 \\ 0.0 & 0.024 & 0.0 \\ 0.0 & 0.0 & 0.003 \end{bmatrix}$
10	Lower right arm	$\begin{bmatrix} 0.018 & 0.0 & 0.0 \\ 0.0 & 0.018 & 0.0 \\ 0.0 & 0.0 & 0.001 \end{bmatrix}$

TABLE A.45 (continued)

Ninety-Fifth-Percentile (95%) Female Body Segment Principal Inertia Matrices (in kg m²)

Body Segment Number	Name	Inertia Matrix (in kg m²) Relative to Body Frame Principal Directions		
11	Right hand	$\begin{bmatrix} 0.003 & 0.0 & 0.0 \\ 0.0 & 0.001 & 0.0 \\ 0.0 & 0.0 & 0.001 \end{bmatrix}$		
12	Upper right leg	$\begin{bmatrix} 0.100 & 0.0 & 0.0 \\ 0.0 & 0.100 & 0.0 \\ 0.0 & 0.0 & 0.025 \end{bmatrix}$		
13	Lower right leg	$\begin{bmatrix} 0.007 & 0.0 & 0.0 \\ 0.0 & 0.007 & 0.0 \\ 0.0 & 0.0 & 0.001 \end{bmatrix}$		
14	Right foot	$\begin{bmatrix} 0.001 & 0.0 & 0.0 \\ 0.0 & 0.006 & 0.0 \\ 0.0 & 0.0 & 0.007 \end{bmatrix}$		
15	Upper left leg	$\begin{bmatrix} 0.100 & 0.0 & 0.0 \\ 0.0 & 0.100 & 0.0 \\ 0.0 & 0.0 & 0.025 \end{bmatrix}$		
16	Lower left leg	$\begin{bmatrix} 0.007 & 0.0 & 0.0 \\ 0.0 & 0.007 & 0.0 \\ 0.0 & 0.0 & 0.001 \end{bmatrix}$		
17	Left foot	$\begin{bmatrix} 0.001 & 0.0 & 0.0 \\ 0.0 & 0.006 & 0.0 \\ 0.0 & 0.0 & 0.007 \end{bmatrix}$		

Glossary

Abduction: spreading the legs apart (opposite of adduction; see Figure 2.18)

Absolute velocity: velocity in an inertial reference frame

Acceleration: see Section 8.4

Acetabulum: rounded socket supporting the head of the femur

Acromion: a pointed projection forming the tip of the shoulder

Active forces: forces exerted on a body such as gravity or contact forces; also called applied forces, as opposed to passive (or inertia) forces (see Section 12.1)

Addition theorem for angular velocity: see Section 8.7.2

Adduction: bring the legs together from a spread position (opposite of abduction; see Figure 2.18)

Afferent: leading to, as a nerve impulse to the brain or spinal cord

Angular velocity: see Section 8.6

Angular velocity matrix: the matrix whose dual vector is the angular velocity (see Equation 8.122)

Anterior: toward the front (opposite of posterior)

Aorta: the large blood vessel providing blood flow out of the heart to the body

Applied forces: forces exerted on a body such as gravity or contact forces; also called active forces, as opposed to inertia (or passive) forces (see Section 12.1)

Arachnoid: fibrous tissue between the brain and dura mater

Avulsion: a traumatic tearing of body tissue

Axilla: armpit

Ball-and-socket joint: a spherical joint with three rotational degrees of freedom

Barber poling: a vehicle occupant moving up relative to a seat belt, particularly a shoulder belt (see Section 15.7)

Beltrami–Mitchell equations: see compatibility equations

Biomechanics: mechanics applied with living systems

Biosystem: a living system

Block multiplication of matrices: multiplication of partitioned matrices with the submatrices treated as elements (see, e.g., Equation 3.101)

Bone maintenance: bone remodeling

Bound vector: a vector acting along a specific line of action (also called a bound vector)

Bryant angles: dextral rotation angles (1-2-3 rotation sequence angles) of a body relative to a reference frame (see Section 8.8)

Cadence: in walking, the number of steps per unit time

Cancellous bone: soft, spongy bone, also called trabecular bone (see Section 7.2)

Carpal: referring to the wrist

Cartilage: a relatively stiff, fibrous connective tissue between bones as in the knees and chest (sternum)

Caudal: referring to the tail

Center of gravity: see mass center

Central inertia dyadic: the dyadic of a body computed relative to the mass center G (see Section 10.4)

Cerebrum: the largest portion of the brain

Cervical: pertaining to the neck

Characteristics of a vector: magnitude (length or norm) and direction (orientation and sense)

Cockpit: vehicle occupant space

Collagen: a tough, fibrous tissue providing strength to bone, ligament, tendon, and cartilage

Column matrix, column array: a matrix with only one column

Compact bone: hard or cortical bone (see Section 7.2)

Compatibility equations: constraint equations to insure consistency in solutions of stress equilibrium equations (see Section 5.7)

Component: a vector or scalar contributing to the sum (or resultant) of a vector (see Section 3.2.5)

Component: one of the addends in a vector sum; a scalar coefficient of a unit vector where a vector is expressed in terms of unit vectors

Compression: a force or stress tending to shorten or decrease the volume of a body

Concussion: loss of consciousness

Configuration graph: see Section 8.8

Conformable matrices: for two matrices, where the number of rows of the first is equal to the number of columns of the second

Constraint equations: analytical expressions resulting from position, movement, or motion constraints on a multibody system (a human body model) (see Sections 13.5 and 13.6)

Convulsion: violent erratic muscle contraction

Coronal plane: a vertical plane dividing the human body into anterior and posterior parts, also called the frontal plane (see Figure 2.8)

Cortical bone: hard or compact bone (see Section 7.2)

Costal: referring to the ribs

Coup/contrecoup injury: head injury occurring at and opposite the site of traumatic impact

Couple: a force system with a zero resultant but a nonzero moment about some point (see Section 4.5.2)

Crash-victim simulator: a computer program or software to model the kinematics and dynamics of a vehicle occupant during a crash

Cutaneous: referring to the skin

D'Alembert's principle: a variant of Newton's second law stating simply that $\mathbf{F} + \mathbf{F}^* = 0$ where $\mathbf{F}^*$ is defined as $-m\mathbf{a}$ with $\mathbf{F}$ being an applied force on a particle, m is the mass of the particle, and $\mathbf{a}$ is the acceleration of the particle in an inertial (or Newtonian) reference frame

Degloving: the traumatic tearing away of skin and flesh from the bones

Degree of freedom: a measure of the manner of possible movement of a particle, body, or system

Deltoid: shoulder muscle

Dermis: the middle or central layer of skin

Dextral rotation angles: so-called Bryant angles (1-2-3 rotation sequence angles) of a body relative to a reference frame (see Section 8.8)

Diagonal matrix: a square matrix with nonzero elements on its diagonal but zero elements off the diagonal

Diaphysis: shaft of a long bone (see Figure 2.23)

Diffuse axonal injury (DAI): scattered microdamage to brain tissue

Digits: fingers or toes

Dimensions of a matrix: the number of rows and columns of a matrix

Dorsiflexion: raising the toe of the foot upward (opposite of plantar flexion; see Figure 2.20)

Double support: the condition in walking when two feet are on the walking surface, as opposed to single support, where only one foot is on the walking surface

Dual vector: a vector forming the nondiagonal entries of a skew-symmetric matrix as in the angular velocity matrix of Equation 8.122

Dummy force method: applying a concentrated force in the strain energy method, at a point, to determine the displacement at that point, with the magnitude of the force being then set to zero

Dura mater: thick outermost layer of the brain

Dyad: the result of a dyadic product (see Section 3.4)

Dyadic: a sum of dyads (see Equation 3.45)

Dyadic product: a means of multiplying vectors resulting in dyads (see Equation 3.42)

Dyadic transpose: a dyadic obtained by interchanging the rows and columns of the array of its components

Dyadic/vector product: a product of a dyadic and a vector producing a vector (see Equations 3.72 and 3.78)

Dynamics: combined study of kinematics, inertia, and kinetics

Efferent: leading away from, as a nerve impulse away from the brain or spinal cord

Eigenvalue: the parameter λ in the expression $A\mathbf{x} = \lambda\mathbf{x}$ (see Equation 3.135); also called principal value

Eigenvalues of inertia: see eigenvalues; the eigenvalues of the inertia matrix

Eigenvector: the vector $\mathbf{x}$ in the expression $A\mathbf{x} = \lambda\mathbf{x}$ (see Equation 3.135); also called principal vector

Elastic modulus: proportionality constant in linear stress–strain equations; also known as Young's modulus of elasticity (see Equation 5.64)

Elements of a matrix: the numbers, or variables, making up a matrix (see Section 3.6)

Engineering shear strain: see Equation 5.18

Epidermis: outermost layer of the skin

Epiligament: a ligament sheath

Epiphysis: rounded, enlarged end of a long bone (see Figure 2.23)

Epitendon: a tendon sheath

Equivalent force systems: force systems with equal resultants and equal moments about some point (see Section 4.5.3)

Euler angles: angles generated by a 1-3-1 or a 3-1-3 rotation sequence of a body relative to a reference frame (see Section 8.8)

Euler parameters: a set of four parameters defining the orientation of a body (see Section 8.16)

Eversion: rotating the soles of the feet outward so as to cause a varus leg position (opposite of inversion; see Figure 2.21)

Extension: the straightening of a limb or the backward movement of the head/neck (opposite of flexion; see, e.g., Figures 2.13 through 2.15)

Femur: thigh bone (see Figure 2.22)

Fibrillation: erratic muscle contraction (see also twitching)

Fibula: smaller of the two lower leg bones

Finite-segment modeling: representing a system by lumped masses or individual bodies (see Figure 2.3); also called lumped-mass modeling

First moment: the product of a particle mass and the position vector locating the particle. First moments are used to locate mass centers as in Section 9.2

Flexion: the bending of a limb or the forward bending of the head/neck (chin to chest, opposite of extension; see, e.g., Figures 2.13 through 2.15)

Flexural stress: stress arising due to bending of a structure

Foramen: an opening through a bone

Force: a push or pull (see Section 4.1)

Free vector: a vector (such as a unit vector) that is not restricted to a specific line or point

Frontal bone: the front and top of the skull (see Figure 2.24)

Frontal plane: see coronal plane

Gait: the procedure of walking; walking analysis

Generalized coordinates: variables describing the degrees of freedom of multibody systems or of human body models

Generalized forces: forces (and moments) projected onto partial velocity (and partial angular velocity) vectors (see Section 12.2)

Generalized speed: a linear combination of generalized coordinate derivatives (see Section 11.6)

Gross modeling: representing a system as a whole

Gross-motion simulator: a whole-body model as in Figure 6.1

Haversian system: a set of concentric cylindrical layers of bone containing an inner blood supply; also called an osteon

Heel strike: in walking, the beginning of the support phase of a leg

Hematoma: bruising or bleeding between the brain and skull or elsewhere in the body

Hinge joint: a pin or revolute joint with one rotational degree of freedom (see Figure 2.28)

Horizontal plane: see transverse plane

Humerus: upper arm bone (see Figure 2.22)

Identity dyadic: a dyadic whose scalar components are the same as the values of the Kronecker delta function (see Equations 3.48 and 3.49)

Identity matrix: a matrix whose diagonal elements are one and whose off-diagonal elements are zero (see Section 3.6.2)

Inertia: mass and mass distribution; resistance to movement

Inertia dyadic: a dyadic composed from inertia vectors and unit vectors (see Equation 10.17)

Inertia forces: forces due to accelerations; also called passive forces, as opposed to applied (or active) forces (see Section 12.1)

Inertia matrix: a matrix whose elements are moments and products of inertia (see Equation 10.18); also, the matrix whose elements are components of an inertia dyadic

Inertia vector: a vector useful in determining inertia properties of particles and rigid bodies, as defined by Equation 10.3; also called second-moment vector

Inertial reference frame: a reference frame where Newton's laws are valid; also called a Newtonian reference frame

Inferior: toward the feet or downward (opposite of superior)

Injury: failure or damage of the tissue of a biosystem

Integrated seat belt: a seat belt webbing with the shoulder belt portion anchored to the seat backrest

Inverse dyadics: dyadics that, multiplied together, produce the identity dyadic (see Equations 3.57 through 3.60)

Inverse matrix: a matrix whose product with a matrix produces an identity matrix (see Equation 3.95)

Inversion: rotating the soles of the feet inward so as to cause a valgus leg position (opposite eversion; see Figure 2.21)

Isometric: a muscle contraction that does not produce movement

Isotonic: a constant force muscle contraction

Isotropic material: a material having the same properties in all directions

Kane's equations: equations (analogous to Lagrange's equations or Newton's laws) that are ideally suited for dynamic analyses of human body models (see Section 13.1)

Kinematics: a study of motion and movement

Kinesiology: the study of human movement and the associated mechanics thereof

Kinetics: a study of forces and force systems

Kronecker delta function: a double-index function with values 1 and 0 depending on whether the indices are equal or not equal (see Equation 3.23); also, elements of the identity matrix

Lateral: toward the outside, away from the median (or sagittal) plane (opposite of medial)

Lateral collateral ligament: a ligament on the lateral side of the knee between the femur and fibula

Ligament: cord or cable connecting bone to bone

Load limiter: a device or seat belt design intended to limit peak seat belt forces during a vehicle crash

Lower body array: an array of numbers defining the connection configurations of a multibody system (see Section 6.2)

Lumbar: lower back

Lumped-mass modeling: representing a system by finite segments or individual bodies (lumped masses) (see, e.g., Figure 10.1); also called finite-segment modeling

Mandible: jaw bone

Mass center: a point of a body, or within a set of particles, for which the sum of the first moments of the particles of the body, or set, is zero; also called center of gravity

Mathematical shear strain: see Equation 5.17

Matrix: a structured array of numbers (see Section 3.6)

Matrix transpose: a matrix resulting from the interchange of the rows and columns of a given matrix (see Section 3.6.3)

Maxilla: upper jaw bone

Medial: toward the middle or median (sagittal) plane (opposite of lateral)

Medial collateral ligament: a ligament on the medial side of the knee between the femur and tibia

Median plane: see sagittal plane

Melanin: a dark pigment providing color to the skin

Membrane analogy: a means of modeling the torsional response of noncircular cross-sectional rods; also called soap film analogy (see Section 5.17)

Metacarpal: refers to the bones of the hand

Metatarsal: refers to the bones of the feet

Modeling: bone formation or bone growth

Modeling: mathematical representation of a physical body or system; a simplified, and usually reduced size, object or structure representing a body or system

Modulus of rigidity: see shear modulus

Moment of inertia: the projection of the second-moment vector (inertia vector) along the direction used to form the second-moment vector, as defined by Equation 10.5 (see also Equation 10.14 for a geometric interpretation)

Negative force system: if two force systems S_1 and S_2 when superimposed result in a zero system, then S_1 and S_2 are said to be negative to each other

Neutral axis: a median axis of a beam that is neither in tension nor compression as the beam is bent

Newtonian reference frame: a reference frame where Newton's laws are valid; also called an inertial reference frame

Normal strain: a strain component normal to or perpendicular to a surface (see also simple strain)

Normal stress: a stress component at a point of a surface directed normal to or along the perpendicular to the surface

Occipital: refers to the back of the head

Occipital bone: the rear part of the skull (see Figure 2.24)

Orbital: refers to the eye socket

Order: dimension of a matrix. For a square matrix, the order is the number of rows (or columns) of the matrix

Orthogonal complement array: an array C, which when premultiplied by a given array B, produces a zero array, that is, $BC = 0$; orthogonal complement arrays are used in processing constraint equation arrays (see Sections 13.7 and 13.8)

Orthogonal dyadic: a dyadic whose inverse is equal to its transpose (see Equation 3.61)

Orthogonal matrix: a matrix whose inverse is equal to its transpose

Osteoblast: a bone-forming cell

Osteoclast: a bone-resorbing cell

Osteocyte: bone cell

Osteon: the primary structural component of cortical (compact) bone; also called the Haversian system

Osteoporosis: a bone disease (porous bones) characterized by low bone mineral density

Parallelogram law: geometric representation of a vector addition (see Figure 3.5)

Parietal bone: the side and rear upper part of the skull (see Figure 2.24)

Partial angular velocity: the vector coefficient in an angular velocity vector of a generalized coordinate derivative or of a generalized speed (see Section 11.4)

Partial velocity: the vector coefficient in a velocity vector of a generalized coordinate derivative or of a generalized speed (see Section 11.10)

Partitioned matrix: a matrix divided into submatrices

Passive forces: forces due to accelerations; also called inertia forces, as opposed to "active" (or applied) forces (see Section 12.1)

Pectoral: refers to the breast or chest

Performance: activity of a living system

Permutation symbol: a three-index function having values 1, −1, and 0, depending upon whether the indices are cyclic, anticyclic, or nondistinct (see Equation 3.36)

Pia mater: fibrous membrane beneath the arachnoid and on the outer surface of the brain

Pin joint: a hinge or revolute joint with one rotational degree of freedom (see Figure 2.28)

Pitch motion: leaning or tilting forward or backward (see Section 15.2)

Plantar flexion: pushing the toe of the foot downward, as in accelerating a vehicle (opposite of dorsiflexion; see Figure 2.20)

Poisson's ratio: an elastic constant that is a measure of the transverse contraction of a body being elongated or transverse expansion of a body being shortened; also known as the transverse contraction ratio (see Equation 5.69)

Porpoising: a vehicle occupant moving up relative to a seat belt, particularly a lap belt (see Section 15.7)

Posterior: toward the rear or back (opposite of anterior)

Pretensioner: a device for eliminating slack and for tightening a seat belt in the beginning of a motor vehicle crash

Principal direction: the direction of an eigenvector or of a principal vector

Principal value: the parameter λ in the expression $A\mathbf{x} = \lambda\mathbf{x}$ (see Equation 3.135)

Principal vector: the vector $\mathbf{x}$ in the expression $A\mathbf{x} = \lambda\mathbf{x}$ (see Equation 3.135)

Product of inertia: the projection of the second-moment vector (inertia vector) along a direction different than that used to form the second-moment vector, as defined by Equation 10.4 (see also Equation 10.15 for a geometric interpretation)

Pronation: axial rotation of the forearm so that the palm of the hand faces downward (opposite of supination; see Figure 2.17)

Prone: lying on the stomach

Pseudoinverse of a matrix: a matrix that for a singular matrix has properties analogous to that of an inverse (see Section 3.6.18)

Quasicoordinate: a variable (generally nonexisting) whose derivative is a generalized speed (see Section 11.6)

Radius: anterior lower arm bone

Radius of gyration: a geometric quantity (a length) used for determining moment of inertia (see Section 10.5)

Rank: the dimension of the largest nonsingular submatrix of a matrix

Reference configuration: an orientation and positioning of the bodies of the human frame so that the local coordinate axes are all mutually aligned and also aligned with the global axes of the torso

Rehabilitation: recovery from injury or disease

Remodeling: bone resorption and subsequent reformation, sometimes called bone maintenance

Resultant: the sum of vectors (see Section 3.2.5)

Revolute joint: a pin, or hinge, joint with one rotational degree of freedom (see Figure 2.28)

Roll motion: rotation about a spinal axis (see Section 15.2)

Rotation dyadic: a dyadic that when multiplied with a vector rotates the vector about a given line through a specified angle (see Section 8.15)

Row matrix, row array: a matrix with any one row

Sacrum: a fusion of five vertebrae at the base of the spine; the tailbone

Sagittal plane: the midplane dividing the human body into its left and right parts, also called the median plane (see Figure 2.8)

Scalar product: a means of multiplying vectors resulting in a scalar (see Equation 3.19)

Scalar triple product: a product of three vectors producing a scalar (see Equation 3.63)

Second moment of area: a geometric property of beam cross sections, sometimes regarded as an area moment of inertia (see Equation 5.105)

Second-moment vector: a vector useful in determining inertia properties of particles and bodies, as defined by Equation 10.3; also called inertia vector

Shear modulus: proportionality constant in linear shear stress–shear strain relations (see Equation 5.68); also known as the modulus of elasticity in shear or the modulus of rigidity

Shear stress: a stress component at a point of a surface directed tangent to the surface

Simple angular velocity: the angular velocity of a body rotating about a fixed axis (see Equation 8.19)

Simple couple: a couple with only two forces (equal in magnitude, but oppositely directed; see Section 4.5.2)

Simple strain: intuitively an average deformation or a change in length per unit length; see also normal strain (see Section 5.2 and Equation 5.18)

Simple stress: uniaxial or normal stress (see Section 5.1)

Single support: the condition in walking where only one foot is on the walking surface, as opposed to "double support," where two feet are on the walking surface

Singular matrix: a matrix whose determinant is zero

Slider joint: a simple translation joint (see Figure 2.29)

Sliding vector: a vector acting along a specific line of action (also called a bound vector)

Soap film analogy: a means of modeling the torsional response of noncircular cross-sectional rods; also called membrane analogy (see Section 5.17)

Spacewalk: an astronaut moving in space outside a satellite

Sphenoid bone: the frontal base of the skull (see Figure 2.24)

Spherical joint: a ball-and-socket joint with three rotational degrees of freedom

Sprain: the stretching or yielding of a ligament or tendon

Square matrix: a matrix with the same numbers of rows and columns

Step length: in walking, the distance in the direction of walking of two successive footprints (one from each foot)

Sternum: cartilage forming the breast bone

Strain: see Section 5.2

Stress: the limiting value of a force component per unit area as the area shrinks to zero (see Section 5.1)

Stress vector: the vector of stress components acting at a point of a given surface (see Section 5.1)

Stride: in walking, the time required to complete both the support phase and the swing phase for a leg

Stride length: in walking, the distance between two successive footprints of the same leg

Subcutaneous: beneath the dermis (skin)

Submarining: a vehicle occupant sliding under a seat belt, particularly a lap belt; alternatively, a lap belt moving up on an occupant's torso (see Section 15.7)

Submatrix: a matrix formed by eliminating rows and/or columns from a given matrix

Summation convention: a rule that repeated indices represents a sum over the range of the index (see Equations 3.16 and 3.17)

Superior: toward the head or upward (opposite of inferior)

Supination: axial rotation of the forearm so that the palm of the hand faces upward (opposite of pronation; see Figure 2.17)

Supine: lying on the back

Swing: in walking, the movement of a nonsupporting leg (see Section 15.4.1)

Symmetric dyadic: a dyadic equal to its transpose (see Equation 3.52)

Symmetric matrix: a matrix that is equal to its transpose (see Section 3.6.5)

Synovial fluid: a lubricating fluid in joints and tendon sheaths

Temporal bone: the side and base of the skull (see Figure 2.24)

Tendon: cord or cable connecting muscle to bone

Tension: a force or stress tending to elongate or enlarge a body

Tetanic: erratic muscle contraction or spasm

Thoracic: referring to the chest

Tibia: largest of the two lower leg bones (see Figure 2.22)

Tissue: biological material (see Section 7.1)

Toe off: in walking, the ending of the support phase of a leg

Torque: the moment of a couple (see Section 4.5.2)

Trabecula: a "little beam" bone structure in cancellous or soft bone

Trabecular bone: soft or spongy bone; also called cancellous bone (see Section 7.2)

Trace: the sum of the diagonal elements of a square matrix

Transformation matrix: a matrix whose elements are used to relate vector components referred to different unit vector bases, as defined and illustrated in Equations 3.160 through 3.168

Transition movement: a movement successively employing different muscle groups

Transverse contraction ratio: see Poisson's ratio

Transverse plane: a horizontal plane dividing the human body into its upper and lower parts (see Figure 2.8)

Triple-vector product: a product of three vectors producing a vector (see Equations 3.69 and 3.70)

Twitching: erratic muscle contraction or spasm

Ulna: posterior lower arm bone

Uniaxial strain: 1D strain

Uniaxial stress: 1D stress

Unit vector: a vector with magnitude one (see Section 3.2.2)

Valgus: leg position with the knees separated more than the feet (bowlegged, opposite of varus; see Figure 2.19)

Varus: leg position with the knees close together (knock-kneed, opposite of valgus; see Figure 2.19)

Vector: intuitively, a directed line segment with the characteristics of magnitude, orientation, and sense (see Section 3.1). More formally, vectors are elements of vector spaces [12,15,20].

Vector product: a means of multiplying vectors resulting in a vector (see Equation 3.31)

Vector–vector: a dyadic (see Section 3.4)

Velocity: see Section 8.3

Wolff's law: in short, form follows function (see Section 7.1)

Yaw motion: leaning or tilting side to side (see Section 15.2)

Young's modulus of elasticity: see elastic modulus

Zero dyadic: a dyadic whose scalar components are zero

Zero force system: a force system with a zero resultant and a zero moment about some point (see Section 4.5.1)

Zero matrix: a matrix, all of whose elements are zero (see Section 3.6.1)

Zero vector: a vector with magnitude zero (see Section 3.2.3)

Bibliography

Bellman, R., *Introduction to Matrix Analysis*, 2nd edn., McGraw-Hill, New York, 1970.

Borisenko, A. I. and I. E. Tarapov (trans. R. A. Silverman), *Vector and Tensor Analysis*, Prentice Hall, Englewood Cliffs, NJ, 1968.

Bronson, R., *Matrix Methods—An Introduction*, Academic Press, New York, 1970.

Carter, D. R. and G. S. Beaupre, *Skeletal Function and Form*, Cambridge University Press, Cambridge, U.K., 2001.

Cochran, G. V. B., *A Primer of Orthopaedic Biomechanics*, Churchill Livingstone, New York, 1982.

Cramer, G. D. and S. A. Darby, *Basic and Clinical Anatomy of the Spine, Spinal Cord, and ANS*. Mosby Corporation, St. Louis, MO, 1995.

Hannon, P. and K. Knapp, *Forensic Biomechanics*, Lawyers & Judges Publishing Company, Tucson, AZ, 2006.

Karamcheti, K., *Vector Analysis and Cartesian Tensors*, Holden Day, San Francisco, CA, 1967.

Knudson, D., *Fundamentals of Biomechanics*, 2nd edn., Springer, New York, 2007.

Kreighbaum, E. and K. M. Barthels, *Biomechanics—A Qualitative Approach for Studying Human Movement*, 4th edn., Allyn & Bacon, Boston, MA, 1996.

McConnell, A. J., *Application of Tensor Analysis*, Dover, New York, 1957.

Ortega, J. M., *Matrix Theory*, Plenum, New York, 1987.

Peterson, D. R. and J. D. Bronzino (Eds.), *Biomechanics Principles and Applications*, CRC Press, Boca Raton, FL, 2008.

Polsson, B. O. and S. N. Bhatia, *Tissue Engineering*, Pearson/Prentice Hall, Upper Saddle River, NJ, 2004.

Ritter, A. B., S. Reisman, and B. B. Michnaik, *Biomedical Engineering Principles*, CRC Press, Taylor & Francis, Boca Raton, FL, 2005.

Sokolnikoff, I. S., *Tensor Analysis—Theory and Applications*, John Wiley & Sons, New York, 1951.

Wills, A. P., *Vector Analysis with an Introduction to Tensor Analysis*, Dover, New York, 1958.

Index